Lecture Notes in Computer Science   9538

Commenced Publication in 1973
Founding and Former Series Editors:
Gerhard Goos, Juris Hartmanis, and Jan van Leeuwen

More information about this series at http://www.springer.com/series/7407

Zsuzsanna Lipták · William F. Smyth (Eds.)

Combinatorial Algorithms

26th International Workshop, IWOCA 2015
Verona, Italy, October 5–7, 2015
Revised Selected Papers

 Springer

Editors
Zsuzsanna Lipták
Department of Computer Science
University of Verona
Verona
Italy

William F. Smyth
Department of Computing and Software
McMaster University
Hamilton
Canada

ISSN 0302-9743   ISSN 1611-3349   (electronic)
Lecture Notes in Computer Science
ISBN 978-3-319-29515-2   ISBN 978-3-319-29516-9   (eBook)
DOI 10.1007/978-3-319-29516-9

Library of Congress Control Number: 2016931327

LNCS Sublibrary: SL1 – Theoretical Computer Science and General Issues

Printed on acid-free paper

This Springer imprint is published by SpringerNature
The registered company is Springer International Publishing AG Switzerland

Preface

This volume contains revised versions of papers presented at the 26th International Workshop on Combinatorial Algorithms (IWOCA 2015), held October 5–7, 2015, in Verona, Italy.

IWOCA 2015 continued a long and well-established tradition of encouraging high-quality research in theoretical computer science and providing an opportunity to bring together specialists and young researchers working in the area. The IWOCA conference series grew out of a 17-year history of the Australasian Workshop on Combinatorial Algorithms (AWOCA). Previous AWOCA and IWOCA meetings have been held in Australia, Indonesia, South Korea, Japan, Czech Republic, Canada, UK, India, France, and the USA.

We solicited high-quality papers in the broad area of combinatorial algorithms. Topics included: algorithms and data structures (including sequential, parallel, distributed, approximation, probabilistic, randomized, and on-line algorithms); algorithms on strings and graphs; applications (bioinformatics, music analysis, networking, and others); combinatorics on words; combinatorial enumeration; combinatorial optimization; complexity theory; computational biology; compression and information retrieval; cryptography and information security; decompositions and combinatorial designs; discrete and computational geometry; graph drawing and labelling; graph theory.

The Program Committee decided to accept 30 papers, out of a total of 90 submissions. One paper was later withdrawn by the authors. Each submission received at least three, and at most seven reviews. Papers were submitted and reviewed using the EasyChair online system. Authors of accepted papers come from 19 countries, on four continents (Asia, Europe, North America, South America).

The scientific program included three invited lectures, given by:

– Béla Bollobás on "Monotone Cellular Automata"
– Frank Ruskey on "Recent Results about Venn Diagrams"
– Esko Ukkonen on "Identifiability of a String from Its Substrings"

We thank the invited speakers for accepting our invitation and for their excellent presentations at the conference.

The program also included an open problem session, chaired by Martin Milanič and Romeo Rizzi, where seven open problems were presented. These can be found at the open problem collection of IWOCA at http://iwoca.org. This year for the first time, we had a Best Student Paper Award, sponsored by the European Association for Theoretical Computer Science (EATCS). The Program Committee decided to assign this award to the paper "The k-Leaf Spanning Tree Problem Admits a Klam Value of 39" by Meirav Zehavi.

We thank all authors who submitted their work for consideration to IWOCA 2015. We wish to thank the Program Committee and the 153 external reviewers, whose many thorough reviews helped us select the papers to be presented at IWOCA 2015. The success of the scientific program is due to their hard work.

We also thank the EATCS (European Association for Theoretical Computer Science), the EATCS Italian Chapter, and the AICA (Associazione Italiana per l'Informatica ed il Calcolo Automatico), for their support of the conference.

IWOCA 2015 was organized by the Department of Computer Science of the University of Verona, whose administrative and financial support we gratefully acknowledge.

November 2015   Zsuzsanna Lipták
                                                                        William F. Smyth

Organization

Steering Committee

Costas Iliopoulos   King's College London, UK
Mirka Miller   University of Newcastle, Australia
William F. Smyth   McMaster University, Canada

Program Committee

Donald Adjeroh	West Virginia University, USA
Hideo Bannai	Kyushu University, Japan
Paola Bonizzoni	Università di Milano-Bicocca, Italy
Gerth Stølting Brodal	Aarhus University, Denmark
Sebastian Böcker	Friedrich Schiller University Jena, Germany
Charles Colbourn	Arizona State University, USA
Jiří Fiala	Charles University, Czech Republic
Dalibor Froncek	University of Minnesota Duluth, USA
Luisa Gargano	Università di Salerno, Italy
Roberto Grossi	Università di Pisa, Italy
Pinar Heggernes	University of Bergen, Norway
Costas Iliopoulos	King's College London, UK
Ralf Klasing	CNRS and University of Bordeaux, France
Christian Komusiewicz	TU Berlin, Germany
Jan Kratochvíl	Charles University, Czech Republic
Dieter Kratsch	Université de Lorraine - Metz, France
Gad M. Landau	University of Haifa, Israel
Thierry Lecroq	University of Rouen, France
Zsuzsanna Lipták	Università di Verona, Italy – Co-chair
Giovanni Manzini	University of Eastern Piedmont, Italy
Martin Milanič	University of Primorska, Slovenia
Mirka Miller	University of Newcastle, Australia
Ian Munro	University of Waterloo, Canada
Wendy Myrvold	University of Victoria, Canada
Gonzalo Navarro	University of Chile, Chile
Kunsoo Park	Seoul National University, South Korea
Solon Pissis	King's College London, UK
Simon Puglisi	University of Helsinki, Finland
M. Sohel Rahman	Bangladesh University of Engineering and Technology, Bangladesh
Romeo Rizzi	Università di Verona, Italy

Frank Ruskey   University of Victoria, Canada
William F. Smyth   McMaster University, Canada – Co-chair

Problem Session Co-chairs

Gabriele Fici   Università di Palermo, Italy
Martin Milanič   University of Primorska, Slovenia
Romeo Rizzi   Università di Verona, Italy
Oliver Schaudt   University of Cologne, Germany

Organizing Committee

Ferdinando Cicalese   Università di Verona, Italy
Giuditta Franco   Università di Verona, Italy
Rosalba Giugno   Università di Catania, Italy
Zsuzsanna Lipták   Università di Verona, Italy
Gabriele Pozzani   Università di Verona, Italy
Simon Puglisi   University of Helsinki, Finland
Romeo Rizzi   Università di Verona Italy
Ugo Solitro   Università di Verona, Italy

Student Volunteers

Daniele Castrovilli   Università di Verona, Italy
Carlo Comin   Università di Trento, Italy
Andrei Florea   Università di Verona, Italy
Giovanni Pasquini   Università di Verona, Italy
Francesca Zerbato   Università di Verona, Italy
Stefania Zubbi   Università di Verona, Italy

Additional Reviewers

Albano, Alexandre
Amit, Mika
Anholcer, Marcin
Araujo-Pardo, Gabriela
Asahiro, Yuichi
Badkobeh, Golnaz
Baisya, Dipankar Ranjan
Bampas, Evangelos
Bampis, Evripidis
Bannister, Michael J.
Beal, Richard
Belmonte, Rémy
Berglin, Edvin
Bevern, René Van
Bodlaender, Hans L.
Böckenhauer, Hans-Joachim
Bonamy, Marthe
Bong, Novi Herawati
Bonifaci, Vincenzo
Bousquet, Nicolas
Bruner, Marie-Louise
Bujtás, Csilla
Calamoneri, Tiziana
Casel, Katrin

Christodoulakis, Manolis
Comin, Carlo
Conte, Alessio
Cordasco, Gennaro
Cordova, Joshimar
Çela, Eranda
Damaschke, Peter
De Agostino, Sergio
Della Vedova, Gianluca
Drake, Brian
Dregi, Markus Sortland
Egidi, Lavinia
Elbassioni, Khaled
Elsässer, Robert
Engler, Martin
Eppstein, David
Fabila-Monroy, Ruy
Faro, Simone
Fernau, Henning
Ferrada, Hector
Fleischauer, Markus
Froese, Vincent
Gagie, Travis
Ganian, Robert
Giannopoulou, Archontia
Giaquinta, Emanuele
Giroire, Frederic
Godbole, Anant
Golovach, Petr
González Yero, Ismael
Gourves, Laurent
Grigorious, Cyriac
Habib, Michel
Hell, Pavol
Hossain, Md. Iqbal
Hsieh, Sun-Yuan
Hufsky, Franziska
Hyyrö, Heikki
I, Tomohiro
Idrees, Samah
Ilie, Lucian
Inenaga, Shunsuke
Iqbal, Sumaiya
Irvine, Veronika
Kammer, Frank
Kelarev, Andrei

Kempa, Dominik
Kiyomi, Masashi
Klavzar, Sandi
Koike Quintanar, Sergio Hiroki
Konvalinka, Matjaz
Koster, Arie
Krattenthaler, Christian
Kärkkäinen, Juha
La Croix, Michael
Lefebvre, Arnaud
Lopez-Ortiz, Alejandro
Mahoney, Thomas
Mamino, Marcello
Mampentzidis, Konstantinos
Manea, Florin
Mansour, Toufik
Manuel, Paul
Marino, Andrea
Mary, Arnaud
Matsuoka, Yoshiaki
Mehrabi, Ali D.
Mercas, Robert
Mertzios, George
Mishna, Marni
Mohamed, Manal
Mömke, Tobias
Nakashima, Yuto
Nayeem, Muhammad Ali
Nichterlein, André
Nishimoto, Takaaki
Ota, Takahiro
Pagli, Linda
Papadopoulos, Charis
Parreau, Aline
Perez-Roses, Hebert
Peters, Daniel
Petreschi, Rossella
Pinho, Armando
Pinzon, Yoan
Pirola, Yuri
Pisanti, Nadia
Prencipe, Giuseppe
Prieur-Gaston, Elise
Pruesse, Gara
Pudwell, Lara
Pérez-Lantero, Pablo

R, Sundara Rajan
Rafiey, Arash
Rahman, Md. Khaledur
Rescigno, Adele
Rizzi, Raffaella
Rozenberg, Liat
Rudolf, Gabor
Russo, Luis M.S.
Ryan, Joe
Samal, Robert
Sawada, Joe
Schauer, Joachim
Scheubert, Kerstin
Schieber, Baruch
Schiermeyer, Ingo
Schweitzer, Pascal
Schwiegelshohn, Uwe
Shalom, Mordechai
Sorge, Manuel

Stephen, Sudeep
Storandt, Sabine
Strothmann, Thim
Todinca, Ioan
Ulfarsson, Henning
Uno, Takeaki
Vaccaro, Ugo
van Duijn, Ingo
van Leeuwen, Erik Jan
Vatter, Vincent
Venturini, Rossano
Werneck, Renato
White, Walton Timothy James
Williams, Aaron
Yamanaka, Katsuhisa
Yi, Eunjeong
Zaccaria, Simone
Žitnik, Arjana

Invited Talks

Monotone Cellular Automata

Béla Bollobás

University of Cambridge, UK, and
University of Memphis, TN, USA

Cellular automata, introduced in the 1940s by von Neumann, are interacting particle systems. In its simple form, we have a set of sites arranged in a grid-like fashion, with each site in one of finitely many states. Starting with such an 'initial configuration', at each time-step the system updates itself according to some fixed *local* rule: each site goes into a state that depends only on the states of a few nearby sites. Examples include the Ising model of ferromagnetism and simple models of the brain.

A cellular automaton with states 0 and 1, say, is *monotone* if every site in state 1 remains in state 1 forever. One of the simplest monotone cellular automata is *bootstrap percolation with infection parameter r*, introduced in 1979 by Chalupa, Leith and Reich. This process is an oversimplified model of the spread of an infection on a graph (with 0 meaning 'healthy' and 1 'infected'), in which a healthy site gets infected if it has at least r infected neighbours. By now, there is a huge literature of bootstrap percolation, with most of the early results due to probabilists, statistical physicists, and computer scientists, and many recent results proved by combinatorialists. I shall present some basic facts about bootstrap percolation, and will describe some important theorems due to Aizenman, Lebowitz, Cerf, Manzo, Cirillo and Holroyd, culminating in some substantial results I have proved with Balogh, Duminil-Copin and Morris.

Recently, with Smith and Uzzell, I introduced a far-reaching generalization of bootstrap percolation on lattices and lattice-like finite graphs. The only assumptions we made about such a process is that it is local, homogeneous and monotone. Surprisingly, much can be proved about these very general processes on $\mathbb{Z}^2$; in particular, Smith, Uzzell, Balister, Przykucki and I classified them into three classes, and proved much about the critical probability in each class. Very recently, Duminil-Copin, Morris, Smith and I have proved much more precise results about the most important class in the classification.

In my lecture, *aimed at non-specialists*, I shall give a brief introduction to some aspects of cellular automata. I shall assume *very little* and will keep my lecture *simple*.

Recent Results About Venn Diagrams

Frank Ruskey

University of Victoria, Canada

An n-Venn diagram is a collection of n simple closed curves in the plane that divide it into 2^n non-empty regions, one unique region per possible intersection of the interiors/exteriors of the curves. If the curves lie in general position; i.e., so that at most 2 curves intersect at a point then it is unknown whether rotationally symmetric diagrams exist for every prime n (the primality of n being an easily proved necessary condition). However, if curves can intersect at 3 or more curves, rotationally symmetric diagrams exist for prime n, and the proof relies on a modification of the classic symmetric chain decomposition of the Boolean lattice. In this talk this proof and later developments, such as the enumeration of symmetric Venn diagrams, will be surveyed. Additional open problems in the area of Venn diagrams will be discussed; e.g., can a new curve always be added to a Venn diagram to get a new Venn diagram?

Identifiability of a String from Its Substrings

Esko Ukkonen[1]

University of Helsinki, Finland

A classic algorithmic challenge in biological sequence analysis, the genome assembly problem asks one to reconstruct the DNA sequence from the short fragments (called reads) that a DNA sequencing instrument samples from the original sequence. The reads are produced in massive amounts but with some reading errors.

The talk discusses the exact version of the problem in which the reads are noiseless. Given a collection F of reads (substrings) sampled from an unknown target string T, the problem is to reconstruct T from F. If F covers the entire T several times and if the repeated substrings of T are contiguously covered by the reads in F, the reconstruction becomes straightforward: Just superpose the reads as suggested by their matching suffixes and prefixes. If, however, there are repeats that are longer than the reads – as is often the case for DNA sequences – the reconstruction becomes ambiguous because there may be several different reconstructions suggested by the reads.

We demonstrate that identifying a unique solution is possible from F, if F is the full k-mer spectrum of T and T does not contain any 3-repeats of length $k-1$ and not any interleaved pair of 2-repeats of length $k-1$. A finite-state automaton-like representation of the pairwise overlaps of the reads is introduced such that the unique identifiability of T reduces to the uniqueness of the Eulerian path in this representation. Generalization for more realistic F with variable-length reads and non-uniform coverage is considered.

References

1. Bresler, G., Bresler, M., Tse, D.: Optimal assembly for high throughput shotgun sequencing. BMC Bioinform. **14**(Suppl. 5), S18 (2013)
2. Myers, E.W.: The fragment assembly string graph. Bioinformatics **21**(Suppl. 2), ii79–ii85 (2005)
3. Pevzner, P.A., Tang, H., Waterman, M.S.: An Eulerian path approach to DNA fragment assembly. PNAS **98**, 9748–9753 (2001)
4. Ukkonen, E.: Approximate string-matching with q-grams and maximal matches. Theor. Comput. Sci. **92**, 191–211 (1992)

---

[1]Supported by European Commission grant SYSCOL (UE7-SYSCOL-258236).

Contents

Speeding Up Cover Time of Sparse Graphs Using Local Knowledge

Mohammed Amin Abdullah[1]($\boxtimes$), Colin Cooper[2], and Moez Draief[1,3]

[1] Mathematical and Algorithmic Sciences Laboratory,
Huawei Technologies Co., Shenzhen, China
mohammed.abdullah@huawei.com

[2] Department of Informatics, King's College London, London, UK
colin.cooper@kcl.ac.uk

[3] Imperial College London, London, UK
m.draief@imperial.ac.uk

Abstract. We analyse the cover time of a random walk on a random graph of a given degree sequence. Weights are assigned to the edges of the graph using a certain type of scheme that uses only local degree knowledge. This biases the transitions of the walk towards lower degree vertices. We demonstrate that, with high probability, the cover time is at most $(1 + o(1))\frac{d-1}{d-2}8n \log n$, where d is the minimum degree. This is in contrast to the precise cover time of $(1 + o(1))\frac{d-1}{d-2}\frac{\theta}{d}n \log n$ (with high probability) given in [1] for a simple (i.e., unbiased) random walk on the same graph model. Here θ is the average degree and since the ratio θ/d can be arbitrarily large, or go to infinity with n, we see that the scheme can give an unbounded speed up for sparse graphs.

Keywords: Random walks · Random graphs · Network exploration

1 Introduction

A simple random walk $\mathcal{W}_u = \mathcal{W}_u(t)$, $t = 0, 1, \ldots$ on a graph G starting from a vertex u is a sequence of movements from one vertex to another where at each step an edge is chosen uniformly at random from the set of edges incident on the current vertex, and then transitioned to next vertex. Various quantities of interest related to the behaviour of the walk can be studied. For example, the *hitting time* $\mathbf{H}[u, v]$ of v is the expected number of steps until $\mathcal{W}_u$ visits v for the first time. That is, $\mathbf{H}[u, v] = \mathbf{E}[\min\{t \in \mathbb{N}_0 : \mathcal{W}_u(t) = v\}]$ (note, by definition, $\mathbf{H}[u, u] = 0$). The *maximum hitting time* is $\max_{u,v} \mathbf{H}[u, v]$.

Another quantity of interest, and the primary focus of this paper, is the *cover time* $\mathbf{COV}[G]$: denoting by $\mathbf{COV}_u[G]$ the expected time it takes $\mathcal{W}_u$ to visit every vertex, $\mathbf{COV}[G] = \max_u \mathbf{COV}_u[G]$.

For simple random walks, asymptotically tight bounds for cover time were given by [6,7]:

$$(1 + o(1))n \log n \leq \mathbf{COV}[G] \leq (1 + o(1))\frac{4}{27}n^3,$$

© Springer International Publishing Switzerland 2016
Z. Lipták and W.F. Smyth (Eds.): IWOCA 2015, LNCS 9538, pp. 1–12, 2016.
DOI: 10.1007/978-3-319-29516-9_1

and these lower and upper bounds are met by the complete graph and the lollipop graph respectively.

It is also natural to define random walks on a weighted graph $G = (V, E, w)$, where w is a function mapping edges to strictly positive values $w : E \to \mathbb{R}^+$. A *weighted* random walk on a vertex u transitions an edge (u, v) with probability $w(u, v)/w(u)$. Simple random walks are a special case where w is a constant function, and we may refer to them as *unweighted* walks.

The study of random walks on general weighted graphs is less developed than the special case of unweighted graphs, and it is not difficult to formulate many open questions on their behaviour. In particular, what bounds exist for hitting times and cover time? This was addressed in part by [8]. The investigation is framed as follows. For a graph G, let $\mathcal{P}(G)$ denote the set of all transition probability matrices for G, that is, stochastic matrices that respect the graph structure, i.e. if $P \in \mathcal{P}(G)$ is a transition matrix on G, we have $P_{u,v} \neq 0$ if and only if $(u, v) \in E$.

For $P \in \mathcal{P}(G)$, let $H_G(P)$ denote the maximum hitting time in G with transition matrix P, and $C_G(P)$ similarly for cover time. Let

$$H_G = \inf_{P \in \mathcal{P}(G)} H_G(P) \quad \text{and} \quad C_G = \inf_{P \in \mathcal{P}(G)} C_G(P).$$

Note that if for a graph G one knows a spanning tree T_G, a transition matrix P can be constructed that gives a simple random walk on T_G, and ignores all other edges of G. By a "twice round the spanning tree" argument of the type employed in [3], this implies a $O(n^2)$ upper bound on H_G and C_G.

In [8], it is shown that for a path graph P_n, any transition matrix will have $\Omega(n^2)$ maximum hitting time (and therefore, cover time). This, in conjunction with the spanning tree argument, implies $\Theta(n^2)$ for both H_G and C_G.

One can also ask the question about the minimum local topological information on the graph G that is always sufficient to construct a transition matrix that 'achieves' this upper bound for both H_G and C_G. Our goal is to devise a particular weighting scheme that gives $O(n^2)$ maximum hitting time for any graph. In [8], the transition probability of edge $e = (u, v)$ is defined as follows:

$$P_{u,v} = \begin{cases} \dfrac{1/\sqrt{d(v)}}{\sum_{w \in N(u)} 1/\sqrt{d(w)}} & \text{if } v \in N(u) \\ 0 & \text{otherwise} \end{cases}$$

where $d(v)$ is the degree of v and $N(v)$ is the neighbour set of v.

We will refer to this as the *Ikeda* scheme. It results in an $O(n^2 \log n)$ upper bound on the cover time for any connected n-vertex graph G. The rationale behind this scheme is that, at a high degree vertex, the biased walk transition favours low degree neighbours, speeding up their exploration and addressing the shortcoming of simple random walks for which low degree nodes are hard to reach.

In the algorithmic context of graph exploration, simple random walks are generally considered to have the benefit of not requiring information beyond what is needed to choose the next edge **uar**. Generally, this implies that a token making the walk can be assumed to know the degree of the vertex it is currently on, but no more information about the structure of the graph. In the Ikeda scheme, information required in addition to the vertex degree, is the degrees of neighbouring vertices.

Whilst the speed up given by the Ikeda scheme is clear for graphs such as the lollipop, which has a cover time of $\Theta(n^3)$, it is not clear how much of an advantage it gives over the simple random walk for sparse graphs. Clearly for regular graphs there can be no difference, but what of graphs that have some variation in vertex degree but are still sparse and perhaps even fairly homogeneous? The main aim of this paper is to answer this question for a different local weighting scheme: for $G = (V, E)$, assign each edge (u, v) weight $w(u, v) = 1/\min\{d(u), d(v)\}$ (equivalently, each edge is assigned *resistance* $r(u, v) = \min\{d(u), d(v)\}$). This weighting scheme defines the following transition matrix of a weighted random walk:

$$P_{u,v} = \begin{cases} \dfrac{w(u,v)}{\sum_{w \in N(u)} w(u,w)} & \text{if } v \in N(u) \\ 0 & \text{otherwise} \end{cases} \tag{1}$$

where $w(u, v) = 1/\min\{d(u), d(v)\}$.

We may, as a matter of convenience, say that $w(u, v) = 0$ if $(u, v) \notin E$ in calculations of transition probabilities. We call this scheme the *minimum degree* (or *min-deg*) scheme. It uses limited local graph information as the Ikeda scheme and provides similar general bounds on the hitting time and cover time of $O(n^2)$ and $O(n^2 \log n)$ respectively. Additionally, however, we show that it can provided an arbitrarily large or unbounded speed up for sparse graphs.

Notation and Terminology

For a graph $G = (V, E)$, let $n = |V|$ and $m = |E|$. Asymptotic quantities are with respect to n. A sequence of events $(\mathcal{E}_n)_n$ on probability spaces indexed by the number of vertices n occurs *with high probability* (**whp**) if $\mathbf{Pr}(\mathcal{E}_n) \to 1$ as $n \to \infty$. For a vertex $v \in V$, $d_v = d(v)$ is the degree of v and $N(v)$ is the set of v's neighbours. For a random walk $\mathcal{W}_u$ on a (weighted or unweighted) graph G, the stationary distribution, should it exist, is denoted by π, and $\pi_v = \pi(v)$ is the stationary probability for vertex v. We use the phrases "weighted random walk on a graph G" and "random walk on a weighted graph G" interchangeably. We shall introduce further notation as needed.

Due to space constraints, some proofs are omitted.

2 General Bounds for the Minimum Degree Weighting Scheme

In this section, we prove an upper bound of $O(n^2)$ and $O(n^2 \log n)$ on the hitting time and cover time, respectively, for the minimum degree scheme with transition matrix (1).

The proof of $O(n^2)$ hitting time in [8] applies the following, generally useful lemma (proof omitted).

Lemma 1. *For any connected graph G and any pair of vertices $u, v \in V$, let $\rho = (x_0, x_1, \ldots, x_\ell)$ where $x_0 = u$ and $x_\ell = v$ be a shortest path between u and v. Then*

$$\sum_{i=0}^{\ell} d(x_i) \leq 3n$$

where $d(x)$ is the degree of vertex x.

In addition, we have that

Lemma 2. *For the minimum degree scheme, let $w(G) = \sum_{u \in V} \sum_{v \in N(u)} w(u, v)$. Then*

$$n \leq w(G) \leq 2n. \tag{2}$$

Consequently,

Theorem 1. *For a graph $G = (V, E, w)$ under the min-deg weighting scheme, $\mathbf{H}[u, v] \leq 6n^2$ for any pair of vertices $u, v \in V$.*

By Matthews' technique [13], we obtain

Corollary 1. $\mathbf{COV}[G] = O(n^2 \log n)$.

The authors of [8] conjecture that their weighting scheme in fact gives an $O(n^2)$ upper bound on cover time. To our knowledge, no weighting scheme has been shown to meet an $O(n^2)$ bound on all simple, connected and undirected graphs. We believe that our weighting scheme provides a similar bound and conjecture so:

Conjecture 1. The minimum degree weighting scheme has $O(n^2)$ cover time on all graphs G.

3   Random Graphs of a Given Degree Sequence

From here on we study a sequence of random graphs on n vertices, where n goes to infinity.

Define $\mathcal{G}(\mathbf{d}_n)$ to be the set of connected simple graphs on the vertex set $V = [n]$ and with degree sequence $\mathbf{d}_n = (d_1^{(n)}, d_2^{(n)}, \ldots, d_n^{(n)})$ where $d_i^{(n)} = d(i)$ is the degree of vertex $i \in V$. Clearly, restrictions on degree sequences are required in order for the model to make sense. An obvious one is that the sum of the degrees in the sequence cannot be odd. Even then, not all degree sequences are *graphical* and not all graphical sequences can produce simple graphs. Take for example the two vertices v and w where $d_v = 3$ and $d_w = 1$. In order to study this model, we restrict the degree sequences to those which are *nice* and graphs which have nice degree sequences are termed the same. The precise definition is given below.

Let
$$\omega = \omega_n = \log \log \log n.$$

For a degree sequence $\mathbf{d}_n$, Let $d = d(\mathbf{d}_n) = d_1^{(n)}$ be the minimum, $\theta = \theta(\mathbf{d}_n)$ the average and $\Delta = \Delta(\mathbf{d}_n) = d_n^{(n)}$ the maximum of the entries in $\mathbf{d}_n$. Let $n_d = \sum_{i=1}^{n} \mathbf{1}_{\{d(i)=d\}}$, that is, the total number of entries in $\mathbf{d}_n$ with value d. We emphasise that d can grow with n – it need not be a fixed integer.

A sequence $(\mathbf{d}_n)_n$ of degree sequences is *nice* if the following conditions are satisfied: For each $\mathbf{d}_n$, **(i)** $n\theta$ is even, **(ii)** $d \geq 3$, **(iii)** $\Delta \leq \omega^{1/4}$, and **(iv)** for some constant $\alpha \in (0,1]$, $n_d/n \to \alpha$ as $n \to \infty$.

Condition **(ii)** ensures the graph is connected (**whp**), and condition **(iii)** is required for our proofs in subsequent lemmas.

To understand condition **(iv)** consider that without it, the sequence of degree sequences could result in sequences of random graphs that have wildly different cover times. As such we may not have convergence. The condition itself is fairly liberal – we do not require the that the degree sequence as a whole converges to a fixed distribution, nor even that d converges to some fixed constant.

Examples of nice sequences/graphs are: Any d-regular graph where $d \leq \omega^{1/4}$; a graph where a positive fraction of the vertices have bounded degree at least 3 and the rest have unbounded degree at most $\omega^{1/4}$; a truncated power-law graph with minimal degree at least 3 and maximal degree at most $\omega^{1/4}$.

In [1], the authors prove the following asymptotic result on the cover time of simple random walks on nice graphs:

Theorem 2 ([1]). *Let $(\mathbf{d}_n)_n$ be nice and let G be chosen **uar** from $\mathcal{G}(\mathbf{d}_n)$. Then **whp**,*

$$\mathbf{COV}[G] = (1 + o(1))\frac{d-1}{d-2}\frac{\theta}{d} n \log n, \tag{3}$$

where d is the effective minimum degree and θ is the average degree.

The *effective minimum degree* is the smallest integer d which satisfies condition **(iv)** above. It coincides with the minimum degree in our context.

We prove the following:

Theorem 3. *Let $(\mathbf{d}_n)_n$ be nice and let G be chosen **uar** from $\mathcal{G}(\mathbf{d}_n)$. Weight the edges of G with the min-deg weighting scheme, that is, for an edge (u,v), assign it weight $w(u,v) = 1/\min\{d(u), d(v)\}$. Denote the resulting graph G_w. Then **whp**,*

$$\mathbf{COV}[G_w] \leq (1 + o(1))\frac{d-1}{d-2} 8n \log n. \tag{4}$$

where d is the minimum degree.

Note that the assumptions on the degree sequence allow for the ratio θ/d to be unbounded. As such, the ratio of the min-deg cover time to the simple cover time, that is, the *speed up*, can be unbounded.

Typical Graphs. Our analysis requires that graphs G taken from $\mathcal{G}(\mathbf{d}_n)$ have certain structural properties. The subset of graphs $\mathcal{G}'(\mathbf{d}_n)$ having these properties form a large proportion of $\mathcal{G}(\mathbf{d}_n)$, in fact, $|\mathcal{G}'(\mathbf{d}_n)|/|\mathcal{G}(\mathbf{d}_n)| = 1 - n^{-\Omega(1)}$ when $(\mathbf{d}_n)_n$ is nice [1]. We term graphs in $\mathcal{G}'(\mathbf{d}_n)$ for $(\mathbf{d}_n)_n$ nice as *typical*, so a graph G drawn **uar** from $\mathcal{G}(\mathbf{d}_n)$ will be typical **whp**.

We need not list all the properties of typical graphs, but we shall use their useful consequences, amongst which are that they are connected, simple, and non-bipartite.

4 Convergence to the Stationary Distribution

In this section we begin with a brief overview of results on Markov chains and random walks on (weighted) graphs. For details, we refer the reader to, e.g., [2,10,11].

Since a weighted random walk on $G = (V, E, w)$ is a reversible Markov chain, we can apply standard results for these types of processes. For example, if G is non-bipartite, then the walk converges to a stationary distribution π, where $\pi_u = \pi(u) = w(u)/w(G)$.

Furthermore, the rate of convergence is related to the *absolute spectral gap* - the difference between the largest eigenvalue, 1 and second largest (in magnitude) eigenvalue λ_* of the probability transition matrix of the walk. Specifically, if $P_u^{(t)}(v) = \mathbf{Pr}(\mathcal{W}_u(t) = v)$ then

$$|P_u^{(t)}(v) - \pi_v| \leq \sqrt{\frac{\pi_v}{\pi_u}} \lambda_*^t. \tag{5}$$

If the walk is made *lazy*, that is, if we append a looping probability of $1/2$ and scale all other transition probabilities accordingly, then the largest eigenvalue remains 1 and second eigenvalue λ_2 is guaranteed to be the second largest in absolute terms. We can then apply the following result, proved independently in [9,12]:

Theorem 4 ([9], [12]). *Let λ_2 be the second largest eigenvalue of a reversible, aperiodic transition matrix $\mathbf{P}$. Then*

$$\frac{\Phi^2}{2} \leq 1 - \lambda_2 \leq 2\Phi \tag{6}$$

where Φ is the conductance.

Corollary 2.

$$|P_u^{(t)}(v) - \pi_v| \leq \sqrt{\frac{\pi_v}{\pi_u}} \left(1 - \frac{\Phi^2}{2}\right)^t. \tag{7}$$

Conductance is defined as follows:

Definition 1 (Conductance). *Let $\mathcal{M}$ be an irreducible, aperiodic Markov chain on some state space Ω. Let the stationary distribution of $\mathcal{M}$ be π with $\pi(x)$ denoting the stationary probability of $x \in \Omega$. Let P be the transition* *matrix for* $\mathcal{M}$. *For* $x, y \in \Omega$ *let* $Q(x, y) = \pi(x)P_{x,y}$ *and for sets* $A, B \subseteq \Omega$, *let* $Q(A, B) = \sum_{x \in A, y \in B} Q(x, y)$. *The* conductance *of* $\mathcal{M}$ *is the quantity*

$$\Phi = \Phi(\mathcal{M}) = \min_{\substack{S \subseteq \Omega \\ \pi(S) \leq 1/2}} \frac{Q(S, \overline{S})}{\pi(S)} \tag{8}$$

where $\pi(S) = \sum_{x \in S} \pi(x)$, *and* $\overline{S} = \Omega \setminus S$.

For a graph G weighted by function w, we write $\Phi(G_w)$ for the conductance of the weighted random walk on G.

Making the walk lazy halves the conductance and doubles the important quantity R_v, which we shall define and elaborate upon below. It also doubles the cover time.

In fact, we do not need to maintain a lazy walk all the time, but will do so only for the duration of the *mixing time* T which we define as follows:

$$T = \omega^2 \log n. \tag{9}$$

Informally, the mixing time is how long it takes for the distribution of a Markov chain to be close to the stationary distribution. After the mixing time, we can revert to the non-lazy walk. It will be seen that the lazy steps during the mixing time will have negligible impact on the asymptotic cover time, since, being poly-logarithmic, it is short compared to other quantities such as hitting time which are linear in n and dominate over it.

More precisely, we show below that for most nice graphs, for any $t \geq T$

$$|P_u^{(t)}(x) - \pi_v| \leq \frac{1}{n^3}, \tag{10}$$

for any vertices u and v in G. This is a corollary of the following lemma:

Lemma 3. *Let* $(d_n)_n$ *be nice and let* G *be chosen* **uar** *from* $\mathcal{G}(d_n)$. *Let* G_w *be* G *weighted with the min-deg weighting scheme. Then* $\Phi(G_w) \geq 1/(100\Delta)$ **whp**, *where* Δ *is the maximum degree.*

We will consider the condition $\Phi(G_w) \geq 1/(100\Delta)$ to be one of the typical properties.

Corollary 3. *For a random walk on a weighted typical graph* G, *we have for* $t \geq T$,

$$|P_u^{(t)}(v) - \pi_v| \leq n^{-3},$$

where $P_u^{(t)}(v)$ *is the probability that the minimum degree random walk is at node* v *at time* t, *given that it started at node* u.

5 First Visit Lemma

The hitting time from the stationary distribution, $\mathbf{H}[\pi, v] = \sum_{u \in V} \pi_u \mathbf{H}[u, v]$, can be expressed as $\mathbf{H}[\pi, v] = Z_{v,v}/\pi_v$, where

$$Z_{v,v} = \sum_{t=0}^{\infty} (P_v^{(t)}(v) - \pi_v), \tag{11}$$

see e.g. [2]. For a (weighted or unweighted) random walk $\mathcal{W}_v$, starting from v define

$$R_v(T) = \sum_{t=0}^{T-1} P_v^{(t)}(v). \tag{12}$$

Thus R_v is the expected number of returns made by $\mathcal{W}_v$ to v during the mixing time, in the graph G. We note that $R_v \geq 1$, as $P_v^{(0)}(v) = 1$.

Let $D(t) = \max_{u,x} |P_u^{(t)}(x) - \pi_x|$. As $\pi_x \geq 1/n^2$ for any vertex of a simple graph, (10) implies that $D(t) \leq \pi_x$ for all $x \in V$ if $t \geq T$.

Lemma 4. *For a random walk $\mathcal{W}_u$ on a graph G (weighted or unweighted), suppose T satisfies (10). Let vertex $v \in V$ be such that $T\pi_v = o(1)$, and $\pi_v < 1/2$, then*

$$\mathbf{H}[\pi, v] = (1 + o(1)) \frac{R_v(T)}{\pi_v}. \tag{13}$$

Let $\boldsymbol{A}_t(v)$ denote the event that $\mathcal{W}_u$ does not visit v in steps $0, ..., t$. We next derive a crude upper bound for $\mathbf{Pr}(\boldsymbol{A}_t(v))$ in terms of $\mathbf{H}[\pi, v]$.

Lemma 5. *For a random walk $\mathcal{W}_u$ on a graph G (weighted or unweighted), suppose T satisfies (10), then*

$$\mathbf{Pr}(\boldsymbol{A}_t(v)) \leq \exp\left(\frac{-(1 - o(1))\lfloor t/\tau_v \rfloor}{2} \right),$$

where $\tau_v = T + 2\mathbf{H}[\pi, v]$.

In order to apply Lemma 5, we shall need to show that the conditions of Lemma 4 are satisfied, and we will need to bound R_v. We start off by bounding the stationary distribution:

Lemma 6. *For a vertex u,*

$$\frac{1}{2n} \leq \pi_u \leq \frac{d(u)}{n}. \tag{14}$$

Corollary 4. $T\pi_u = o(1)$.

We see that for nice degree sequences, the conditions of Lemma 4 are satisfied. It remains to bound R_v, the expected number of returns in the mixing time.

5.1   The Number of Returns in the Mixing Time

Let Γ_v denote the subgraph of G induced by all vertices within distance ω of v. From [1], and in conjunction with the restriction $\Delta \le \omega^{1/4}$, we have the following, which we shall consider to be a typical property:

Proposition 1 ([1]). *With high probability, Γ_v is either a tree or has a unique cycle.*

Let Γ_v° be the set of vertices in Γ_v that are at distance ω from v.

Lemma 7. *Suppose G_w is typical and weighted with the min-degree scheme. Let $\mathcal{W}_v^*$ denote the (weighted) walk on Γ_v starting at v with Γ_v° made into absorbing states. Assume further that there are no cycles in Γ_v°. Let $R_v^* = \sum_{t=0}^\infty r_t^*$ where r_t^* is the probability that $\mathcal{W}_v^*$ is at vertex v at time t. Then*

$$R_v = R_v^* + O(\sqrt{\omega} e^{-\Omega(\sqrt{\omega})}).$$

We apply this in the proof of the following lemma:

Lemma 8. *Suppose $\Delta \le \omega^{1/4}$, G_w is typical and weighted with the min-degree scheme. Let v be a vertex in G_w.*

   (a) *If Γ_v is a tree, $R_v \le \frac{d-1}{d-2} + O(\sqrt{\omega} e^{-\Omega(\sqrt{\omega})})$.*
   (b) *$R_v \le \frac{d}{d-2}\frac{d-1}{d-2} + O(\sqrt{\omega} e^{-\Omega(\sqrt{\omega})}) \le 6 + O(\sqrt{\omega} e^{-\Omega(\sqrt{\omega})})$.*

5.2   The Number of Vertices Not Locally Tree-Like

We wish to bound the number of vertices v that are not locally tree-like, i.e., for which Γ_v has a cycle.

Lemma 9. *Suppose $G \in \mathcal{G}(\mathbf{d})$ is drawn **uar**. With probability at least $1 - n^{-\Omega(1)}$, the number of vertices not locally tree-like is at most $n^{1/10}$.*

6   Upper Bound on the Cover Time

Let the random variable c_u be the time taken by the (weighted) random walk $\mathcal{W}_u$ starting from vertex u to visit every vertex of a connected (weighted) graph G. Let U_t be the number of vertices of G which have not been visited by $\mathcal{W}_u$ by step t. We note the following:

$$\mathbf{COV}_u[G] = \mathbf{E}[c_u] = \sum_{t>0} \mathbf{Pr}(c_u \ge t), \qquad (15)$$

$$\mathbf{Pr}(c_u \ge t) = \mathbf{Pr}(c_u > t-1) = \mathbf{Pr}(U_{t-1} > 0) \le \min\{1, \mathbf{E}[U_{t-1}]\}. \qquad (16)$$

Recall that $\mathbf{A}_s(v)$ is the event that vertex v has not been visited by time s. It follows from (15), (16) that

$$\mathbf{COV}_u[G] \le t+1+\sum_{s\ge t} \mathbf{E}[U_s] = t+1+\sum_{v}\sum_{s\ge t} \mathbf{Pr}(\mathbf{A}_s(v)). \qquad (17)$$

We use Lemmas 4 and 5, which hold for weighted random walks (see Chap. 2, General Markov Chains, in [2] for justification of (11) and the inequality $D(s + t) \leq 2D(s)D(t)$. All other expressions in the proofs hold for weighted random walks). Thus,

$$\mathbf{Pr}(\boldsymbol{A}_t(v)) \leq \exp\left(\frac{-(1 - o(1))\lfloor t/\tau_v \rfloor}{2}\right),$$

where $\tau_v = T + 2\mathbf{H}[\pi, v]$ and $\mathbf{H}[\pi, v] = (1 + o(1))R_v/\pi_v$.

Hence, for a given v,

$$\sum_{s \geq t} \mathbf{Pr}(\boldsymbol{A}_s(v)) \leq \sum_{s \geq t} \exp\left(\frac{-(1 - o(1))\lfloor s/\tau_v \rfloor}{2}\right)$$

$$\leq \tau_v \sum_{s \geq \lfloor t/\tau_v \rfloor} \exp\left(\frac{-(1 - o(1))s}{2}\right)$$

$$\leq 3\tau_v \exp\left(\frac{-(1 - o(1))}{2}\left\lfloor \frac{t\pi_v}{T\pi_v + (1 + o(1))2R_v} \right\rfloor\right).$$

Since $T\pi_v = o(1)$ and $\pi_v \geq 1/2n$ from (14), we get

$$\sum_{s \geq t} \mathbf{Pr}(\boldsymbol{A}_s(v)) \leq 3\tau_v \exp\left(\frac{-(1 - o(1))}{2}\left\lfloor \frac{t}{(1 + o(1))4nR_v} \right\rfloor\right)$$

Let $t = t^* = (1 + \epsilon)8\frac{d-1}{d-2}n \log n$ where $\epsilon \to 0$ sufficiently slowly. Then

$$\sum_{s \geq t} \mathbf{Pr}(\boldsymbol{A}_s(v)) \leq 3\tau_v \exp\left(-(1 + \Theta(\epsilon))\frac{d-1}{d-2}\frac{\log n}{R_v}\right) \tag{18}$$

We partition the double sum $\sum_v \sum_{s \geq t} \mathbf{Pr}(\boldsymbol{A}_s(v))$ from (17) into

$$\sum_{v \in V_A} \sum_{s \geq t} \mathbf{Pr}(\boldsymbol{A}_s(v)) + \sum_{v \in V_B} \sum_{s \geq t} \mathbf{Pr}(\boldsymbol{A}_s(v))$$

where V_A are locally tree-like and V_B are not.

If v is locally tree-like, then using Theorem 8 (a), the RHS of (18) is bounded by

$$3\tau_v n^{-(1+\Theta(\epsilon))} = 3(T + 2(1 + o(1))R_v/\pi_v)n^{-(1+\Theta(\epsilon))} = O(1)n^{-\Theta(\epsilon)}.$$

Thus,

$$\sum_{v \in V_A} \sum_{s \geq t} \mathbf{Pr}(\boldsymbol{A}_s(v)) \leq O(1)n^{1-\Theta(\epsilon)} = o(t). \tag{19}$$

For any v (i.e., including those not locally tree-like), (18) is bounded by

$$3\tau_v n^{-(1+\Theta(\epsilon))\frac{d-1}{6(d-2)}} \leq O(1)n^{1-(1+\Theta(\epsilon))\frac{d-1}{6(d-2)}} \tag{20}$$

Using Lemma 9 to sum the bound (20) over all non locally tree-like vertices, we get

$$\sum_{v \in V_B} \sum_{s \geq t} \mathbf{Pr}(A_s(v)) \leq O(1) n^{\frac{1}{10} + 1 - (1 + \Theta(\epsilon)) \frac{d-1}{6(d-2)}} = O(n^{\frac{1}{2}}) = o(t). \qquad (21)$$

Hence, combining (17), (19) and (21) for $t = t^*$, Theorem 3 follows.

Compare this with (3), we see that the speed up,

$$S = \frac{\mathbf{COV}[G]}{\mathbf{COV}[G_w]} = \Omega(\theta),$$

Therefore $S \to \infty$ as $n \to \infty$ if $\theta \to \infty$ as $n \to \infty$. That is, we can have an unbounded speed up.

We conjecture that the following tighter bound holds:

Conjecture 2. Equation (4) can be replaced by

$$\mathbf{COV}[G_w] \leq (1 + o(1)) \frac{d-1}{d-2} n \log n.$$

References

1. Abdullah, M., Cooper, C., Frieze, A.M.: Cover time of a random graph with given degree sequence. Discrete Math. **312**(21), 3146–3163 (2012)
2. Aldous, D., Fill, J.: Reversible Markov Chains and Random Walks on Graphs. http://stat-www.berkeley.edu/pub/users/aldous/RWG/book.html
3. Aleliunas, R., Karp, R.M., Lipton, R.J., Lovasz, L., Rackoff, C.: Random walks, universal traversal sequences, and the complexity of maze problems. In: Proceedings of the 20th Annual IEEE Symposium on Foundations of Computer Science, pp.218–223 (1979)
4. Chandra, A.K., Raghavan, P., Ruzzo, W.L., Smolensky, R., Tiwari, P.: The electrical resistance of a graph captures its commute and cover times. Comput. Complex. **6**, 312–340 (1997)
5. Doyle, P.G., Snell, J.L.: Random Walks and Electrical Networks (2006)
6. Feige, U.: A tight upper bound for the cover time of random walks on graphs. Random Struct. Algorithms **6**, 51–54 (1995)
7. Feige, U.: A tight lower bound for the cover time of random walks on graphs. Random Struct. Algorithms **6**, 433–438 (1995)
8. Ikeda, S., Kubo, I., Yamashita, M.: The hitting and the cover times of random walks on finite graphs using local degree information. Theor. Comput. Sci. **410**, 94–100 (2009)
9. Jerrum, M., Sinclair, A.: Approximating the permanent. SIAM J. Comput. **18**, 1149–1178 (1989)
10. Levin, D.A., Peres, Y., Wilmer, E.L.: Markov Chains and Mixing Times. AMS Press (2008)
11. Lovász, L.: Random walks on graphs: a survey. In: Miklós, D., Sós, V.T., Szonyi, T. (eds.) Combinatorics, Paul Erdős is Eighty, vol. 2, János Bolyai Mathematical Society, Budapest, pp. 353–398 (1996)

12. Lawler, G.F., Sokal, A.D.: Bounds on the L^2 spectrum for Markov chains and Markov processes: a generalization of Cheeger's inequality. Tran. Amer. Math. Soc. **309**, 557–580 (1988)
13. Matthews, P.: Covering problems for Markov chains. Ann. Prob. **16**, 1215–1228 (1988)

Minimum Activation Cost Edge-Disjoint Paths in Graphs with Bounded Tree-Width

Hasna Mohsen Alqahtani[✉] and Thomas Erlebach

Department of Computer Science, University of Leicester, Leicester, UK
{hmha1,terlebach}@leicester.ac.uk

Abstract. In *activation network* problems we are given a directed or undirected graph $G = (V, E)$ with a family $\{f_{uv} : (u, v) \in E\}$ of monotone non-decreasing activation functions from D^2 to $\{0, 1\}$, where D is a constant-size subset of the non-negative real numbers, and the goal is to find activation values $x_v \in D$ for all $v \in V$ of minimum total cost $\sum_{v \in V} x_v$ such that the activated set of edges satisfies some connectivity requirements. We propose an algorithm that optimally solves the *minimum activation cost of k edge-disjoint st-paths* (st-MAEDP) problem in $O(|V||D|^{tw+1}tw^3(k+1)^{(tw+3)2^{(tw+3)}}(tw+3)^{2(tw+3)+3})$ time for graphs with treewidth bounded by a constant tw.

1   Introduction

The *activation network* setting can be defined as follows. We are given a directed or undirected graph $G = (V, E)$ together with a family $\{f_{uv} : (u, v) \in E\}$ of monotone non-decreasing activation functions from D^2 to $\{0, 1\}$, where D is a constant-size subset of the non-negative real numbers, such that the activation of an edge depends on the chosen values from the domain D at its endpoints. An edge $(u, v) \in E$ is *activated* for chosen values x_u and x_v if $f_{uv}(x_u, x_v) = 1$, and the activation function f_{uv} is called *monotone non-decreasing* if $f_{uv}(x_u, x_v) = 1$ implies $f_{uv}(y_u, y_v) = 1$ for any $y_u \geq x_u$, $y_v \geq x_v$. The objective of activation network problems is to find activation values $x_v \in D$ for all $v \in V$ such that the total activation cost $\sum_{v \in V} x_v$ is minimized and the activated set of edges satisfies some connectivity requirements. Activation problems generalize several problems studied in the network literature such as power optimization, minimum broadcast tree and installation cost optimization. Several activation problems have been studied for arbitrary graphs in the recent research literature. Unfortunately, many of these problems are computationally hard. Obtaining a polynomial-time approximation algorithm is a typical approach to deal with an NP-hard problem. Another important approach is studying the problem on graph classes with nice decomposition properties such as bounded treewidth graphs to determine efficient algorithms. Panigrahi [8] shows that the fundamental problem of finding *minimum activation k edge-disjoint st-paths* (st-MAEDP) is NP-hard. It is an interesting area of research to investigate this problem for graphs with treewidth bounded by a constant tw. In this paper, we focus on developing a

© Springer International Publishing Switzerland 2016
Z. Lipták and W.F. Smyth (Eds.): IWOCA 2015, LNCS 9538, pp. 13–24, 2016.
DOI: 10.1007/978-3-319-29516-9_2

polynomial-time algorithm that optimally solves the st-MAEDP for graphs of bounded treewidth. (Note that k is not a constant but part of the input.)

Related Work. Activation network problems were first introduced by Panigrahi in [8] and various of these problems have been investigated in [1,2,6–8]. The *minimum activation st-path* (MAP) problem is optimally solvable in polynomial-time [8]. However, the *minimum spanning activation tree* (MSpAT) problem is NP-hard to approximate within a factor of $o(\log n)$ and there exists a $O(\log n)$-approximation algorithm for this problem [8]. Activation network problems generalize several problems studied in the network literature such as power optimization problems, which are modelled by a graph $G = (V, E)$ where each edge (u, v) in G is assigned a threshold power requirement θ_{uv}. For undirected graphs, an edge (u, v) is activated for chosen values x_u and x_v if each of these values is at least θ_{uv}, and it is activated if $x_u \geq \theta_{uv}$ for the directed case. The *minimum power st-path* (MPP) problem can be solved in polynomial time for both directed and undirected graphs [5]. For directed graphs, the *minimum power k node-disjoint st-paths* problem is also optimally solvable in polynomial time [4,11]. However, the *minimum power k edge-disjoint st-paths* problem is unlikely to admit even a polylogarithmic approximation algorithm for both the directed and undirected variants [4].

There is a $2k$-approximation algorithm for the st-MAEDP problem and a 2-approximation algorithm for the *minimum activation node-disjoint st-paths* (st-MANDP) problem [6]. In [1], we have studied the st-MAEDP and st-MANDP problems when $k = 2$, denoted by st-MA2EDP and st-MA2NDP respectively. We proved that a ρ-approximation algorithm for the st-MA2NDP problem implies a ρ-approximation algorithm for the st-MA2EDP problem. In the same paper, we obtained a 1.5-approximation algorithm for the st-MA2NDP problem and hence for the st-MA2EDP problem. The problems st-MAEDP and st-MANDP for the restricted version of activation networks with $|D| = 2$ and a single activation function for all edges have also been studied in [1]. Under this restriction, the st-MANDP problem is optimally solvable in polynomial time for arbitrary k (except for one case of the activation function, in which we require $k = 2$). The st-MAEDP problem, however, remains NP-hard [1,8]. So far these problems for an arbitrary constant-size D and fixed $k \geq 2$ are not known to be NP-hard. Recently, in [2], we considered the st-MANDP and the problem of finding *minimum activation cost node-disjoint paths* (MANDP) between k disjoint terminal pairs, $(s_1, t_1), \ldots, (s_k, t_k)$, for graphs of bounded treewidth. We proposed algorithms that optimally solve the st-MANDP problem in polynomial-time and the MANDP problem in linear-time for graphs with bounded treewidth [2]. There exists a polynomial-time algorithm that optimally solves the st-MA2EDP problem for graphs of bounded treewidth [1,2]. Other relevant work, applications and motivations of activation network problems have been addressed in [1,2,6–8].

Our Results. We develop a polynomial-time algorithm for the st-MAEDP problem for graphs with treewidth bounded by tw. Our algorithm efficiently combines an edge-coloring scheme with dynamic programming over a *nice tree-decomposition* (see Sect. 2). The edge-coloring scheme was introduced in [12] to develop a polynomial-time algorithm that solves the *minimum shared-edge kst-paths* (MSEP) problem (finding k paths between s and t with minimum number of shared edges among the paths) for graphs of bounded treewidth. The MSEP is a generalization of the problem of finding k edge disjoint st-paths.

The rest of the paper is organized as follows. In Sect. 2 we recall some definitions and results of the class of graphs with bounded treewidth. Then in Sect. 3, we present a polynomial-time algorithm that solves the st-MAEDP problem optimally for graphs with treewidth bounded by tw. We conclude the paper by stating some open problems in Sect. 4.

2 Preliminaries

In this paper we consider the class of graphs of bounded treewidth. A graph $G = (V, E)$ has treewidth tw if it has a *tree-decomposition* of width tw [9]. The *tree-decomposition* concept is defined as follows.

Definition 1. *Given a graph $G = (V, E)$, a tree $\mathcal{T} = (I, F)$ and a family $\mathcal{X} = \{X_i\}_{i \in I}$ of subsets of V (called bags). The pair $(\mathcal{X}, \mathcal{T})$ is called a tree-decomposition of G if it satisfies the following conditions:*

- $V = \bigcup_{i \in I} X_i$.
- *For every edge $(v, w) \in E$, there exists an $i \in I$ with $v \in X_i$ and $w \in X_i$.*
- *For every vertex $v \in V$, the nodes $i \in I$ with $v \in X_i$ form a subtree of $\mathcal{T}$.*

The width of $(\mathcal{X}, \mathcal{T})$ is the number $\max_{i \in I} |X_i| - 1$. The treewidth tw of the graph G is the minimum width among all possible tree-decompositions of the graph.

Theorem 1 ([3]). *For any fixed tw, there exists a linear-time algorithm that checks whether a given graph $G = (V, E)$ has treewidth at most tw, and if so, outputs a tree-decomposition $(\mathcal{X}, \mathcal{T})$ of G with width at most tw.*

Definition 2. *A tree-decomposition $(\mathcal{X}, \mathcal{T})$ is called a* nice *tree-decomposition, if $\mathcal{T}$ is a binary tree rooted at some $r \in I$ that satisfies the following:*

- *Each node is either a leaf, or has exactly one or two children.*
- *Let $i \in I$ be a leaf. Then $X_i = \{u, v\}$ for some $(u, v) \in E$.*
- *For every edge $(u, v) \in E$, there is exactly one leaf $i \in I$ such that $u, v \in X_i$. (We say that the edge (u, v) is associated with that leaf $i \in I$).*
- *Let $j \in I$ be the only child of $i \in I$, then either $X_i = X_j \cup \{v\}$ or $X_i = X_j \setminus \{v\}$. The node i is called an introduce node or forget node, respectively.*
- *Let $j, j' \in I$ be the two child nodes of a node $i \in I$, then $X_j = X_{j'} = X_i$. The node i is called a join node of $\mathcal{T}$.*

Scheffler presented in [10] a special tree-decomposition that follows the structure of a nice tree-decomposition as defined above but with no restriction on the size of leaf bags (i.e., it does not require leaf bags to be of size 2). We call that type of tree-decomposition a Scheffler-type nice tree-decomposition. Any given tree-decomposition for a graph $G = (V, E)$ with treewidth at most tw can be easily converted into a Scheffler-type nice tree-decomposition of width tw and size $O(|V|)$ in linear time, if tw is a fixed constant [10]. One can also transform a Scheffler-type nice tree-decomposition to have leaves with bags of size 2 in linear time. Therefore, a nice tree-decomposition with leaf bags of size 2 and width tw can be constructed from any given tree-decomposition of the same treewidth. However, each leaf node of a Scheffler-type nice tree-decomposition may produce $O(tw^3)$ new nodes in the construction of leaves with size 2. The resulting tree-decomposition, therefore, has $O(|V|tw^3)$ nodes.

Throughout this paper, we consider simple undirected graphs $G = (V, E)$ given as an input with a nice tree-decomposition of width at most tw. We define X_i^+ to be the set of all vertices in X_j for all nodes $j \in I$ such that $j = i$ or j is a descendant of i. We denote by G_i^+ a partial graph of G. For a leaf node i, G_i^+ is the subgraph of G with vertex set X_i and the edge of G that is associated with i. For a non-leaf node i, G_i^+ is the graph that is the union of G_j^+ over all children j of i. Note that the graph G_r^+ for the root r of the tree-decomposition is equal to G.

3   Minimum Activation Cost k Edge-Disjoint st-Paths

Given are an activation network $G = (V, E)$ and a pair of source and destination vertices $s, t \in V$. In this section, we consider the st-MAEDP problem where the goal is to find activation values $\{x_v : v \in V\}$ of minimum total cost $\sum_{v \in V} x_v$ such that the activated set of edges contains k edge-disjoint st-paths $\mathcal{P}^{st} = \mathcal{P}_1, \ldots, \mathcal{P}_k$. We present a polynomial-time algorithm that solves the st-MAEDP problem optimally in the case of graphs of bounded tree-width using dynamic programming techniques. The algorithm follows a bottom-up approach to compute a number of possible sub-solutions per nice tree-decomposition node $i \in I$. It is easy to compute the sub-solutions for a leaf node $i \in I$ because the partial graph G_i^+ consists of two vertices that are connected by an edge in G associated with i. For a non-leaf node, we use the information previously computed for its children. The algorithm also constructs a table tab_i to store the computed information for each node $i \in I$.

We use an edge-coloring scheme to compute the sub-solutions per tree node. Let $C = \{0, 1, \ldots, k\}$ be a set of colors. We consider a coloring $f_i : E(G_i^+) \to C$ for the graph G_i^+. For each color $c \in C$, we define $G_i^+(f_i, c)$ to be the subgraph of G_i^+ induced by the edges with color c. Each color $c \in C \setminus \{0\}$ represents the edges used by $\mathcal{P}_c$. Denote by $P(X)$ the set of all possible partitions of the set X. We define $C(X_i) = (\mathcal{Y}_1, \ldots, \mathcal{Y}_k)$ to be a *color vector* on X_i such that $\mathcal{Y}_c \in P(X_i \cup \{s, t\})$ for all $c \in C \setminus \{0\}$. $P_{st}(X_i)$ denotes the set $P(X_i \cup \{s, t\})$. A color vector $C(X_i) = (\mathcal{Y}_1, \ldots, \mathcal{Y}_k)$ on X_i is called *active* if G_i^+ has a coloring f_i such that every element of the partition $\mathcal{Y}_c$, for each $c \in C \setminus \{0\}$, is a set resulting from the intersection between $X_i \cup \{s, t\}$ and the vertex set of a connected component of $G_i^+(f_i, c)$ (see Fig. 1 for an example of an active color vector). $\mathcal{Y}(X_i, f_i, c)$ denotes the partition $\mathcal{Y}_c$ of $X_i \cup \{s, t\}$. This color vector concept was introduced in [12] for developing a polynomial-time algorithm that optimally

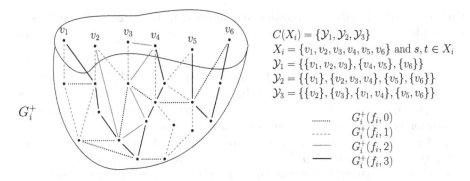

$$C(X_i) = \{\mathcal{Y}_1, \mathcal{Y}_2, \mathcal{Y}_3\}$$
$$X_i = \{v_1, v_2, v_3, v_4, v_5, v_6\} \text{ and } s,t \in X_i$$
$$\mathcal{Y}_1 = \{\{v_1, v_2, v_3\}, \{v_4, v_5\}, \{v_6\}\}$$
$$\mathcal{Y}_2 = \{\{v_1\}, \{v_2, v_3, v_4\}, \{v_5\}, \{v_6\}\}$$
$$\mathcal{Y}_3 = \{\{v_2\}, \{v_3\}, \{v_1, v_4\}, \{v_5, v_6\}\}$$

$$\cdots \cdots \quad G_i^+(f_i, 0)$$
$$\text{-----} \quad G_i^+(f_i, 1)$$
$$\text{——} \quad G_i^+(f_i, 2)$$
$$\text{——} \quad G_i^+(f_i, 3)$$

Fig. 1. An example of an active color vector

solves the MSEP problem for graphs with bounded treewidth. One simple way to compute the sub-solutions is by storing a color vector and an activation-value function per each row of table tab_i. However, the number of color vectors per node i can only be bounded by $(tw + 3)^{k(tw+3)}$ because $|X_i| \leq tw + 1$, $|X_i \cup \{s,t\}| \leq tw + 3$ and $|P_{st}(X_i)| \leq (tw + 3)^{tw+3}$. Therefore, the number of color vectors is not always polynomially bounded. Hence the size of the table tab_i is also not always polynomially bounded. Therefore we define a mapping $\gamma_i : P_{st}(X_i) \rightarrow \{0, 1, \ldots, k\}$ to be a *count* on X_i to obtain a polynomial-time algorithm to optimally solve the st-MAEDP problem. We say that the count γ_i on X_i is an *active count* if G_i^+ has a coloring f_i with a color vector $C(X_i) = (\mathcal{Y}_1, \ldots, \mathcal{Y}_k)$ such that $\gamma_i(A) = |\{c \in C \setminus \{0\} : A = \mathcal{Y}_c\}|$ for each $A \in P_{st}(X_i)$. For any active count γ_i, it is clear that $\sum_{A \in P_{st}(X_i)} \gamma_i(A) = k$. In Fig. 1, the counts would be 1 for all partitions $A \in \{\mathcal{Y}_1, \mathcal{Y}_2, \mathcal{Y}_3\} \subseteq P_{st}(X_i)$ and 0 otherwise. Consider a solution $\mathcal{P} = \mathcal{P}_1, \ldots, \mathcal{P}_k$ for the st-MAEDP problem. Let $\mathcal{P}^i = \mathcal{P}[G_i^+]$ be the induced solution in a partial graph G_i^+ (i.e., the set of vertices and edges that are both in $\mathcal{P}$ and in G_i^+). Since the interaction between $\mathcal{P}^i$ in G_i^+ and the rest of the graph happens only in vertices of X_i, we can consider an activation-value function $\Lambda_i : X_i \rightarrow D$ and a count $\gamma_i : P_{st}(X_i) \rightarrow \{0, 1, \ldots, k\}$ to represent $\mathcal{P}^i$ in G_i^+. The idea of using counts instead of color vectors is based on [12].

3.1   Processing the Tree Decomposition

For a tree node $i \in I$, the table tab_i has multiple rows and each row represents a solution of a unique combination of a count γ_i and a function of activation values Λ_i. Let $val(\gamma_i, \Lambda_i)$ denote the minimum cost value of an assignment of activation values for G_i^+ which satisfies the restrictions Λ_i and activates an edge-colored subgraph of G_i^+ that satisfies the count γ_i. The value $val(\gamma_i, \Lambda_i)$ is also stored in tab_i. We compute the sub-solution tables starting at the leaves towards the root.

Leaf. Let $i \in I$ be a leaf, $X_i = \{u, v\}$. Let (γ_i, Λ_i) be any row of tab_i. We distinguish the following cases and define the value $val(\gamma_i, \Lambda_i)$ for each case.

Each case corresponds to a possible sub-solution in G_i^+. Recall that G_i^+ is a single edge. The sub-solution's cost $val(\gamma_i, \Lambda_i)$ is set to $\sum_{v \in X_i} \Lambda_i(v)$ if one of the following cases applies:

- $\gamma_i(A') = k - 1$ for $A' = \{\{u\}, \{v\}, \{s\}, \{t\}\}$ and $\gamma_i(A'') = 1$ for $A'' = \{\{u, v\}, \{s\}, \{t\}\}$ and $\gamma_i(A) = 0$ for all $A \in P_{st}(X_i) \setminus \{A', A''\}$, where $f_{uv}(\Lambda_i(u), \Lambda_i(v)) = 1$ and $s, t \notin X_i$. Intuitively, this means that the sub-solution is a path with one edge not containing s or t.
- $\gamma_i(A') = k - 1$ for $A' = \{\{u\}, \{v\}\}$ and $\gamma_i(A'') = 1$ for $A'' = \{\{u, v\}\}$ and $\gamma_i(A) = 0$ for all $A \in P_{st}(X_i) \setminus \{A', A''\}$, where $f_{uv}(\Lambda_i(u), \Lambda_i(v)) = 1$ and $s, t \in X_i$. Intuitively, this means that the sub-solution is a path with one edge containing s and t.
- $\gamma_i(A') = k - 1$ for $A' = \{\{u\}, \{v\}, \{s\}\}$ and $\gamma_i(A'') = 1$ for $A'' = \{\{u, v\}, \{s\}\}$ and $\gamma_i(A) = 0$ for all $A \in P_{st}(X_i) \setminus \{A', A''\}$, where $f_{uv}(\Lambda_i(u), \Lambda_i(v)) = 1$ and $s \notin X_i$ and $t \in X_i$. Intuitively, the sub-solution is a path with one edge and one endpoint equal to t. (The roles of s and t can be exchanged.)
- $\gamma_i(A') = k$ for $A' = \{\{u\}, \{v\}, \{s\}, \{t\}\}$ and $\gamma_i(A) = 0$ for all $A \in P_{st}(X_i) \setminus \{A'\}$, where $s, t \notin X_i$. Intuitively, this means that the sub-solution has no edges.
- $\gamma_i(A') = k$ for $A' = \{\{u\}, \{v\}\}$ and $\gamma_i(A) = 0$ for all $A \in P_{st}(X_i) \setminus \{A'\}$, where $s, t \in X_i$. Intuitively, this means that the sub-solution has no edges.
- $\gamma_i(A') = k$ for $A' = \{\{u\}, \{v\}, \{s\}\}$ and $\gamma_i(A) = 0$ for all $A \in P_{st}(X_i) \setminus \{A'\}$, where $s \notin X_i$ and $t \in X_i$. Intuitively, this means that the sub-solution has no edges. (The roles of s and t can be exchanged.)

In these cases we construct an edge-colored subgraph of G_i^+ plus activation-values that may be part of a global solution. If none of the above cases applies, $val(\gamma_i, \Lambda_i) = +\infty$.

Introduce. Let $i \in I$ be an introduce node, and $j \in I$ its only child. We have $X_j \subset X_i$, $|X_i| = |X_j| + 1$ and let v be the additional vertex in X_i. The vertex v is isolated since i does not introduce edges in G_i^+. For every row (γ_j, Λ_j) in tab_j, there are $|D|$ rows in tab_i such that for all $A \in P_{st}(X_j)$ and all $u \in X_i \setminus \{v\}$, $\gamma_i(A \cup \{\{v\}\}) = \gamma_j(A)$ if $v \notin \{s, t\}$ and $\gamma_i(A) = \gamma_j(A)$ if $v \in \{s, t\}$ and $\Lambda_i(u) = \Lambda_j(u)$. The sub-solution's cost $val(\gamma_i, \Lambda_i)$ for these rows is set to $val(\gamma_j, \Lambda_j) + \Lambda_i(v)$.

Forget. Let $i \in I$ be a forget node, and $j \in I$ its only child. We have $X_i \subset X_j$, $|X_j| = |X_i| + 1$ and let v be the discarded vertex. Let A^{-v} be the partition $A \in P_{st}(X_j)$ after removing the vertex v if $v \notin \{s, t\}$ and A^{-v} equals the partition A if $v \in \{s, t\}$. Note that $A^{-v} \in P_{st}(X_i)$ for all $A \in P_{st}(X_j)$. For $A \in P_{st}(X_i)$, we say that $W(A)$ is the set of all partitions $B \in P_{st}(X_j)$ such that $B^{-v} = A$ (i.e., $W(A) = \{B \in P_{st}(X_j) : B^{-v} = A\}$). For each row (γ_i, Λ_i), we consider all (γ_j, Λ_j) such that for all $u \in X_i$ and $A \in P_{st}(X_i)$, $\Lambda_i(u) = \Lambda_j(u)$ and $\gamma_i(A) = \sum_{B \in W(A)} \gamma_j(B)$. The sub-solution's cost $val(\gamma_i, \Lambda_i)$ is the minimum of $val(\gamma_j, \Lambda_j)$ over all these (γ_j, Λ_j).

Join. Let $i \in I$ be a join node, and $j, j' \in I$ its two children. We have $X_i = X_j = X_{j'}$. We call a mapping $\beta_i : P_{st}(X_i) \times P_{st}(X_i) \to \{0, 1, \ldots, k\}$ a pair-count on X_i. We define β_i to be an active pair-count if G_i^+ has a coloring f_i such that $\beta_i(A_j, A_{j'}) = |\{c \in C : A_j = \mathcal{Y}(X_j, f_j, c) \text{ and } A_{j'} = \mathcal{Y}(X_{j'}, f_{j'}, c)\}|$ for each $A_j, A_{j'} \in P_{st}(X_i)$ where f_j and $f_{j'}$ are the restriction of f_i to $E(G_j^+)$ and $E(G_{j'}^+)$, respectively. Let $val(\beta_i, \Lambda_i)$ denote the minimum cost value of an assignment of activation values for G_i^+ which satisfies the restrictions Λ_i and activates an edge-colored subgraph of G_i^+ that satisfies the pair-count β_i. The algorithm computes all $val(\beta_i, \Lambda_i)$ of all β_i on X_i from all pairs of sub-solutions (γ_j, Λ_j) and $(\gamma_{j'}, \Lambda_{j'})$ such that both have the same activation-value function $(\Lambda_j(u) = \Lambda_{j'}(u)$ for all $u \in X_i)$ and the pair of active counts γ_j and $\gamma_{j'}$ satisfy the following conditions:

C_1. $\gamma_j(A_j) = \sum_{A \in P_{st}(X_i)} \beta_i(A_j, A)$ for all $A_j \in P_{st}(X_i)$.
C_2. $\gamma_{j'}(A_{j'}) = \sum_{A \in P_{st}(X_i)} \beta_i(A, A_{j'})$ for all $A_{j'} \in P_{st}(X_i)$.

The value $val(\beta_i, \Lambda_i)$ is set to be the summation value of the pair of sub-solutions (γ_j, Λ_j) and $(\gamma_{j'}, \Lambda_{j'})$ that satisfy the above conditions minus the activation cost of X_i. To determine the pair (γ_i, Λ_i) that corresponds to the pair (β_i, Λ_i) of pair-count and an activation-value function, we construct a bipartite graph for each pair of partitions with a pair-count greater than 0 as follows. For each pair $A_j, A_{j'} \in P_{st}(X_i)$ where $\beta_i(A_j, A_{j'}) \geq 1$, we construct a bipartite graph $H_{\beta_i}^{(A_j, A_{j'})} = (A_j \cup A_{j'}, E_{\beta_i}^{(A_j, A_{j'})})$ with partite sets A_j and $A_{j'}$, where the vertices $a_j \in A_j$ and $a_{j'} \in A_{j'}$ are joined by an edge in $E_{\beta_i}^{(A_j, A_{j'})}$ iff $a_j \cap a_{j'} \neq \emptyset$. Assume that $D_1, D_2, \ldots, D_b$ are the connected components of the graph $H_{\beta_i}^{(A_j, A_{j'})}$. We define $\mathcal{U}(A_j, A_{j'})$ to be the family of vertex sets $\{\bigcup_{v \in D_l} v : 1 \leq l \leq b\}$. We set the value $val(\gamma_i, \Lambda_i)$ to be the minimum $val(\beta_i, \Lambda_i)$ over all (β_i, Λ_i) such that for each $A_i \in P_{st}(X_i)$:

$$\gamma_i(A_i) = \sum \beta_i(A_j, A_{j'})$$

where the summation above is taken over all pairs $A_j, A_{j'} \in P_{st}(X_i)$ such that $A_i = \mathcal{U}(A_j, A_{j'})$.

Extracting the Solution at the Root. The algorithm checks all the pairs (γ_r, Λ_r) of the root bag X_r such that for all $A \in P_{st}(X_r)$ where $\gamma_r(A) \geq 1$, there is a set in A containing both s and t. In this case (γ_r, Λ_r) corresponds to a feasible solution. The output of the algorithm is the minimum cost value among all the feasible solutions obtained at the root. For each row (γ_i, Λ_i) of bag X_i, we store the rows of X_i's children that were used in the calculation of $val(\gamma_i, \Lambda_i)$. Computing the optimum solution is possible by traversing top-down in the decomposition tree to the leaves (traceback) to get the activation values, and then running a maximum flow algorithm on the unit-capacity graph of the activated edges to get the k edge-disjoint st-paths.

3.2   Analysis

The following lemmas analyse the running time of the algorithm and show that we can efficiently compute an optimal solution for the st-MAEDP problem for graphs with bounded treewidth. Let an instance of the problem be given by an activation network $G = (V, E)$ with treewidth bounded by tw and terminals $s, t \in V$. Let $\mathcal{P}^{OPT}$ represent an optimal solution for this instance. We use $\mathcal{C}(\mathcal{P}^i)$ to denote the activation cost of a sub-solution $\mathcal{P}^i$ in a partial graph G_i^+.

Lemma 1. *The st-MAEDP algorithm requires $O(|V||D|^{tw+1}(k+1)^{(tw+3)^{2(tw+3)}} (tw+3)^{2(tw+3)+3}tw^3)$ time.*

Proof. The running time of the algorithm depends on the size of the tables and the combination of tables during the bottom-up traversal. For each set of vertices X, there are at most $|X|^{|X|}$ possible partitions. Therefore, for each node i, $|P_{st}(X_i)| \leq (|X_i| + 2)^{(|X_i|+2)} \leq (tw+3)^{(tw+3)}$. That means there are at most $(k+1)^{(tw+3)^{(tw+3)}}$ possible active counts γ on X_i. The table tab_i in a processed bag X_i contains no more than $(k+1)^{(tw+3)^{(tw+3)}}|D|^{tw+1}$ rows corresponding to the possible active counts γ and the $|D|$ possible activation values for each vertex of X_i. Consider all possible row combinations with equal activation functions for two tables for a join node. Since there are at most $(tw+3)^{2(tw+3)}$ possible pairs of partitions, there are at most $(k+1)^{(tw+3)^{2(tw+3)}}$ possible active pair-counts on X_i. For each pair-count β_i, there is a pair of active counts γ_j and $\gamma_{j'}$ such that the conditions C_1 and C_2 are satisfied. Computing γ_j and $\gamma_{j'}$ from β_i takes $O((tw+3)^{2(tw+3)})$ time. Therefore, the computation of the value $val(\beta, \Lambda)$ for each combination of pair-count β and activation function Λ needs $O((tw+3)^{2(tw+3)})$ time. Thus we see that all pairs (β, Λ) and $val(\beta, \Lambda)$ can be computed in $O((k+1)^{(tw+3)^{2(tw+3)}}(tw+3)^{2(tw+3)}|D|^{tw+1})$ time. Since $|A| \leq tw+3$ for all $A \in P_{st}(X_i)$, the bipartite graph $H_{\beta_i}^{(A_j, A_{j'})}$ defined in the join node contains no more than $(tw+3)^2$ edges and can be constructed in $O((tw+3)^3)$ time. Therefore, for each β_i, one can compute the active count γ_i that satisfies $\gamma_i(A_i) = \sum \beta_i(A_j, A_{j'})$ where the summation is taken over all $A_j, A_{j'} \in P_{st}(X_i)$ such that $A_i = \mathcal{U}(A_j, A_{j'})$ and update the value $val(\gamma_i, \Lambda_i)$ to be equal to $val(\beta_i, \Lambda_i)$ if $val(\beta_i, \Lambda_i) \leq val(\gamma_i, \Lambda_i)$ in $O((tw+3)^{2(tw+3)}(tw+3)^3)$ time. Thus, we can compute all $val(\gamma, \Lambda)$ of all combinations of active count γ and activation-value function Λ on X_i in time $O((k+1)^{(tw+3)^{2(tw+3)}}(tw+3)^{2(tw+3)+3}|D|^{tw+1})$. Since the tree-decomposition T has $O(|V|tw^3)$ nodes, one can compute all the tables for all nodes in $O(|V||D|^{tw+1}tw^3(k+1)^{(tw+3)^{2(tw+3)}}(tw+3)^{2(tw+3)+3})$ time.   □

Lemma 2. *For any processed bag X_i, let $\mathcal{P}_i^{OPT}$ be the induced solution of $\mathcal{P}^{OPT}$ in G_i^+ and $(\gamma_i^{OPT}, \Lambda_i^{OPT})$ be the corresponding count and activation values, then*

$$val(\gamma_i^{OPT}, \Lambda_i^{OPT}) \leq \mathcal{C}(\mathcal{P}_i^{OPT}) \tag{1}$$

Proof. We use induction over the tree decomposition to prove that the value of $(\gamma_i^{OPT}, \Lambda_i^{OPT})$ is at most the activation cost of $\mathcal{P}_i^{OPT}$. The base case are the leaf nodes where the hypothesis clearly holds. Let us assume that the induction hypothesis holds for all descendants of bag X_i. We want to prove that the induction hypothesis holds also for X_i and that can be proven by showing that the hypothesis holds for all different types of the bag X_i.

Introduce. Let us assume that i is an introduce node and j its only child and let v be the additional vertex in X_i. The statement (1) holds for X_j because it is a child of the bag X_i. Let $(\gamma_j^{OPT}, \Lambda_j^{OPT})$ be the corresponding active count and activation values of $\mathcal{P}_j^{OPT}$. The following statement holds:

$$val(\gamma_j^{OPT}, \Lambda_j^{OPT}) \leq \mathcal{C}(\mathcal{P}_j^{OPT})$$

$(\gamma_i^{OPT}, \Lambda_i^{OPT})$ and $(\gamma_j^{OPT}, \Lambda_j^{OPT})$ both agree on the activation values for all vertices in $X_i \setminus \{v\}$. The additional vertex v is an isolated vertex in the induced graph G_i^+ and the one vertex set $\{v\}$ is in A for all $A \in P_{st}(X_i)$ such that $\gamma_i^{OPT}(A) \geq 1$. Since $\Lambda_i^{OPT}(v) \in D$ is the activation value of the vertex v, then:

$$\begin{aligned} val(\gamma_i^{OPT}, \Lambda_i^{OPT}) &= val(\gamma_j^{OPT}, \Lambda_j^{OPT}) + \Lambda_i^{OPT}(v) \\ &\leq \mathcal{C}(\mathcal{P}_j^{OPT}) + \Lambda_i^{OPT}(v) \\ &= \mathcal{C}(\mathcal{P}_i^{OPT}) \end{aligned}$$

The statement (1) holds for an introduce node.

Forget. Assume that i is a forget node and j its only child and let v be the discarded vertex in X_i. The statement (1) holds for X_j because it is a child of the bag X_i. Let $(\gamma_j^{OPT}, \Lambda_j^{OPT})$ be the corresponding active count and activation values of $\mathcal{P}_j^{OPT}$. The following statement holds:

$$val(\gamma_j^{OPT}, \Lambda_j^{OPT}) \leq \mathcal{C}(\mathcal{P}_j^{OPT})$$

$(\gamma_i^{OPT}, \Lambda_i^{OPT})$ and $(\gamma_j^{OPT}, \Lambda_j^{OPT})$ both agree on activation values for all vertices in $X_j \setminus \{v\}$. The discarded vertex v is either an isolated vertex or part of a connected component in the induced solution $\mathcal{P}_j^{OPT}$ and that means the value $val(\gamma_j^{OPT}, \Lambda_j^{OPT})$ is one of the values that has been considered when calculating $val(\gamma_i^{OPT}, \Lambda_i^{OPT})$. Then:

$$val(\gamma_i^{OPT}, \Lambda_i^{OPT}) \leq val(\gamma_j^{OPT}, \Lambda_j^{OPT}) \leq \mathcal{C}(\mathcal{P}_j^{OPT}) = \mathcal{C}(\mathcal{P}_i^{OPT})$$

Then the statement (1) holds for a forget node.

Join. Assume that i is a join node and j and j' its children. Let $(\gamma_i^{OPT}, \Lambda_i^{OPT})$ be the corresponding active count and activation values of the induced solution $\mathcal{P}_i^{OPT}$ in G_i^+. $\mathcal{P}_i^{OPT}$ is the union of its children sub-solutions $\mathcal{P}_j^{OPT}$ and $\mathcal{P}_{j'}^{OPT}$. Let $(\gamma_j^{OPT}, \Lambda_j^{OPT})$ and $(\gamma_{j'}^{OPT}, \Lambda_{j'}^{OPT})$ be the corresponding active count and activation values of $\mathcal{P}_j^{OPT}$ and $\mathcal{P}_{j'}^{OPT}$, respectively. The statement (1) holds for X_j and $X_{j'}$ because they are children of the bag X_i. For all $l \in \{j, j'\}$ the following statement holds:

$$val(\gamma_l^{OPT}, \Lambda_l^{OPT}) \leq \mathcal{C}(\mathcal{P}_l^{OPT})$$

Let $(\beta_i^{OPT}, \Lambda_i^{OPT})$ be the corresponding active pair-count and activation values of the induced solution $\mathcal{P}_i^{OPT}$ in G_i^+ which satisfies conditions C_1 and C_2 for $(\gamma_j^{OPT}, \Lambda_j^{OPT})$ and $(\gamma_{j'}^{OPT}, \Lambda_{j'}^{OPT})$. We know that $\mathcal{P}_i^{OPT}$ is a sub-solution in G_i^+, therefore, $val(\beta_i^{OPT}, \Lambda_i^{OPT})$ is one of the values that has been considered when computing $val(\gamma_i^{OPT}, \Lambda_i^{OPT})$. Then:

$$
\begin{aligned}
val(\gamma_i^{OPT}, \Lambda_i^{OPT}) &\leq val(\beta_i^{OPT}, \Lambda_i^{OPT}) \\
&= val(\gamma_j^{OPT}, \Lambda_j^{OPT}) + val(\gamma_{j'}^{OPT}, \Lambda_{j'}^{OPT}) - \sum_{v \in X_i} \Lambda_i^{OPT}(v) \\
&\leq \mathcal{C}(\mathcal{P}_j^{OPT}) + \mathcal{C}(\mathcal{P}_{j'}^{OPT}) - \sum_{v \in X_i} \Lambda_i^{OPT}(v) \\
&= \mathcal{C}(\mathcal{P}_i^{OPT})
\end{aligned}
$$

Then the induction hypothesis holds for a join node.   □

Lemma 3. *For any processed bag X_i, any pair (γ_i, Λ_i) where $val(\gamma_i, \Lambda_i) = c_i < \infty$ corresponds to an edge-coloring plus activation-values $\mathcal{P}^i = (f_i, \Lambda_i^+)$ where $f_i : E(G_i^+) \to \{0, \ldots, k\}$ and $\Lambda_i^+ : X_i^+ \to D$ with the following properties:*

- *The active count of $\mathcal{P}^i$ in X_i is γ_i.*
- *The activation values of $\mathcal{P}^i$ in X_i are Λ_i.*
- *The total activation cost in X_i^+ is c_i.*

Proof. We prove by induction that for any bag X_i there exists an edge-coloring and activation-values $\mathcal{P}^i$ with the above properties. The base case are the leaf nodes where the hypothesis clearly holds. Let us assume that the induction hypothesis holds for all the descendants of bag X_i. We want to prove that the induction hypothesis holds also for X_i and that can be proved by showing that the hypothesis holds for all different types of the bag X_i.

Introduce. Assume that i is an introduce node and j its only child and v the additional vertex in X_i. Let (γ_i, Λ_i) be some entry with $val(\gamma_i, \Lambda_i) = c_i < \infty$. The induction hypothesis holds for X_j. Let (γ_j, Λ_j) be the corresponding active count and activation function that have been used for the calculation of $val(\gamma_i, \Lambda_i)$. Then (γ_j, Λ_j) corresponds to an edge-coloring and activation-values $\mathcal{P}^j$ that satisfies the above properties. From the algorithm, (γ_i, Λ_i) and (γ_j, Λ_j) both agree on the activation values for all vertices in $X_i \setminus \{v\}$. Moreover, the additional vertex v is an isolated vertex in the induced graph G_i^+ and the one vertex set $\{v\}$ is in A for all $A \in P_{st}(X_i)$ such that $\gamma_i(A) \geq 1$. The value of (γ_i, Λ_i) is equal to the value of (γ_j, Λ_j) added to the activation value of the vertex v. The union of the isolated vertex v and $\mathcal{P}^j$ is an edge-coloring with activation-values that satisfies all properties of the induction hypothesis. Thus, the induction hypothesis holds for an introduce node.

Forget. Assume that i is a forget node and j its only child and v the discarded vertex. Let (γ_i, Λ_i) be some entry with $val(\gamma_i, \Lambda_i) = c_i < \infty$. The induction hypothesis holds for X_j. Let (γ_j, Λ_j) be the corresponding active count and activation function that have been used for the calculation of $val(\gamma_i, \Lambda_i)$. Then (γ_j, Λ_j) corresponds to an edge-coloring and activation-values $\mathcal{P}^j$ that satisfies the above properties. From the algorithm, (γ_j, Λ_j) has the minimum value cost among all possible rows of X_j that can produce (γ_i, Λ_i) and both agree on activation values for all vertices in $X_j \setminus \{v\}$. The discarded vertex v is part of the edge-coloring $\mathcal{P}^j$. Therefore (γ_i, Λ_i) corresponds to the edge-coloring and activation-values $\mathcal{P}^j$ and we can set $\mathcal{P}^i = \mathcal{P}^j$. The induction hypothesis holds for a forget node.

Join. Assume that i is a join node and j and j' its children. Let (γ_i, Λ_i) be some entry with $val(\gamma_i, \Lambda_i) = c_i < \infty$. Since $val(\gamma_i, \Lambda_i) = c_i < \infty$, then there is a pair of active pair-count and activation function (β_i, Λ_i) such that $val(\gamma_i, \Lambda_i) = val(\beta_i, \Lambda_i)$ and for each $A_i \in P_{st}(X_i)$, $\gamma_i(A_i) = \sum \beta_i(A_j, A_{j'})$ where the summation is taken over all pairs $A_j, A_{j'} \in P_{st}(X_i)$ satisfying $A_i = \mathcal{U}(A_j, A_{j'})$. Let (γ_j, Λ_j) and $(\gamma_{j'}, \Lambda_{j'})$ be the pairs that have been used for the calculation of $val(\beta_i, \Lambda_i)$ which satisfy the conditions C_1 and C_2. We know that $val(\gamma_j, \Lambda_j) < \infty$ and $val(\gamma_{j'}, \Lambda_{j'}) < \infty$ and the induction hypothesis holds for X_j and $X_{j'}$. Therefore, (γ_j, Λ_j) and $(\gamma_{j'}, \Lambda_{j'})$ correspond to edge-colorings with activation-values $\mathcal{P}^j = (f_j, \Lambda_j^+)$ and $\mathcal{P}^{j'} = (f_{j'}, \Lambda_{j'}^+)$, respectively. For each pair of partitions A_j and $A_{j'}$ where $\gamma_j(A_j) = r_j$, $\gamma_{j'}(A_{j'}) = r_{j'}$ and $\beta_i(A_j, A_{j'}) = r > 0$, there are r_l colors for partition A_l in $\mathcal{P}^l = (f_l, \Lambda_l)$ for $l \in \{j, j'\}$. Therefore, we choose r unused colors $z_1^j, z_2^j, \ldots, z_r^j$ for A_j in $\mathcal{P}^j$ and r unused colors $z_1^{j'}, z_2^{j'}, \ldots, z_r^{j'}$ for $A_{j'}$ in $\mathcal{P}^{j'}$ and then recolor both z_w^j in $G_j^+(f_j, z_w^j)$ and $z_w^{j'}$ in $G_{j'}^+(f_{j'}, z_w^{j'})$ with a new color $z_w^{jj'}$ for all $w \in \{1, \ldots, r\}$. The colors $z_1^j, z_2^j, \ldots, z_r^j$ in $\mathcal{P}^j$ and colors $z_1^{j'}, z_2^{j'}, \ldots, z_r^{j'}$ in $\mathcal{P}^{j'}$ are now marked as used. Note that there are always enough unused colors because the conditions C_1 and C_2 are satisfied. After recoloring $\mathcal{P}^j$ and $\mathcal{P}^{j'}$ and since for each $A_i \in P_{st}(X_i)$, $\gamma_i(A_i) = \sum \beta_i(A_j, A_{j'})$ where the summation is taken over all pairs $A_j, A_{j'} \in P_{st}(X_i)$ satisfying $A_i = \mathcal{U}(A_j, A_{j'})$, it follows that the active count of the union $\mathcal{P}^j \cup \mathcal{P}^{j'}$ in X_i is γ_i. Moreover, (γ_i, Λ_i), (γ_j, Λ_j) and $(\gamma_{j'}, \Lambda_{j'})$ all have the same activation-value function $(\Lambda_i(u) = \Lambda_j(u) = \Lambda_{j'}(u)$ for all $u \in X_i)$. That means the activation values of the union $\mathcal{P}^j \cup \mathcal{P}^{j'}$ in X_i are Λ_i. The algorithm also computes the cost value of (γ_i, Λ_i) as follows:

$$val(\gamma_i, \Lambda_i) = val(\beta_i, \Lambda_i) = val(\gamma_j, \Lambda_j) + val(\gamma_{j'}, \Lambda_{j'}) - \sum_{v \in X_i} \Lambda_i(v)$$

Since $val(\gamma_j, \Lambda_j)$ and $val(\gamma_{j'}, \Lambda_{j'})$ are the total activation costs in X_j^+ and $X_{j'}^+$, respectively, then the summation of these activation costs minus the activation values for all $u \in X_i$ is the total activation cost in X_i^+. We can set $\mathcal{P}^i$ to be the edge-coloring and activation values $\mathcal{P}^j \cup \mathcal{P}^{j'}$ that satisfies the properties of the lemma. The induction hypothesis holds for a join node. $\square$

We obtain the following theorem by combining the above lemmas.

Theorem 2. *The st-MAEDP problem for graphs with treewidth bounded by tw can be solved optimally in $O(|V||D|^{tw+1}tw^3(k+1)^{(tw+3)2^{(tw+3)}}(tw+3)^{2(tw+3)+3})$ time.*

Corollary 1. *For any fixed k, the st-MAEDP problem for graphs of bounded treewidth can be solved optimally in linear-time FPT parameterized by the treewidth tw.*

4 Conclusion

To the best of our knowledge, the st-MAEDP problem for graphs with treewidth bounded by tw considered here has not been addressed before. We established an algorithm that solves the problem optimally in $O(|V||D|^{tw+1}tw^3(k + 1)^{(tw+3)^{2(tw+3)}} (tw+3)^{2(tw+3)+3})$ time. Our algorithm also solves the st-MAEDP problem when $k = 2$ in linear-time and this is an improvement over the cubic algorithm obtained in [2]. It would be interesting if one can obtain a faster or even linear-time algorithm for the st-MAEDP problem in graphs with treewidth bounded by a constant.

References

1. Alqahtani, H.M., Erlebach, T.: Approximation algorithms for disjoint st-paths with minimum activation cost. In: Spirakis, P.G., Serna, M. (eds.) CIAC 2013. LNCS, vol. 7878, pp. 1–12. Springer, Heidelberg (2013)
2. Alqahtani, H.M., Erlebach, T.: Minimum activation cost node-disjoint paths in graphs with bounded treewidth. In: Geffert, V., Preneel, B., Rovan, B., Štuller, J., Tjoa, A.M. (eds.) SOFSEM 2014. LNCS, vol. 8327, pp. 65–76. Springer, Heidelberg (2014)
3. Bodlaender, H.L.: A linear time algorithm for finding tree-decompositions of small treewidth. In: STOC 1993. ACM, pp. 226–234 (1993)
4. Hajiaghayi, M.T., Kortsarz, G., Mirrokni, V.S., Nutov, Z.: Power optimization for connectivity problems. In: Jünger, M., Kaibel, V. (eds.) IPCO 2005. LNCS, vol. 3509, pp. 349–361. Springer, Heidelberg (2005)
5. Lando, Y., Nutov, Z.: On minimum power connectivity problems. In: Arge, L., Hoffmann, M., Welzl, E. (eds.) ESA 2007. LNCS, vol. 4698, pp. 87–98. Springer, Heidelberg (2007)
6. Nutov, Z.: Survivable network activation problems. In: Fernández-Baca, D. (ed.) LATIN 2012. LNCS, vol. 7256, pp. 594–605. Springer, Heidelberg (2012)
7. Nutov, Z.: Approximating Steiner networks with node-weights. SIAM J. Comput. **39**(7), 3001–3022 (2010)
8. Panigrahi, D.: Survivable network design problems in wireless networks. In: 22nd Annual SIAM Symposium on Discrete Algorithms, pp. 1014–1027. ACM (2011)
9. Robertson, N., Seymour, P.D.: Graph minors. XIII. The disjoint paths problem. J. Comb. Theory, Ser. B **63**, 65–110 (1995)
10. Scheffler, P.: A practical linear time algorithm for disjoint paths in graphs with bound tree-width. Technical report 396, Dept. Mathematics, Technische Universität Berlin (1994)
11. Srinivas, A., Modiano, E.: Finding minimum energy disjoint paths in wireless ad-hoc networks. Wireless Netw. **11**(4), 401–417 (2005)
12. Ye, Z.Q., Li, Y.M., Lu, H.Q., Zhou, X.: Finding paths with minimum shared edges in graphs with bounded treewidth. In: The 2013 World Congress in Computer Science, Computer Engineering, and Applied Computing, FCS 2013, pp. 40–47 (2013). ISBN: 1601322429

A Fast Scaling Algorithm for the Weighted Triangle-Free 2-Matching Problem

Stepan Artamonov[1][⊠] and Maxim Babenko[2]

[1] Moscow State University, Moscow, Russia
swatmad@gmail.com
[2] National Research University Higher School of Economics (HSE), Moscow, Russia
maxim.babenko@gmail.com

Abstract. A *perfect 2-matching* in an undirected graph $G = (V, E)$ is a function $x \colon E \to \{0, 1, 2\}$ such that for each node $v \in V$ the sum of values $x(e)$ on all edges e incident to v equals 2. If $\operatorname{supp}(x) = \{e \in E \mid x(e) \neq 0\}$ contains no triangles, then x is called *triangle-free*. Polyhedrally, triangle-free 2-matchings are harder than 2-matchings, but easier than usual 1-matchings.

Concerning the weighted case, Cornuéjols and Pulleyblank devised a combinatorial strongly-polynomial algorithm that finds a perfect triangle-free 2-matching of minimum cost. A suitable implementation of their algorithm runs in $O(VE \log V)$ time.

In case of integer edge costs in the range $[0, C]$, for both 1- and 2-matchings some faster scaling algorithms are known that find optimal solutions within $O(\sqrt{V}\alpha(E, V) \log V E \log(VC))$ and $O(\sqrt{V}E \log(VC))$ time, respectively, where α denotes the inverse Ackermann function. So far, no efficient cost-scaling algorithm is known for finding a minimum-cost perfect triangle-free 2-matching. The present paper fills this gap by presenting such an algorithm with time complexity of $O(\sqrt{V}E \log V \log(VC))$.

1 Introduction

1.1 Basic Notation and Definitions

We shall use some standard graph-theoretic notation throughout the paper. For an undirected graph G we denote its sets of nodes and edges by $V(G)$ and $E(G)$, respectively. Unless stated otherwise, we allow parallel edges and loops in graphs. A subgraph of G induced by a subset $U \subseteq V(G)$ is denoted by $G[U]$. For $U \subseteq V(G)$, the set of edges with one end in U and the other in $V(G) - U$ is denoted by $\delta(U)$; for $U = \{u\}$ the latter notation is shortened to $\delta(u)$. Also $\gamma(U)$ denotes the set of edges with both endpoints in U. For an arbitrary set W and a function $f \colon W \to \mathbb{R}$ we denote its *support set* by $\operatorname{supp}(f) = \{w \in W \mid f(w) \neq 0\}$. For an arbitrary subset $W' \subseteq W$ we write $f(W')$ to denote $\sum_{w \in W'} f(w)$.

© Springer International Publishing Switzerland 2016
Z. Lipták and W.F. Smyth (Eds.): IWOCA 2015, LNCS 9538, pp. 25–37, 2016.
DOI: 10.1007/978-3-319-29516-9_3

The following objects will be of primary interest throughout the paper:

Definition 1. *Given an undirected graph G, a* 2-matching *in G is a function* $x \colon E(G) \to \{0, 1, 2\}$ *such that* $x(\delta(v)) \leq 2$ *for all* $v \in V(G)$. *If* $x(\delta(v)) = 2$ *for all* $v \in V(G)$, *then* x *is called* perfect. *A vertex* v *is called* free *from* x, *if* $x(\delta(v)) = 0$. *If* $\text{supp}(x)$ *contains no triangles, then* x *is called* triangle-free.

Consider some non-negative real valued edge costs $c \colon E(G) \to \mathbb{R}_+$. Then a natural combinatorial problem consists in finding a perfect triangle-free 2-matching x with minimum total cost $c \cdot x$. For this problem Cornuéjols and Pulleyblank [CP80a] devised a combinatorial polynomial algorithm. While they did not specifically aim for the best time bound, their algorithm could be implemented to run in $O(VE \log V)$ time pretty easily (hereinafter in complexity bounds we identify sets with their cardinalities).

1.2   Related Work and Our Contribution

Now let edge costs be integers in the range $[0, C]$. The problem of finding a perfect triangle 2-matchings of minimum cost is closely related to some other problems in matching theory, for which faster cost-scaling algorithms are known.

First, we may allow triangles in $\text{supp}(x)$ and start looking for a perfect 2-matching of minimum cost. This problem is trivially reducible to minimum cost perfect bipartite matching; a classical algorithm [GT89], which employs cost scaling and blocking augmentations, solves it in $O(\sqrt{V}E \log(VC))$ time.

Second, in Definition 1 we may replace $x(\delta(v)) \leq 2$ by $x(\delta(v)) \leq 1$ and get the usual notion of *1-matchings*. For general graphs G, a sophisticated algorithm from [GT91] solves the minimum-cost perfect matching problem within $O(\sqrt{V\alpha(E, V)} \log V E \log(VC))$ time.

Apart from the primal-dual algorithm given in [CP80a], there are no known methods for solving the weighted perfect triangle-free 2-matching problem. For the case of simple triangle-free 2-matchings (when $x(e) \leq 1$ for $e \in E$), Kobayashi gave a polynomial time algorithm [Kob14]. Also, some relevant prior art exists for the unweighted case, where the goal is to find a matching with maximum *size* $x(E(G))$.

For unweighted 2-matchings (or, equivalently, 1-matchings in bipartite graphs) Hopcroft and Karp devised an $O(\sqrt{V}E)$ time algorithm [HK73] (by use of *clique compression*, the latter bound was improved to $O(\sqrt{V}E \log_V(V^2/E))$ in [FM95]). Later, a conceptually similar but much more involved $O(\sqrt{V}E)$-time algorithm [MV80] for matchings in general graphs was devised (and its running time was similarly improved to $O(\sqrt{V}E \log_V(V^2/E))$ in [GK04]).

Concerning unweighted triangle-free 2-matchings, Cornuéjols and Pulleyblank [CP80b] gave a natural augmenting path algorithm; with a suitable implementation its time complexity is $O(VE)$. To match the latter with the complexity of 1- and 2-matchings, [BGR10] proposed a method that reduces the problem to a pair of maximum 1-matching computations. Unfortunately, this approach does not seem to extend to weighted problems.

Hence, no efficient cost scaling algorithm for the perfect triangle-free 2-matching problem seems to be known. In this paper we present one with the time complexity of $O(\sqrt{V}E \log V \log(VC))$.

At the high level, our approach follows the standard avenue of cost-scaling matching algorithms. It employs bit scaling, and for any current edge costs runs an alternating sequence of primal steps (involving blocking augmentations) and dual steps (adjusting reduced costs and thus enabling primal steps to progress). Some nontrivial combinatorial ingredients, however, are needed to make it work.

2 Preliminaries

2.1 Basic 2-Matchings

Let x be a 2-matching. It is easy to see that the set $\{e \in E(G)|x(e) = 1\}$ forms a collection of disjoint paths and cycles. Call a 2-matching x *basic* if such collection consists solely of odd cycles. An odd cycle C with $x(e) = 1$ for every $e \in E(C)$ is called an *odd 1-cycle*. The following fact is obvious:

Proposition 1. *Let x be a perfect triangle-free 2-matching. Then there exists a basic perfect triangle-free 2-matching x' of the same or smaller cost.*

Hereinafter, if not stated otherwise, all 2-matchings are assumed to be basic.

2.2 Triangle Clusters

A set of three pairwise-adjacent edges $\{\{u, v\}, \{v, w\}, \{w, u\}\} \subseteq E(G)$ (for some distinct $u, v, w \in V(G)$) will be called a *triangle*. By a *triangle cluster* we mean a connected graph whose edges partition into disjoint triangles such that any two triangles have at most one node in common, and if such a node exists, then it is an articulation point of the cluster. For triangle-free 2-matchings, triangle clusters play the role analogous to that of factor-critical subgraphs for 1-matchings (cf. [CP80a]).

2.3 LP Formulation

Nice LP descriptions are known for both 1- and 2-matchings. For perfect 2-matchings, Definition 1 provides the needed set of linear constraints capturing the convex hull of the family of perfect 2-matchings.

For perfect 1-matchings, the equalities

$$x(\delta(v)) = 1 \quad \text{for all } v \in V(G) \tag{1}$$

are generally not sufficient; as shown by Edmonds [Edm65], one must add the following *odd-subset* conditions:

$$x(\gamma(U)) \le \frac{1}{2}(|U| - 1) \quad \text{for all subsets } U \subseteq V(G) \text{ of odd size.} \tag{2}$$

Let $\mathcal{T}(G)$ be the set of all triangles in G. For the family of perfect triangle-free 2-matchings, its polyhedral description is proved in [CP80a] to be

$$
\begin{aligned}
x(\delta(v)) &= 2 \quad \text{for all } v \in V(G) \\
x(T) &\leq 2 \quad \text{for all } T \in \mathcal{T}(G).
\end{aligned} \tag{3}
$$

Note that (3) is in some sense simpler than (1) and (2) but is harder than the description of perfect 2-matchings. Hence, one would expect an algorithm for perfect triangle 2-matchings to be somewhat simpler than that of Gabow and Tarjan for perfect 1-matchings. As we show later, this is indeed the case.

Taking (3) into account, we establish the LP formulation of our problem; namely, for $x \colon E(G) \to \mathbb{R}_+$

$$
\begin{aligned}
&\text{minimize } c \cdot x \\
&\text{subject to } x(\delta(v)) = 2 \text{ for all } v \in V(G) \\
&\qquad\qquad x(T) \leq 2 \quad \text{for all } T \in \mathcal{T}(G)
\end{aligned} \tag{4}
$$

For an edge $e = \{u, v\} \in E(G)$ we denote $\mathcal{T}(e) := \{T \in \mathcal{T}(G) \mid e \in T\}$ (the set of triangles intersecting e). To construct the dual problem, we introduce vertex and triangle *potentials* $\pi_V \colon V(G) \to \mathbb{R}$ and $\pi_{\mathcal{T}} \colon \mathcal{T}(G) \to \mathbb{R}_+$ (which are dual to node and triangle constraints in (4)). To simplify the notation, we typically combine π_V and $\pi_{\mathcal{T}}$ into a single mapping π (minding that $\pi(T) \geq 0$ should hold for all $T \in \mathcal{T}(G)$, while $\pi(v)$ is of an arbitrary sign for $v \in V(G)$). Given π, the *reduced costs* $c_\pi(e)$ are defined for all edges $e = \{u, v\} \in E(G)$ by $c_\pi(e) := c(e) + \pi(u) + \pi(v) + \pi(\mathcal{T}(e))$. Now the dual problem to (4) is as follows:

$$
\begin{aligned}
&\text{maximize } -2 \cdot (\pi(V(G)) + \pi(\mathcal{T}(G))) \\
&\text{subject to } c_\pi(e) \geq 0 \text{ for all } e \in E(G)
\end{aligned} \tag{5}
$$

Fix arbitrary feasible primal and dual solutions x and π, respectively. Then x will be called *triangle-consistent* w.r.t. π if $\pi(T) > 0$ implies $x(T) = 2$ for all $T \in \mathcal{T}(G)$.

3   Algorithm

3.1   Overview of Cost Scaling

Let G be an undirected graph without loops or parallel edges endowed with integer edge costs $c \colon E(G) \to \{0, \ldots, C\}$. We introduce the following relaxation of the complementary slackness conditions (cf. [GT89]):

Definition 2. *Given a 2-matching x and feasible duals π, we call x 1-feasible if $x(e) > 0$ implies $c_\pi(e) \in \{0, 1\}$ for all $e \in E(G)$. A 2-matching x is called 1-optimal if it is perfect, 1-feasible and triangle-consistent w.r.t. π.*

Lemma 1. *Let edge costs c be divisible by k for some integer $k > |V|$. Then any 1-optimal triangle-free 2-matching has, in fact, the minimum cost among all perfect triangle-free 2-matchings.*

Using Lemma 1, we multiply all edge costs by $|V| + 1$ and search for a 1-optimal triangle-free 2-matching. Now, we initially start with $c \equiv 0$, and then perform $O(\log(VC))$ *scaling phases*. Each phase "opens" one new bit of edge costs, i.e. sets $c(e) := 2c(e) + \delta(e)$ where $\delta(e) \in \{0, 1\}$. Totally there are $O(\log(VC))$ phases, and the goal of a phase is to compute a 1-optimal solution and the duals for these updated costs.

3.2 Scaling Phase

Each scaling phase works with some fixed edge costs c, maintains a graph $\widehat{G}$ obtained from G by contracting some vertex-disjoint triangle clusters, a feasible set of duals π, and a 1-feasible 2-matching x in $\widehat{G}$ (possibly containing triangles in $\mathrm{supp}(x)$). Here by *contracting* a triangle cluster C we mean first removing $E(C)$ and then replacing $V(C)$ with a new vertex v_C, such that for every edge e, all its endpoints that were in $V(C)$ are now replaced with v_C. Note that for a contracted cluster C $G[V(C)]$ may contain some edges outside of C. These edges turn into loops. Also, we do not merge parallel edges during contractions.

For a vertex $v \in V(\widehat{G})$ we denote by $\mathrm{cl}(v)$ the corresponding triangle cluster in G that was contracted into v. For $v \in V(G)$ not affected by contractions we assume $\mathrm{cl}(v)$ to be the degenerate triangle cluster consisting of just v itself. For an edge $e \in E(\widehat{G})$ we denote by $\varphi(e)$ its preimage in G, i.e. the edge e' in G corresponding to e in $\widehat{G}$. The latter is well-defined since we do not merge parallel edges.

Also, note that while we do not define π on $V(\widehat{G})$, the definition of 1-optimal 2-matchings involves reduced costs, which are well-defined for $e \in E(\widehat{G})$ by $c_\pi(e) := c_\pi(\varphi(e))$. Since $\widehat{G}$ can contain loops, x can also be positive on these loops. For a loop e at v we assume that e counts twice in $\delta(v)$, and if $x(e) > 0$, then we ensure that $x(e) = 1$, and regard e as an odd 1-cycle of length 1.

The following properties of $\widehat{G}$, π and x will be maintained:

(INV1) For each contracted triangle cluster C and $e \in E(C)$, one has $c_\pi(e) \in \{0, 1\}$.

(INV2) Each triangle T in G with $\pi(T) > 0$ belongs to some contracted cluster.

(INV3) For each $v \in V(G)$, $2\pi(v)$ is integer; for each $T \in \mathcal{T}(G)$, $\pi(T)$ is integer.

(INV4) For each 1-cycle Γ in x, its preimage $\varphi(\Gamma) = \{\varphi(e') \mid e' \in E(C)\}$ in G does not form a triangle.

Note that for a 1-cycle Γ in $\widehat{G}$, if $\varphi(\Gamma)$ forms a triangle in G, then Γ is a triangle in $\widehat{G}$, while the converse is not true.

A phase finishes when x becomes perfect.

Lemma 2. *Let $\widehat{G}$, π, and x satisfy conditions (INV1)–(INV4). Suppose additionally that x is perfect. Then x can be extended in $O(E)$ time into a 1-optimal triangle-free 2-matching in G.*

Let us explain how a phase starts. We get certain previous edge costs c', feasible duals π', a 1-optimal primal solution x' and a contracted graph $\widehat{G}'$, obeying (INV1)–(INV4) from the previous phase (for the very first phase one may assume that $\widehat{G}' = G$, $\pi' = 0$, $c' = 0$, $x' = 0$).

During scaling we set $c := 2c' + \delta$, where δ can only have values 0 and 1. We initially define $\pi := 2\pi'$ (for both vertices and triangles) and $\widehat{G} := \widehat{G}'$. This way $c_\pi \cdot x' = O(V)$ by 1-feasibility of x'.

Note that $c'(e) \in \{0, 1\}$ holds for edges belonging to the contracted clusters. Now $c(e) \in \{0, 1, 2, 3\}$ holds for these edges, possibly violating (INV1). To fix this, a certain NORMALIZE-TRIANGLES routine is run (explained in detail in Subsect. 3.6) to decrease the reduced costs of edges inside the contracted triangles, where appropriate, and ensure that all of (INV1)–(INV3) are true.

It remains to construct a 1-optimal 2-matching x for the current costs. This is done by MATCH procedure described in the upcoming subsection.

3.3   Match Routine

We start with $x = 0$ (i.e. discard the primal solution from the previous phase), and then gradually improve x until it becomes perfect. During MATCH duals π are updated and some triangles in G are expanded and contracted while preserving (INV1)–(INV3). The goal is to construct a 1-optimal 2-matching x in $\widehat{G}$ satisfying (INV4). This is done by a series of blocking augmentations and dual adjustments similar to [GT89].

Before proceeding further, let us introduce the notion of *augmenting paths* as follows. Consider an edge-simple path $P = v_0 e_0 v_1 e_1 \ldots v_k e_k v_{k+1}$ in $\widehat{G}$ viewed as an alternating sequence of vertices and edges. We assume that $v_0, \ldots, v_k$ are pairwise-distinct, i.e. P can only contain a single repeated vertex v_{k+1}.

Suppose that v_0 is a free vertex and $e_0, \ldots, e_l$ is the maximal prefix of P consisting of edges e with $x(e) = 0$ and $x(e) = 2$ (alternatively). Also, suppose that: (i) $l = k$ and v_{k+1} is free; or (ii) $l < k$ and $x(e_{l+1}) = \ldots = x(e_k) = 1$; or (iii) $l = k$, $x(e_k) = 0$, and $v_{k+1} = v_i$ for some even $i < k + 1$. In cases (ii) and (iii) P consists of a vertex-simple prefix denoted by $T(P)$ and a vertex-simple cycle denoted by $C(P)$.

Given P as above, one can turn x into a 2-matching x' of greater size in a usual manner; see Fig. 1 for examples. Moreover, in cases (i) and (ii) the resulting x' obeys (INV4). We call P *regular* in case (i) and *cycle-breaking* in case (ii).

Case (iii) is more tricky, since x' may violate (INV4). In particular, if $\varphi(C(P))$ is a triangle in G, then one cannot augment along such P but can contract $C(P)$, joining triangle clusters $\mathrm{cl}(v)$ for $v \in V(C(P))$. On the other hand, if $\varphi(C(P))$ does not form a triangle in G, then x' obeys (INV4); we call such P *cycle-forming*.

Definition 3. *For a 2-matching x and duals π edge $e \in E(\widehat{G})$ is called eligible if: (i) $x(e) = 1$ and $c_\pi(e) \in \{0, 1\}$; or (ii) $x(e) = 2$ and $c_\pi(e) = 1$; or (iii) $x(e) = 0$ and $c_\pi(e) = 0$. A path P is called eligible if every edge $e \in E(P)$ is eligible.*

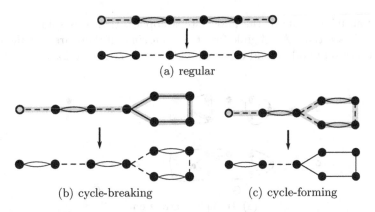

(a) regular

(b) cycle-breaking   (c) cycle-forming

Fig. 1. Examples of various types of augmenting paths (shaded) together with augmentation results. Dotted lines correspond to edges e with $x(e) = 0$, solid lines correspond to edges e with $x(e) = 1$, doubled lines correspond to edges e with $x(e) = 2$.

Note that augmenting along an eligible cycle-forming augmenting path yields an odd 1-cycle consisting of eligible edges, and such augmentation will be the only way to produce an odd 1-cycle in the 2-matching we construct, hence edge e with $x(e) = 1$ will always be eligible.

During the search we construct a certain rooted forest $\mathcal{F}$. We call a vertex $v \in V(\mathcal{F})$ *odd* if the path from v to its root has an odd length, and *even* otherwise. Then $\mathcal{F}$ is called *eligible* if: (i) the roots of $\mathcal{F}$ are exactly all the vertices in $\widehat{G}$ that are free from x; (ii) each edge $e \in E(\mathcal{F})$ is eligible; (iii) for a non-root vertex v the edge e to its parent has $x(e) = 0$ if v is odd, and $x(e) = 2$ if v is even; (iv) no vertex $v \in V(\mathcal{F})$ belongs to an odd 1-cycle of x. We now describe a routine $\text{GROW}(\mathcal{F}, e)$ that takes an eligble edge $e = \{v, u\}$, where $v \in V(\mathcal{F})$, and performs at most one the following steps:

1. Enlarges $\mathcal{F}$ by adding e; or
2. Modifies $\widehat{G}$ and $\mathcal{F}$ by contracting a triangle containing e; or
3. Claims the existence of an eligible x-augmenting path in $\widehat{G}$.

Suppose v is odd. If $x(e) = 2$, then we add e to $\mathcal{F}$; if $x(e) \neq 2$, then we do nothing. Now let v be even. Since either v is a root or the edge e' to its parent has $x(e') = 2$, we have $x(e) = 0$. If $u \notin V(\mathcal{F})$, then since every free vertex belongs to $\mathcal{F}$, either u belongs to a 1-cycle of x (and an eligible cycle-breaking augmenting path is found), or there exists an edge e' incident to u with $x(e') = 2$. In the latter case e is added to $\mathcal{F}$. Finally, let $u \in V(\mathcal{F})$. If u is odd, then we do nothing. Otherwise, there are two possibilities. If v and u are from distinct trees of $\mathcal{F}$, then an eligible regular augmenting path is found. Else, $\mathcal{F} + e$ contains a cycle C. If $\varphi(C)$ is a triangle in G, then C is contracted (see Fig. 2(b)). Otherwise, an eligible cycle-forming augmenting path is found (see Fig. 2(a)). It is straighforward to see that (INV1) and (INV4) are preserved.

If for an edge e one of the above steps applies, then e is said to be *valid*; otherwise, e is *invalid*. An eligible forest $\mathcal{F}$ is *maximal* if there are no valid edges, and for every odd node v and every triangle $T \in \mathcal{T}(\mathrm{cl}(v))$ we have $\pi(T) > 0$.

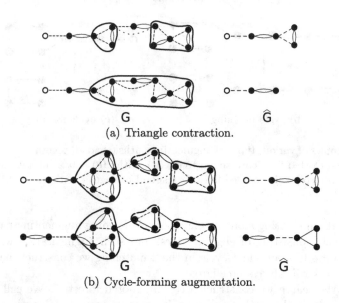

(a) Triangle contraction.

(b) Cycle-forming augmentation.

Fig. 2. Some possible cases appearing in GROW. The algorithm only deals with x in $\widehat{G}$ but the corresponding extensions of x to G are also shown.

MATCH consists of a sequence of *iterations*, each proceeding as follows. It runs BLOCKING-AUGMENT routine that, given a graph $\widehat{G}$, a 2-matching x in $\widehat{G}$, and duals π satisfying (INV1)–(INV4), performs a series of augmentations along eligible augmenting paths and contractions of triangles, and returns a maximal eligible forest $\mathcal{F}$ for the resulting x. If x is perfect, then MATCH stops. Otherwise, it invokes DUAL routine to adjust the duals π while maintaining (INV1)–(INV4) and the 1-feasibility of x, and obtain at least one eligible x-augmenting path; then MATCH starts a new iteration. Both BLOCKING-AUGMENT and DUAL are described in the upcoming subsections.

3.4   Blocking-Augment Routine

We now show how to perform a series of augmentations and contractions, and to construct a maximal eligible forest. Since the duals remain unchanged, we only need to care about eligible edges e with $c_\pi(e) \in \{0, 1\}$.

Stage I: First, we uncontract all triangles T with $\pi(T) = 0$ (see [CP80a, pp.152–153, Sect. 4] for the explanation of uncontractions). Then, we find a maximal ("blocking") sequence of eligible regular or cycle-forming x-augmenting paths. For each such path, we augment x and iterate. This step is carried out with the help of a slight modification of the blocking augmentation algorithm from [GT91, Sect. 8]. This algorithm runs a depth-first search and grows a search tree starting from free vertices. The only difference is that when we examine an edge $\{u, v\}$ from the current vertex u, we claim that an augmenting path is found not only if v is free, but also if v belongs to an odd 1-cycle of x. In the latter case we "break" this cycle, making its eligible edges available for the traversal. Since the DFS is purely local, this change does not affect its correctness.

Stage II: Now we "block" along cycle-forming augmenting paths. We build an eligible forest $\mathcal{F}$ applying GROW operations in a DFS-like manner. Initially $\mathcal{F}$ consists of all free vertices (regarded as roots). We also maintain another *dormant* forest $\mathcal{F}'$, which will be formed by detaching certain subtrees of $\mathcal{F}$ rooted at odd vertices. This $\mathcal{F}'$ has the same notion of even and odd vertices. Finally, $\mathcal{P}$ will denote a set of vertex-disjoint eligible cycle-forming x-augmenting paths found so far, and $U := \bigcup_{P \in \mathcal{P}} V(P)$ (initially both $\mathcal{P}$ and U are empty).

For a vertex $v \in V(\mathcal{F})$, an edge $e = \{v, u\}$ is called *feasible* if it is eligible and either v even or $x(e) = 2$. Our DFS at v will scan all such feasible edges e. Following the usual notation, we call a vertex v *active* if the DFS has entered v, and *scanned* if the DFS has finished examining all feasible edges incident to v.

We maintain the following invariants:

Lemma 3. *During the above DFS*

(i) *there are no eligible regular or cycle-breaking x-augmenting paths in $\widehat{G}$;*
(ii) *there are no eligible edges between an even vertex of $\mathcal{F}$ and an even vertex of $\mathcal{F}'$;*
(iii) *there are no eligible edges between scanned even vertices of $\mathcal{F}$.*

Let v be the current vertex, r be the root of the tree containing v. If v is odd then v does not belong to an odd 1-cycle (by eligibility of $\mathcal{F}$). Hence, v has at most one incident edge $e = \{v, u\}$ with $x(e) = 2$, and if e is feasible then we perform GROW$(\mathcal{F}, e)$ and continue our DFS at the newly added even vertex u.

Now let v be even, $e = \{v, u\}$ be a feasible edge we currently examine. If $u \in U$, we skip e. Otherwise, let $u \notin U$. If $u \notin V(\mathcal{F}) \cup V(\mathcal{F}')$, then we again perform GROW$(\mathcal{F}, e)$. Note that this GROW cannot find an eligible cycle-breaking augmenting path by Lemma 3(i), so e is added to $\mathcal{F}$, and we continue the DFS at the new odd vertex u. If u is an odd vertex of $\mathcal{F}$ or $\mathcal{F}'$, we do nothing; we also claim that u cannot be an even vertex of $\mathcal{F}'$ (see Lemma 3(ii)). Finally, if u is an even vertex of $\mathcal{F}$, we claim that u belongs to the same subtree as v, since otherwise there have been a regular x-augmenting path (contradicting Lemma 3(i)). Hence $\mathcal{F} + e$ contains a cycle C. If $\varphi(C)$ forms a triangle in G, then GROW performs a contraction, forming a new vertex v' and applying the corresponding change to $\mathcal{F}$. After that, v' becomes an active even vertex, and the DFS continues at v'. Note that some feasible edges that are incident to v' may have already been examined. More precisely, for a vertex $v_0 \in V(C)$ which was contracted into v, if an edge $f = \{v_0, w\}$ was examined from v_0, then its image in the new graph $f' = \{v', w\}$ is also regarded as examined.

Now suppose that $\varphi(C)$ is not a triangle in G, then an eligible cycle-forming augmenting path P is found. We add P to $\mathcal{P}$ and halt the DFS from r. Let $\mathcal{F}_r$ be the tree of $\mathcal{F}$ rooted at r. We remove the vertices $V(P)$ from $\mathcal{F}_r$, thus partitioning $\mathcal{F}_r$ into disjoint subtrees rooted at some odd vertices. Then, for each such tree T rooted at w we add an artificial vertex r_T and connect it with w by an artificial edge. This way T is now rooted at r_T, and the parity of vertices is preserved. Finally, the subtree at r_T is moved to the dormant forest $\mathcal{F}'$.

Stage III: We extend $\mathcal{F}$ as follows. Note that $U, V(\mathcal{F}), V(\mathcal{F}')$ are pairwise disjoint. We examine all eligible edges $e = \{v, u\} \in \delta(V(\mathcal{F}))$, where v is an even vertex of $\mathcal{F}$. If $u \in U$, then we add e to $\mathcal{F}$. If $u \in V(\mathcal{F}')$, then u is odd by Lemma 3(ii); we add e to $\mathcal{F}$ moving u together with its subtree from $\mathcal{F}'$ to $\mathcal{F}$. These steps are iterated until there are no such edges from even vertices of $\mathcal{F}$.

At last, we augment x along every path in $\mathcal{P}$.

Lemma 4. *The above algorithm constructs a maximum eligible forest $\mathcal{F}$.*

3.5   Dual Routine

We now describe how to change π. Let $\mathcal{F}$ be the maximal eligible forest obtained from BLOCKING-AUGMENT routine. By *shifting* a vertex $v \in V(\widehat{G})$ by δ we mean the following: if v is not contracted, then increase $\pi(v)$ by δ; if v is contracted, then increase $\pi(u)$ by δ for each $u \in V(\mathrm{cl}(v))$ and decrease $\pi(T)$ by 2δ for each $T \in \mathcal{T}(\mathrm{cl}(v))$. We update the duals π by shifting all even vertices by $-\varepsilon$ and shifting all odd vertices by $+\varepsilon$. The maximality of $\mathcal{F}$ implies that this update is possible for some $\varepsilon > 0$; moreover, we choose the maximum possible ε. The non-negativity of reduced costs and duals corresponding to triangles imply a certain upper bound for ε.

Lemma 5. *If $\varepsilon = +\infty$, then there are no triangle-free 2-matchings in G. If ε is finite, then it is positive and 2ε is integer.*

One can see that the dual transformation does not change the values of $c_\pi(e)$ for edges $e \in E(\mathcal{F})$, keeps π feasible and satisfying (INV3), keeps x 1-feasible, and keeps $\mathcal{F}$ eligible w.r.t. π and x. Moreover, the choice of ε ensures that after the update at least of the following applies: either there exists a triangle T in $\mathrm{cl}(v)$ for some odd vertex v such that $\pi(T) = 0$ holds after the update, or there exists a new valid edge, which can be used to grow $\mathcal{F}$ further. In the former case we uncontract T and insert the relevant eligible edges into $\mathcal{F}$ (see again [CP80a, pp.152–153, Sect. 4]) We perform a sequence of GROW steps until $\mathcal{F}$ becomes maximal again or an x-augmenting path is found. In the former case we repeat the dual adjustments; in the latter case DUAL stops.

3.6   Normalize-Triangles Routine

Here we explain how to "fix" duals π in order to ensure that (INV1) holds after doubling the costs and increasing some of them by 1. The algorithm updates dual variables and possibly expands some of the contracted triangles as follows.

Consider a contracted triangle cluster C. For a triangle T and an edge $e \in T$, denote by opp-$v_T(e)$ the vertex v of this triangle that is not incident to e. Also let opp-cl$_T(e)$ be the connected component (again regarded as a triangle cluster) of C containing opp-$v_T(e)$ that arises from deleting the edges of T from C. Let e be the edge of T with $c_\pi(e) \in \{2,3\}$. Fix a positive integer ε and update the duals by decreasing $\pi(T)$ by ε, increasing $\pi(u)$ by ε for all $u \in V(\text{opp-cl}_T(e))$ and decreasing $\pi(T')$ by 2ε for all $T' \in \mathcal{T}(\text{opp-cl}_T(e))$. Note that this transformation does not change c_π for $e' \in E(C) - \{e\}$ and decreases $c_\pi(e)$ by ε. It also obviously preserves (INV3). However for some triangle T' we may get $\pi(T') = 0$ or $\pi(T') = -1$ after the change. In the first case we uncontract T'. In the second case we reset $\pi(T') := 0$ and then uncontract T'. This increases $c_\pi(e)$ on the edges of T' by 1 but the edges of T are unaffected by this operation.

The above transformations are applied to all relevant edges e and lead to the updated duals π obeying (INV1)–(INV3). With a suitable implementation using ET-trees [HK95], NORMALIZE-TRIANGLES runs in $O(E + V \log V)$ time. Also the following fact (employed in Sect. 4) is true:

Lemma 6. *Let π be the duals obtained by NORMALIZE-TRIANGLES. There exists a perfect triangle-free 2-matching x' in G such that x' is triangle-consistent w.r.t. π and obeys $c_\pi \cdot x' = O(V)$.*

4 Complexity Analysis

NORMALIZE-TRIANGLES takes $O(E + V \log V)$ time, BLOCKING-AUGMENT and DUAL run in $O((V + E) \log V)$ time, where the $\log V$ factor comes from supporting contractions and uncontractions, which can be done using Link/cut trees [ST83]. Hence, it remains to bound the number of iterations in MATCH.

Lemma 7. *The number of iterations of MATCH routine is $O(\sqrt{V})$.*

Proof. Consider the duals π_0 and the graph $\widehat{G}_0$ right after NORMALIZE-TRIANGLES. By Lemma 6 there exists a perfect triangle-free 2-matching x_0 in G such that x_0 is triangle-consistent w.r.t. π_0 and $c_{\pi_0} \cdot x_0 = O(V)$. For each $v \in V(\widehat{G}_0)$ pick a single vertex in cl(v) (arbitrarily) and denote the resulting set of vertices by F_0.

Now let π_1 be the duals, $\widehat{G}_1$ be the current graph, and x'_1 be the current 1-feasible 2-matching in $\widehat{G}_1$ obtained after the i-th iteration. Then $V(\text{cl}(v)) \cap F_0 \neq \emptyset$ holds for each free w.r.t. x'_1 vertex v. Pick a single one vertex in $V(\text{cl}(v)) \cap F_0$ (again arbitrarily) for all free v and denote the resulting set by F_1. Let Δ be the sum of all ε-s in dual adjustments performed so far; then $\pi(v)$ was decreased exactly by Δ for all $v \in F_1$. This follows as each $v \in F_1$ was appearing in some (possibly trivial) cl(v') for a free (and thus even) vertex v' during all iterations up to the i-th one, and thus $\pi(v)$ was receiving decreases by ε in each DUAL.

We claim that $|F_1| \cdot \Delta = O(V)$. Assuming this is true, the proof follows by the usual case splitting. Indeed, by Lemmas 5 and 4 each run of the DUAL routine increases Δ by at least $1/2$. Hence, after $O(\sqrt{V})$ iterations we get $|F_1| = O(\sqrt{V})$, and it remains to run $O(\sqrt{V})$ augmentations to finish MATCH.

Let x_1 be the extension of x_1' into contracted triangle clusters. Consider $\Lambda := (x_1 - x_0) \cdot (c_{\pi_1} - c_{\pi_0})$. Then $\Lambda = x_1 \cdot c_{\pi_1} + x_0 \cdot c_{\pi_0} - x_0 \cdot c_{\pi_1} - x_1 \cdot c_{\pi_0} \leq x_1 \cdot c_{\pi_1} + x_0 \cdot c_{\pi_0} \leq O(V) + O(V) = O(V)$ due to non-negativity of reduced costs and the 1-feasibility of x_1. On the other hand, for arbitrary duals π denote $\vartheta\pi = c_\pi - c$. Then $\Lambda = (x_1 - x_0)(\vartheta\pi_1 - \vartheta\pi_0) = x_1 \cdot \vartheta\pi_1 + x_0 \cdot \vartheta\pi_0 - x_1\vartheta \cdot \pi_0 - x_0 \cdot \vartheta\pi_1$. Since x_0 is perfect and triangle-consistent w.r.t. π_0, $x_0 \cdot \vartheta\pi_0 = 2\pi_0(V(G)) + 2\pi_0(\mathcal{T}(G))$. Also, since x_1 is basic and is triangle-consistent w.r.t. π_1 we similarly get $x_1 \cdot \vartheta\pi_1 = 2\pi_1(V(G) - F_1) + 2\pi_1(\mathcal{T}(G))$.

Now, note that $x_1 \cdot \vartheta\pi_0 = \sum_{v \in V(G)} x_1(\delta(v))\pi_0(v) + \sum_{T \in \mathcal{T}(G)} x_1(T)\pi_0(T) = 2\pi_0(V(G) - F_1) + \sum_{T \in \mathcal{T}(G)} x_1(T)\pi_0(T) \leq 2\pi_0(V(G) - F_1) + 2\pi_0(\mathcal{T}(G))$. Similarly, $x_0 \cdot \vartheta\pi_1 \leq 2\pi_1(V(G)) + 2\pi_1(\mathcal{T}(G))$, and hence $\Lambda \geq 2\pi_0(F_1) - 2\pi_1(F_1)$. Since each dual adjustment decreases $\pi(v)$ for all $v \in F_1$ by ε, $2\pi_0(F) - 2\pi_1(F) \geq 2|F_1|\Delta$. Combining this with $\Lambda = O(V)$ we get the desired bound $|F_1|\Delta = O(V)$.   □

We conclude with

Theorem 1. *For a graph G and integer edge costs $c: E(G) \to [0, C]$, a minimum-cost perfect triangle-free 2-matching can be found in $O(E\sqrt{V}\log V \log(VC))$ time.*

References

[BGR10] Babenko, M.A., Gusakov, A., Razenshteyn, I.P.: Triangle-free 2-matchings revisited. Discrete Math. Alg. Appl. **2**, 643–654 (2010)

[CP80a] Cornuéjols, G., Pulleyblank, W.: A matching problem with side conditions. Discrete Math. **29**, 135–159 (1980)

[CP80b] Cornuéjols, G., Pulleyblank, W.R.: Perfect triangle-free 2-matchings. Math. Program. Stud. **13**, 1–7 (1980)

[Edm65] Edmonds, J.: Maximum matching and a polyhedron with 0, 1 vertices. J. Res. Natl. Bur. Stan. **69B**, 125–130 (1965)

[FM95] Feder, T., Motwani, R.: Clique partitions, graph compression and speeding-up algorithms. J. Comput. Syst. Sci. **51**, 261–272 (1995)

[GK04] Goldberg, A.V., Karzanov, A.V.: Maximum skew-symmetric flows and matchings. Math. Program. **100**(3), 537–568 (2004)

[GT89] Gabow, H.N., Tarjan, R.E.: Faster scaling algorithms for network problems. SIAM J. Comput. **18**, 1013–1036 (1989)

[GT91] Gabow, H.N., Tarjan, R.E.: Faster scaling algorithms for general graph matching problems. J. ACM **38**, 815–853 (1991)

[HK73] Hopcroft, J.E., Karp, R.M.: An $n^{5/2}$ algorithm for maximum matchings in bipartite graphs. SIAM J. Comput. **2**(4), 225–231 (1973)

[HK95] Henzinger, M.R., King, V.: Randomized dynamic graph algorithms with polylogarithmic time per operation. In: Proceeding of 27th ACM Symposium on Theory of Computing, pp. 519–527 (1995)

[Kob14] Kobayashi, Y.: Triangle-free 2-matchings and m-concave functions on jump systems. Discrete Appl. Math. **175**, 35–42 (2014)

[MV80] Micali, S., Vazirani, V.: An $O(\sqrt{V} \cdot E)$ algorithm for finding maximum matching in general graphs. In: Proceedings of 21st IEEE Symposium on Foundations of Computer Science, pp. 248–255 (1980)

[ST83] Sleator, D.D., Tarjan, R.E.: A data structure for dynamic trees. J. Comput. Syst. Sci. **26**(3), 362–391 (1983)

1-Page and 2-Page Drawings with Bounded Number of Crossings per Edge

Carla Binucci[1]([📧]), Emilio Di Giacomo[1], Md. Iqbal Hossain[2], and Giuseppe Liotta[1]

[1] Università degli Studi di Perugia, Perugia, Italy
{carla.binucci,emilio.digiacomo,giuseppe.liotta}@unipg.it
[2] Bangladesh University of Engineering and Technology, Dhaka, Bangladesh
mdiqbalhossain@cse.buet.ac.bd

Abstract. A 2-page drawing of a graph is such that the vertices are drawn as points along a line and each edge is a circular arc in one of the two half-planes defined by this line. If all edges are in the same half-plane, the drawing is called a 1-page drawing. We want to compute 1-page and 2-page drawings of planar graphs such that the number of crossings per edge does not depend on the number of the vertices. We show that for any constant k, there exist planar graphs that require more than k crossings per edge in either a 1-page or a 2-page drawing. We then prove that if the vertex degree is bounded by Δ, every planar 3-tree has a 2-page drawing with a number of crossings per edge that only depends on Δ. Finally, we show a similar result for 1-page drawings of partial 2-trees.

1 Introduction

A *k-page book embedding* (also known as *stack layout*) of a planar graph G is a crossing-free drawing of G where the vertices are represented as points along a line called *spine* and each edge is a circular arc in one of k half-planes bounded by the spine; each such half-plane is called a *page*. The minimum number of pages to compute a book embedding of a planar graph G is the *book thickness* of G. It is known that not all planar graphs have book thickness two; in fact, Bernhart and Kainen proved that a planar graph has page number two if and only if it is sub-Hamiltonian [4]. On the other hand, Yannakakis showed that every planar graph has book thickness at most four [13].

As a consequence, if we want to compute a drawing of a planar graph such that all vertices are points of a line and each edge is a circular arc inside one of at most two half-planes, edge crossings are unavoidable. If the edges are restricted in exactly one half-plane, we talk about 1-page drawings; otherwise, we talk about 2-page drawings. Since testing a graph for sub-Hamiltonicity is an NP-complete problem [6], the result by Bernhart and Kainen implies that minimizing the

---

Research supported in part by the MIUR project AMANDA "Algorithms for MAssive and Networked DAta", prot. 2012C4E3KT_001.

© Springer International Publishing Switzerland 2016
Z. Lipták and W.F. Smyth (Eds.): IWOCA 2015, LNCS 9538, pp. 38–51, 2016.
DOI: 10.1007/978-3-319-29516-9_4

number of edge crossings in a 2-page drawing is also NP-complete. Two recent papers by Bannister et al. [3] and by Bannister and Eppstein [2] show that the problem is fixed parameter tractable with respect to various graph parameters, such as cyclomatic number and treewidth.

In this paper we study the problem of computing 1-page and 2-page drawings of planar graphs such that the number of crossings per edge is bounded by a function that does not depend on the size of the graph. We prove the following theorems about planar graphs with bounded treewidth and bounded vertex degree.

Theorem 1. *Let G be a planar 3-tree with maximum degree Δ and n vertices. There exists an $O(n)$-time algorithm that computes a 2-page drawing of G with at most 2Δ crossings per edge. Also, for every integer constant k, there exist infinitely many planar 3-trees that do not admit a 2-page drawing with at most k crossings per edge.*

Theorem 2. *Let G be a partial 2-tree with maximum degree Δ and n vertices. There exists an $O(n)$-time algorithm that computes a 1-page drawing of G with at most Δ^2 crossings per edge. Also, for every integer constant k, there exist infinitely many partial 2-trees that do not admit a 1-page drawing with at most k crossings per edge.*

Related Literature and Paper Organization. 1-page and 2-page drawings are among the oldest and more common graph drawing conventions and they have received different names during the years. For example, they were studied in the 60s under the name of *network permutations* (see, e.g., [11]); they were called *linear embeddings* in the 90s (see, e.g., [10]); they were introduced in the InfoVis community less than fifteen years ago with the name of *arc diagrams* (see, e.g., [12]).

Also, the contribution of this paper can be related with a fertile research stream in graph drawing devoted to computing drawings where some edge crossing configurations are forbidden. In particular, a graph is said to be k-planar if it has a drawing where each edge is crossed at most k times. Theorems 1 and 2 compute k-planar 1-page and 2-page drawings where k is a function of Δ. Recent results about k-planar graphs and drawings include [1,8,9].

Preliminary definitions are given in Sect. 2. Theorem 1 is proved in Sect. 3, while Theorem 2 is proved in Sect. 4. Open problems are discussed in Sect. 5.

2 Preliminaries

A *drawing* Γ of a graph $G = (V, E)$ is a mapping of the vertices in V to points of the plane and of the edges in E to Jordan arcs connecting their corresponding endpoints but not passing through any other vertex. Γ is a *planar drawing* if no edge is crossed; it is a *k-planar drawing* if each edge is crossed at most k times. A planar drawing of a graph partitions the plane into topologically connected regions, called *faces*. The unbounded region is called the *outer face*. A *planar embedding* of a planar graph is an equivalence class of planar drawings that define the same set of faces. Two drawings with the same planar embedding have the same circular ordering of the edges around each vertex. A planar graph together with a planar embedding is called a *plane* graph. A plane graph is *maximal* if all its faces are triangles. Given a plane graph, we denote by $C_{x,y,z}$ the cycle formed by three mutually adjacent vertices x, y, and z, and by $G(C_{x,y,z})$ the subgraph of G consisting of $C_{x,y,z}$ and all the vertices inside it.

The concept of planar embedding can be extended to k-planar drawings as follows. Given a k-planar drawing Γ we can planarize it by replacing each crossing with a dummy vertex. A k-*planar embedding* of a k-planar graph is an equivalence class of k-planar drawings whose planarized versions have the same planar embedding. Notice that two drawings with the same k-planar embedding have the same circular ordering of the edges around each vertex, the same set of crossings, and the same circular ordering of the edges around each crossing. An *outer k-planar embedding* is a k-planar embedding with all vertices on the outer face. Analogously an *outerplanar embedding* is a planar embedding with all vertices on the outer face. An *outer k-planar graph* (respectively *outerplanar graph*) is a graph that admits an outer k-planar embedding (respectively an *outerplanar embedding*).

A 2-*page drawing* of a graph G is a drawing Γ of G such that the vertices lie on a horizontal line ℓ, called *spine*, and each edge is drawn as a semicircle in one of the two half-planes defined by ℓ. Each of these half-planes is called a *page*. If all edges of Γ are on a single page, Γ is called a 1-page drawing. If each edge of Γ has at most k crossings, we call Γ a k-*planar2-page (1-page) drawing*. Figure 1(a) shows an example of 1-planar 2-page drawing.

A graph is *connected* if every pair of vertices of G is connected by a path. A k-*connected* graph G is such that removing any $k-1$ vertices leaves G connected; 2-connected and 1-connected graphs are also called *biconnected*, and *simply connected* graphs, respectively. A vertex whose removal disconnects the graph is called a *cut-vertex*. Hence, a connected graph is biconnected if it has no cutvertices. Given a simply connected graph G, the maximal subgraphs of G not containing a cutvertex are the *biconnected components* of G. Notice that a biconnected component of a connected graph is either a biconnected subgraph or a single edge.

A graph is *Hamiltonian* if it has a simple cycle containing all its vertices. A k-*planar Hamiltonian augmentation* of a k-planar graph G is a k-planar Hamiltonian multigraph[1] G' obtained by adding edges to G and possibly changing the k-planar embedding of G. A k-planar Hamiltonian augmentation such that no edge of the Hamiltonian cycle has crossings is called an *uncrossed k-planar Hamiltonian augmentation*. The following lemma establishes the connection between k-planar 2-page drawings and uncrossed k-planar Hamiltonian augmentations.

---

[1] A multigraph is a graph that can have multiple edges between the same pair of vertices.

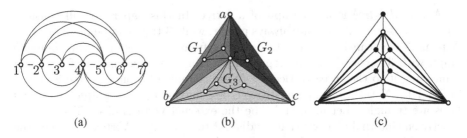

Fig. 1. (a) A 1-planar 2-page drawing. (b) A plane 3-tree. The three subgraphs G_1, G_2 and G_3 are highlighted. (c) An example of a graph G used in the proof of Lemma 4.

Lemma 1. *Let G be a planar graph. G admits a k-planar 2-page drawing if and only if G admits an uncrossed k-planar Hamiltonian augmentation.*

Proof. Consider a k-planar 2-page drawing Γ of G and let $v_1, v_2, \ldots, v_n$ be the vertices of G in the linear order they appear along the spine ℓ of Γ. For every $i = 1, 2, \ldots, n-1$, if the edge (v_i, v_{i+1}) does not exist in G, we add it to G (in either page); analogously, we add the edge (v_n, v_1) if it does not exist in G (in either page). Let G' be the resulting graph. The edges $(v_1, v_2), (v_2, v_3), \ldots, (v_{n-1}, v_n), (v_n, v_1)$ form a Hamiltonian cycle H of G'. Since each edge (v_i, v_{i+1}) connects vertices that are consecutive along ℓ and every edge is either above or below ℓ, the edge (v_i, v_{i+1}) is not crossed. Analogously the edge (v_n, v_1) is not crossed because it connects the first vertex along ℓ to the last one. Thus, no edge of H is crossed and G' is an uncrossed k-planar Hamiltonian augmentation of G.

Assume now that G admits an uncrossed k-planar Hamiltonian augmentation G' and let H be the Hamiltonian cycle of G'. Since no edge of H is crossed, every edge not in H is either completely inside H or completely outside H in the k-planar embedding of G'. We compute a k-planar 2-page drawing as follows. We order the vertices of G' (and therefore of G) along ℓ according to the order in which they appear along H (starting from an arbitrary vertex). The edges of H and the edges inside H are drawn below ℓ and the remaining edges (i.e., those outside H) are drawn above ℓ. It is easy to see that two edges cross in the defined 2-page drawing if and only if they cross in the k-planar embedding of G'. It follows that each edge is crossed at most k times in the computed 2-page drawing. □

The following lemma can be proven analogously to the previous one.

Lemma 2. *Let G be a planar graph. G admits a k-planar 1-page drawing if and only if G admits an uncrossed outer k-planar Hamiltonian augmentation.*

Let $k > 0$ be a given integer. A *k-tree* is a graph recursively defined as follows: (i) The complete graph with k vertices is a k-tree. (ii) A k-tree with $n+1$ vertices ($n \geq k$) can be obtained from a k-tree H with n vertices by adding one vertex and making it adjacent to a clique of size k of H.

A *partial k-tree* is a subgraph of a k-tree. In this paper we will consider 2-trees and 3-trees. 2-trees are always planar, while 3-trees can be planar or not. A plane 3-tree can be constructed starting from a 3-cycle (i.e., the complete graph with 3 vertices) and by repeatedly adding a vertex inside a face f and connecting it to the three vertices of f. An example of a plane 3-tree is shown in Fig. 1(b). Notice that a plane 3-tree G is a maximal planar graph, i.e., all its faces are triangles. Let a, b, and c be the external vertices of G. There exists exactly one internal vertex r that is adjacent to a, b, and c. Vertex r is called the *representative vertex* of G. Furthermore, the three subgraphs $G_1 = G(C_{a,b,r})$, $G_2 = G(C_{c,a,r})$, and $G_3 = G(C_{b,c,r})$ are plane 3-trees (see Fig. 1(b)). A *partial plane 3-tree* is a subgraph of a plane 3-tree.

3   Plane 3-Trees

In this section, we first prove that, for every constant $k > 0$, there exist infinitely many (partial) plane 3-trees that do not admit a k-planar 2-page drawing. Let G be a graph and let h be a positive integer. We define the *h-extension* of G as the graph G^* constructed by attaching h paths of length 2 to each edge of G. It is easy to see that the h-extension of a plane 3-tree is a partial plane 3-tree for every positive integer h. We have the following lemma. The proof is omitted for space reasons.

Lemma 3. *Let h be a positive integer, and let G be a planar graph. In any h-planar drawing of the $3h$-extension G^* of G, there are no two edges of G that cross each other.*

Using the previous lemma we can prove the following.

Lemma 4. *Let $k > 0$ be any given integer. For every $n > 4$ there exits a partial plane 3-tree with $27kn + 3n - 54k - 4$ vertices that does not admit a k-planar 2-page drawing.*

Proof. Let G' be a plane 3-tree with n vertices and let G be the plane 3-tree obtained by adding a vertex inside each face of G' (and connecting it to the three vertices of the face). The vertices of G shared with G' are called *white* vertices, the remaining ones are called *black* vertices (see Fig. 1(c) for an example). Since G' has $2n - 4$ faces, then G has n white vertices and $2n - 4$ black vertices. Let G^* be the $3k$-extension of G. Notice that G^* has $27kn + 3n - 54k - 4$ vertices. We prove that G^* does not admit a k-planar 2-page drawing. Based on Lemma 1 it is sufficient to show that G^* does not admit an uncrossed k-planar Hamiltonian augmentation.

Suppose as a contradiction, that G^* has an uncrossed k-planar Hamiltonian augmentation $\overline{G}$ and let H be the corresponding Hamiltonian cycle. Let u and v be two vertices of G^* that are two black vertices of G. By Lemma 3, the subgraph of G^* coinciding with G is planarly embedded in the considered k-planar embedding of $\overline{G}$. This means that u is inside a triangle τ_u of white vertices and v is inside another triangle τ_v of white vertices. If there is no vertex between u and v in H, then H must cross a white edge (i.e. an edge with two white endvertices) in order to move from τ_u to τ_v, but this is not possible because the edges of H do not have crossings. Thus, in H there must be at least one white vertex between any two black vertices. This means that the number of white vertices must be at least the number of black vertices. Since the black vertices are $2n - 4$, the white vertices are n, and $n > 4$, this is not possible. □

Since Lemma 4 states that for plane 3-tree it is not always possible to obtain a 2-page drawing with a constant number of crossings per edge, we investigate whether such a drawing exists with a number of crossings that, although not constant, does not depend on the size of the graph. We show that every plane 3-tree G admits a 2Δ-planar 2-page drawing, where Δ is the maximum vertex degree of G. Based on Lemma 1 it is sufficient to show that G admits an uncrossed 2Δ-planar Hamiltonian augmentation. To this aim, we describe an algorithm, which we call PLANE3TREEEMBEDDER, that computes an uncrossed 2Δ-planar Hamiltonian augmentation with some additional properties.

Let G be a plane 3-tree of maximum vertex degree Δ. Let a, b, and c be the external vertices of G (in counterclockwise order) and, if G has more than three vertices, let r be the representative vertex of G. Let $G_1 = G(C_{a,b,r})$, $G_2 = G(C_{c,a,r})$, and $G_3 = G(C_{b,c,r})$ (see Fig. 1(b)). We denote by Δ_i $(i = 1, 2, 3)$ the maximum vertex degree in G_i, by $\delta(v)$ the degree of vertex v in G, and by $\delta_i(v)$ the degree of vertex v in G_i. Let $e_1 = (u, v)$ and $e_2 = (v, w)$ be two external edges of G arbitrarily chosen (i.e. $u, v, w \in \{a, b, c\}$). We prove that G admits an uncrossed k-planar Hamiltonian augmentation with a Hamiltonian cycle H such that the following properties hold:

P1: $k \le 2\Delta$, i.e., each edge is crossed at most 2Δ times. In particular, the edges of the external face of G have no crossings, while the edges incident to a, b, and c have at most $\Delta + \delta(a)$, $\Delta + \delta(b)$, and $\Delta + \delta(c)$ crossings, respectively.

P2: e_1 and e_2 belong to H;

P3: Let H' be the path obtained from H by removing the external vertices of G. If H' is not empty, let z_1 and z_2 be the end-vertices of H'. Then the edges (z_1, u) and (z_2, w) belong to H, and the edges (z_1, v) and (z_2, v) are not crossed (see Fig. 2).

If G has 3 vertices, H coincides with G and the three properties **P1**, **P2**, and **P3** trivially hold.

If G has more than 3 vertices, then it can be decomposed into the three graphs $G_1 = G(C_{a,b,r})$, $G_2 = G(C_{c,a,r})$, and $G_3 = G(C_{b,c,r})$ (see Fig. 1(b)). Without loss of generality assume that e_1 is the edge (a, b) and that e_2 is the edge (b, c), i.e., assume that $u = a$, $v = b$, and $w = c$ (the other cases are analogous). The subgraph G_1 recursively admits a k_1-planar Hamiltonian augmentation with a Hamiltonian cycle H_1 that contains the edges (a, b) and (a, r) and satisfies the properties **P1-P3**. Analogously, G_2 recursively admits a k_2-planar Hamiltonian augmentation with a Hamiltonian

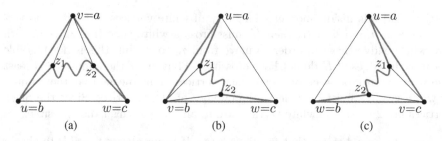

Fig. 2. Illustration of property **P3**. The Hamiltonian cycle is shown with bold edges. The thin edges are not crossed. (a) $e_1 = (b, a)$, $e_2 = (a, c)$. (b) $e_1 = (a, b)$, $e_2 = (b, c)$. (c) $e_1 = (a, c)$, $e_2 = (c, b)$.

cycle H_2 that contains the edges (a, r) and (r, c) and satisfies the properties **P1-P3**. Finally, G_3 recursively admits a k_3-planar Hamiltonian augmentation with a Hamiltonian cycle H_3 that contains the edges (r, c) and (c, b) and satisfies the properties **P1-P3**.

We now describe how the algorithm PLANE3TREEEMBEDDER constructs H starting from H_1, H_2, and H_3. From each cycle H_i (for $i = 1, 2, 3$) it constructs a path H_i' as follows.

Construction of H_1'. If G_1 has three vertices (see Fig. 3(a)), then H_1 coincides with G_1 and H_1' consists of the only edge (a, b). If G_1 has more than three vertices (see Fig. 3(b)), then by Property **P3** (see Fig. 2(a)) there exists a vertex z_1' such that the edge (z_1', b) belongs to H_1 and the edge (z_1', a) has no crossings. Also, there exists a vertex z_2' such that (z_2', r) belongs to H_1. H_1' is obtained from H_1 by: (i) removing the vertex r (and therefore the edges (z_2', r) and (a, r)); (ii) removing the edge (z_1', b); and (iii) adding the edge (z_1', a). Notice that in both cases H_1' starts at b and contains all vertices of G_1 except r. The last vertex is either a or z_2'.

Construction of H_2'. If G_2 has three vertices (see Fig. 3(c)), then H_2 coincides with G_2 and H_2' consists of the only vertex r. If G_2 has more than three vertices (see Fig. 3(d)), then by Property **P3** (see Fig. 2(b)) there exists a vertex z_1'' such that the edge (z_1'', c) belongs to H_2 and the edge (z_1'', r) has no crossings. Also, there exists a vertex z_2'' such that (z_2'', a) belongs to H_2. H_2' is obtained from H_2 by: (i) removing the vertices a and c (and therefore the edges (a, z_2''), (a, r), (c, r), and (z_1'', c)); and (ii) adding the edge (z_1'', r). Notice that in both cases H_2' ends at r and contains all vertices of G_2 except a and c. The first vertex is either r itself or z_2''.

Construction of H_3'. If G_3 has three vertices (see Fig. 3(e)), then H_3 coincides with G_3 and H_3' consists of the two edges (r, c) and (c, b). If G_3 has more than three vertices (see Fig. 3(f)), then by Property **P3** (see Fig. 2(c)) there exists a vertex z_1''' such that the edge (z_1''', b) belongs to H_3 and the edge (z_1''', c) has no crossings. Also, there exists a vertex z_2''' such that (z_2''', r) belongs to H_3. H_3' is obtained from H_3 by: (i) removing the edge (c, r); (ii) removing the edge (z_1''', b);

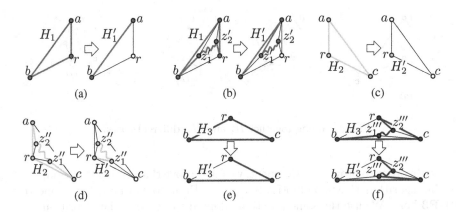

Fig. 3. Obtaining H_i' from H_i ($i = 1, 2, 3$). (a)–(b) G_1. (c)–(d) G_2. (e)–(f) G_3.

and (iii) adding the edge (z_1''', c). Notice that in both cases H_3' starts at r, ends at b, and contains all vertices of G_3.

Once H_1', H_2', and H_3' have been constructed, algorithm PLANE3TREE EMBEDDER glues them together to form H. H_2' and H_3' have the vertex r in common so they can be glued without any further modification. In order to glue H_1' and H_2' the last vertex of H_1' is connected to the first vertex of H_2' with an edge e^*. We have the following cases:

The last vertex of H_1' is a and the first vertex of H_2' is r. (see Fig. 4(a)).
In this case $e^* = (a, r)$. Since it is not crossed, it can be added to H.

The last vertex of H_1' is a and the first vertex of H_2' is z_2''. (see Fig. 4(b)).
In this case the edge $e^* = (a, z_2'')$ was an edge of H_2 (by property **P3**) and therefore it is not crossed. Thus, it can be added to H.

The last vertex of H_1' is z_2' and the first vertex of H_2' is r. (see Fig. 4(c)).
In this case the edge $e^* = (z_2', r)$ was an edge of H_1 (by property **P3**) and therefore it is not crossed. Thus, it can be added to H.

The last vertex of H_1' is z_2' and the first vertex of H_2' is z_2''. (see Fig. 4(d)).
In this case the edge $e^* = (z_2', z_2'')$ does not exist in G. We can add it to G creating one crossing with the edge (a, r). Notice that this crossing involves an edge of H, i.e., the edge (z_2', z_2''). Since we want no crossings on the edges of H, we reroute the edge (a, r) (shown bold in Fig. 4(d)) so that it crosses all the edges of G_2 incident to c (except (a, c) and (c, r)) and no other edge of G_2.

In summary the cycle H consists of: (i) the path H_1'; (ii) the edge e^*; (iii) the path H_2'; and (iv) the path H_3'. The following lemma proves the correctness of the algorithm PLANE3TREEEMBEDDER.

Lemma 5. *Let G be a plane 3-tree. Algorithm PLANE3TREEEMBEDDER computes an uncrossed k-planar Hamiltonian augmentation of G with a Hamiltonian cycle that satisfies properties **P1-P3**.*

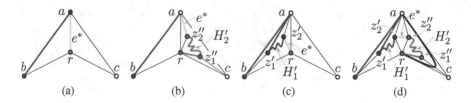

Fig. 4. Gluing together H_1' and H_2': different cases.

Proof. In order to prove the statement, we must prove that the cycle H computed by the algorithm PLANE3TREEEMBEDDER is Hamiltonian and that properties **P1-P3** hold. We use the same notation adopted to describe the algorithm.

H **contains all vertices of** G. All the vertices of G distinct from a, b, c, and r belong to either H_1', or H_2', or H_3'. Vertex a belongs to H_1'. Vertex b belongs to H_1' and H_3' (it has degree one in both paths and therefore it has degree two in H). Vertex c belongs to H_3'. Vertex r belongs to H_2' and H_3' (it has degree one in both paths and therefore it has degree two in H).

Property P1 holds. We first prove that the edges of H are not crossed. Suppose first that the edge (a, r) is not rerouted. In this case the edges crossed in G were already crossed in G_1, G_2, or G_3. The edges of H are either edges of H_1, H_2, and H_3, in which case they are not crossed by induction, or they are the edges (z_1', a), (z_1'', r), or (z_1''', c). By property **P3** these edges are not crossed. Suppose now that we reroute the edge (a, r). In this case there can be edges that were not crossed in G_2 and that get crossed in G. These edges are (a, r) and all the edges incident to c except (a, c) and (r, c). Denote by E^* the edges that get additional crossings. The edge (a, r) is not part of H by construction. The edges incident to c crossed by (a, r) are not part of H, too. Namely, the edges of H_2 incident to c are either (a, c) and (r, c), which are not in H (and are not crossed), or (z_1'', c) and (r, c), which are also not in H.

Consider now the edges not in H. Each edge that is not in E^* has the same number of crossings that it had in G_1, G_2, or G_3 and therefore it satisfies Property **P1** by induction. Let $\chi^{(a,r)}$ be the number of crossings that are created on the edge (a, r) by the rerouting. $\chi^{(a,r)}$ is less than $\delta_2(c)$ and therefore we have $\chi^{(a,r)} \leq \delta_2(c) \leq \Delta \leq \Delta + \delta(c) \leq 2\Delta$. Consider now an edge $e \in E^*$. Let χ_2^e be the number of crossings that the edge e has in G_2. By induction, χ_2^e is at most $\Delta_2 + \delta_2(c)$. The number of crossings that the edge e has in G is $\chi^e = \chi_2^e + 1 \leq \Delta_2 + \delta_2(c) + 1 \leq \Delta + \delta_2(c) + 1$. Since $\delta(c) \geq \delta_2(c) + 1$, we have that $\chi^e \leq \Delta + \delta(c) \leq 2\Delta$.

Property P2 holds. The edge $e_1 = (b, a)$ belongs to H_1', while the edge $e_2 = (b, c)$ belongs to H_3'. Thus they both belong to H.

Property P3 holds. The vertex z_1 coincides with z_1''' and the vertex z_2 coincides with z_1'. Since the 2Δ-planar embedding of G_1 and G_3 is not changed (we only changed the 2Δ-planar embedding of G_2), property **P3** holds because it held for G_1 and G_3. □

Proof of Theorem 1. The existence of a 2-page drawing with 2Δ crossings per edge follows from Lemmas 1 and 5. The existence of infinitely many planar 3-trees that do not admit a 2-page drawing with a constant number of crossings per edge follows from Lemma 4 and by the fact that a graph does not admit a k-planar 2-page drawing if it contains a subgraph that does not admit such a drawing. The proof of the time complexity of PLANE3TREEEMBEDDER is omitted for space reasons.

4   Partial 2-Trees

We start by proving that, for every constant $k > 0$, there exist infinitely many (partial) 2-trees that do not admit a k-planar 1-page drawing.

Lemma 6. *Let $k > 0$ be any given integer. For every $n > 4$ there exists a 2-tree with $6kn + n - 9k$ vertices that does not admit a k-planar 1-page drawing.*

Proof. Let G be a 2-tree with n vertices that is not outerplanar and let G^* be the $3k$-extension of G. Since G has $2n-3$ edges, G^* has $6kn+n-9k$ vertices. We show that G^* does not admit an uncrossed outer k-planar Hamiltonian augmentation. By Lemma 2, this implies that G^* does not admit a k-planar 1-page drawing. If G^* has an uncrossed outer k-planar Hamiltonian augmentation $G^\star$, then, by Lemma 3, the subgraph of G^* coinciding with G is planarly embedded in the outer k-planar embedding of $G^\star$. But this means that G is outerplanar, which is not true by definition. □

We now prove that every partial 2-tree admits a Δ^2-planar 1-page drawing. It is known that a graph G is a partial 2-tree if and only if every biconnected component of G is a series-parallel graph [5]. A *series-parallel* graph G is a multigraph with two distinguished vertices, called the *source s* and the *sink t* of G, which is recursively defined as follows: (i) A single edge (s, t) is a series-parallel graph with source s and sink t; (ii) Given two series-parallel graphs G_1 and G_2 with sources s_1 and s_2, respectively and sinks t_1 and t_2, respectively, then:

- The graph obtained by identifying t_1 with s_2 is a series-parallel graph with source s_1 and sink t_2. This operation is called *series composition*.
- The graph obtained by identifying the two sources s_1 and s_2 and identifying the two sinks t_1 and t_2 is a series-parallel graph with source $s_1 = s_2$ and sink $t_1 = t_2$. This operation is called *parallel composition*.

The series-parallel graphs as defined above, are often called *two-terminal series-parallel graphs* [7]. The source and the sink of a series-parallel graph are called its *poles*. We only consider series-parallel graphs without multiple edges. Let G be a series-parallel graph. In the following we denote by s and t the poles of G, by Δ the maximum vertex degree of G, by $\delta(v)$ the degree of a vertex v in G, and by h_s (respectively h_t) the number of times that s (respectively t) is a pole of a parallel composition in the construction of G. If G is the series or parallel composition of two graphs G_1 and G_2, we denote by s_i, t_i, Δ_i, $\delta_i(v)$, h_{i,s_i}, and h_{i,t_i} the analogous elements for G_i ($i = 1, 2$). We have the following technical lemma whose proof is omitted for space reasons.

Lemma 7. *Let G be a series-parallel graph. Then $h_s \le \Delta$ and $h_t \le \Delta$.*

We now prove that every series-parallel graph admits an uncrossed outer Δ^2-planar Hamiltonian augmentation. By Lemma 2 this implies that every biconnected partial 2-tree admits a Δ^2-planar 1-page drawing. Let G be a series-parallel graph. We describe an algorithm, called SPEMBEDDER, that computes an uncrossed outer k-planar Hamiltonian augmentation of G with a Hamiltonian cycle H with the following properties:

PA: $k \le \Delta^2$, i.e., each edge is crossed at most Δ^2 times. In particular, the edges of H have no crossings and the edges incident to s and t have at most $h_s \cdot \Delta$ and $h_t \cdot \Delta$ crossings, respectively. Notice that by Lemma 7, $h_s \cdot \Delta \le \Delta^2$ and $h_t \cdot \Delta \le \Delta^2$.
PB: (s,t) belongs to H;

If G has a single edge (s,t), then H is obtained by adding a dummy edge in parallel between s and t and the properties **PA–PB** trivially hold. If G has more than one edge, then G is either the series or the parallel composition of two series-parallel graphs G_1 and G_2. By induction, each G_i ($i = 1, 2$) has an uncrossed outer k_i-planar Hamiltonian augmentation with Hamiltonian cycle H_i such that the properties **PA–PB** hold. Let $s_i = v_{i,1}, v_{i,2}, \ldots, v_{i,n_i} = t_i$ be the vertices of G_i in the order they appear along H_i.

Series composition. We remove the edge (s_i, t_i) from H_i ($i = 1, 2$), thus obtaining a path H_i'. Notice that the last vertex of H_1' is t_1 and the first vertex of H_2' is s_2. In G these two vertices coincide, and therefore H_1' and H_2' are joined together in G. We add the edge (s_1, t_2), thus obtaining the cycle H: $s_1 = v_{1,1}, v_{1,2}, \ldots, v_{1,n_1-1}, t_1 = s_2, v_{2,2}, \ldots, v_{2,n_2} = t_2$. All the other edges of G are embedded inside H (see Fig. 5(b)).

Parallel composition. Assume that G is obtained as a parallel composition of G_1 and G_2. We remove the vertex t_1 (and therefore (s_1, t_1) and (v_{1,n_1-1}, t_1)) from H_1 and the vertex s_2 (and therefore $(s_2, v_{2,2})$ and (s_2, t_2)) from H_2. We then add the edges $(v_{1,n_1-1}, v_{2,2})$ and (s_1, t_2), and obtain the cycle H: $s_1 = v_{1,1}, v_{1,2}, \ldots, v_{1,n_1-1}, v_{2,2}, \ldots, v_{2,n_2} = t_2$. All the other edges of G are embedded inside H (see Fig. 5(c)).

Lemma 8. *Let G be a series-parallel graph. Algorithm SPEMBEDDER computes an uncrossed outer k-planar Hamiltonian augmentation with a Hamiltonian cycle that satisfies properties **PA–PB**.*

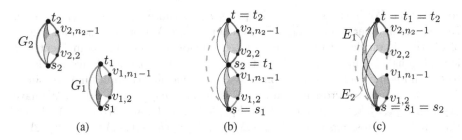

Fig. 5. Algorithm SPEMBEDDER. (a) G_1 and G_2. (b) Series composition. (c) Parallel composition. The bold edges represent the Hamiltonian cycles H_1, H_2, and H. The dashed edges are (possibly) dummy edges.

Sketch of proof. It is immediate to see that the cycle H computed by the algorithm SPEMBEDDER is Hamiltonian and that the constructed embedding is outer k-planar (for some k). We must prove that properties **PA–PB** hold. Property **PB** holds by construction both for series and parallel composition. Thus we concentrate on property **PA**. We use the same notation adopted to describe the algorithm (see also Fig. 5). Due to space reasons, full details are omitted here.

Series composition. It is easy to see that no crossing has been introduced on the edges of G_1 and of G_2 with the series composition and that the only added edge, i.e., edge (s, t), has no crossing. Using this fact, we can prove property **PA**. Let e be an edge of H. If $e = (s, t)$ then it is not crossed as shown above. If e belongs to G_1 (respectively G_2), then it belongs to H_1 (respectively H_2) and therefore is not crossed. Let e be an edge of G_1 (respectively G_2) not in H and that is not incident to s or t. By induction e has $\Delta_1^2 \le \Delta^2$ (respectively $\Delta_2^2 \le \Delta^2$) crossings. Consider now an edge e of G_1 that is incident to s (the case when e belongs to G_2 and/or is incident to t is analogous). By induction, e has at most $h_{1,s_1} \cdot \Delta_1$ crossings. Since $\Delta_1 \le \Delta$ and $h_{1,s_1} = h_s$, the number of crossings of e is at most $h_s \cdot \Delta$.

Parallel composition. It is easy to see that the only edges that get additional crossings with the parallel composition, are those connecting s to vertices of G_2 (denote them as E_2) and those connecting t to the vertices of G_1 (denote them as E_1), and that the two added edges $(v_{1,n_1-1}, v_{2,2})$ and (s_1, t_2) do not cross any other edge. We now prove property **PA**. Let e be an edge of H. If $e = (s, t)$ or $e = (v_{1,n_1-1}, v_{2,2})$ then it is not crossed as shown above. If e belongs to G_1, then it belongs to H_1 and does not belong to E_1. Therefore it is not crossed. Analogously, if e belongs to G_2, then it belongs to H_2 and does not belong to E_2. Also in this case it is not crossed. Let e be an edge of G_1 (respectively G_2) not belonging to H and that is not incident to s or t. By induction e has $\Delta_1^2 \le \Delta^2$ (respectively $\Delta_2^2 \le \Delta^2$) crossings. Consider now an edge e of G_1 that is incident to s (the case when e belongs to G_2 and is incident to t is analogous). By induction, e has at most $h_{1,s_1} \cdot \Delta_1$ crossings. Since e belongs to G_1 and is incident to s, then it does not belong to E_1 and has the same crossings it had before the composition. We have $\Delta_1 \leq \Delta$ and $h_{1,s_1} \leq h_s$ (because $h_s = h_{1,s_1} + h_{2,s_2} + 1$). Hence, the number of crossings of e is at most $h_s \cdot \Delta$. Consider now an edge e of G_2 that is incident to s (the case when e belongs to G_1 and is incident to t is analogous). By induction, e has at most $h_{2,s_2} \cdot \Delta_2$ crossings. Since e belongs to G_2 and is incident to s, it belongs to E_2 and has $\delta_1(t_1)$ additional crossings. Thus, the number of crossings on e in G is at most $h_{2,s_2} \cdot \Delta_2 + \delta_1(t_1)$. We have $\Delta_2 \leq \Delta$, $h_{2,s_2} \leq h_s - 1$ (because $h_s = h_{1,s_1} + h_{2,s_2} + 1$), and $\delta_1(t_1) \leq \Delta$. Hence, the number of crossings of e is $h_{2,s_2} \cdot \Delta_2 + \delta_1(t_1) \leq (h_s - 1) \cdot \Delta + \Delta = h_s \cdot \Delta$. $\square$

Proof of Theorem 2. The existence of graphs that do not admit a k-planar 1-page drawing, for every constant k, follows from Lemma 6, while Lemma 8 implies that every biconnected 2-tree has a Δ^2-planar 1-page drawing. The proof of extension to simply connected 2-trees and of the time complexity are omitted for space reasons.

5 Open Problems

The results of this paper suggest several open problems. For example: (1) Does the 2Δ bound hold for partial plane 3-trees? And for general (partial) 3-trees? (2) Can the Δ^2 bound for 2-trees be reduced? Notice that a lower bound of $\lceil \frac{\Delta-3}{2} \rceil$ holds. Namely, consider a 2-tree G consisting of $\Delta - 1$ triangles sharing an edge. In any 1-page drawing of G, there exists at least one edge with at least $\lceil \frac{\Delta-3}{2} \rceil$ crossings. (3) Can we prove a bound that only depends on Δ for general planar graphs?

References

1. Alam, M.J., Brandenburg, F.J., Kobourov, S.G.: Straight-line grid drawings of 3-connected 1-planar graphs. In: Wismath, S., Wolff, A. (eds.) GD 2013. LNCS, vol. 8242, pp. 83–94. Springer, Heidelberg (2013)
2. Bannister, M.J., Eppstein, D.: Crossing minimization for 1-page and 2-page drawings of graphs with bounded treewidth. In: Duncan, C., Symvonis, A. (eds.) GD 2014. LNCS, vol. 8871, pp. 210–221. Springer, Heidelberg (2014)
3. Bannister, M.J., Eppstein, D., Simons, J.A.: Fixed parameter tractability of crossing minimization of almost-trees. In: Wismath, S., Wolff, A. (eds.) GD 2013. LNCS, vol. 8242, pp. 340–351. Springer, Heidelberg (2013)
4. Bernhart, F., Kainen, P.C.: The book thickness of a graph. J. Comb. Theor. Ser. B **27**(3), 320–331 (1979)
5. Bodlaender, H.L.: A partial k-arboretum of graphs with bounded treewidth. Theor. Comput. Sci. **209**(12), 1–45 (1998)
6. Chung, F.R.K., Leighton, F.T., Rosenberg, A.L.: Embedding graphs in books: a layout problem with applications to VLSI design. SIAM J. Discrete Math. **8**(1), 33–58 (1987)
7. Di Battista, G., Eades, P., Tamassia, R., Tollis, I.G.: Graph Drawing: Algorithms for the Visualization of Graphs. Prentice-Hall, Upper Saddle River (1999)

8. Di Giacomo, E., Didimo, W., Liotta, G., Montecchiani, F.: Area requirement of graph drawings with few crossings per edge. Comput. Geom. **46**(8), 909–916 (2013)
9. Di Giacomo, E., Liotta, G., Montecchiani, F.: Drawing outer 1-planar graphs with few slopes. In: Duncan, C., Symvonis, A. (eds.) GD 2014. LNCS, vol. 8871, pp. 174–185. Springer, Heidelberg (2014)
10. Masuda, S., Nakajima, K., Kashiwabara, T., Fujisawa, T.: Crossing minimization in linear embeddings of graphs. IEEE Trans. Comput. **39**(1), 124–127 (1990)
11. Nicholson, T.: Permutation procedure for minimising the number of crossings in a network. Proc. Inst. Electr. Eng. **115**(1), 21–26 (1968)
12. Wattenberg, M.: Arc diagrams: visualizing structure in strings. In: InfoVis 2002, pp. 110–116. IEEE Computer Society (2002)
13. Yannakakis, M.: Embedding planar graphs in four pages. J. Comput. Syst. Sci. **38**(1), 36–67 (1989)

Longest Common Extensions in Partial Words

Francine Blanchet-Sadri[1]([✉]), Rachel Harred[1], and Justin Lazarow[2]

[1] Department of Computer Science, University of North Carolina,
P.O. Box 26170, Greensboro, NC 27402–6170, USA
{blanchet,r_harred}@uncg.edu
[2] Department of Mathematics, University of Texas at Austin,
1 University Station C1200, Austin, TX 78712–0233, USA
jlazarow@utexas.edu

Abstract. The *Longest Common Extension* of a pair of positions (i, j) in a string, or word, is the longest substring starting at i and j. The *LCE problem* considers a word and a set of pairs of positions and computes for each pair in the set, the longest common extension starting at both positions in the pair. This problem finds applications in matching with don't-care characters, approximate string searching, finding all exact or approximate tandem repeats, to name a few. From a practical point of view, Ilie et al. (Journal of Discrete Algorithms, 2010) looked for simple and efficient algorithms for the LCE problem. In this paper, we extend their analyses to partial words, strings with don't-cares or holes. In this context, we compute the *Longest Common Compatible Extension* of each pair of positions (i, j) in a partial word, i.e., the longest substrings starting at i and j that are *compatible*. We show that our results match with those of total words (partial words without holes). We find that one of the simplest algorithms for implementing the LCE problem is optimal on average in this case.

1 Introduction

The *Longest Common Extension* of a pair of positions (i, j) in a string, or word, is the longest substring starting at both i and j, i.e., the longest common prefix of the suffixes that start at i and j. Given as input a word and a set of pairs of positions, the *LCE problem* outputs for each pair in the set, the longest common extension starting at both positions in the pair. Variants of this problem appear in the literature and find important applications in matching with don't-care characters [6], approximate string searching [13,14,17], finding all exact or approximate tandem repeats [7,12,15], to name a few. The LCE problem can be optimally solved by first preprocessing the input word in linear time in its length so that the longest common extension of each pair of positions be computed in constant time. A first approach is based on the constant-time computation of Lowest Common Ancestors in the suffix tree of the input word [1,2,8,18] while a second approach is based on Range Minimum Queries in the suffix array of the input word [1,2,5,16].

© Springer International Publishing Switzerland 2016
Z. Lipták and W.F. Smyth (Eds.): IWOCA 2015, LNCS 9538, pp. 52–64, 2016.
DOI: 10.1007/978-3-319-29516-9_5

From a practical point of view, Ilie, Navarro and Tinta [9] looked for simple and efficient algorithms for the LCE problem. They observed that the average of the LCE values is very small; they actually calculated the limit of this average, over any given alphabet size, when the word's length goes to infinity. As an important consequence, they gave algorithms that solve the LCE problem in constant time without any preprocessing. These algorithms are the best in practice with respect to both time and space. By computing the longest common extensions required in Landau and Vishkin's approximate string searching algorithm [14] with their simplest LCE algorithm, Ilie et al. obtained an algorithm that runs 13 to 20 times faster.

Blanchet-Sadri and Lazarow [3] extended suffix trees to partial words, strings with don't-cares or holes, by introducing a suffix directed acyclic graph, with *compatibility links*, that exhibits all the suffixes while preserving the *Longest Common Compatible Extension* of each pair of positions (i, j), i.e., the longest substrings starting at i and j that are *compatible*. They gave an optimal $O(nh+n)$ time and space algorithm for constructing the suffix dag of a given partial word w of length n with an arbitrary number of holes h over a fixed alphabet by modifying Weiner's algorithm [19]. With $O(nh + n)$ preprocessing time, finding the longest common compatible extension of a given pair of positions in w requires constant time. Later, using ideas from suffix arrays, alignment techniques, and dynamic programming [10,11], Crochemore et al. [4] provided an algorithm for the LCCE problem, over any integer alphabet, with a slightly better runtime. Their simpler data structure can be constructed in $O(n\mu + n)$ time and space, where μ is the number of blocks of consecutive holes in w.

In this paper, we extend the analyses of Ilie et al. [9] to partial words. We compute the longest common compatible extension of each pair of positions (i, j) in any given partial word. We show that our results match with those of total words (partial words without holes). We find that one of the simplest algorithms for implementing the LCE problem is optimal on average in this case.

The contents of our paper are as follows: In Sect. 2, we recall a few concepts on partial words such as compatibility and we introduce a few new concepts related to the LCE problem. In Sect. 3, we estimate the average value of the longest common compatible extensions over all h-hole partial words of a given length over an alphabet of a given size. In Sect. 4, we show how to compute the longest common compatible extensions in the h-hole case. Finally in Sect. 5, we conclude with some remarks.

2 Preliminaries

Let Σ be an alphabet. We assume, unless otherwise stated, that the cardinality of Σ, denoted by $\mathtt{card}(\Sigma)$ or $|\Sigma|$, is at least two. A *partial word* over Σ is a sequence of characters from the extended alphabet $\Sigma_\diamond = \Sigma \cup \{\diamond\}$, where the character $\diamond$ represents a don't-care or a *hole*. A *total word* is a partial word with no holes. We denote by $|w|$ the *length* of w, or the number of characters in w. We denote by ε the empty word, i.e., the word of length zero. We denote by Σ^*

(respectively, $\Sigma_\diamond^*$) the set of total words (respectively, partial words) over Σ and by Σ^n (respectively, $\Sigma_\diamond^n$) the set of total words (respectively, partial words) of length n over Σ.

The character at position i of a partial word w is denoted $w[i]$, with position numbers starting at 1. A *factor* of w is a consecutive sequence of characters in w. We denote by $w[i..j]$ the factor of w starting at position i and ending at position j. A *prefix* of w is a factor of the form $w[1..j]$, while a *suffix* is of the form $w[i..|w|]$.

We denote by $H(w)$ the set of positions that are holes in w. Two partial words u and v of equal length over Σ are *compatible*, denoted by $u \uparrow v$, if $u[i] = v[i]$ for every $i \notin H(u) \cup H(v)$. For example, if $w = ab\diamond bba\diamond\diamond baab$, a partial word with three holes over the alphabet $\{a, b\}$, then $w[1..4] = ab\diamond b \uparrow w[6..9] = a\diamond\diamond b$. A *square* in a partial word w is a factor of the form uv, with u and v compatible.

A *strong period* of a partial word w is a positive integer p such that $w[i] = w[j]$ for all i, j such that $i \equiv j \mod p$ and $i, j \notin H(w)$. A *weak period* of w is a positive integer p such that $w[i] = w[i + p]$ for all i such that $i, i + p \notin H(w)$.

Define the *longest common compatible extension* of a pair of positions (i, j) in a partial word w as the longest factor of w starting at i that is compatible with a factor of w starting at j. When $\mathrm{card}(H(w)) = h$, denote by $\mathrm{LCCE}_w(i, j)$ the length of the longest common compatible extension of (i, j) in w. Returning to our example, when $w = ab\diamond bba\diamond\diamond baab$ we have $\mathrm{LCCE}_w(1, 6) = 4$. Also denote by $\mathrm{Avg\_LCCE}(n, \sigma, h)$ the average length of the longest common compatible extension in all partial words of length n over an alphabet of cardinality σ with exactly h holes.

3   Computing the Average LCCE in the h-Hole Case

The total word case was studied in [9]. We study the h-hole partial word case here, for $h \geq 1$. We wish to determine the average value of the LCCE over all h-hole partial words of a given length n over an arbitrary σ-letter alphabet Σ.

The following hold:

$$\mathrm{Avg\_LCCE}(n, \sigma, h)$$

$$= \frac{1}{\binom{n}{h}\sigma^{n-h}} \sum_{w \in \Sigma_\diamond^n, |H(w)|=h} \left(\frac{1}{\binom{n}{2}} \sum_{1 \leq i < j \leq n} \mathrm{LCCE}_w(i, j) \right)$$

$$= \frac{1}{\binom{n}{h}\binom{n}{2}\sigma^{n-h}} \sum_{k=1}^{n-1} k \sum_{1 \leq i < j \leq n-k+1} \mathrm{card}(\{w \mid |H(w)| = h, \mathrm{LCCE}_w(i, j) = k\}).$$

$$(1)$$

Determining this average value in the partial word case requires much more combinatorial work than determining the corresponding average in the total word case. To make things easier, we split our analysis of this average value into six cases (we do this to emphasize six cases based on the positioning of i and j).

We denote the set of partial words of length n over the σ-letter alphabet Σ matching the given k value of the longest common compatible extension starting at the i and j positions and having exactly h holes by $K^c_{i,j,k,\sigma,h}(n)$, where c is used to denote a specific case (there are six cases numbered from $c = 1$ to $c = 6$ covered in Lemmas 1–6 below). We say that the pair (i,j) *satisfies* c if it satisfies the conditions stipulated by case c. More precisely,

$$K^c_{i,j,k,\sigma,h}(n) = \{w \mid w \in \Sigma^n_\diamond, |H(w)| = h, \text{LCCE}_w(i,j) = k, \text{ and } (i,j) \text{ satisfies } c\}.$$

Note that if $w \in K^c_{i,j,k,\sigma,h}(n)$ then $w[i..i+k-1] \uparrow w[j..j+k-1]$. We also adopt the notation

$$N^c_{k,\sigma,h}(n) = \text{card}(\{(i,j) \mid (i,j) \text{ satisfies } c \text{ in some } w \in K^c_{i,j,k,\sigma,h}(n)\}).$$

Let us now introduce our cases.

Lemma 1 ($c = 1$: $j \leq n - k$ and $j - i > k$). *The following hold:*

$$\text{card}(K^1_{i,j,k,\sigma,h}(n)) = \sum_{m=0}^{h} \sum_{\ell=0}^{\lfloor \frac{m}{2} \rfloor} A \binom{n - 2k - 2}{h - m} (\sigma - 1)\sigma^{n-k-1-h+m-\ell}, \qquad (2)$$

where $A = \binom{k}{\ell}\binom{k-\ell}{m-2\ell}2^{m-2\ell}$, and

$$N^1_{k,\sigma,h}(n) = \frac{1}{2}(n - 2k)(n - 2k - 1). \qquad (3)$$

Proof. We have $w[i+k] \not\uparrow w[j+k]$, so neither is a hole. We have $\sigma(\sigma-1)$ choices of letters for $w[i+k]$ and $w[j+k]$ since they are not equal. Assume there are m holes in $w[i..i+k-1]w[j..j+k-1]$ for some m, $0 \leq m \leq h$. So there are $h - m$ holes in $u = w[1..i-1]w[i+k+1..j-1]w[j+k+1..n]$. We have $\binom{n-2k-2}{h-m}$ possible position sequences for the $h - m$ holes to be placed and $\sigma^{n-2k-2-h+m}$ choices of letter sequences for the remaining positions in u. This count totals to $\binom{n-2k-2}{h-m}$ $(\sigma - 1)\sigma^{n-2k-1-h+m}$ choices for $w[1..i-1]w[i+k..j-1]w[j+k..n]$.

Next, let us count the number of choices for $v = w[i..i+k-1]w[j..j+k-1]$. Align the positions in $w[i..i+k-1]$ and $w[j..j+k-1]$ to create k columns (the top row representing $w[i..i+k-1]$ and the bottom row representing $w[j..j+k-1]$). When we distribute the m holes, some of these columns may end up containing a pair of holes. Say there are ℓ columns that contain a pair of holes, so $m - 2\ell$ of the remaining $k - \ell$ columns contain only one hole appearing in the top row or the bottom row. This count totals to

$$\sum_{\ell=0}^{\lfloor \frac{m}{2} \rfloor} \binom{k}{\ell}\binom{k - \ell}{m - 2\ell}2^{m-2\ell}\sigma^{k-\ell}$$

for v. Combining these counts gives us Eq. (2). $\square$

Lemma 2 ($c = 2$: $j = n - k + 1$ **and** $j - i > k$).   *The following hold:*

$$\mathrm{card}(K^2_{i,j,k,\sigma,h}(n)) = \sum_{m=0}^{h} \sum_{\ell=0}^{\lfloor \frac{m}{2} \rfloor} A \binom{n - 2k}{h - m} \sigma^{n-k-h+m-\ell}, \qquad (4)$$

where $A = \binom{k}{\ell}\binom{k-\ell}{m-2\ell} 2^{m-2\ell}$, *and*

$$N^2_{k,\sigma,h}(n) = n - 2k. \qquad (5)$$

Let s be a finite sequence of positive integers in some discrete interval of integers $[1..q]$. Then g_s is defined as the number of non-empty gaps created by s. For example, if $s = 125$ in $[1..9]$, then $g_s = 2$ counting the gap 34 and the gap 6789.

Lemma 3 ($c = 3$: $j \leq n - k$ **and** $j - i < k$).   *The following hold:*

$$\mathrm{card}(K^3_{i,j,k,\sigma,h}(n)) = \sum_{m=0}^{h} C \binom{i - 1 + n - j - k}{h - m} (\sigma - 1)\sigma^{i-1+n-j-k-h+m}, \quad (6)$$

where $C = \sum_{(s_1,\ldots,s_{j-i}) \in S}(\sigma^{g_{s_1}} \cdots \sigma^{g_{s_{j-i}}})$ *with S being the set of tuples of the form* $(s_1, \ldots, s_{j-i})$ *that satisfy the conditions:*

- $s_1, \ldots, s_r$ *are subsequences of* $[1..q']$ *and* $s_{r+1}, \ldots, s_{j-i}$ *of* $[1..q]$,
- $|s_1| = m_1, \ldots, |s_{j-i}| = m_{j-i}$ *and* $m_1 + \cdots + m_{j-i} = m$,
- s_ℓ *does not contain* $\frac{k-\ell+1}{j-i} + 1$ *for* $\ell \in \{1, \ldots, j - i\}$ *satisfying* $\ell \equiv (k + 1)$ mod $(j - i)$,

when $q = \lfloor \frac{j+k-i}{j-i} \rfloor$, $q' = \lceil \frac{j+k-i}{j-i} \rceil$, $r = (j + k - i)$ mod $(j - i)$, *and*

$$N^3_{k,\sigma,h}(n) = \frac{1}{2}(k - 2)(k - 1) + (k - 1)(n - 2k + 1). \qquad (7)$$

Proof. Note that $w[i + k] = w[j + (i + k - j)]$ nor $w[j + k]$ can be a hole so that we have $\sigma(\sigma - 1)$ choices of letters for $w[i + k]$ and $w[j + k]$. Assume there are m holes in $w[i..j + k - 1]$ for some m, $0 \leq m \leq h$. So there are $h - m$ holes in $u = w[1..i - 1]w[j + k + 1..n]$. We have $\binom{i-1+n-j-k}{h-m}$ possible position sequences for the $h - m$ holes to be placed and $\sigma^{i-1+n-j-k-h+m}$ choices of letter sequences for the remaining positions in u. This count totals to $\binom{i-1+n-j-k}{h-m}(\sigma - 1)\sigma^{i-1+n-j-k-h+m}$ choices for $w[1..i - 1]w[j + k..n]$.

Next, let us count the number of choices for $v = w[i..j + k - 1]$. We have $k - i + j - 1$ choices for where to put the m holes since $w[i + k]$ cannot be a hole. Align the positions in v to create $j - i$ columns. Let $j + k - i = q(j - i) + r$ where $0 \leq r < j - i$. So $q = \lfloor \frac{j+k-i}{j-i} \rfloor$ and $r = (j + k - i)$ mod $(j - i)$. Let $q' = \lceil \frac{j+k-i}{j-i} \rceil$. There are r of the $j - i$ columns that have q' elements each and the $j - i - r$ remaining columns have q elements each. When we distribute the m holes, say there are m_1 holes placed in Column 1, m_2 in Column 2, and so on.

So we have the restrictions $m_1 + \cdots + m_{j-i} = m$, $m_1, \ldots, m_r \in [0..q']$, and $m_{r+1}, \ldots, m_{j-i} \in [0..q]$. Note that $w[i..j + k - 1]$ is weakly $(j - i)$-periodic or each of the $j - i$ columns is weakly 1-periodic, i.e., each column can be viewed as a partial word of length q' or q with the holes creating gaps (each non-empty gap is assigned one of the σ letters). In any given column, say s denotes the subsequence of $[1..q']$ or $[1..q]$ representing the positions with holes, and let g_s denote the number of non-empty gaps created by the holes in the column. This column will then be associated with σ^{g_s} choices of letters for the gaps.

Let $\ell \in \{1, \ldots, j - i\}$ be such that $i + k - i + 1 = k + 1 \equiv \ell \mod (j - i)$, i.e., ℓ is the column number where position $(i + k)$ falls into. The first entry of column 1 is $i = (i - 1) + 1$, so the first entry of column ℓ is $(i - 1) + \ell$. Then if $s_1, \ldots, s_r$ are subsequences of $[1..q']$ and $s_{r+1}, \ldots, s_{j-i}$ are subsequences of $[1..q]$ with the restriction that s_ℓ does not contain $\frac{(i+k)-(i-1)-\ell}{j-i} + 1 = \frac{k-\ell+1}{j-i} + 1$ (this is because $w[i + k] \neq \diamond$) and $|s_1| = m_1, \ldots, |s_{j-i}| = m_{j-i}$, we get that this count totals to

$$\sum_{(s_1, \ldots, s_{j-i}) \in S} \left(\sigma^{g_{s_1}} \cdots \sigma^{g_{s_{j-i}}} \right)$$

for v. Combining both counts gives us Eq. (6). $\qquad \square$

Lemma 4 ($c = 4$: $j = n - k + 1$ and $j - i < k$). *The following hold:*

$$\mathbf{card}(K^4_{i,j,k,\sigma,h}(n)) = \sum_{m=0}^{h} D \binom{i-1}{h-m} \sigma^{i-1-h+m}, \qquad (8)$$

where $D = \sum_{(s_1, \ldots, s_{j-i}) \in T} (\sigma^{g_{s_1}} \cdots \sigma^{g_{s_{j-i}}})$ with T being the set of tuples of the form $(s_1, \ldots, s_{j-i})$ that satisfy the conditions:

- $s_1, \ldots, s_r$ *are subsequences of $[1..q']$ and $s_{r+1}, \ldots, s_{j-i}$ of $[1..q]$,*
- $|s_1| = m_1, \ldots, |s_{j-i}| = m_{j-i}$ *and $m_1 + \cdots + m_{j-i} = m$,*

when $q = \lfloor \frac{j+k-i}{j-i} \rfloor$, $q' = \lceil \frac{j+k-i}{j-i} \rceil$, and $r = (j + k - i) \mod (j - i)$, and

$$N^4_{k,\sigma,h}(n) = k - 1. \qquad (9)$$

Let us illustrate Lemma 4 with an example. Consider the parameters $n = 10$, $h = 3$, $\sigma = 3$, $k = 4$, and $(i, j) = (4, 7)$. Let us calculate $\mathbf{card}(K^4_{i,j,k,\sigma,h}(n))$. For $m = 0$, the only possibility for (m_1, m_2, m_3) is $(0, 0, 0)$. So $s_1 = s_2 = s_3 = \varepsilon$, and $\sigma^{g_{s_1}} = \sigma^{g_{s_2}} = \sigma^{g_{s_3}} = \sigma$. This totals to $\sigma^{g_{s_1}} \sigma^{g_{s_2}} \sigma^{g_{s_3}} \binom{i-1}{h-m} \sigma^{i-1-h+m} = 27 \binom{4-1}{3-0} \sigma^{4-1-3+0} = 27$.

For $m = 1$, we have three possibilities for (m_1, m_2, m_3): $(1, 0, 0)$, $(0, 1, 0)$, and $(0, 0, 1)$. So the possibilities for $s_1 s_2 s_3$ along with their associated $\sigma^{g_{s_1}} \sigma^{g_{s_2}} \sigma^{g_{s_3}}$ are: $1\varepsilon\varepsilon$ with $\sigma\sigma\sigma = 27$, $2\varepsilon\varepsilon$ with $\sigma^2\sigma\sigma = 81$, $3\varepsilon\varepsilon$ with $\sigma\sigma\sigma = 27$, $\varepsilon 1\varepsilon$ with $\sigma\sigma\sigma = 27$, $\varepsilon 2\varepsilon$ with $\sigma\sigma\sigma = 27$, $\varepsilon\varepsilon 1$ with $\sigma\sigma\sigma = 27$, and $\varepsilon\varepsilon 2$ with $\sigma\sigma\sigma = 27$. This totals to $243 \binom{4-1}{3-1} \sigma^{4-1-3+1} = 2187$.

For $m = 2$, we have six possibilities for (m_1, m_2, m_3):

$$(2, 0, 0), (0, 2, 0), (0, 0, 2), (1, 1, 0), (1, 0, 1), (0, 1, 1).$$

So the possibilities for $s_1 s_2 s_3$ along with their associated $\sigma^{g_{s_1}} \sigma^{g_{s_2}} \sigma^{g_{s_3}}$ are: $12\varepsilon\varepsilon$, $13\varepsilon\varepsilon$, and $23\varepsilon\varepsilon$ totalling $(3\sigma)\sigma\sigma = 81$, $\varepsilon 12\varepsilon$ with $\sigma\sigma^0\sigma = 9$, $\varepsilon\varepsilon 12$ with $\sigma\sigma\sigma^0 = 9$, 11ε, 12ε, 21ε, 22ε, 31ε, and 32ε totalling $(\sigma + \sigma^2 + \sigma)(2\sigma)\sigma = 270$, $1\varepsilon 1$, $1\varepsilon 2$, $2\varepsilon 1$, $2\varepsilon 2$, $3\varepsilon 1$, and $3\varepsilon 2$ also totalling 270, $\varepsilon 11$, $\varepsilon 12$, $\varepsilon 21$, and $\varepsilon 22$ totalling $\sigma(2\sigma)(2\sigma) = 108$. This totals to $747\binom{4-1}{3-2}\sigma^{4-1-3+2} = 20169$.

For $m = 3$, we have eight possibilities for (m_1, m_2, m_3):

$$(3,0,0), (2,1,0), (2,0,1), (1,2,0), (1,1,1), (1,0,2), (0,2,1), (0,1,2).$$

This totals to $999\binom{4-1}{3-3}\sigma^{4-1-3+3} = 26973$. Thus, $\text{card}(K^4_{i,j,k,\sigma,h}(n)) = 49356$.

Lemma 5 ($c = 5$: $j \leq n - k$ and $j - i = k$). *The following hold:*

$$\text{card}(K^5_{i,j,k,\sigma,h}(n)) = \sum_{m=0}^{h} \sum_{\ell=0}^{\lfloor \frac{m}{2} \rfloor} B\binom{n - 2k - 1}{h - m}(\sigma - 1)\sigma^{n-k-1-h+m-\ell}, \quad (10)$$

where $B = \left[\binom{k-1}{\ell}\binom{k-1-\ell}{m-2\ell-1}2^{m-2\ell-1} + \binom{k-1}{\ell}\binom{k-1-\ell}{m-2\ell}2^{m-2\ell}\right]$, *and*

$$N^5_{k,\sigma,h}(n) = n - 2k. \quad (11)$$

Proof. This is the case when $w[i..j+k-1]$ is a square. We have $w[i+k] \not\curlyvee w[j+k]$, so there are $\sigma - 1$ choices for $w[j+k]$. Note that neither $w[i+k]$ nor $w[j+k]$ is a hole. Assume there are m holes in the square for some m, $0 \leq m \leq h$. So there are $h-m$ holes in $u = w[1..i-1]w[j+k+1..n]$. We have $\binom{n-2k-1}{h-m}$ choices of placement for the $h - m$ holes and $\sigma^{n-2k-1-h+m}$ choices for the non-holes in u. This gives a count of $\binom{n-2k-1}{h-m}(\sigma - 1)\sigma^{n-2k-1-h+m}$ choices for $w[1..i-1]w[j+k..n]$.

Next, let us count the number of choices for the square $w[i..j+k-1]$. Align the positions in $w[i..i+k-1]$ and $w[j..j+k-1]$ to create k columns (the top row representing $w[i..i+k-1]$ and the bottom row representing $w[j..j+k-1]$). When we distribute the m holes, some of these columns, except for the first, may end up containing a pair of holes. Say there are ℓ columns that contain a pair of holes. There are two possibilities for the first column: the position in the top row is a hole, in which case $m - 2\ell - 1$ of the remaining $k - 1 - \ell$ columns contain only one hole appearing in the top row or the bottom row, or the position in the top row of the first column is not a hole, in which case $m - 2\ell$ of the remaining $k - 1 - \ell$ columns contain only one hole appearing in the top row or the bottom row. Thus, the total number of choices for the square is

$$\sum_{\ell=0}^{\lfloor \frac{m}{2} \rfloor} \left[\binom{k-1}{\ell}\binom{k-1-\ell}{m-2\ell-1}2^{m-2\ell-1} + \binom{k-1}{\ell}\binom{k-1-\ell}{m-2\ell}2^{m-2\ell}\right]\sigma^{k-\ell}.$$

Combining both counts gives us Eq. (10). $\qquad \square$

Lemma 6 ($c = 6$: $j = n - k + 1$ and $j - i = k$). *The following hold:*

$$\text{card}(K^6_{i,j,k,\sigma,h}(n)) = \sum_{m=0}^{h} \sum_{\ell=0}^{\lfloor \frac{m}{2} \rfloor} A\binom{n - 2k}{h - m}\sigma^{n-k-h+m-\ell}, \quad (12)$$

where $A = \binom{k}{\ell}\binom{k-\ell}{m-2\ell}2^{m-2\ell}$, and

$$N_{k,\sigma,h}^6(n) = 1. \tag{13}$$

When $j \leq n - k$, we can verify that

$$N_{k,\sigma,h}^1(n) + N_{k,\sigma,h}^3(n) + N_{k,\sigma,h}^5(n) = \binom{n-k}{2},$$

and when $j = n - k + 1$, that

$$N_{k,\sigma,h}^2(n) + N_{k,\sigma,h}^4(n) + N_{k,\sigma,h}^6(n) = n - k.$$

Thus we count exactly as many pairs (i, j) there can be. To total all such partial words and (i, j) pairs, we must sum up the product of the number of partial words matching the case for fixed i and j with the number of (i, j) pairs matching the case. With exception to cases 3 and 4, the number of partial words matching the case for fixed i and j is independent of i and j. Thus if we denote $P_{i,j,k,\sigma,h}^c(n)$ as the total number of choices of partial words and pairs (i, j), we get

$$P_{i,j,k,\sigma,h}^c(n) = N_{k,\sigma,h}^c(n)\,\mathsf{card}(K_{i,j,k,\sigma,h}^c(n))$$

for all c except $c = 3$ or $c = 4$. For $c = 3$, we obtain that $P_{i,j,k,\sigma,h}^3(n)$ equals

$$\sum_{j=2}^{k-1}\left(\sum_{i=1}^{j-1}\left(\sum_{m=0}^{h}C\binom{i-1+n-j-k}{h-m}\right)(\sigma-1)\sigma^{i-1+n-j-k-h+m}\right)$$
$$+\sum_{j=k}^{n-k}\left(\sum_{i=j-k+1}^{j-1}\left(\sum_{m=0}^{h}C\binom{i-1+n-j-k}{h-m}\right)(\sigma-1)\sigma^{i-1+n-j-k-h+m}\right),$$

while for $c = 4$, we obtain that $P_{i,j,k,\sigma,h}^4(n)$ equals

$$\sum_{j=n-k+1}^{n-k+1}\left(\sum_{i=n-2k+2}^{n-k}\left(\sum_{m=0}^{h}D\binom{i-1}{h-m}\right)\sigma^{i-1-h+m}\right),$$

where C and D are as in Lemmas 3 and 4. We then have the following theorem by rewriting (1).

Theorem 7. *The following holds:*

$$\mathsf{Avg\_LCCE}(n,\sigma,h) = \frac{1}{\binom{n}{h}\binom{n}{2}\sigma^{n-h}}\sum_{k=1}^{n-1}k\left(\sum_{c=1}^{6}P_{i,j,k,\sigma,h}^c(n)\right),$$

where

$$P^1_{i,j,k,\sigma,h}(n) = \tfrac{1}{2}(n-2k)(n-2k-1)$$
$$\sum_{m=0}^{h}\sum_{\ell=0}^{\lfloor\frac{m}{2}\rfloor} A\binom{n-2k-2}{h-m}(\sigma-1)\sigma^{n-k-1-h+m-\ell},$$

$$P^2_{i,j,k,\sigma,h}(n) = (n-2k)\sum_{m=0}^{h}\sum_{\ell=0}^{\lfloor\frac{m}{2}\rfloor} A\binom{n-2k}{h-m}\sigma^{n-k-h+m-\ell},$$

$$P^3_{i,j,k,\sigma,h}(n) = \sum_{j=2}^{k-1}\left(\sum_{i=1}^{j-1}\left(\sum_{m=0}^{h} C\binom{i-1+n-j-k}{h-m}(\sigma-1)\sigma^{i-1+n-j-k-h+m}\right)\right)$$
$$+\sum_{j=k}^{n-k}\left(\sum_{i=j-k+1}^{j-1}\right.$$
$$\left.\left(\sum_{m=0}^{h} C\binom{i-1+n-j-k}{h-m}(\sigma-1)\sigma^{i-1+n-j-k-h+m}\right)\right),$$

$$P^4_{i,j,k,\sigma,h}(n) = \sum_{j=n-k+1}^{n-k+1}\left(\sum_{i=n-2k+2}^{n-k}\left(\sum_{m=0}^{h} D\binom{i-1}{h-m}\sigma^{i-1-h+m}\right)\right),$$

$$P^5_{i,j,k,\sigma,h}(n) = (n-2k)\sum_{m=0}^{h}\sum_{\ell=0}^{\lfloor\frac{m}{2}\rfloor} B\binom{n-2k-1}{h-m}(\sigma-1)\sigma^{n-k-1-h+m-\ell},$$

$$P^6_{i,j,k,\sigma,h}(n) = \sum_{m=0}^{h}\sum_{\ell=0}^{\lfloor\frac{m}{2}\rfloor} A\binom{n-2k}{h-m}\sigma^{n-k-h+m-\ell},$$

where A, B, C and D are as in Lemmas 1, 5, 3 and 4, respectively.

We next show that asymptotically, the average length of the longest common compatible extension between any two positions in a partial word with h holes over a σ-letter alphabet is $\frac{1}{\sigma-1}$.

Theorem 8. *For fixed σ and h, $\lim_{n\to\infty} \mathrm{Avg\_LCCE}(n,\sigma,h) = \frac{1}{\sigma-1}$.*

Proof. Let $S(n) = \sum_{1\le c\le 6} P^c_{i,j,k,\sigma,h}(n)$. We wish to calculate

$$\lim_{n\to\infty} \mathrm{Avg\_LCCE}(n,\sigma,h) = \lim_{n\to\infty}\frac{\sum_{k=1}^{n-1} kS(n)}{\binom{n}{h}\binom{n}{2}\sigma^{n-h}} = \sum_{1\le c\le 6}\lim_{n\to\infty}\frac{\sum_{k=1}^{n-1} kP^c_{i,j,k,\sigma,h}(n)}{\binom{n}{h}\binom{n}{2}\sigma^{n-h}}.$$

So let us look at the six limits when n goes to infinity, which can be calculated using Theorem 7. For $c = 1$, we get

$$\lim_{n\to\infty}\frac{\sum_{k=1}^{n-1} kP^1_{i,j,k,\sigma,h}(n)}{\binom{n}{h}\binom{n}{2}\sigma^{n-h}}$$
$$= \lim\frac{\frac{(\sigma-1)}{2\sigma}\sum_{k=1}^{n-1}\frac{k}{\sigma^k}(n-2k)(n-2k-1)\left(\sum_{m=0}^{h}\binom{n-2k-2}{h-m}(2\sigma)^m\left(\sum_{\ell=0}^{\lfloor\frac{m}{2}\rfloor}\frac{1}{(4\sigma)^\ell}\binom{k}{\ell}\binom{k-\ell}{m-2\ell}\right)\right)}{\binom{n}{h}\binom{n}{2}}$$
$$= \frac{1}{(\sigma-1)}.$$

To see this, the denominator, $\binom{n}{h}\binom{n}{2}$, is $O(n^{h+2})$. As to the numerator, the third sum, $\sum_{\ell=0}^{\lfloor\frac{m}{2}\rfloor}\frac{1}{(4\sigma)^\ell}\binom{k}{\ell}\binom{k-\ell}{m-2\ell}$, is $O(k^m)$. Each summand in the second sum is then $O((n+k)^{h-m}k^m)$. So the first sum involves expressions that are big-O of expressions like

$$n^{h-m-p+p_1} \sum\nolimits_{k=1}^{n-1} \frac{k^{m+p+p_2+1}}{\sigma^k},$$

where $m \in [0..h]$, $p \in [0..h-m]$, $p_1, p_2 \in [0..2]$, and $p_1+p_2 \in [1..2]$. So to calculate the limit, we must calculate the limit of weighted (by constants) expressions like

$$\frac{1}{n^{m+p-p_1+2}} \sum\nolimits_{k=1}^{n-1} \frac{k^{m+p+p_2+1}}{\sigma^k},$$

where $m \in [0..h]$, $p \in [0..h-m]$, $p_1, p_2 \in [0..2]$, and $p_1 + p_2 \in [1..2]$. Note that p_1 refers to the number of n's selected from $(n - 2k)(n - 2k - 1)$ and p_2 refers to the number of k's. Since each such $\lim_{n\to\infty} \sum_{k=1}^{n-1} \frac{k^{m+p+p_2+1}}{\sigma^k}$ is $O(1)$, we obtain

$$\lim_{n\to\infty} \frac{1}{n^{m+p-p_1+2}} \sum_{k=1}^{n-1} \frac{k^{m+p+p_2+1}}{\sigma^k} = 0,$$

when $m + p - p_1 + 2 > 0$. So we need to consider the expressions where $m + p - p_1 + 2 = 0$, which occur exactly when $m = p = 0$, $p_1 = 2$, and $p_2 = 0$. Thus, since each of $\lim_{n\to\infty} \sum_{k=1}^{n-1} \frac{k}{\sigma^k}, \ldots, \lim_{n\to\infty} \sum_{k=1}^{n-1} \frac{k^{h+1}}{\sigma^k}$ is $O(1)$, we obtain

$$\begin{aligned}
&\lim_{n\to\infty} \frac{\sum_{k=1}^{n-1} k P^1_{i,j,k,\sigma,h}(n)}{\binom{n}{h}\binom{n}{2}\sigma^{n-h}} \\
&= \lim_{n\to\infty} \frac{\frac{(\sigma-1)}{2\sigma} \sum_{k=1}^{n-1} \frac{k}{\sigma^k} n^2 \binom{n-2k-2}{h}}{\binom{n}{h}\binom{n}{2}} \\
&= \lim_{n\to\infty} \frac{\frac{(\sigma-1)}{\sigma} \sum_{k=1}^{n-1} \frac{k}{\sigma^k} n^2 (n - 2k - 2) \cdots (n - 2k - 2 - h + 1)}{n(n-1)\cdots(n-h+1)n(n-1)} \\
&= \frac{(\sigma-1)}{\sigma} \lim_{n\to\infty} \sum_{k=1}^{n-1} \frac{k}{\sigma^k} = \frac{(\sigma-1)}{\sigma} \frac{\sigma}{(\sigma-1)^2}.
\end{aligned}$$

We can show that the limits related to $c \in \{2, \ldots, 6\}$ are all 0. Thus, the only term in $\frac{\sum_{k=1}^{n-1} k S(n)}{\binom{n}{h}\binom{n}{2}\sigma^{n-h}}$ that does not go to zero is from the $c = 1$ part, which goes to $\frac{1}{\sigma-1}$. Thus, $\lim_{n\to\infty}$ Avg $\_$LCCE$(n, \sigma, h) = \frac{1}{\sigma-1}$.   □

4   Computing the LCCE in the h-Hole Case

Theorem 8 shows that, on average, the length of the longest common compatible extension between any two positions in a partial word with h holes over a σ-letter alphabet will be $\frac{1}{\sigma-1}$. Thus we will only need to make $\frac{\sigma}{\sigma-1}$ comparisons on average when computing the longest common compatible extension between two suffixes. Thus we replace equality with compatibility in the DIRECTCOMP algorithm from [9] to find an optimal algorithm on average.

Algorithm 1. DIRECTCOMPPARTIAL(w, i, j)

$k \leftarrow 0$
while $(w[i + k] \uparrow w[j + k])$ **and** $(j + k \leq n)$ **do**
  $k \leftarrow k + 1$
return k

Figure 1 illustrates the average LCCE for some parameters n, σ, and h.

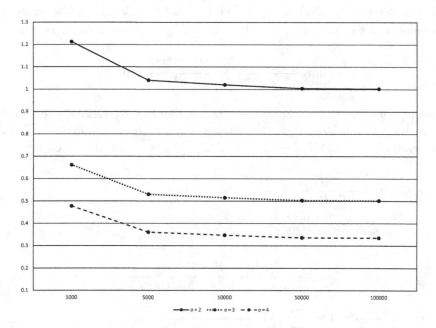

Fig. 1. y-axis represents the average $\mathtt{Avg\_LCCE}(n, \sigma, h)$ for number of holes $h = 50$ and alphabet size $\sigma \in \{2, 3, 4\}$; x-axis represents length $n \in \{1,000, \ldots, 100,000\}$

5   Conclusion

In this paper, we designed a simple and efficient algorithm for implementing the LCCE problem in constant time, that requires no preprocessing and that is optimal on average, for partial words with h holes. We did so by extending the analyses of Ilie et al. [9] to such partial words. We observed that the average of the LCCE values is very small; we actually calculated the limit of this average, over any given alphabet size, when the partial word's length goes to infinity. Estimating the average value of the LCCE over all partial words of a given length, with an arbitrary number of holes over an alphabet of a given size, could be used to improve the time and space efficiencies of algorithms on partial words.

Acknowledgements. Project sponsored by the National Security Agency under Grant Number H98230-15-1-0232. The United States Government is authorized to reproduce and distribute reprints notwithstanding any copyright notation herein. This manuscript is submitted for publication with the understanding that the United States Government is authorized to reproduce and distribute reprints. This material is based upon work supported by the National Science Foundation under Grant No. DMS–1060775.

References

1. Bender, M.A., Farach-Colton, M.: The LCA problem revisited. In: Gonnet, G.H., Viola, A. (eds.) LATIN 2000. LNCS, vol. 1776, pp. 88–94. Springer, Heidelberg (2000)
2. Berkman, O., Vishkin, U.: Recursive star-free parallel data structure. SIAM J. Comput. **22**, 221–242 (1993)
3. Blanchet-Sadri, F., Lazarow, J.: Suffix trees for partial words and the longest common compatible prefix problem. In: Dediu, A.-H., Martín-Vide, C., Truthe, B. (eds.) LATA 2013. LNCS, vol. 7810, pp. 165–176. Springer, Heidelberg (2013)
4. Crochemore, M., Iliopoulos, C.S., Kociumaka, T., Kubica, M., Langiu, A., Radoszewski, J., Rytter, W., Szreder, B., Waleń, T.: A note on the longest common compatible prefix problem for partial words (2013). arxiv:1312.2381v1
5. Fischer, J., Heun, V.: Theoretical and practical improvements on the RMQ-problem, with applications to LCA and LCE. In: Lewenstein, M., Valiente, G. (eds.) CPM 2006. LNCS, vol. 4009, pp. 36–48. Springer, Heidelberg (2006)
6. Gusfield, D.: Algorithms on Strings, Trees, and Sequences. Cambridge University Press, Cambridge (1997)
7. Gusfield, D., Stoye, J.: Linear time algorithms for finding and representing all the tandem repeats in a string. J. Comput. Syst. Sci. **69**, 525–546 (2004)
8. Harel, D., Tarjan, R.E.: Fast algorithms for finding nearest common ancestors. SIAM J. Comput. **13**(2), 338–355 (1984)
9. Ilie, L., Navarro, G., Tinta, L.: The longest common extension problem revisited and applications to approximate string searching. J. Discrete Algorithms **8**(4), 418–428 (2010)
10. Kärkkäinen, J., Sanders, P.: Simple linear work suffix array construction. In: Baeten, J.C.M., Lenstra, J.K., Parrow, J., Woeginger, G.J. (eds.) ICALP 2003. LNCS, vol. 2719, pp. 943–955. Springer, Berlin (2003)
11. Kasai, T., Lee, G.H., Arimura, H., Arikawa, S., Park, K.: Linear-time longest-common-prefix computation in suffix arrays and its applications. In: Amir, A., Landau, G.M. (eds.) CPM 2001. LNCS, vol. 2089, pp. 181–192. Springer, Heidelberg (2001)
12. Landau, G., Schmidt, J.P., Sokol, D.: An algorithm for approximate tandem repeats. J. Comput. Biol. **8**, 1–18 (2001)
13. Landau, G., Vishkin, U.: Introducing efficient parallelism into approximate string matching and a new serial algorithm. In: STOC 1986, pp. 220–230. ACM Press (1986)
14. Landau, G., Vishkin, U.: Fast parallel and serial approximate string matching. J. Algorithms **10**, 157–169 (1989)
15. Main, M.G., Lorentz, R.J.: An O(nlog n) algorithm for finding all repetitions in a string. J. Algorithms **5**(3), 422–432 (1984)

16. de Castro Miranda, R., Ayala-Rincón, M.: A modification of the Landau-Vishkin algorithm computing longest common extensions via suffix arrays. In: Setubal, J.C., Verjovski-Almeida, S. (eds.) BSB 2005. LNCS (LNBI), vol. 3594, pp. 210–213. Springer, Heidelberg (2005)
17. Myers, G.: An O(nd) difference algorithm and its variations. Algorithmica **1**, 251–266 (1986)
18. Schieber, B., Vishkin, U.: On finding lowest common ancestors: simplification and parallelization. SIAM J. Comput. **17**, 1253–1262 (1988)
19. Weiner, P.: Linear pattern matching algorithm. SWAT **1973**, 1–11 (1973)

Adding Isolated Vertices Makes Some Online Algorithms Optimal

Joan Boyar[1]([⊠]) and Christian Kudahl[1]

Department of Mathematics and Computer Science,
University of Southern Denmark, Odense, Denmark
{joan,kudahl}@imada.sdu.dk

Abstract. An unexpected difference between online and offline algorithms is observed. The natural greedy algorithms are shown to be worst case online optimal for ONLINE INDEPENDENT SET and ONLINE VERTEX COVER on graphs with "enough" isolated vertices, Freckle Graphs. For ONLINE DOMINATING SET, the greedy algorithm is shown to be worst case online optimal on graphs with at least one isolated vertex. These algorithms are not online optimal in general. The online optimality results for these greedy algorithms imply optimality according to various worst case performance measures, such as the competitive ratio. It is also shown that, despite this worst case optimality, there are Freckle graphs where the greedy independent set algorithm is objectively less good than another algorithm.

It is shown that it is NP-hard to determine any of the following for a given graph: the online independence number, the online vertex cover number, and the online domination number.

1 Introduction

This paper contributes to the larger goal of better understanding the nature of online optimality, greedy algorithms, and different performance measures for online algorithms. The graph problems ONLINE INDEPENDENT SET, ONLINE VERTEX COVER and ONLINE DOMINATING SET, which are defined below, are considered in the *vertex-arrival* model, where the vertices of a graph, G, are revealed one by one. When a vertex is revealed (we also say that it is "requested"), its edges to previously revealed vertices are revealed. At this point, an algorithm irrevocably either accepts the vertex or rejects it. This model is well-studied (see for example, [6, 8–10, 12, 13, 15]).

We show that, for some graphs, an obvious Greedy algorithm for each of these problems performs less well than another online algorithm and thus is not online optimal. However, this Greedy algorithm performs (at least in some sense) at least as well as any other online algorithm for these problems, as long as the graph has enough isolated vertices. Thus, in contrast to the case with

Supported in part by the Villum Foundation, the Stibo-Foundation, and the Danish Council for Independent Research, Natural Sciences.

Z. Lipták and W.F. Smyth (Eds.): IWOCA 2015, LNCS 9538, pp. 65–76, 2016.
DOI: 10.1007/978-3-319-29516-9_6

offline algorithms, adding isolated vertices to a graph can improve an algorithm's performance, even making it "optimal".

For an online algorithm for these problems and a particular sequence of requests, let S denote the set of accepted vertices, which we call a *solution*. When all vertices have been revealed (requested and either accepted or rejected by the algorithm), S must fulfill certain conditions:

- In the ONLINE INDEPENDENT SET problem, S must form an independent set. That is, no two vertices in S may have an edge between them. The goal is to maximize $|S|$.
- In the ONLINE VERTEX COVER problem, S must form a vertex cover. That is, each edge in G must have at least one endpoint in S. The goal is to minimize $|S|$.
- In the ONLINE DOMINATING SET problem, S must form a dominating set. That is, each vertex in G must be in S or have a neighbor in S. The goal is to minimize $|S|$.

If a solution does not live up to the specified requirement, it is said to be infeasible. The score of a feasible solution is $|S|$. The score of an infeasible solution is ∞ for minimization problems and $-\infty$ for maximization problems. Note that for ONLINE DOMINATING SET, it is not required that S form a dominating set at all times. It just needs to be a dominating set when the whole graph has been revealed. If, for example, it is known that the graph is connected, the algorithm might reject the first vertex since it is known that it will be possible to dominate this vertex later.

In Sect. 2, we define the greedy algorithms for the above problems, along with concepts analogous to the online chromatic number of Gyárfás et al. [7] for the above problems, giving a natural definition of optimality for online algorithms. In Sect. 3, we show that greedy algorithms are not in general online optimal for these problems. In Sect. 4, we define Freckle Graphs, which are graphs which have "enough" isolated vertices to make the greedy algorithms online optimal. In Sect. 5, it is shown that the online optimality results for these greedy algorithms imply optimality according to various worst case performance measures, such as the competitive ratio. In Sect. 6, it is shown that, despite this worst case optimality, there is a family of Freckle graphs where the greedy independent set algorithm is objectively less good than another algorithm. Various NP-hardness results concerning optimality are proven in Sect. 7. There are some concluding remarks in the last section.

2   Algorithms and Preliminaries

For each of the three problems, we define a greedy algorithm.

- In ONLINE INDEPENDENT SET, GIS accepts a revealed vertex, v, iff no neighbors of v have been accepted.

- In ONLINE VERTEX COVER, GVC accepts a revealed vertex, v, iff a neighbor of v has previously been revealed but not accepted.
- In ONLINE DOMINATING SET, GDS accepts a revealed vertex, v, iff no neighbors of v have been accepted.

For an algorithm ALG, we define $\overline{\text{ALG}}$ be the algorithm that simulates ALG and accepts exactly those vertices that ALG rejects. This defines a bijection between ONLINE INDEPENDENT SET and ONLINE VERTEX COVER algorithms. Note that GVC $= \overline{\text{GIS}}$.

For a graph, G, an ordering of the vertices, ϕ, and an algorithm, ALG, we let ALG($\phi(G)$) denote the score of ALG on G when the vertices are requested in the order ϕ. We let $|G|$ denote the number of vertices in G.

For minimization problems, we define:

$$\text{ALG}(G) = \max_{\phi} \text{ALG}(\phi(G))$$

That is, ALG(G) is the highest score ALG can achieve over all orderings of the vertices in G.

For maximization problems, we define:

$$\text{ALG}(G) = \min_{\phi} \text{ALG}(\phi(G))$$

That is, ALG(G) is the lowest score ALG can get over all orderings of the vertices in G.

Observation 1. *Let ALG be an algorithm for* ONLINE INDEPENDENT SET. *Let a graph, G, with n vertices be given. Now, $\overline{\text{ALG}}$ is an* ONLINE VERTEX COVER *algorithm and* $\text{ALG}(G) + \overline{\text{ALG}}(G) = n$.

The equality $\text{ALG}(G) + \overline{\text{ALG}}(G) = n$ holds, since a worst ordering of G for ALG is also a worst ordering for $\overline{\text{ALG}}$.

In considering online algorithms for coloring, [7] defines the online chromatic number, which intuitively is the best result (minimum number of colors) any online algorithm can be guaranteed to obtain for a particular graph (even when the graph, but not the ordering, is known in advance). We define analogous concepts for the problems we consider, defining for every graph a number representing the best value any online algorithm can achieve. Note that in considering all algorithms, we include those which know the graph in advance. Of course, when the graph is known, the order in which the vertices are requested is not known to an online algorithm, and the label given with a requested vertex does not necessarily correspond to its label in the known graph: The subgraph revealed up to this point might be isomorphic to more than one subgraph of the known graph and it could correspond to any of these subgraphs.

Let $I^O(G)$ denote the *online independence number* of G. This is the largest number such that there exists an algorithm, ALG, for ONLINE INDEPENDENT SET with $\text{ALG}(G) = I^O(G)$. Similarly, let $V^O(G)$, the *online vertex cover number*, be the smallest number such that there exists an algorithm, ALG, for ONLINE

Algorithm 1. IS-STAR, an online optimal algorithm for independent set for S_n

1: **for** request to vertex v **do**
2:     **if** v is the first vertex **then**
3:        reject v
4:     **else if** v is the second vertex and it has an edge to the first **then**
5:        reject v
6:     **else if** v has more than one neighbor already **then**
7:        reject v
8:     **else**
9:        accept v

VERTEX COVER with $\text{ALG}(G) = V^O(G)$. Also let $D^O(G)$, the *online domination number*, be the smallest number such that there exists an algorithm, ALG, for ONLINE DOMINATING SET with $\text{ALG}(G) = D^O(G)$.

The same relation between the online independence number and the online vertex cover number holds as between the independence number and the vertex cover number.

Observation 2. *For a graph, G with n vertices, we have $I^O(G) + V^O(G) = n$.*

3   Non-optimality of Greedy Algorithms

We start by motivating the other results in this paper by showing that the greedy algorithms are not optimal in general. In particular, they are not optimal on the star graphs, S_n, $n \geq 3$, which have a center vertex, s, and n other vertices, adjacent to s, but not to each other.

The algorithm, IS-STAR, does much better than GIS for the independent set problem on star graphs:

Theorem 1. *For a star graph, S_n, IS-STAR$(S_n) = n - 1$ and GIS$(S_n) = 1$.*

Proof. An intuitive way of thinking of IS-STAR is that it accepts a vertex if it is not possible that it is the center vertex, s. Since IS-STAR never accepts s, it produces an independent set. For every ordering of the vertices, IS-STAR will reject the first vertex. If the first vertex is s, it will reject the second vertex. Otherwise, it will reject s when it comes. Thus, IS-STAR$(S_n) = n - 1$. On the other hand, GIS$(G) = 1$, since it will accept s if it is requested first.   □

Since $n - 1 > 1$ for $n \geq 3$, we can conclude that GIS is not an optimal online algorithm for all graph classes.

Corollary 1. *For ONLINE INDEPENDENT SET, there exists an infinite family of graphs, S_n for $n \geq 3$, and an online algorithm, IS-STAR, such that GIS$(S_n) <$ IS-STAR(S_n).*

To show that GVC is not an optimal algorithm for ONLINE VERTEX COVER, we consider $\overline{\text{IS-STAR}}$.

Corollary 2. *For* ONLINE VERTEX COVER, *there exists an infinite family of graphs,* S_n *for* $n \geq 3$, *and an online algorithm,* $\overline{\text{IS-STAR}}$, *such that* $\overline{\text{IS-STAR}}(S_n) < \text{GVC}(S_n)$.

Proof. Using Observation 1 and Theorem 1, we have that $\overline{\text{IS-STAR}}(S_n) = n + 1 - \text{IS-STAR}(S_n) = 2$ and $\text{GVC}(S_n) = n + 1 - \text{GIS}(S_n) = n$. $\qquad\square$

Finally, for ONLINE DOMINATING SET, we have a similar result.

Corollary 3. *For* ONLINE DOMINATING SET, *there exists an infinite family of graphs,* S_n *for* $n \geq 3$, *and an online algorithm,* $\overline{\text{IS-STAR}}$, *such that* $\overline{\text{IS-STAR}}(S_n) < \text{GDS}(S_n)$.

Proof. Requesting s last ensures that GDS accepts n vertices, It can never accept all $n + 1$ vertices, so $\text{GDS}(S_n) = n$. On the other hand, $\overline{\text{IS-STAR}}(S_n) = 2$ (as in the proof of Corollary 2). We note that a vertex cover is also a dominating set in connected graphs. This means that $\overline{\text{IS-STAR}}$ always produces a dominating set in S_n. $\qquad\square$

4   Optimality of Greedy Algorithms on Freckle Graphs

For a graph, G, we let

- k denote the number of isolated vertices,
- G' denote the graph induced by the non-isolated vertices,
- n denote the number of vertices in G',
- $b(G')$ be a maximum independent set in G', and
- $s(G')$ be a minimum inclusion-maximal independent set in G' (that is, a smallest independent set such that including any vertex in the set would cause it to no longer be independent).

Note that $|s(G')|$ is also known as the *independent domination number* of G' (see [1] for more information).

Using this notation, we define the following class of graphs.

Definition 1. *A graph, G, is a Freckle Graph if* $k + |s(G')| \geq I^O(G')$.

Note that all graphs where at least half the vertices are isolated are Freckle Graphs. If the definition was changed to this (which might be less artificial), the results presented here would still hold, but our definition gives stronger results. The name comes from the idea that such a graph in many cases has a lot of isolated vertices (freckles). Furthermore, any graph can be turned into a Freckle Graph by adding enough isolated vertices. Note that a complete graph is a Freckle Graph. To make the star graph, S_n, a freckle graph, we need to add $n - 2$ isolated vertices. The results below show that GIS and GVC are online optimal on all Freckle Graphs.

Theorem 2. *For any algorithm,* ALG, *for* ONLINE INDEPENDENT SET, *and for any Freckle Graph, G,* $\text{GIS}(G) \geq \text{ALG}(G)$.

Proof. First, we note that GIS will accept the k isolated vertices. In G', it will accept an inclusion-maximal independent set. Since we take the worst ordering, it accepts $|s(G')|$ vertices. We get $\text{GIS}(G) = k + |s(G')|$. Now we describe an adversary strategy which ensures that an arbitrary algorithm, ALG, accepts at most $k + |s(G')|$ vertices.

The adversary starts by presenting $k + |s(G')|$ isolated vertices. Either ALG accepts $|s(G')|$ vertices or it rejects k vertices.

If ALG accepts $|s(G')|$ vertices, the adversary decides that they are exactly those in $s(G')$. This means that ALG will accept no other vertices in G'. Thus, it accepts at most $k + s(G')$ vertices.

If ALG rejects k vertices, the adversary decides that they are the k isolated vertices. The remaining graph is now G'. By definition, ALG cannot accept more than $I^O(G')$ vertices on the worst vertex ordering.

Since $k + |s(G')| \geq I^O(G')$ (because G is a Freckle Graph), ALG accepts no more than $k + |s(G')|$ vertices.   □

Corollary 4. *For any Freckle Graph, G, $\text{GIS}(G) = I^O(G)$.*

Corollary 5. *For any algorithm, ALG, for* ONLINE VERTEX COVER, *and for any Freckle Graph, G, $\text{GVC}(G) \leq \text{ALG}(G)$.*

Proof. This follows from Theorem 2, Observation 1, and the fact that $\text{GVC} = \overline{\text{GIS}}$.   □

Corollary 6. *For any Freckle Graph, G, $\text{GVC}(G) = V^O(G)$.*

For ONLINE DOMINATING SET something similar holds.

Theorem 3. *For any algorithm, ALG, for* ONLINE DOMINATING SET *and for any graph, G, with at least one isolated vertex, $\text{GDS}(G) \leq \text{ALG}(G)$.*

Proof. Recall that k denotes the number of isolated vertices in G, and G' denotes the subgraph of G induced by the non-isolated vertices. Note that GDS always produces an independent set. Thus, GDS accepts at most $k + |b(G')|$ vertices; it accepts exactly the k isolated vertices and the vertices in $b(G')$ if these are presented first.

Let an algorithm, ALG, be given. The adversary can start by presenting $k + |b(G')|$ isolated vertices. If at least one of these vertices is not accepted by ALG, the adversary can decide that this was in fact an isolated vertex, which can now no longer be dominated. Thus, $\text{ALG}(G) = \infty$. If ALG accepts all the presented vertices, it gets a score of at least $k + |b(G')|$.   □

Corollary 7. *For a any graph, G, with an isolated vertex, $\text{GDS}(G) = D^O(G)$.*

5 Implications for Worst Case Performance Measures

Do the results from the previous section mean that GIS is a good algorithm for ONLINE INDEPENDENT SET if the input graph is known to be a Freckle Graph?

The answer to this depends on how the performance of online algorithms is measured. In general, the answer is yes, if a measure, that only considers the worst case, is used.

The most commonly used performance measure for online algorithms is *competitive analysis* [14]. For maximization problems, an algorithm, ALG, is said to be c-competitive if there exists a constant, b, such that for any input sequence, I, $\text{OPT}(I) \leq c\text{ALG}(I) + b$ where $\text{OPT}(I)$ is the score of the optimal offline algorithm. For minimization problems, we require that $\text{ALG}(I) \leq c\text{OPT}(I) + b$. The competitive ratio of ALG is inf $\{c : \text{ALG is } c\text{-competitive}\}$. For *strict competitive analysis*, the definition is the same, except there is no additive constant.

Another measure is *on-line competitive analysis* [6], which was introduced for online graph coloring. The definition is the same as for competitive analysis except that $\text{OPT}(I)$ is replaced by $\text{OPTON}(I)$, which is score of the best online algorithm that knows the requests in I but not their ordering. For graph problems, this means the vertex-arrival model is used, as in this paper. The algorithm is allowed to know the final graph.

Corollary 8. *For* ONLINE INDEPENDENT SET *on Freckle Graphs, no algorithm has a smaller competitive ratio, strict competitive ratio, or on-line competitive ratio than GIS.*

Proof. Let ALG be a c-competitive algorithm for some c. Theorem 2 implies that GIS is also c-competitive. This argument also holds for the strict competitive ratio and the on-line competitive ratio.   □

Corollary 9. *For* ONLINE VERTEX COVER *on Freckle Graphs, no algorithm has a smaller competitive ratio, strict competitive ratio, or on-line competitive ratio than GVC.*

Corollary 10. *For* ONLINE DOMINATING SET *on the class of graphs with at least one isolated vertex, no algorithm has a smaller competitive ratio, strict competitive ratio, or on-line competitive ratio than GDS.*

Similar results hold for relative worst order analysis [3]. According to relative worst order analysis, for minimization problems in this graph model, one algorithm, A, is at least as good as another algorithm, B, on a graph class, if for all graphs G in the class, $A(G) \leq B(G)$. The inequality is reversed for maximization problems. It follows from the definitions that if an algorithm is optimal with respect to on-line competitive analysis, it is also optimal with respect to relative worst-order analysis. This was observed in [4]. Thus, the above results show that the three greedy algorithm in the corollaries above are also optimal on Freckle Graphs, under relative worst order analysis.

6 A Subclass of Freckle Graphs Where Greedy is Not Optimal

Although these greedy algorithms are optimal with respect to some worst case measures, this does not mean that these greedy algorithms are always the best choice for *all* Freckle Graphs. There is a subclass of Freckle Graphs where another algorithm is objectively better than GIS, and bijective analysis and average analysis [2] reflect this.

Theorem 4. *There exists an infinite class of Freckle Graphs* $\tilde{G} = \{G_n \mid n \geq 2\}$ *and an algorithm* Almost-GIS *such that for all* $n \geq 2$ *the following holds:*

$$\forall \phi \; \text{Almost-GIS}(\phi(G_n)) \geq \text{GIS}(\phi(G_n))$$
$$\exists \phi \; \text{Almost-GIS}(\phi(G_n)) > \text{GIS}(\phi(G_n))$$

Proof. Consider the graph $G_n = (V, E)$, where

$$V = \{x_1, x_2, \ldots, x_n, y_1, y_2, \ldots, y_n, z, u_1, u_2, \ldots, u_n, v_1, v_2, \ldots, v_{n-1}\}$$
$$E = \{(x_i, y_i), (y_i, z), (z, u_i) \mid 1 \leq i \leq n\}.$$

Fig. 1 shows the graph G_4.

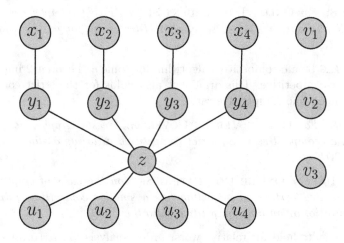

Fig. 1. The graph G_4.

We start by noting that G_n is a Freckle Graph since:

- The smallest maximal independent set, which does not contain isolated vertices, has size $n + 1$.
- The maximum independent set, which does not contain isolated vertices, has size $2n$.
- The number of isolated vertices is $n - 1$.

The algorithm, Almost-GIS, is identical to GIS, except that it rejects a vertex if it already has two neighbors when it is presented. Note that it never matters when $v_1, \ldots, v_{n-1}$ are requested since they will all be accepted by GIS and Almost-GIS. Consider any ordering of the vertices of G where GIS and `Almost-GIS` do not accept the same independent set. There must exist a first vertex, w, which is accepted by one of the algorithms and rejected by the other. By definition of the algorithms, it must be the case that w is rejected by `Almost-GIS` and accepted by `GIS`. It must hold that w has two neighbors, which have not been accepted by either algorithm. This can only happen if $w = z$ and the two neighbors are y_i and y_j where x_i and x_j have already been presented and accepted by both algorithms and no u_k have been presented yet. In this case, z is accepted by `GIS` and rejected by `Almost-GIS`. However, $u_1, \ldots, u_n$ are accepted by `Almost-GIS` and rejected by `GIS`. Since $n \geq 2$ and since both `GIS` and `Almost-GIS` accept exactly one of x_i and y_i, $1 \leq i \leq n$, we get that on every ordering, ϕ, where `GIS` and `Almost-GIS` accept a different independent set, `Almost-GIS`$(\phi(G)) < $ `GIS`$(\phi(G))$. Such an ordering always exists (the ordering $x_1, \ldots, x_n, y_1, \ldots, y_n, z, u_1, \ldots, u_n, v_1, \ldots, v_{n-1}$ achieves this). $\square$

Competitive analysis, on-line competitive analysis, and relative worst order ratio do not identify `Almost-GIS` as a better algorithm than `GIS` on the class of graphs $\tilde{G}$ defined in the proof of Theorem 4. There are, however, other measures which do this. Bijective analysis and average analysis [2] are such measures. Let I_n be the set of all input sequences of length $4n$. Since we are considering the rather restricted graph class $\tilde{G}$, I_n denotes all orderings of the vertices in G_n (since these are the only inputs of length $4n$). For an algorithm A to be considered better than another algorithm B for a maximization problem, it must hold for sufficiently large n that there exists a bijection $f : I_n \to I_n$ such that the following holds:

$$\forall I \in I_n \ A(I) \geq B(f(I))$$
$$\exists I \in I_n \ A(I) > B(f(I))$$

Theorem 5. *`Almost-GIS` is better than `GIS` on the class $\tilde{G}$ according to bijective analysis.*

Proof. We let the bijection f be the identity and the result follows from Theorem 4. $\square$

Average analysis is defined such that if one algorithm is better than another according to bijective analysis, it is also better according to average analysis. Thus, `Almost-GIS` is better than `GIS` on the class $\tilde{G}$ according to average analysis.

Note that `Almost-GIS` is not an optimal algorithm for all Freckle Graphs. The class of graphs, $K_{n,n}$, for $n \geq 2$, consisting of complete bipartite graphs with n vertices in each side of the partition, is a class where `Almost-GIS` can behave very poorly. Note that on these graphs, `GIS` is optimal and always finds an independent set of size n, which is optimal, so these graphs are Freckle Graphs, even though they have no isolated vertices. If the first request to `Almost-GIS` is a vertex from one side of the partition and the next two are from the other side of the partition, `Almost-GIS` only accepts one vertex, not n.

7 Complexity of Determining the Online Independence Number, Vertex Cover Number, and Domination Number

Given a graph, G, it is easy to check if it has an isolated vertex and apply Theorem 3. However, Theorem 2 and Corollary 5 might not be as easy to apply, because it is not obvious how one can check if a graph is a Freckle Graph ($k + |s(G')| \geq I^O(G')$). In some cases, this is easy. For example, any graph where at least half the vertices are isolated is a Freckle Graph. We leave the hardness of recognizing Freckle Graphs as an open problem, but we show a hardness result for deciding if $I^O(G) \leq q$.

Theorem 6. *Given $q \in \mathbb{N}$ and a graph, G, deciding if $I^O(G) \leq q$ is NP-hard.*

Proof. Note that it is NP-complete to determine if the minimum maximal independent set of a graph, $G = (V, E)$, has size at most L, for an integer L [5]. To reduce from this problem, we create $\tilde{G} = (\tilde{V}, E)$ which is the same as G, but has $|V|$ extra isolated vertices, and a bound $\tilde{L} = L + |V|$. $\tilde{G}$ is a Freckle Graph, since $|V| \geq I^O(G)$. By Corollary 4, $\text{GIS}(\tilde{G}) = I^O(\tilde{G})$. Since $\text{GIS}(\tilde{G}) = |s(G)| + |V|$, the original graph, G, has a minimum maximal independent set of size L, if and only if $\tilde{G}$ has online independence number at most $\tilde{L}$.   □

The hardness of computing the online independence number implies the hardness of computing the online vertex cover number.

Corollary 11. *Given $q \in \mathbb{N}$ and a graph, G, deciding if $V^O(G) \geq q$ is NP-hard.*

Proof. This follows from Observation 2 and Theorem 6.   □

Theorem 7. *Given $q \in \mathbb{N}$ and a graph, G, deciding if $D^O(G) \geq q$ is NP-hard.*

Proof. We make a reduction from INDEPENDENT SET. In INDEPENDENT SET, a graph, G and an $L \in \mathbb{N}$ is given. It is a yes-instance if and only if there exists an independent set of size at least L. We reduce instances of INDEPENDENT SET, (G,L), to instances of ONLINE DOMINATING SET, $(\tilde{G},\tilde{L})$, such that there exists an independent set in G of size at least L if and only if $D^O(\tilde{G}) \geq \tilde{L}$. The reduction is very simple. We let $\tilde{G}$ be the graph which consists of G with one additional isolated vertex. We set $\tilde{L} = L + 1$. Assume first that any independent set in G has size at most $L-1$. This means that any independent set in $\tilde{G}$ has size at most L. Since GDS produces an independent set, it will accept at most $L < \tilde{L}$ vertices. Assume now that there is an independent size of at least L in G. Then, there exists an independent set of size at least $L + 1$ in $\tilde{G}$. If these vertices are presented first, GDS will accept them. From Theorem 3, we get that no algorithm for ONLINE DOMINATING SET can do better (since $\tilde{G}$ has an isolated vertex), which means that $D^O(\tilde{G}) \geq \tilde{L}$.   □

Theorem 8. *Given $q \in \mathbb{N}$ and a graph, G, the problem of deciding if $I^O(G) \leq q$ is in PSPACE.*

Proof. Let $q \in \mathbb{N}$ and a graph, $G = (V, E)$, be given. We sketch an algorithm that uses only polynomial space which decides if $I^O(G) \leq q$. We view the problem as a game between the adversary and the algorithm where the algorithm wins if it gets an independent set of size at least q. A move for the adversary is revealing a vertex along with edges to a subset of the previous vertices such that the resulting graph is an induced subgraph of G. These are possible to enumerate since induced subgraph can be solved in polynomial space. A move by the algorithm is accepting or rejecting that vertex. We make two observations: The game has only polynomial length (each game has length $2|V|$), and it is always possible in polynomial space to enumerate the possible moves from a game state. Thus, an algorithm can traverse the game tree using depth first search and recursively compute for each game state if the adversary or the algorithm has a winning strategy. □

Similar proofs can be used to show that the problems of deciding if $V^O(G) \geq q$ and $D^O(G) \geq q$ are in PSPACE as well. It remains open whether these problems are NP-complete, PSPACE-complete, or neither. In [11] it was shown that determining the online chromatic number is PSPACE-complete if the graph is pre-colored.

8   Concluding Remarks

A strange difference between online and offline algorithms is observed: Adding isolated vertices to a graph can change an algorithm from not being optimal to being optimal (according to many measures). This is even more surprising for vertex cover than for independent set, since in the offline case, adding isolated vertices to a graph can improve its approximation ratio in the case of the independent set problem. It is hard to see how adding isolated vertices to a graph could in any way help an offline algorithms for vertex cover.

Note that GIS = GDS. This means that for Freckle Graphs with at least one isolated vertex, GIS is an algorithm which solves both online independent set (a maximization problem) and online dominating set (a minimization problem) online optimally. This is quite unusual, since the independent sets and dominating sets it will find in the worst case can be quite different for the same graph.

It is tempting to believe that the greedy algorithms considered here are the unique optimal algorithms for these problems. Is this true?

As mentioned earlier, the NP-hardness results presented here do not answer the question as to how hard it is to recognize Freckle Graphs. This is left as an open problem.

We have shown it to be NP-hard to decide if $I^O(G) \leq q$, $V^O(G) \geq q$, and $D^O(G) \geq q$, but there is nothing to suggest that these problems are contained in NP. They are in PSPACE, but it is left as an open problem if they are NP-complete, PSPACE-complete or somewhere in between.

Acknowledgments. The authors would like to thank Lene Monrad Favrholdt for interesting and helpful discussions.

References

1. Allan, R.B., Laskar, R.: On domination and independent domination numbers of a graph. Discrete Math. **23**(2), 73–76 (1978)
2. Angelopoulos, S., Dorrigiv, R., López-Ortiz, A.: On the separation and equivalence of paging strategies. In: 18th ACM-SIAM Symposium on Discrete Algorithms (SODA), pp. 229–237 (2007)
3. Boyar, J., Favrholdt, L.M.: The relative worst order ratio for online algorithms. ACM Trans. Algorithm **3**(2), 22 (2007)
4. Boyar, J., Favrholdt, L.M., Medvedev, P.: The relative worst order ratio of online bipartite graph coloring, Unpublished manuscript
5. Garey, M.R., Johnson, D.S.: Computers and Intractability: A Guide to the Theory of NP-Completeness. W. H. Freeman & Co., New York (1979). Under Dominating Set on p. 190
6. Gyárfás, A., Király, Z., Lehel, J.: On-line competitive coloring algorithms. Tech. rep. TR-9703-1, Institute of Mathematics at Eötvös Loránd University (1997). http://www.cs.elte.hu/tr97/
7. Gyárfás, A., Király, Z., Lehel, J.: On-line 3-chromatic graphs I. Triangle-free graphs. SIAM J. Discrete Math. **12**, 385–411 (1999)
8. Gyárfás, A., Lehel, J.: First fit and on-line chromatic number of families of graphs. ARS Combinatoria **29C**, 168–176 (1990)
9. Halldórsson, M.M.: Online coloring known graphs. In: 10th ACM-SIAM Symposium on Discrete Algorithms (SODA), pp. 917–918 (1999)
10. Halldórsson, M.M., Iwama, K., Miyazaki, S., Taketomi, S.: Online independent sets. Theoret. Comput. Sci. **289**(2), 953–962 (2002)
11. Kudahl, C.: Deciding the on-line chromatic number of a graph with pre-coloring is PSPACE-complete. In: Paschos, V.T., Widmayer, P. (eds.) CIAC 2015. LNCS, vol. 9079, pp. 313–324. Springer, Heidelberg (2015)
12. Lehel, A.G.J., Kiraly, Z.: On-line graph coloring and finite basis problems. Combinatorics: Paul Erdos is Eighty **1**, 207–214 (1993)
13. Lovász, L., Saks, M., Trotter, W.: An on-line graph coloring algorithm with sublinear performance ratio. Discrete Math. **75**(13), 319–325 (1989)
14. Sleator, D.D., Tarjan, R.E.: Amortized efficiency of list update and paging rules. Commun. ACM **28**(2), 202–208 (1985)
15. Vishwanathan, S.: Randomized online graph coloring. J. Algorithms **13**(4), 657–669 (1992)

Gray Codes for AT-Free Orders via Antimatroids

Jou-Ming Chang, Ton Kloks, and Hung-Lung Wang[✉]

Institute of Information and Decision Sciences,
National Taipei University of Business, Taipei, Taiwan
{spade,antonius,hlwang}@ntub.edu.tw

Abstract. The AT-free order is a linear order of the vertices of a graph the existence of which characterizes AT-free graphs. We show that all AT-free orders of an AT-free graph can be generated in $O(1)$ amortized time.

1 Convex Geometries and Elimination Orders

Antimatroids are the warp and woof for various algorithms that generate elimination orders.

Definition 1. *A set system is a pair $(V, \mathcal{F})$ where V is a finite set and where $\mathcal{F}$ is a collection of subsets of V. The set system is accessible if each nonempty $X \in \mathcal{F}$ has an element $x \in X$ such that $X - x \in \mathcal{F}$. A set system $(V, \mathcal{F})$ is an* <u>antimatroid</u> *if it is accessible and closed under unions.*

Definition 2. *A* <u>convexity space</u> *is a pair $(V, \mathcal{N})$ where V is a finite set and $\mathcal{N}$ a collection of subsets such that*

(i) $\varnothing \in \mathcal{N}$ and $V \in \mathcal{N}$ and
(ii) $\mathcal{N}$ is closed under intersections.

The elements of $\mathcal{N}$ are called convex. For any subset $X \subseteq V$ one defines the <u>convex hull</u> $\sigma(X)$ as the intersection of all the convex sets that contain X.

Definition 3. *A convexity space is a* <u>convex geometry</u> *if it satisfies the*

> ANTI-EXCHANGE PROPERTY:
>
> *If $X \subseteq V$, and y and z are two points outside $\sigma(X)$, then*
>
> $$y \in \sigma(X + z) \quad \Rightarrow \quad z \notin \sigma(X + y).$$

Proposition 1. *A set system $(V, \mathcal{F})$ is an antimatroid if and only if $(V, \mathcal{F}^\star)$ is a convex geometry, where*

$$\mathcal{F}^\star = \{\, V - X \mid X \in \mathcal{F} \,\}.$$

T. Kloks—This author thanks the institute for their hospitality and support.

Z. Lipták and W.F. Smyth (Eds.): IWOCA 2015, LNCS 9538, pp. 77–87, 2016.
DOI: 10.1007/978-3-319-29516-9_7

Here are some examples of convex geometries. Let $G = (V, E)$ be a graph.

1. Call a subset $X \subseteq V$ convex if all chordless paths between vertices in X are contained in X. Let $\mathcal{M}$ be the collection of convex sets. The set system $(V, \mathcal{M})$ is a convex geometry if and only if G is chordal.
2. Call X convex if all geodesics between vertices in X are contained in X. Let $\mathcal{G}$ be the collection of convex sets. Then $(V, \mathcal{G})$ is a convex geometry if and only if G is Ptolemaic.
3. Call a set X convex if every even-chorded path whose endpoints are in X is contained in X. Let $\mathcal{S}$ be the collection of convex sets. Then $(V, \mathcal{S})$ is a convex geometry if and only if G is strongly chordal.
4. Call X convex when every induced P_3 with ends in X is contained in X. Let $\mathcal{P}$ be the collection of convex sets. Then $(V, \mathcal{P})$ is a convex geometry if and only if G is HHDA-free, that is, weak bipolarizable.

The following theorem serves as an example of Gray codes for some elimination orders mentioned above, and as an outline of the technique used in this paper to prove our major result. For a detailed proof we refer to [3, 9–11].

Theorem 1. *The elimination orders in the list below can be generated in constant amortized time:*

(a) Perfect elimination orders for chordal graphs.
(b) Simple elimination orders for strongly chordal graphs.
(c) Semi-perfect elimination orders for HHDA-free graphs.
(d) Linear extensions of any poset.

Proof (Sketch). The generic algorithm for generating these elimination orders was obtained by Pruesse and Ruskey [9] (see also [11]).

A simple language is a pair $(V, \mathcal{L})$ where V is a finite set, called the alphabet of letters, and $\mathcal{L}$ is a collection of simple words. A word is a finite sequence of letters. A word is simple if each letter occurs at most once. If $\alpha \in \mathcal{L}$ then denote by $\tilde{\alpha}$ the set of letters that occur in α. A language antimatroid is a simple language $(V, \mathcal{L})$ such that $\varnothing \in \mathcal{L}$ and $\mathcal{L}$ satisfies the exchange axiom:

> if $\alpha, \beta \in \mathcal{L}$ and if $\tilde{\alpha} \not\subseteq \tilde{\beta}$ then there is a letter $x \in \tilde{\alpha} \backslash \tilde{\beta}$ such that $\beta x \in \mathcal{L}$.

The basic words of a language antimatroid is the set of words of maximal length. There is a 1-1 correspondence between antimatroids and antimatroid languages (see, e.g., [1]).

For example, the collection of perfect elimination orders of a chordal graph G form the basic words of a language antimatroid (see, e.g., [1]).

Let $(V, \mathcal{L})$ be a language antimatroid. A transposition oracle is an algorithm to decide, for a basic word $\alpha \in \mathcal{L}$ and a given transposition of two adjacent letters, whether the new word is a basic word of $\mathcal{L}$. If $\alpha x y \beta$ is a basic word, then a transposition of the (adjacent) letters x and y is the operation that produces the word $\alpha y x \beta$.

Pruesse and Ruskey prove in their paper [9, Theorem 5] that, if $(V, \mathcal{L})$ is a language antimatroid with an $O(1)$ transposition oracle, then the basic words can be generated in $O(1)$ amortized time, such that each word differs from the next by no more than two transpositions.   □

Asteroidal Sets. Asteroidal sets were introduced in [7] as follows (see also, e.g., [8]). (The definition appeared also, earlier, in [12] where it was used to characterize certain subclasses of chordal graphs.)

Definition 4. *Let G be a graph. A set $A \subseteq V(G)$ is an asteroidal set if for each $a \in A$, the set $A - a$ is contained in one component of $G - N[a]$, where $N[a]$ denotes the closed neighborhood of a.*

Asteroidal sets of cardinality three are *asteroidal triples*. They were used by Lekkerkerker and Boland to characterize interval graphs. Many classes of graphs that one encounters often in practical situations, are AT-free, for example, interval graphs, permutation graphs, and cocomparability graphs. Many NP-complete problems, e.g., independent set, domination and 3-coloring, become polynomial when one restricts the graphs to a class with a bounded asteroidal number.

Recently, elimination orders, called AT-*free orders*, were obtained for graphs with bounded asteroidal number [6]. We describe the result for AT-*free graphs*, which are graphs without an asteroidal triple. An example is given in Fig. 1. In Sect. 2 we show that these AT-free orders are the basic words of a language antimatroid. In Sect. 3, We show that all AT-free orders of a graph can be listed in $O(1)$ amortized time.

2   A Convex Geometry

Throughout this section, let G be a connected AT-free graph.

Definition 5. *For two vertices x and y let the interval $I(x, y)$ be defined as the set of vertices $z \in V(G)$ for which there is a chordless x, z-path that avoids $N[y]$ and a chordless y, z-path that avoids $N[x]$.*

Let $\mathcal{I}$ be the set of triples (x, z, y) for which $z \in I(x, y)$. Notice that

$$(x, z, y) \in \mathcal{I} \quad \Rightarrow \quad (y, z, x) \in \mathcal{I} \text{ and } x, y \text{ and } z \text{ are pairwise distinct.} \quad (1)$$

Now $\mathcal{I}$ is a <u>strict betweenness</u>, in the sense of Chvátal [5]. According to [2] (see also [5]), any such ternary relation $\mathcal{I}$ defines a convexity space $(V, \mathcal{N})$ with

$$\mathcal{N} = \{ K \subseteq V \mid \{a, c\} \subseteq K \quad \text{and} \quad (a, b, c) \in \mathcal{I} \quad \Rightarrow \quad b \in K \}. \quad (2)$$

Since there are no triples $(a, b, a) \in \mathcal{I}$ with $b \neq a$, all singletons $\{a\} \in \mathcal{N}$.

Definition 6. *Let $X \subseteq V$. A vertex $b \in X$ is an <u>extreme</u> point of X if there are no vertices a and c in X such that $b \in I(a, c)$.*

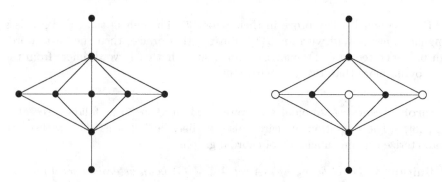

Fig. 1. The AT-free graph on the left has twelve triples $(x, z, y) \in \mathcal{I}$, that is, ordered triples (x, y, z) with $z \in I(x, y)$; the horizontal P_5 entails two such triples and there are ten triples consisting of the two pendants and one vertex of the horizontal P_5. Notice that the two vertices of degree 6 are not in any triple of $\mathcal{I}$ (neither one is in an independent set of cardinality 3). There are four extreme points, namely the two pendants and their neighbors. It is *not* true that for every triple $(x_1, x_2, x_3) \in \mathcal{I}$ there is an extreme point $\bar{x}_3$ with $(x_1, x_2, \bar{x}_3) \in \mathcal{I}$. For the AT-free graph on the left it *is* true that for every triple $(x_1, x_2, x_3) \in \mathcal{I}$ there exist extreme points $\bar{x}_1$ and $\bar{x}_3$ such that $(\bar{x}_1, x_2, \bar{x}_3) \in \mathcal{I}$. Figure 2 shows an AT-free graph in which this does not hold. Compare [5, Theorems 1 and 3]. The graph on the right has white vertices as an asteroidal triple.

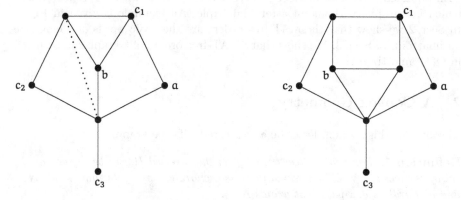

Fig. 2. The graph on the left is an example of an AT-free graph with a crossing pair. The strict betweenness does not satisfy Eq. (3). On the left, the extreme points are c_3 and its neighbor, c_1 and the common neighbor of c_2 and c_1. Although $(a, b, c_2) \in \mathcal{I}$, there are no two extreme vertices that form a triple with b in the middle. In the graph on the right, $(c_1, b, c_3) \in \mathcal{I}$, that is, b is between two extremes.

The set of extreme points of $X \subseteq V$ is denoted by $\mathrm{ex}(X)$. We note that the concept of extreme points is inherited from that in dealing with convex geometries. For simplicity, we use the same notation when there is no danger of misunderstanding.

Remark 1. Chvátal shows in [5, Theorem 1] the equivalence of the following two statements for any strict betweenness $\mathcal{B}$ on a finite ground set V.

(i) For all $X \subseteq V$ and all x_1, x_2 and $x_3 \in X$ with $(x_1, x_2, x_3) \in \mathcal{B}$, there exist $\bar{x}_1 \in \mathrm{ex}(X)$ and $\bar{x}_3 \in \mathrm{ex}(X)$ such that $(\bar{x}_1, x_2, \bar{x}_3) \in \mathcal{B}$.

(ii)

$$(a, b, c_2) \in \mathcal{B} \quad \text{and} \quad (c_1, c_2, c_3) \in \mathcal{B} \quad \Rightarrow$$
$$(a, b, c_1) \in \mathcal{B} \quad \text{or} \quad (a, b, c_3) \in \mathcal{B} \quad \text{or} \quad (c_1, b, c_3) \in \mathcal{B}. \quad (3)$$

Figure 2 shows an AT-free graph in which the strict betweenness does not satisfy (3).

Monophonic convexity in chordal graphs and strict order betweenness in posets satisfy the following condition which is stronger than the Condition (3), mentioned above.

$$(a, b, c_2) \in \mathcal{B} \quad \text{and} \quad (c_1, c_2, c_3) \in \mathcal{B} \quad \Rightarrow \quad (a, b, c_1) \in \mathcal{B} \quad \text{or} \quad (a, b, c_3) \in \mathcal{B}. \quad (4)$$

Chvátal proves in [5, Theorem 3] that (4) is equivalent with the statement that, for all $X \subseteq V$ and for all $\{x_1, x_2, x_3\} \subseteq X$,

$$(x_1, x_2, x_3) \in \mathcal{B} \quad \Rightarrow \quad \exists_{\bar{x}_3 \in \mathrm{ex}(X)} \ (x_1, x_2, \bar{x}_3) \in \mathcal{B}. \quad (5)$$

Figure 1 shows an AT-free graph in which the strict betweenness $\mathcal{I}$ does not satisfy (4).

Notation. For a subset X of vertices we write $N(X)$ for the set of vertices in $V \setminus X$ that have a neighbor in X. Similarly, we use $N(x)$ to denote the open neighborhood of x, consisting of vertices adjacent to x. For nonadjacent vertices x and y, let $C^y(x)$ denote the component of $G - N[y]$ that contains x.

Definition 7. *A pair of triples* $(a, b, c_2) \in \mathcal{I}$ *and* $(c_1, c_2, c_3) \in \mathcal{I}$ *is a* <u>crossing</u> *if*

(i) c_3 *is in a component* D *of* $G - N[b]$, $D \neq C^b(a)$ *and* $D \neq C^b(c_2)$;
(ii) $c_1 \in N(b) \cap N(a)$;
(iii) $N(D) \subseteq N(a) \cap N(c_2)$.

When a convexity space is defined by a strict betweenness that satisfies Eq. (3) then it defines a convex geometry [5, Lemma 2]. The Carathéodory number of such a convex geometry is at most 2.[1] However, there are strict betweenness relations that do not satisfy Eq. (3) and that, nonetheless, define convex geometries with Carathéodory number 2. In his paper Chvátal gives an example with 4 elements of a convex geometry with Carathéodory number 2 which does not satisfy Eq. (3). The strict betweenness relation for AT-free graphs also defines a convex geometry. The example in Fig. 2 shows that the Carathéodory number can be more than 2. The following lemma characterizes the AT-free graphs that give rise to convex geometries that satisfy Eq. (3).

---

[1] A convexity space has <u>Carathéodory number</u> k if k is the smallest integer satisfying the following property. If $x \in \sigma(X)$ for a set X then there exists a $X' \subseteq X$ with $|X'| \leq k$ and $x \in \sigma(X')$.

Lemma 1. *Assume that $\mathcal{I}$ does not have a crossing pair of triples. Then $\mathcal{I}$ satisfies* (3), *that is,*

$$(a, b, c_2) \in \mathcal{I} \quad and \quad (c_1, c_2, c_3) \in \mathcal{I} \quad \Rightarrow$$
$$(a, b, c_1) \in \mathcal{I} \quad or \quad (a, b, c_3) \in \mathcal{I} \quad or \quad (c_1, b, c_3) \in \mathcal{I}.$$

Proof. Assume that c_1 and c_3 are both in $C^b(c_2)$. Notice that there is a path from c_1 to b that avoids $N[a]$. This follows since $(a, b, c_2) \in \mathcal{I}$ implies that there is a b, c_2-path that avoids $N[a]$ and, since $c_1 \in C^b(c_2)$ and $a \notin C^b(c_2)$, there is also a c_1, c_2-path avoiding $N[a]$.

If $N(C^b(a)) \not\subseteq N(c_1)$ then $(a, b, c_1) \in \mathcal{I}$, since in that case there is an a, b-path avoiding $N[c_1]$. By symmetry we may also assume that $N(C^b(a)) \subseteq N(c_3)$. Thus,

$$N(C^b(a)) \subseteq N(c_1) \cap N(c_3).$$

On the other hand, $N(C^b(a)) \not\subseteq N(c_2)$, since by assumption $(a, b, c_2) \in \mathcal{I}$ and so there is an a, b-path avoiding $N(c_2)$. This contradicts the assumption that $(c_1, c_2, c_3) \in \mathcal{I}$, since there is a c_1, c_3-path avoiding $N[c_2]$.

Assume that $c_1 \in C^b(a)$ and $c_3 \in C^b(c_2)$. Since $(c_1, c_2, c_3) \in \mathcal{I}$, we have that

$$N(C^b(a)) \not\subseteq N(c_3).$$

Thus there exists an a, b-path that avoids $N[c_3]$. Notice

$$(a, b, c_2) \in \mathcal{I} \quad \Rightarrow \quad N(C^b(c_2)) \not\subseteq N(a).$$

So, there is a b, c_3-path avoiding $N[a]$. This proves that $(a, b, c_3) \in \mathcal{I}$.

Assume that $c_1 \in C^b(a)$ and $c_3 \in C^b(a)$. Since $(c_1, c_2, c_3) \in \mathcal{I}$ there must be paths from c_1 and c_3 to vertices in $N(C^b(a))$ that avoids each other's neighborhood. This implies that $(c_1, b, c_3) \in \mathcal{I}$.

Assume that $c_1 \in C^b(a)$ and that c_3 is in some component D of $G - N[b]$, $D \neq C^b(a)$ and $D \neq C^b(c_2)$. Since $(c_1, c_2, c_3) \in \mathcal{I}$, we have that

$$N(C^b(a)) \not\subseteq N(c_3) \quad and \quad N(D) \not\subseteq N(c_1).$$

This implies that $(c_1, b, c_3) \in \mathcal{I}$.

Assume $c_1 \in C^b(a)$ and $c_3 \in N(b)$. There exists an a, b-path that avoids $N(c_2)$. Since $c_1 \in C^b(a)$, there is a c_1, b-path that avoids $N(c_2)$. If we extend this path to c_3 we find a c_1, c_3-path that avoids $N(c_2)$. This contradicts the assumption that $(c_1, c_2, c_3) \in \mathcal{I}$.

Assume that c_1 and c_3 are both neighbors of b. Since $(c_1, c_2, c_3) \in \mathcal{I}$, c_1 and c_3 must be in different components of $G - N[c_2]$. This is not possible, since c_1 and c_3 have a common neighbor $b \notin N[c_2]$.

Assume that $c_1 \in C^b(c_2)$ and that $c_3 \in N(b)$. Since $(a, b, c_2) \in \mathcal{I}$ we have that $N(C^b(a)) \not\subseteq N(c_2)$. Since $(a, b, c_2) \in \mathcal{I}$ and $c_1 \in C^b(c_2)$, there exists a b, c_1-path that avoids $N[a]$. This implies that, if $N(C^b(a)) \not\subseteq N(c_1)$ then $(a, b, c_1) \in \mathcal{I}$. So, now assume that $N(C^b(a)) \subseteq N(c_1)$. This contradicts that $(c_1, c_2, c_3) \in \mathcal{I}$, since there is a c_1, c_3-path that avoids $N[c_2]$.

Assume that $c_1 \in C^b(c_2)$ and that c_3 is in a component D of $G - N[b]$, $D \neq C^b(a)$ and $D \neq C^b(c_2)$. The fact that $(a, b, c_2) \in \mathcal{I}$ implies that there is a b, c_2-path that avoids $N[a]$. Then, since $c_1 \in C^b(c_2)$, there is a b, c_1-path that avoids $N[a]$. Now, if $N(C^b(a)) \not\subseteq N(c_1)$, then there is a, b-path that avoids $N[c_1]$, and with the previous observation this implies that $(a, b, c_1) \in \mathcal{I}$. So, now assume that

$$N(C^b(a)) \subseteq N(c_1).$$

Because $(c_1, c_2, c_3) \in \mathcal{I}$, there exists a c_2, c_3-path that avoids $N[c_1]$. This implies that $N(D) \not\subseteq N(c_1)$ and so there exists a c_3, b-path that avoids $N[c_1]$. If $N(C^b(a)) \not\subseteq N(c_3)$ then there is a c_1, b-path that avoids $N[c_3]$, implying that $(c_1, b, c_3) \in \mathcal{I}$. Henceforth assume that

$$N(C^b(a)) \subseteq N(c_3).$$

Now, we must also have that $N(C^b(a)) \subseteq N[c_2]$, otherwise there would be a c_1, c_3-path avoiding $N[c_2]$, contradicting $(c_1, c_2, c_3) \in \mathcal{I}$. Thus, summarizing, we have that

$$N(C^b(a)) \subseteq N(c_1) \cap N(c_2) \cap N(c_3).$$

But then there can be no a, b-path avoiding $N[c_2]$, contradicting $(a, b, c_2) \in \mathcal{I}$.

Assume there is a component D of $G - N[b]$, $D \neq C^b(a)$ and $D \neq C^b(c_2)$, with $\{c_1, c_3\} \subset D$. Since $(c_1, c_2, c_3) \in \mathcal{I}$ there exist paths from c_1 and c_3 to vertices in $N(D)$ avoiding each other's neighborhood. This implies $(c_1, b, c_3) \in \mathcal{I}$.

A similar argument applies in the case where there are two components D_1 and D_3 of $G - N[b]$, both different from $C^b(a)$ and $C^b(c_2)$, with $c_1 \in D_1$ and $c_3 \in D_3$. Namely, because $(c_1, c_2, c_3) \in \mathcal{I}$, there must exist paths from c_1 and c_3 to $N(D_1)$ and $N(D_3)$ avoiding each other's neighborhood, implying that $(c_1, b, c_3) \in \mathcal{I}$.

For the final case, assume that $c_1 \in N(b)$ and that c_3 is in a component D of $G - N[b]$, where $D \neq C^b(a)$ and $D \neq C^b(c_2)$. Notice that

$$N(C^b(a)) \not\subseteq N(c_3) \quad \text{and} \quad N(D) \not\subseteq N(a) \quad \Rightarrow \quad (a, b, c_3) \in \mathcal{I}.$$

Assume this is not the case. First assume that $N(C^b(a)) \subseteq N(c_3)$. We have that

$$(a, b, c_2) \in \mathcal{I} \quad \Rightarrow \quad N(C^b(a)) \not\subseteq N(c_2).$$

Now there exists a c_1, c_3-path avoiding $N[c_2]$ (via $N(C^b(a)) \backslash N(c_2)$ and b), contradicting the assumption that $(c_1, c_2, c_3) \in \mathcal{I}$. Now assume that $N(D) \subseteq N(a)$. If $N(D) \not\subseteq N(c_2)$, then there is a c_3, c_1-path avoiding $N[c_2]$, a contradiction. Finally, assume that

$$N(D) \subseteq N(a) \cap N(c_2).$$

Since there is a c_2, c_3-path that avoids $N[c_1]$, $N(D) \not\subseteq N[c_1]$. This implies an a, c_2-path avoiding $N[c_1]$. Since $N(C^b(a)) \not\subseteq N(c_2)$, there exists a a, c_1-path avoiding $N[c_2]$. There exists a c_2, b-path that avoids $N[a]$. If $c_1 \notin N(a)$, there is a c_2, c_1-path avoiding $N[a]$. Then $\{c_1, c_2, a\}$ is an asteroidal triple. So we have that $c_1 \in N(a)$. Then the pair of triples (a, b, c_2) and (c_1, c_2, c_3) is crossing, contradicting the assumption that $\mathcal{I}$ has no crossing pairs.

This proves the lemma. □

The following corollary follows via some results of [5].

Corollary 1. *Let G be a connected* AT-*free graph. Assume that $\mathcal{I}$ contains no crossing pair of triples. Let $(V, \mathcal{N})$ be the convexity space as defined in* (2). *Then $(V, \mathcal{N})$ is a convex geometry.*

Proof. According to [5, Lemma 2], the claim is implied by Lemma 1.   □

For an AT-free graph, the strict betweenness defined in Definition 5 and Eq. (1) defines a convex geometry, as shown later in Theorem 2. To prove that, we need the following lemma, which gives a sufficient condition for a convexity space being a convex geometry. For $X \subseteq V$, let $\sigma^0(X) = X$. For $i \geq 1$, let

$$\sigma^i(X) = \{x \mid \exists_{a,b \in \sigma^{i-1}(X)} \, (a, x, b) \in \mathcal{I}\} \setminus \cup_{0 \leq j < i} \, \sigma^j(X).$$

For $x \in X$, the *index* of x with respect to X is i if $x \in \sigma^i(X)$.

Lemma 2. *Let $(V, \mathcal{N})$ be a convexity space. For $X \subseteq V$, $\{y, z\} \subseteq V$, and $\{y, z\} \cap \sigma(X) = \varnothing$, if*

$$\exists_{a' \in \sigma(X)} \, (a', y, z) \in \mathcal{I} \quad \text{implies} \quad z \notin \sigma(X + y),$$

then $(V, \mathcal{N})$ is a convex geometry.

Proof. The lemma can be proved by induction on the index of y with respect to $\sigma(X) + z$. Details will appear in the full version of this paper.   □

Theorem 2. *Let G be a connected* AT-*free graph. The strict betweenness defined in Definition 5 and Eq. (1) defines a convex geometry with convex sets defined as in Eq. (2).*

Proof. By Lemma 2, it suffices to show that for $X \subseteq V$, $\{y, z\} \subseteq V$, and $\{y, z\} \cap \sigma(X) = \varnothing$,

$$\exists_{x \in \sigma(X)} \, (x, y, z) \in \mathcal{I} \quad \text{implies} \quad z \notin \sigma(X + y),$$

where $V = V(G)$. First, we show that the assumption that

$$(x, y, z) \in \mathcal{I} \quad \text{and} \quad (x', z, y) \in \mathcal{I}$$

leads to a contradiction, where $x' \in \sigma(X)$. Since $(x, y, z) \in \mathcal{I}$, x and z lie in different components of $G - N[y]$. Furthermore, since $(x', z, y) \in \mathcal{I}$, x' and z lie in the same component of $G - N[y]$. This implies that there is an x', z-path that avoids $N[x]$. Since $(x', z, y) \in \mathcal{I}$, x' and y are in different components of $G - N[z]$. Since $(x, y, z) \in \mathcal{I}$, x lies in the same component of $G - N[z]$ as y, and so there exists an x, y-path that avoids $N[x']$. Since $(x', z, y) \in \mathcal{I}$, there exists also a y, z-path that avoids $N[x']$. The concatenation of these two paths gives an x, z-path that avoids $N[x']$. These two observations imply that $z \in I(x, x')$ unless x and x' are adjacent or the same. But that is not the case, since x' is in the same component of $G - N[y]$ as z which is different from the component of $G - N[y]$ that contains x. Thus $z \in I(x, x')$ which implies $z \in \sigma(X)$; a contradiction.

Now we prove that $z \notin \sigma(X + y)$. By the previous case, we may assume that z is not between two elements of $\sigma(X) + y$. Assume that $(p, z, q) \in \mathcal{I}$ and $(y, q, r) \in \mathcal{I}$ for some $p, r \in \sigma(X)$. Then, by Lemma 1, either

$$(p, z, y) \in \mathcal{I} \quad \text{or} \quad (p, z, r) \in \mathcal{I} \quad \text{or} \quad (y, z, r) \in \mathcal{I} \tag{6}$$

or we have a crossing pair. By assumption none of the cases in (6) occurs so we have a crossing pair. Since $(a', y, z) \in \mathcal{I}$, y and z are not adjacent, so we have that p, q and y lie in distinct components of $G - N[z]$. Since $(a', y, z) \in \mathcal{I}$, there is a a', y-path that avoids $N[z]$, that is, a' lies in the same component of $G - N[z]$ as y. Since $(a', y, z) \in \mathcal{I}$, a' and z are in different components of $G - N[y]$. There is a z, p-path that avoids $N[y]$, since $(p, z, q) \in \mathcal{I}$ and $N(C^z(y)) \subseteq N(q)$. This implies that there is a z, p-path that avoids $N[a']$. Also, there is a y, z-path that avoids $N[a']$ which implies that there is a y, p-path that avoids $N[a']$, since p is adjacent to $N(C^z(y))$. This proves that

$$y \in I(a', p) \quad \Rightarrow \quad y \in \sigma(X)$$

which is a contradiction. The remaining cases will appear in the full version of this paper. This proves the theorem. □

3 The Oracle

Definition 8. *A linear order σ of the vertices of a graph G is an* AT-*free order if*

$$(x, z, y) \in \mathcal{I} \quad \Rightarrow \quad z <_\sigma \max \{ x, y \}, \tag{7}$$

where

$$\max \{ x, y \} = \begin{cases} x, & \text{if } y <_\sigma x, \\ y, & \text{otherwise.} \end{cases}$$

Recently, Corneil and Stacho published the following characterization of AT-free graphs [6].

Theorem 3. *A graph is* AT-*free if and only if it has an* AT-*free order.*

Definition 9. *Let $(V, \mathcal{L})$ be a language antimatroid. Let $w = \alpha x y \beta$ be a basic word in which x and y are adjacent letters. A transposition of w with respect to x and y is the operation that produces the word $w' = \alpha y x \beta$.*

Definition 10. *An $O(1)$ transposition oracle is an oracle that determines in constant time if a word w' is a basic word, given that w is a basic word, where w' is a word obtained from w by a transposition of two adjacent letters.*

Pruesse and Ruskey proved the following theorem [9, Theorem 5].

Theorem 4. *Let $(V, \mathcal{L})$ be a language antimatroid with an $O(1)$ transposition oracle. Then the basic words can be generated in constant amortized time such that each word differs from the next by at most two transpositions.*

In the previous section we have shown that an AT-free graph G with its collection of convex sets $\mathcal{N}$ as defined in Definition (2) forms a convex geometry. The basic words of the corresponding language antimatroid are the AT-free orders. In the following theorem we prove that this language antimatroid has an $O(1)$ amortized time transposition oracle.

Theorem 5. *The* AT-*free orders of an* AT-*free graph can be generated in* $O(1)$ *amortized time.*

Proof (sketch). We show that there is an $O(1)$ transposition oracle.

Let σ be an ordering of the vertices such that

$$z \in I(x, y) \quad \Rightarrow \quad z <_\sigma \max\{ x, y \}.$$

Assume that p and q are consecutive vertices in this order, say $p <_\sigma q$, and let σ' be the order obtained from σ by transposing p and q, that is,

$$p <_\sigma q \quad \text{and} \quad q <_{\sigma'} p.$$

Then σ' is <u>not</u> an AT-free order only if there exists a vertex r with

$$p \in I(q, r) \quad \text{and} \quad r <_\sigma p.$$

The algorithm keeps a data structure which maintains for all ordered pairs x and y the value

$$f_\sigma(x, y) = |\{ z \mid x \in I(y, z) \quad z <_\sigma x <_\sigma y \}| \tag{8}$$

Then, using this data structure, the algorithm can decide in $O(1)$ time whether transposing x and y yields an AT-free order, namely when $f_\sigma(x, y) = 0$.

Unfortunately, this oracle has a side effect. Consider $x <_\sigma y <_\sigma z$. After a transposition of x and y,

$$f_{\sigma'}(x, z) = f_\sigma(x, z) + 1 \quad \text{if } y \in I(x, z)$$
$$f_{\sigma'}(y, z) = f_\sigma(y, z) - 1 \quad \text{if } x \in I(y, z).$$

We now make use of an observation made by Sawada [11, Observation 1]. According to this observation updates need only be made for vertices z in a limited range. According to [11, Theorem 13] the basic words can be generated in constant amortized time. We refer to the discussion in [11, Sect. 6.2, after Theorem 15, pp. 85–86] for further details.

This proves the claim. $\square$

Acknowledgments. The authors would like to thank the anonymous reviewers for the comments. Jou-Ming Chang was supported in part by the MOST grant 104-2221-E-114-002-MY3. Hung-Lung Wang was supported in part by the MOST grant 104-2221-E-114-003.

References

1. Björner, A., Ziegler, G.: 8 Introduction to greeds. In: White, N. (ed.) Matroid Applications - Encyclopedia of Mathematics and its Applications, vol. 40, pp. 284–357. Cambridge University Press, Cambridge (1992)
2. Calder, J.: Some elementary properties of interval convexities. J. Lond. Math. Soc. **3**, 422–428 (1971)
3. Chandran, L., Ibarra, L., Ruskey, F., Sawada, J.: Generating and characterizing the perfect elimination orderings of a chordal graph. Theor. Comput. Sci. **307**, 303–317 (2003)
4. Chang, J.-M., Ho, C.-W., Ko, M.-T.: LexBFS-ordering in asteroidal triple-free graphs. In: Aggarwal, A.K., Pandu Rangan, C. (eds.) ISAAC 1999. LNCS, vol. 1741, pp. 163–172. Springer, Heidelberg (1999)
5. Chvátal, V.: Antimatroids, betweenness, convexity. In: Cook, W., Lovász, L., Vygen, J. (eds.) Research Trends in Combinatorial Optimization. Bonn, 2009, pp. 57–64. Springer, Heidelberg (2008)
6. Corneil, D., Stacho, J.: Vertex ordering characterizations of graphs of bounded asteroidal number. J. Graph Theor. **78**, 61–79 (2015)
7. Kloks, T., Kratsch, D., Müller, H.: Asteroidal sets in graphs. In: Möhring, Rolf H. (ed.) WG 1997. LNCS, vol. 1335, pp. 229–241. Springer, Heidelberg (1997)
8. Kloks, T., Wang, Y.: Advances in graph algorithms. Manuscript viXra:1409.0165 (2014)
9. Pruesse, G., Ruskey, F.: Gray codes from antimatroids. Order **10**(3), 239–252 (1993)
10. Pruesse, G., Ruskey, F.: Generating linear extensions fast. SIAM J. Comput. **23**, 373–386 (1994)
11. Sawada, J.: Oracles for vertex elimination orderings. Theor. Comput. Sci. **341**, 73–90 (2005)
12. Walter, J.: Representations of chordal graphs as subtrees of a tree. J. Graph Theor. **2**, 265–267 (1978)

Enumerating Cyclic Orientations of a Graph

Alessio Conte[1]([✉]), Roberto Grossi[1], Andrea Marino[1], and Romeo Rizzi[2]

[1] Università di Pisa, Pisa, Italy
{conte,grossi,marino}@di.unipi.it
[2] Università di Verona, Verona, Italy
rizzi@di.univr.it

Abstract. Acyclic and cyclic orientations of an undirected graph have been widely studied for their importance: an orientation is acyclic if it assigns a direction to each edge so as to obtain a directed acyclic graph (DAG) with the same vertex set; it is cyclic otherwise. As far as we know, only the enumeration of acyclic orientations has been addressed in the literature. In this paper, we pose the problem of efficiently enumerating all the *cyclic* orientations of an undirected connected graph with n vertices and m edges, observing that it cannot be solved using algorithmic techniques previously employed for enumerating acyclic orientations. We show that the problem is of independent interest from both combinatorial and algorithmic points of view, and that each cyclic orientation can be listed with $\tilde{O}(m)$ delay time. Space usage is $O(m)$ with an additional setup cost of $O(n^2)$ time before the enumeration begins, or $O(mn)$ with a setup cost of $\tilde{O}(m)$ time.

1 Introduction

Given an undirected graph $G(V, E)$ with $n = |V|$ vertices and $m = |E|$ edges, an orientation transforms G into a directed graph $\overrightarrow{G}$ by assigning a direction to each edge. That is, an orientation of G is the directed graph $\overrightarrow{G}(V, \overrightarrow{E})$ such that the vertex set V is the same as G, and the edge set $\overrightarrow{E}$ is an orientation of E: exactly one direction between $(u, v) \in \overrightarrow{E}$ and $(v, u) \in \overrightarrow{E}$ holds for any undirected edge $\{u, v\} \in E$. An orientation $\overrightarrow{G}$ is *acyclic* when $\overrightarrow{G}$ does not contain any directed cycles, so $\overrightarrow{G}$ is a directed acyclic graph (DAG); otherwise we say that the orientation $\overrightarrow{G}$ is cyclic.

Acyclic orientations of undirected graphs have been studied in depth. Many results concern the number of such orientations: Stanley [13] shows how the number of acyclic orientations of a graph can be computed by using the chromatic polynomial (a special case of Tutte's polynomial). Other results rely on the so called acyclic orientation game: Alon et al. [2] inquire about the number of edges that have to be examined in order to identify an acyclic orientation of a random

This work has been partially supported by the Italian Ministry of Education, University, and Research (MIUR) under PRIN 2012C4E3KT national research project AMANDA — Algorithmics for MAssive and Networked DAta.

Z. Lipták and W.F. Smyth (Eds.): IWOCA 2015, LNCS 9538, pp. 88–99, 2016.
DOI: 10.1007/978-3-319-29516-9_8

graph G; Pikhurko [11] gives an upper bound on this number of edges in general graphs. The counting problem is known to be #P-complete [9] and enumeration algorithms that list all the acyclic orientations of a graph are given in [3,12].

We consider *cyclic* orientations, which have been also studied from many points of view. Counting them is #P-complete, as doing so provides the number of acyclic ones (the total number of orientations is 2^m, and they are partitioned in these two kinds). In Fisher et al. [6], given a graph G and an acyclic orientation of it, the number of *dependent edges*, i.e. edges generating a cycle if reversed, has been studied. This number of edges implicitly gives a hint on the number of cyclic orientations in a graph.

In this paper we address the problem of enumerating all the cyclic orientations of an undirected graph G. Without loss of generality, we assume that G is connected.

Problem 1. Enumerating the set of all the directed graphs $\overrightarrow{G}$ that are cyclic orientations of a given undirected connected graph G.

We analyze the cost of an enumeration algorithm for Problem 1 in terms of its *setup* cost, meant as the initialization time before the algorithm is able to list the solutions, and its *delay* cost, which is a well-known performance measure bounding the worst-case time between any two consecutively enumerated solutions (e.g. [8]). We are interested in algorithms with guaranteed $\tilde{O}(m)$ delay, where the $\tilde{O}$ notation ignores polylogarithmic multiplicative factors.

A naive solution to Problem 1 uses the fact that enumeration algorithms exist for listing acyclic orientations [3,12]. It enumerates the cyclic orientations by difference, namely, by enumerating all the 2^m possible edge orientations and removing the α acyclic ones. However, this solution does not guarantee any polynomial delay, as the number $\beta = 2^m - \alpha$ of cyclic orientations can be much larger or much smaller than the number α of acyclic ones. For example, a tree with m edges has $\alpha = 2^m$ and $\beta = 0$. On the other extreme of the situation, we have cliques. An oriented clique is also called a *tournament*, and a *transitive tournament* is a tournament with no cycles. A clique of n nodes can generate 2^m different tournaments, out of which exactly $\alpha = n!$ will be transitive tournaments [10]. As 2^m grows faster than $n!$, where $m = \frac{n \cdot (n-1)}{2}$, we have that the ratio $\alpha/\beta = n!/(2^m - n!)$ tends to 0 for increasing n.

To the best of our knowledge, an enumeration algorithm for Problem 1 with guaranteed $\tilde{O}(m)$ delay is still missing. We provide such an algorithm in this paper, namely, listing each cyclic orientation with $\tilde{O}(m)$ delay time. Space usage is $O(m)$ memory cells with a setup cost of $O(n^2)$ time, or $O(mn)$ memory cells with a setup cost of $\tilde{O}(m)$ time. Although Problem 1 could have less applications than enumerating acyclic orientations [3,12], it is a rich source for enumeration and a ground-field for new combinatorial and algorithmic techniques.

In the following, for the sake of clarity, we will call *edges* the elements of E (undirected graph) and *arcs* the elements of $\overrightarrow{E}$ (directed graph). We will assume that the graph in input G is connected and we will denote as $n = |V|$ and $m = |E|$ respectively its number of nodes and edges.

The paper is organized as follows. Section 2 gives an overview of our enumeration algorithm. Section 3 describes the initialization steps, and shows how to reduce the problem from the input graph to a suitable multi-graph that guarantees to have a chordless cycle (hole) of logarithmic size. The latter is crucial to obtain the claimed delay, and can be seen as a form of kernelization [1]. Section 4 shows how to enumerate in the multigraph and obtain the cyclic orientations for the input graph. Section 5 describes how to absorb the setup cost using more space. Finally, some conclusions are drawn in Sect. 6.

2   Algorithm Overview

The intuition behind our algorithm for an undirected connected graph $G(V_G, E_G)$ is the following one. Suppose that G is cyclic, otherwise there are no cyclic orientations. Consider one cycle[1] $C(V_C, E_C)$ in G: we can orient its edges in two ways so that the resulting $\overrightarrow{C}$ is a directed cycle. At this point, *any* orientation of the remaining edges, e.g. those in $E_G \backslash E_C$, will give a cyclic orientation of G. Thus we generate all possible orientations of the edges in $E_G \backslash E_C$, and then assign some suitable orientations to the edges in E_C. This guarantees that we have at least two solutions for each orientation of $E_G \backslash E_C$, namely, setting the orientation of E_C so that $\overrightarrow{C}$ is one of the two possible directed cycles. Yet this is not enough as we could have a cyclic orientation even if $\overrightarrow{C}$ is acyclic.

In general we must consider the following cases. One easy case is that the orientation of $E_G \backslash E_C$ already produces a directed cycle: any orientation of E_C will give a cyclic orientation of G. Another easy case, as seen above, is for the two orientations of E_C such that $\overrightarrow{C}$ is a directed cycle: any orientation of $E_G \backslash E_C$ will give a cyclic orientation of G. It remains the case when the orientations of both $E_G \backslash E_C$ and E_C are individually acyclic: when put together, we might or might not have a directed cycle in the resulting orientation of G. To deal with the latter case, we need to "massage" G and transform it into a multigraph as follows, referring the reader to Algorithm 1.

Algorithm setup is performed as described in Sect. 3. We preprocess G with bridge removal and edge chain compression, recalling that a bridge is an edge whose deletion increases the number of connected components and that a chain is a maximal path of degree-2 nodes. The result is an undirected connected multigraph $M(V_M, E_M)$, where the edges are labeled as *simple* and *chain*. After that, we find a chordless cycle of logarithmic size C in M, called *log-hole* , and remove E_C from E_M, obtaining the labeled multigraph $M'(V_M, E'_M)$, where $E'_M = E_M \backslash E_C$.

Enumerating cyclic orientations, described in Sect. 4, exploits the property (which we will show later) that finding cyclic orientations of G corresponds to finding particular orientations in M', called *extended cyclic orientations*, and of C, called *legal orientations*. In the **for** loop, these orientations of M' and C are enumerated so as to find all the cyclic orientations of G. As we will see for the latter task, it is important to have C of logarithmic size to guarantee our claimed delay.

---

[1] This will actually be a chordless cycle of logarithmic size (called log-hole).

Algorithm 1. Returning all the cyclic orientations of G

Input: An undirected connected graph $G(V, E)$
Output: All the cyclic orientations $\overrightarrow{G}(V, \overrightarrow{E})$

Algorithm setup (Section 3):
Remove bridges and isolated nodes from G
$M(V_M, E_M) \leftarrow$ replace G's maximal paths of degree-2 nodes with chain edges
$C(V_C, E_C) \leftarrow$ a log-hole of M
$M'(V_M, E'_M) \leftarrow$ delete the edges of C from M, i.e. $E'_M = E_M \setminus E_C$

Enumerate cyclic orientations (Section 4):
for each extended orientation $\overrightarrow{M}'$ of multigraph M' **do**
  for each legal orientation $\overrightarrow{C}$ of log-hole C (see Algorithm 2) **do**
    $\overrightarrow{M}''(V_M, \overrightarrow{E}'') \leftarrow$ combine $\overrightarrow{M}'$ and $\overrightarrow{C}$, where $\overrightarrow{E}'' = \overrightarrow{E}'_M \cup \overrightarrow{E}_C$
    Output each of the cyclic orientations $\overrightarrow{G}$ of G corresponding to $\overrightarrow{M}''$

---

3 Algorithm Setup

3.1 Reducing the Problem to Extended Cyclic Orientations

In the following we show how to reduce Problem 1 to an extended version that allows us to neglect bridges and chains.

Bridge Removal. Given an undirected graph $G(V, E)$, since a bridge cannot be included in any cycle of G, we remove all *bridges*. By removing bridges and computing the cyclic orientations in the cleaned graph, we can still generate solutions for the original graph by using both orientations of each bridge, as emphasized by the following lemma, whose proof is straightforward.

Lemma 1. *Let G be a graph and $\{u, v\}$ a bridge. Let G' be the graph G without the edge $\{u, v\}$. The set of all the cyclic orientations of G is composed by the orientations $\overrightarrow{G}'(V, \overrightarrow{E}' \cup \{(u, v)\})$ and $\overrightarrow{G}'(V, \overrightarrow{E}' \cup \{(v, u)\})$, for all the cyclic orientations $\overrightarrow{G}'(V, \overrightarrow{E}')$ of G'.*

For the sake of simplicity, after removing the bridges we remove also isolated nodes (i.e., nodes of degree zero). It is easy to see that all the surviving nodes have degree 2 or greater.

The bridges of a graph can be found in linear time [14]. Finding and removing bridges and removing isolated nodes can be done in $O(m)$ time and space.

The rationale for removing bridges is to have shorter cycles: for example, consider a "necklace" graph with n nodes, for n even, such that $n/2$ nodes form a cycle, and the remaining $n/2$ nodes have degree 1 and are attached to one of the nodes in the cycle, such that each node in the cycle has degree 3 and is connected to one node of degree 1. With the removal of bridges and isolated nodes, the cycle has only nodes of degree 2 and can be compressed as discussed in the next paragraph.

Chain Compression. It consists in finding all the maximal paths $v_1, \ldots, v_k$ where v_i has degree 2 (with $2 \leq i \leq k - 1$), and replacing each of them, called *chain*, with just one edge, called *chain edge*. It is easy to see that this task can be accomplished in $O(m)$ time by traversing the graph G in a DFS fashion from a node of degree ≥ 3. The output is an undirected connected multigraph $M(V_M, E_M)$, where $V_M \subseteq V$ are the nodes of V whose degree is ≥ 3, and E_M are the chain edges plus all the edges in E which are not part of a chain. The latter ones are called simple edges to distinguish them from the chain edges. In the rest of the paper, M will be seen as a multigraph where $|V_M| \geq 4$ and each of the edges has a label in {simple, chain}, since it might contain parallel edges or loops.[2] For this, we define the concept of *extended orientation* as follows.

Definition 1 (Extended Orientation). For a multigraph $M(V_M, E_M)$ having self loops and edges labeled as simple and chain, an *extended orientation* $\overrightarrow{M}(V_M, \overrightarrow{E_M})$ is a directed multigraph whose arc set $\overrightarrow{E_M}$ assigns a direction or *broken* to each edge in E_M: in particular, for any simple edge $\{u, v\} \in E_M$, exactly one direction between (u, v) and (v, u) is assigned; for any chain edge $\{u, v\} \in E_M$, either the edge is broken, or exactly one direction between (u, v) and (v, u) is assigned. A directed cycle in $\overrightarrow{M}$ cannot contain a broken edge.

Broken edges correspond to chain edges that, when expanded as edges of G, do not have an orientation as a directed path. This means that they cannot be traversed in either direction. Note that this situation cannot happen for simple edges. The following lemma holds.

Lemma 2. *If we have an algorithm that lists all the extended cyclic orientations of $M(V_M, E_M)$ with delay $f(|E_M|)$, for some $f : \mathbb{R} \to \mathbb{R}$, then we have an algorithm that lists all the acyclic orientations of the graph $G(V, E)$ with delay $O(f(|E_M|) + |E|)$.*

Proof. For each extended cyclic orientations $\overrightarrow{M}$ we return a set S of cyclic orientations of G: any simple edge e of $\overrightarrow{M}$ maintains the same direction specified by $\overrightarrow{M}$ in all the solutions in S; for each chain c of $\overrightarrow{M}$, we consider the edges corresponding to c in G, say $e_1, e_2, \ldots, e_h$: if c has a direction in $\overrightarrow{M}$, the same direction of c is assigned to all the edges e_j in all the solutions in S; if c has no direction assigned, i.e. *broken*, we have to consider all the possible $2^h - 2$ ways of making the path $e_1, e_2, \ldots, e_h$ broken (these are all the possible ways of directing the edges except the only two directing a path). All the solutions in S differ for the way they replace the chain edges.

Getting extended cyclic orientations in $f(|E_M|)$ delay, iterating over all the chain edges c, and iterating over all the corresponding edges of c assigning the specified directions as explained above, we return acyclic orientations of the graph $G(V, E)$ with delay $O(f(|E_M|) + |E|)$. $\square$

Lemma 2 allows us to concentrate on extended cyclic orientations of the labeled multigraph M rather than on cyclic orientations of G. Conceptually,

---

[2] The degenerate case of M with ≤ 3 nodes can be easily handled.

we have to assign binary values (the orientation) to simple edges and ternary values (the orientation or broken) to chain edges. If we complicate the problem on one side by introducing these multigraphs with chain edges, we have a relevant benefit on the other side, as shown next.

3.2 Logarithmically Bounded Hole

A *logarithmically bounded hole* (hereafter, log-hole) is a chordless cycle whose length is either the *girth* of the graph (i.e. the length of its shortest cycle) or this length plus one.[3]

Given the labeled multigraph obtained in Sect. 3.1, namely $M(V_M, E_M)$, we perform the following two steps.

1. **Finding a log-hole.** Find a log-hole $C(V_C, E_C)$ in $M(V_M, E_M)$.[4]
2. **Removing the log-hole.** Remove the edges in E_C from M, obtaining $M'(V_M, E'_M)$, where $E'_M = E_M - E_C$.

Properties of the log-hole. Since M is a multigraph with self-loops, a log-hole $C(V_C, E_C)$ of M can potentially be a self-loop. In any case, the following well-known result holds.

Lemma 3 (Logarithmic girth [4,5]). *Let $G(V, E)$ be a graph in which every node has degree at least 3. The girth of G is $\leq 2\lceil \log |V| \rceil$.*

Lemma 3 means that the log-hole C of M has length at most $2\lceil \log |V_M| \rceil + 1$, thus motivating our terminology.

The log-hole C can be found by finding the girth, that is performing a BFS on each node of the multigraph M to identify the shortest cycle that contains that node, in time $O(|V_M| \cdot |E_M|)$. By applying the algorithm in [7], which easily extends to multigraphs, we compute C in time $O(|V_M|^2)$: in this case, if chords are present in the found C, in time $O(|C|) = O(\log |V_M|)$ we can check whether C includes a smaller cycle and redefine C accordingly.

4 Enumerating Cyclic Orientations

We now want to list all the cyclic orientations of the input graph G. By Lemma 2 this is equivalent to listing the extended cyclic orientations of the corresponding labeled multigraph $M(V_M, E_M)$ obtained from G by bridge removal and chain compression. We now show that the latter task can be done by suitably combining some orientations from the labeled multigraph $M'(V_M, E'_M)$ and the log-hole $C(V_C, E_C)$ using an algorithm that is organized as follows.

---

[3] Minimum cycle means the cycle having minimum number of edges (e.g. a self loop). Ties are broken arbitrarily.

[4] When computing the log-hole of M, chain edges count just one, like simple edges.

1. **Finding extended orientations.** Enumerate all extended orientations (not necessarily cyclic) $\overrightarrow{M}'$ of the multigraph M'.
2. **Putting back the log-hole.** For each listed $\overrightarrow{M}'(V_M, \overrightarrow{E}'_M)$, consider all the extended orientations $\overrightarrow{C}(V_C, \overrightarrow{E}_C)$ of the log-hole C such that $\overrightarrow{E}'_M \cup \overrightarrow{E}_C$ contains a directed cycle, and obtain the extended cyclic orientations for the multigraph M.

Finding Extended Orientations. This is now an easy task. For each edge $\{u, v\}$ in E'_M that is labeled as simple, both the directions (u, v) and (v, u) can be assigned; if $\{u, v\}$ is labeled as chain, the directions (u, v) and (v, u), and broken can be assigned. Each combination of these decisions produces an extended orientation of $M'(V_M, E_M)$. If there are s simple edges and c chain edges in M', where $s + c = |E'_M|$, this generates all possible $2^s 3^c$ extended orientations. Each of them can be easily listed in $O(|E'_M|)$ delay (actually less, but this is not the dominant cost).

Putting Back the log-hole. For each listed $\overrightarrow{M}'$ we have to decide how to put back the edges of the cycle C, namely, how to find the orientations of C that create directed cycles.

Definition 2. Given the cycle $C(V_C, E_C)$ and $\overrightarrow{M}'(V_M, \overrightarrow{E}'_M)$, we call *legal orientation* $\overrightarrow{C}(V_C, \overrightarrow{E}_C)$ any extended orientation of C such that the resulting multigraph $\overrightarrow{M}''(V_M, \overrightarrow{E}'')$ is cyclic, where $\overrightarrow{E}'' = \overrightarrow{E}'_M \cup \overrightarrow{E}_C$.

The two following cases are possible.

1. $\overrightarrow{M}'$ is cyclic. In this case each edge in E_C can receive any direction, including broken if the edge is a chain edge: each combination of these assignments will produce a legal orientation that will be output.
2. $\overrightarrow{M}'$ is acyclic. Since C is a cycle, there are at least two legal orientations obtained by orienting C as a directed cycle clockwise and counterclockwise. Moreover, adding just an oriented subset of edges $D \subseteq C$ to $\overrightarrow{M}'$ can create a cycle in $\overrightarrow{M}'$: in this case, any orientation of the remaining edges of $C \backslash D$ (including broken for chain edges) will clearly produce a legal orientation.

While the first case is immediate, the second case has to efficiently deal with the following problem.

Problem 2. Given $\overrightarrow{M}'$ acyclic and cycle C, enumerate all the legal orientations $\overrightarrow{C}(V_C, \overrightarrow{E}_C)$ of C.

In order to solve Problem 2, we exploit the properties of C. In particular, we compute the reachability matrix R among all the nodes in V_C, that is, for each pair u, v of nodes in V_C, $R(u, v)$ is 1 if u can reach v in $\overrightarrow{M}$, 0 otherwise. We say that R is *cyclic* whether there exists a pair i, j such that $R(i, j) = R(j, i) = 1$.

This step can be done by performing a BFS in $\overrightarrow{M}'$ from each node in V_C: by Lemma 3 we have $|V_C| < 2\lceil \log |V_M| \rceil + 1$, and so the cost is $O(|E'_M| \cdot \log |V_M|)$ time. Deciding the orientation of the edges and the chain edges in E_C is done with a ternary partition of the search space. Namely, for each edge $\{u, v\}$ in E_C, if $\{u, v\}$ is simple we try the two possible direction assignments, while if it is a chain we also try the broken assignment. In order to avoid dead-end recursive calls, after each assignment we update the reachability matrix R and perform the recursive call only if this partial direction assignment will produce at least one solution: both the update of R and the dead-end check can be done in $O(\log^2 |V_M|)$ (that is, the size of R).

Scheme for Legal Orientations. The steps are shown by Algorithm 2. At the beginning the reachability matrix R is computed and is passed to the recursive routine `LegalOrientations`. At each step, $\overrightarrow{C}'$ is the partial legal orientation to be completed and I is the set of broken edges declared so far. Also, j is the index of the next edge $\{c_j, c_{j+1}\}$ of the cycle C, with $1 \le j \le h$ (we assume $c_{h+1} = c_1$ to close the cycle): if $j = h + 1$ then all the edges of C have been considered and we output the solution $\overrightarrow{C}'$ together with the list I of broken edges in $\overrightarrow{C}'$. Each time the procedure is called we guarantee that the reachability matrix R is updated.

Let $\{u, v\}$ be the next edge of C to be considered: for each possible direction assignment (u, v) or (v, u) of this edge, we have to decide whether we will be able to complete the solution considering this assignment. This is done by trying to add the arc to the current solution. If there is already a cycle, clearly we can complete the solution. Otherwise, we perform a *reachability check* on $\{c_{j+1}, \dots c_{h+1}\}$: it is still possible to create a directed cycle if and only if any two of the nodes in $\{c_{j+1}, \dots c_{h+1}\}$, say c_f and c_g satisfy $R(c_f, c_g) = 1$ or $R(c_g, c_f) = 1$. This condition guarantees that a cycle will be created in the next calls, since we know there are edges in C between c_f and c_g that can be oriented suitably. Finally, when $\{u, v\}$ is a chain, the broken assignment is also considered: R does not need to be updated as the broken edge does not change the reachability of $\overrightarrow{M}'$.

The reachability and cyclicity checks are done by updating and checking the reachability matrix R (and restoring R when needed). Updating R when adding an arc (u, v) corresponds to making v, and all nodes reachable from v, reachable from u and nodes that can reach u. This can be done by simply performing an **or** between the corresponding rows in time $O(\log^2 |V_M|)$, since R is $|C| \times |C|$. The reachability check can be done in $O(\log^2 |V_M|)$ time. The cyclicity (checking whether a cycle has been already created) takes the same amount of time by looking for a pair of nodes x', y' in $\{c_1, \dots c_j\}$ such that $R(x', y') = R(y', x') = 1$.

Lemma 4. *Algorithm 2 outputs in $O(|E'_M| \log |V_M|)$ time the first legal orientation of C, and each of the remaining ones with $O(\log^3 |V_M|)$ delay.*

Proof. Before calling the `LegalOrientations` procedure we have to compute the reachability matrix from scratch and this costs $O(|E'_M| \log |V_M|)$ time. In the following we will bound the delay between two outputs returned by the

---

Algorithm 2. Returning all legal orientations of C

Input: $\overrightarrow{M}'(V_M, \overrightarrow{E}'_M)$ acyclic, a cycle $C(V_C, E_C)$ with $V_C \subseteq V_M$
Output: All the legal orientations $\overrightarrow{C}(V_C, \overrightarrow{E}_C)$

Build the reachability matrix R for the nodes of V_C in $\overrightarrow{M}'$
Let $V_C = \{c_1, \dots c_h\}$, where $c_{h+1} = c_1$ by definition
Execute LegalOrientations $(\overrightarrow{C}'(\emptyset, \emptyset), 1, R, \emptyset)$

Procedure LegalOrientations$(\overrightarrow{C}'(V'_C, \overrightarrow{E}'_C), j, R, I)$
   if $j = h + 1$ **then**
      | output $\overrightarrow{C}'$ and its set I of broken edges
   else
      $u \leftarrow c_j$, $v \leftarrow c_{j+1}$
      $R_1 \leftarrow R$ updated by adding the arc (u, v);
      if R_1 *is cyclic or has positive reachability test on* $\{c_{j+1}, \dots, c_{h+1}\}$ **then**
         LegalOrientations $(\overrightarrow{C}'(V'_C, \overrightarrow{E}'_C \cup \{(u, v)\}), j + 1, R_1, I)$
      $R_2 \leftarrow R$ updated adding the arc (v, u);
      if R_2 *is cyclic or has positive reachability test on* $\{c_{j+1}, \dots, c_{h+1}\}$ **then**
         LegalOrientations $(\overrightarrow{C}'(V'_C, \overrightarrow{E}'_C \cup \{(v, u)\}), j + 1, R_2, I)$
      if $\{u, v\}$ *is a chain edge* **then**
         if R *is cyclic or has positive reachability test on* $\{c_{j+1}, \dots, c_{h+1}\}$
         then
            LegalOrientations $(\overrightarrow{C}', j + 1, R, I \cup \{\{u, v\}\})$

LegalOrientations procedure. Firstly, note that each call produces at least one solution. This is true when $j = 1$ since we have two possible legal orientations of C. Before performing any call at depth j, the caller function checks whether this will produce at least one solution. Only calls that will produce at least one solution are then performed. This means that in the recursion tree, every internal node has at least one child and each leaf corresponds to a solution. Hence the delay between any two consecutive solutions is bounded by the cost of a leaf-to-root path and the cost of a root-to-the-next-leaf path in the recursion tree induced by LegalOrientations. Since the height of the recursion tree is $O(\log |V_M|)$, i.e. the edges of C, and the cost of each recursion node is $O(\log^2 |V_M|)$, we can conclude that the delay between any two consecutive solutions is bounded by $O(\log^3 |V_M|)$. As it can be seen, it is crucial that the size of C is (poly)logarithmic.   □

Lemma 5 (Correctness).

1. *All the extended cyclic orientations of M are output.*
2. *Only extended cyclic orientations of M are output.*
3. *There are no duplicates.*

Proof 1. Any extended cyclic orientation $\overrightarrow{M}$ can be seen as the union $\overrightarrow{M}''$ of $\overrightarrow{M}'$ and $\overrightarrow{C}$, which are two edge disjoint directed subgraphs. Our algorithm enumerates all the extended orientations of M', and, for each of them, all legal extended orientations $\overrightarrow{C}$: if there is a cycle in $\overrightarrow{M}$ involving only edges in E'_M all the extended orientations of C are legal; otherwise just the extended orientations $\overrightarrow{C}$ of C whose arcs create a cycle in $\overrightarrow{M}''$ are legal. Hence any extended cyclic orientation $\overrightarrow{M}$ is output.

2. Any output solution is an extended orientation: each edge in M' and in C has exactly one direction or is broken. Moreover, any output solution contains at least one cycle: if there is not a cycle in M', a cycle is created involving the edges in C.

3. All the extended orientations $\overrightarrow{M} = \overrightarrow{M}''$ in output differ for at least one arc in $\overrightarrow{E}'_M$ or an arc in $\overrightarrow{E_C}$. Hence there are no duplicate solutions.   □

As a result, we obtain delay $\tilde{O}(|E_M|)$.

Lemma 6. *The extended cyclic orientations of $M(V_M, E_M)$ can be enumerated with delay $\tilde{O}(|E_M|)$ and space $O(|E_M|)$.*

Proof. Finding extended orientations $\overrightarrow{M}'$ of M' can be done with $O(|E'_M|)$ delay. Every time a new $\overrightarrow{M}'$ has been generated, we apply Algorithm 2. By Lemma 4 we output the first cyclic orientation $\overrightarrow{M}$ of M with delay $O(|E_M| \log |V_M|)$ and the remaining ones with delay $O(\log^3 |V_M|)$. Hence the maximum delay between any two consecutive solutions is $O(|E'_M| + |E_M| \log |V_M|) = O(|E_M| \log |V_M|) = \tilde{O}(|E_M|)$. The space usage is linear in all the phases: in particular in Algorithm 2 the space is $O(\log^2 |V_M|)$, because of the reachability matrix R, which is smaller than $O(|E_M|)$.   □

Applying Lemmas 2 and 6, and considering the setup cost in Sect. 3 ($|V_M| \le |V|$ and $|E_M| \le |E|$), we can conclude as follows.

Theorem 1. *Algorithm 1 lists all cyclic orientations of $G(V, E)$ with setup cost $O(|V|^2)$, and delay $\tilde{O}(|E|)$. The space usage is $O(|E|)$ memory cells.*

5   Absorbing the Setup Cost

In this section, we show how to modify our approach to get a setup time equal to the delay, requiring space $\Theta(|V| \cdot |E|)$.

Theorem 2. *All cyclic orientations of $G(V, E)$ can be listed with setup cost $\tilde{O}(|E|)$, delay $\tilde{O}(|E|)$, and space usage of $\Theta(|V| \cdot |E|)$ memory cells.*

We use $n = |V|$ and $m = |E|$ for brevity. Let A_1 be the following algorithm that takes $T_1 = O(mn)$ time to generate n solutions, each with $\tilde{O}(m)$ delay, starting from any given cycle of size $\ge \log n$. This cycle is found by performing a BFS on an arbitrary node u, and identifying the shortest cycle C_u containing u.

Now, if $|C_u| < \log n$, since C_u is a log-hole as required, we stop the setup and run the algorithms in the previous sections setting $C = C_u$. The case of interest in this section is when $|C_u| \geq \log n$. We take a cyclic orientation $\overrightarrow{C_u}$ of C_u, and then n arbitrary orientations of the edges in $G\backslash C_u$. The setup cost is $O(m)$ time and we can easily output each solution in $\tilde{O}(m)$ delay. We denote this set of n solutions by Z_1.

Also, let A_2 be the algorithm behind Theorem 1, with a setup cost of $O(mn)$ and $\tilde{O}(m)$ delay (i.e. Algorithm 1). We denote the time taken by A_2 to list the first n solutions, including the $O(mn)$ setup cost, by $T_2 = \tilde{O}(mn)$, and this set of n solutions by Z_2. Since Z_1 and Z_2 can have nonempty intersection, we want to avoid duplicates.

We show how to obtain an algorithm A that lists all the cyclic orientations without duplicates with $\tilde{O}(m)$ setup cost and delay, using $O(mn)$ space. Even though the delay cost of A is larger than that of A_1 and A_2 by a constant factor, the asymptotic complexity is not affected by this constant, and remains $\tilde{O}(m)$.

Algorithm A executes simultaneously and independently the two algorithms A_1 and A_2. Recall that these two algorithms take $T_1 + T_2$ time in total to generate Z_1 and Z_2 with $\tilde{O}(m)$ delay. However those in Z_2 are produced after a setup cost of $O(mn)$. Hence A slows down on purpose by a constant factor c, thus requiring $c(T_1 + T_2)$ time: it has time to find the distinct solutions in $Z_1 \cup Z_2$ and build a dictionary D_1 on the solutions in Z_1. (Since an orientation can be represented as a binary string of length m, a binary trie can be employed as dictionary D_1, supporting each dictionary operation in $O(m)$ time.) During this time, A outputs the n solutions from Z_1 with a delay of $c(T_1 + T_2)/n = \tilde{O}(m)$ time each, while storing the rest of solutions of $Z_2 \backslash Z_1$ in a buffer Q.

After $c(T_1 + T_2)$ time, the situation is the following: Algorithm A has output the n solutions in Z_1 with $\tilde{O}(m)$ setup cost and delay. These solutions are stored in D_1, so we can check for duplicates. We have buffered at most n solutions of $Z_2 \backslash Z_1$ in Q. Now the purpose of A is to continue with algorithm A_2 alone, with $\tilde{O}(m)$ delay per solution, avoiding duplicates. Thus for each solution given by A_2, algorithm A suspends A_2 and waits so that each solution is output in $c(T_1 + T_2)/n$ time: if the solution is not in D_1, A outputs it; otherwise A extracts one solution from the buffer Q and outputs the latter instead. Note that if there are still d duplicates to handle in the future, then Q contains exactly d solutions from $Z_2 \backslash Z_1$ (and Q is empty when $A-2$ completes its execution). Thus, A never has to wait for a non-duplicated solution. The delay is the maximum between $c(T_1 + T_2)/n$ and the delay of A_2, hence $\tilde{O}(m)$. The additional space is dominated by that of Q, namely, $O(mn)$ memory cells to store up to n solutions.

We also have an amortized cost using the lemma below, where $f(x) = \tilde{O}(x)$ and $s = |V|$.

Lemma 7. *Listing all the extended cyclic orientations of $M(V_M, E_M)$ with delay $O(f(|E_M|))$ and setup cost $O(s \cdot |V_M|)$ implies that the average cost per solution is $O(f(|E_M|) + |E_M|)$.*

Proof. We perform a BFS on an arbitrary node u, and identify the shortest cycle $C_u(V_u, E_u)$ that contains u. This costs $O(m)$ time. Note that $C_u(V_u, E_u)$ is a hole (i.e. it has no chords). Note that a minimum cycle in M either is C_u or contains a node in $V_M - V_u$: hence we perform all the BFSs from each node in $V_M - V_u$, as explained in [7] with an overall cost of $O(|V_M| \cdot |V_M - V_u|)$. The number of extended orientations of M is at least $2^{|E_M - E_u|} \geq 2^{|V_M - V_u|}$. Our setup cost is $O(s \cdot |V_M|)$, with $s \leq |V_M|$, and the number of solutions is at least 2^s. The overall average cost per solution is at most $O(2^s \cdot f(|E_M|) + s \cdot |V_M|)/2^s$, which is $O\left(f(|E_M|) + |E_M| \cdot \frac{s}{2^s}\right)$.   □

6   Conclusions

In this paper we considered the problem of efficiently enumerating cyclic orientations of graphs. The problem is interesting from a combinatorial and algorithmic point of view, as the fraction of cyclic orientations over all the possible orientations can be as small as 0 or very close to 1. We provided an efficient algorithm to enumerate the solutions with delay $\tilde{O}(m)$ and overall complexity $\tilde{O}(\alpha \cdot m)$, with α being the number of solutions.

References

1. Abu-Khzam, F.N., Collins, R.L., Fellows, M.R., Langston, M.A., Suters, W.H., Symons, C.T.: Kernelization algorithms for the vertex cover problem: theory and experiments. In: ALENEX/ANALC, vol. 69 (2004)
2. Alon, N., Tuza, Z.: The acyclic orientation game on random graphs. Random Struct. Algorithms 6(2–3), 261–268 (1995)
3. Barbosa, V.C., Szwarcfiter, J.L.: Generating all the acyclic orientations of an undirected graph. Inf. Process. Lett. 72(1), 71–74 (1999)
4. Bollobas, B.: Extremal Graph Theory. Dover Publications Incorporated, New York (2004)
5. Erdős, P., Pósa, L.: On the maximal number of disjoint circuits of a graph. Publ. Math. Debrecen 9, 3–12 (1962)
6. Fisher, D.C., Fraughnaugh, K., Langley, L., West, D.B.: The number of dependent arcs in an acyclic orientation. J. Comb. Theor. Ser. B 71(1), 73–78 (1997)
7. Itai, A., Rodeh, M.: Finding a minimum circuit in a graph. SIAM J. Comput. 7(4), 413–423 (1978)
8. Johnson, D.S., Papadimitriou, C.H., Yannakakis, M.: On generating all maximal independent sets. Inf. Process. Lett. 27(3), 119–123 (1988)
9. Linial, N.: Hard enumeration problems in geometry and combinatorics. SIAM J. Algeb. Discrete Meth. 7(2), 331–335 (1986)
10. Moon, J.: Topics on tournaments. In: Selected Topics in Mathematics. Athena series. Holt, Rinehart and Winston (1968)
11. Pikhurko, O.: Finding an unknown acyclic orientation of a given graph. Comb. Probab. Comput. 19(1), 121–131 (2010)
12. Squire, M.B.: Generating the acyclic orientations of a graph. J. Algorithms 26(2), 275–290 (1998)
13. Stanley, R.P.: Acyclic orientations of graphs. Discrete Math. 5(2), 171–178 (1973)
14. Tarjan, R.E.: A note on finding the bridges of a graph. Inf. Process. Lett. 2(6), 160–161 (1974)

Filling the Complexity Gaps for Colouring Planar and Bounded Degree Graphs

Konrad K. Dabrowski[1], François Dross[2,3], Matthew Johnson[1(✉)],
and Daniël Paulusma[1]

[1] School of Engineering and Computing Sciences,
Durham University, Science Laboratories, South Road, Durham DH1 3LE, UK
{konrad.dabrowski,matthew.johnson2,daniel.paulusma}@durham.ac.uk
[2] Ecole Normale Supérieure de Lyon, Lyon, France
[3] Laboratoire d'Informatique,
de Robotique et de Microélectronique de Montpellier, Montpellier, France
francois.dross@ens-lyon.fr

Abstract. We consider a natural restriction of the LIST COLOURING problem, k-REGULAR LIST COLOURING, which corresponds to the LIST COLOURING problem where every list has size exactly k. We give a complete classification of the complexity of k-REGULAR LIST COLOURING restricted to planar graphs, planar bipartite graphs, planar triangle-free graphs and to planar graphs with no 4-cycles and no 5-cycles. We also give a complete classification of the complexity of this problem and a number of related colouring problems for graphs with bounded maximum degree.

Keywords: List colouring · Choosability · Planar graphs · Maximum degree

1 Introduction

A *colouring* of a graph is a labelling of the vertices so that adjacent vertices do not have the same label. We call these labels *colours*. Graph colouring problems are central to the study of combinatorial algorithms and they have many theoretical and practical applications. A typical problem asks whether a colouring exists under certain constraints, or how difficult it is to find such a colouring. For example, in the LIST COLOURING problem, a graph is given where each vertex has a list of colours and one wants to know if the vertices can be coloured using only colours in their lists. The CHOOSABILITY problem asks whether such list colourings are guaranteed to exist whenever all the lists have a certain size. In fact, an enormous variety of colouring problems can be defined and there is now a vast literature on this subject. For longer introductions to the type of problems we consider we refer to two recent surveys [6,12].

First and last author supported by EPSRC (EP/K025090/1).

Z. Lipták and W.F. Smyth (Eds.): IWOCA 2015, LNCS 9538, pp. 100–111, 2016.
DOI: 10.1007/978-3-319-29516-9_9

In this paper, we are concerned with the computational complexity of colouring problems. For many such problems, the complexity is well understood in the case where we allow every graph as input, so it is natural to consider problems with restricted inputs. We consider a variant of the LIST COLOURING problem, closely related to CHOOSABILITY, and give a complete classification of its complexity for planar graphs and a number of subclasses of planar graphs by combining known results with new results. Some of the known results are for (planar) graphs with bounded degree. We use these results to fill some more complexity gaps by giving a complete complexity classification of a number of colouring problems for graphs with bounded maximum degree.

1.1 Terminology

A *colouring* of a graph $G = (V, E)$ is a function $c : V \rightarrow \{1, 2, \ldots\}$ such that $c(u) \neq c(v)$ whenever $uv \in E$. We say that $c(u)$ is the *colour* of u. For a positive integer k, if $1 \leq c(u) \leq k$ for all $u \in V$, then c is a k-*colouring* of G. We say that G is k-*colourable* if a k-colouring of G exists. The COLOURING problem is to decide whether a graph G is k-colourable for some given integer k. If k is fixed, that is, not part of the input, we obtain the k-COLOURING problem.

A *list assignment* of a graph $G = (V, E)$ is a function L with domain V such that for each vertex $u \in V$, $L(u)$ is a subset of $\{1, 2, \ldots\}$. This set is called the *list* of *admissible* colours for u. If $L(u) \subseteq \{1, \ldots, k\}$ for each $u \in V$, then L is a k-*list assignment*. The *size* of a list assignment L is the maximum list size $|L(u)|$ over all vertices $u \in V$. A colouring c *respects* L if $c(u) \in L(u)$ for all $u \in V$. Given a graph G with a k-list assignment L, the LIST COLOURING problem is to decide whether G has a colouring that respects L. If k is fixed, then we have the LIST k-COLOURING problem. Fixing the size of L to be at most ℓ gives the ℓ-LIST COLOURING problem. We say that a list assignment L of a graph $G = (V, E)$ is ℓ-*regular* if, for all $u \in V$, $L(u)$ contains exactly ℓ colours. This gives us the following problem, which is one focus of this paper. It is defined for each integer $\ell \geq 1$ (note that ℓ is *fixed*; that is, ℓ is not part of the input).

ℓ-REGULAR LIST COLOURING
Instance: a graph G with an ℓ-regular list assignment L.
Question: does G have a colouring that respects L?

A k-*precolouring* of a graph $G = (V, E)$ is a function $c_W : W \rightarrow \{1, 2, \ldots, k\}$ for some subset $W \subseteq V$. A k-colouring c of G is an *extension* of a k-precolouring c_W of G if $c(v) = c_W(v)$ for each $v \in W$. Given a graph G with a precolouring c_W, the PRECOLOURING EXTENSION problem is to decide whether G has a k-colouring that extends c_W. If k is fixed, we obtain the k-PRECOLOURING EXTENSION problem.

The relationships amongst the problems introduced are shown in Fig. 1.

For an integer $\ell \geq 1$, a graph $G = (V, E)$ is ℓ-*choosable* if, for every ℓ-regular list assignment L of G, there exists a colouring that respects L. The corresponding decision problem is the CHOOSABILITY problem. If ℓ is fixed, we obtain the ℓ-CHOOSABILITY problem.

We emphasize that ℓ-REGULAR LIST COLOURING and ℓ-CHOOSABILITY are two fundamentally different problems. For the former we must decide whether

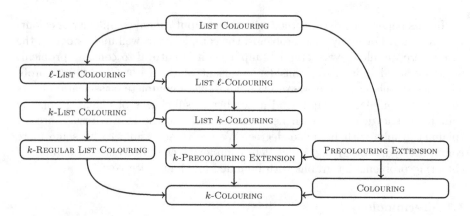

Fig. 1. Relationships between COLOURING and its variants. An arrow from one problem to another indicates that the latter is (equivalent to) a special case of the former; k and ℓ are any two arbitrary integers for which $\ell \geq k$. For instance, k-COLOURING is a special case of k-REGULAR LIST COLOURING. This can be seen by giving the list $L(u) = \{1, \ldots, k\}$ to each vertex u in an instance graph of COLOURING. We also observe that ℓ-REGULAR LIST COLOURING and k-REGULAR LIST COLOURING are not comparable for any $k \neq \ell$.

there exists a colouring that respects a *particular* ℓ-regular list assignment. For the latter we must decide whether or not *every* ℓ-regular list assignment has a colouring that respects it. As we will see later, this difference also becomes clear from a complexity point of view: for some graph classes ℓ-REGULAR LIST COLOURING is computationally easier than ℓ-CHOOSABILITY; whereas, perhaps more surprisingly, for other graph classes, the reverse holds.

For two vertex-disjoint graphs G and H and positive integer k, we let $G + H$ denote the disjoint union $(V(G) \cup V(H), E(G) \cup E(H))$, and kG denote the disjoint union of k copies of G. If G is a graph containing an edge e or a vertex v then $G - e$ and $G - v$ denote the graphs obtained from G by deleting e or v, respectively. If G' is a subgraph of G then $G - G'$ denotes the graph with vertex set $V(G)$ and edge set $E(G) \setminus E(G')$. We let C_n, K_n and P_n denote the cycle, complete graph and path on n vertices, respectively. A wheel is a cycle with an extra vertex added that is adjacent to all other vertices. The wheel on n vertices is denoted W_n; note that $W_4 = K_4$. A graph on at least three vertices is *2-connected* if it is connected and there is no vertex whose removal disconnects it. A *block* of a graph is a maximal subgraph that is connected and cannot be disconnected by the removal of one vertex (so a block is either 2-connected, a K_2 or an isolated vertex). A *block graph* is a connected graph in which every block is a complete graph. A *Gallai tree* is a connected graph in which every block is a complete graph or a cycle. We say that B is a *leaf-block* of a connected graph G if B contains exactly one cut vertex u of G and $B \setminus u$ is a component of $G - u$. For a set of graphs $\mathcal{H}$, a graph G is $\mathcal{H}$-*free* if G contains no induced subgraph isomorphic to a graph in $\mathcal{H}$, whereas G is $\mathcal{H}$-*subgraph-free* if it contains no subgraph isomorphic to a graph in $\mathcal{H}$. The *girth* of a graph is the length of its shortest cycle.

1.2 Known Results for Planar Graphs

We start with a classical result observed by Erdős et al. [9] and Vizing [22].

Theorem 1 ([9,22]). 2-LIST COLOURING *is polynomial-time solvable.*

Garey et al. proved the following result, which is in contrast to the fact that every planar graph is 4-colourable by the Four Colour Theorem [2].

Theorem 2 ([10]). 3-COLOURING *is* NP-*complete for planar graphs of maximum degree 4.*

Next we present results found by several authors on the existence of k-choosable graphs for various graph classes.

Theorem 3. *The following statements hold for k-choosability:*

 (i) *Every planar graph is 5-choosable* [20].
 (ii) *There exists a planar graph that is not 4-choosable* [24].
 (iii) *Every planar triangle-free graph is 4-choosable* [15].
 (iv) *Every planar graph with no 4-cycles is 4-choosable* [16].
 (v) *There exists a planar triangle-free graph that is not 3-choosable* [25].
 (vi) *There exists a planar graph with no 4-cycles, no 5-cycles and no intersecting triangles that is not 3-choosable* [19].
(vii) *Every planar bipartite graph is 3-choosable* [1].

We note that smaller examples of graphs than were used in the original proofs have been found for Theorems 3.(ii) [13], 3.(v) [18] and 3.(vi) [29] and that Theorem 3.(vi) strengthens a result of Voigt [26]. We recall that Thomassen [21] first showed that every planar graph of girth at least 5 is 3-choosable, and that a number of results have since been obtained on 3-choosability of planar graphs in which certain cycles are forbidden; see, for example, [4,7,27,28].

We will also use the following result of Chlebík and Chlebíková.

Theorem 4 ([5]). LIST COLOURING *is* NP-*complete for 3-regular planar bipartite graphs that have a list assignment in which each list is one of $\{1,2\}$, $\{1,3\}$, $\{2,3\}$, $\{1,2,3\}$ and all the neighbours of each vertex with three colours in its list have two colours in their lists.*

1.3 New Results for Planar Graphs

Theorems 1–3 have a number of immediate consequences for the complexity of ℓ-REGULAR LIST COLOURING when restricted to planar graphs. For instance, Theorem 2 implies that 3-REGULAR LIST COLOURING is NP-complete for planar graphs, whereas Theorem 3.(i) shows that 5-REGULAR LIST COLOURING is polynomial-time solvable on this graph class. As such, it is a natural question to determine the complexity for the missing case $\ell = 4$. In this section we settle this missing case and also present a number of new hardness results for ℓ-REGULAR LIST COLOURING restricted to various subclasses of planar graphs. At the end of this section we show how to combine the known results with our new ones to obtain a number of dichotomies (Corollaries 3–6). We deduce some of our new results from two more general theorems, namely Theorems 5 and 6, which we state below; see Sect. 2 for a proof of Theorem 5 (we omitted the proof of Theorem 6).

Theorem 5. *Let* $\mathcal{H}$ *be a finite set of 2-connected planar graphs. Then* 4-REGULAR LIST COLOURING *is* NP-*complete for planar* $\mathcal{H}$-*subgraph-free graphs if there exists a planar* $\mathcal{H}$-*subgraph-free graph that is not 4-choosable.*

Note that the class of $\mathcal{H}$-subgraph-free graphs is contained in the class of $\mathcal{H}$-free graphs. Hence, whenever a problem is NP-complete for $\mathcal{H}$-subgraph-free graphs, it is also NP-complete for $\mathcal{H}$-free graphs.

Combining Theorem 5 with Theorem 3.(ii) yields the following result which, as we will see later, was the only case for which the complexity of k-REGULAR LIST COLOURING for planar graphs was not settled.

Corollary 1. 4-REGULAR LIST COLOURING *is* NP-*complete for planar graphs.*

Theorem 5 has more applications. For instance, consider the non-4-choosable planar graph H from the proof of Theorem 1.7 in [13]. It can be observed that H is W_p-subgraph-free for all $p \geq 8$. Wheels are 2-connected and planar. Hence if $\mathcal{H}$ is any finite set of wheels on at least eight vertices then 4-REGULAR LIST COLOURING is NP-complete for planar $\mathcal{H}$-subgraph-free graphs.

Our basic idea for proving Theorem 5 is to pick a minimal counterexample H with list assignment L (which must exist due to Theorem 3.(ii)). We select an "appropriate" edge $e = uv$ and consider the graph $F' = F - e$. We reduce from an appropriate colouring problem restricted to planar graphs and use copies of F' as a gadget to ensure that we can enforce a regular list assignment. The proof of the next theorem also uses this idea.

Theorem 6. *Let* $\mathcal{H}$ *be a finite set of 2-connected planar graphs. Then* 3-REGULAR LIST COLOURING *is* NP-*complete for planar* $\mathcal{H}$-*subgraph-free graphs if there exists a planar* $\mathcal{H}$-*subgraph-free graph that is not 3-choosable.*

Theorem 6 has a number of applications. For instance, if we let $\mathcal{H} = \{K_3\}$ then Theorem 6, combined with Theorem 3.(v), leads to the following result.

Corollary 2. 3-REGULAR LIST COLOURING *is* NP-*complete for planar triangle-free graphs.*

Theorem 6 can also be used for other classes of graphs. For example, let $\mathcal{H}$ be a finite set of graphs, each of which includes a 2-connected graph on at least five vertices as a subgraph. Let $\mathcal{I}$ be the set of these 2-connected graphs. The graph K_4 is a planar $\mathcal{I}$-subgraph-free graph that is not 3-choosable (since it is not 3-colourable). Therefore, Theorem 6 implies that 3-REGULAR LIST COLOURING is NP-complete for planar $\mathcal{H}$-subgraph-free graphs. We can obtain more hardness results by taking some other planar graph that is not 3-choosable, such as a wheel on an even number of vertices. Also, if we let $\mathcal{H} = \{C_4, C_5\}$ we can use Theorem 6 by combining it with Theorem 3.(vi) to find that 3-REGULAR LIST COLOURING is NP-complete for planar graphs with no 4-cycles and no 5-cycles. We strengthen this result as follows (proof omitted).

Theorem 7. 3-REGULAR LIST COLOURING *is* NP-*complete for planar graphs with no 4-cycles, no 5-cycles and no intersecting triangles.*

Corollaries 1 and 2 and Theorem 7 can be seen as strengthenings of Theorems 3.(ii), 3.(v) and 3.(vi), respectively. Moreover, they complement Theorem 2, which implies that 3-LIST COLOURING is NP-complete for planar graphs, and a result of Kratochvíl [14] that, for planar bipartite graphs, 3-PRECOLOURING EXTENSION is NP-complete. Corollaries 1 and 2 also complement results of Gutner [13] who showed that 3-CHOOSABILITY and 4-CHOOSABILITY are Π_2^p-complete for planar triangle-free graphs and planar graphs, respectively. However, we emphasize that, for special graph classes, it is not necessarily the case that ℓ-CHOOSABILITY is computationally harder than ℓ-REGULAR LIST COLOURING. For instance, contrast the fact that CHOOSABILITY is polynomial-time solvable on $3P_1$-free graphs [11] with our next result (proof omitted).

Theorem 8. 3-REGULAR LIST COLOURING *is* NP-*complete for* $(3P_1, P_1 + P_2)$-*free graphs.*

Our new results, combined with known results, close a number of complexity gaps for the ℓ-REGULAR LIST COLOURING problem. Combining Corollary 1 with Theorems 1, 2 and 3.(i) gives us Corollary 3. Combining Theorem 7 with Theorems 1 and 3.(iv) gives us Corollary 4. Combining Corollary 2 with Theorems 1 and 3.(iii) gives us Corollary 5, whereas Theorems 1 and 3.(vii) imply Corollary 6.

Corollary 3. *Let ℓ be a positive integer. Then ℓ-REGULAR LIST COLOURING, restricted to planar graphs, is* NP-*complete if $\ell \in \{3, 4\}$ and polynomial-time solvable otherwise.*

Corollary 4. *Let ℓ be a positive integer. Then ℓ-REGULAR LIST COLOURING, restricted to planar graphs with no 4-cycles and no 5-cycles and no intersecting triangles, is* NP-*complete if $\ell = 3$ and polynomial-time solvable otherwise (even if we allow intersecting triangles and 5-cycles).*

Corollary 5. *Let ℓ be a positive integer. Then ℓ-REGULAR LIST COLOURING, restricted to planar triangle-free graphs, is* NP-*complete if $\ell = 3$ and polynomial-time solvable otherwise.*

Corollary 6. *Let ℓ be a positive integer. Then ℓ-REGULAR LIST COLOURING, restricted to planar bipartite graphs, is polynomial-time solvable.*

1.4   Known Results for Bounded Degree Graphs

First we present a result of Kratochvíl and Tuza [15].

Theorem 9 ([15]). LIST COLOURING *is polynomial-time solvable on graphs of maximum degree at most* 2.

Brooks' Theorem [3] states that every graph G with maximum degree d has a d-colouring unless G is a complete graph or a cycle with an odd number of vertices. The next result of Vizing [23] generalizes Brooks' Theorem to list colourings.

Theorem 10 ([23]). *Let d be a positive integer. Let $G = (V, E)$ be a connected graph of maximum degree at most d and let L be a d-regular list assignment for G. If G is not a cycle or a complete graph then G has a colouring that respects L.*

And we need another result of Chlebík and Chlebíková [5].

Theorem 11 ([5]). PRECOLOURING EXTENSION *is polynomial-time solvable on graphs of maximum degree* 3.

1.5   New Results for Bounded Degree Graphs

The following result is obtained by making a connection to Gallai trees (proof omitted).

Theorem 12. *Let k be a positive integer. Then k-PRECOLOURING EXTENSION is polynomial-time solvable for graphs of maximum degree at most k.*

We have the following two classifications. The first one is an observation obtained by combining only previously known results, whereas the second one also makes use of our new result.

Corollary 7. *Let d be a positive integer. The following two statements hold for graphs of maximum degree at most d.*

(i) LIST COLOURING *is NP-complete if $d \geq 3$ and polynomial-time solvable if $d \leq 2$.*

(ii) PRECOLOURING EXTENSION *and* COLOURING *are NP-complete if $d \geq 4$ and polynomial-time solvable if $d \leq 3$.*

Proof. We first consider (i). If $d \geq 3$, we use Theorem 4. If $d \leq 2$, we use Theorem 9. We now consider (ii). If $d \geq 4$, we use Theorem 2. If $d \leq 3$, we use Theorem 11.   □

Corollary 8. *Let d and k be two positive integers. The following two statements hold for graphs of maximum degree at most d.*

(i) k-LIST COLOURING *and* LIST k-COLOURING *are NP-complete if $k \geq 3$ and $d \geq 3$ and polynomial-time solvable otherwise.*

(ii) k-REGULAR LIST COLOURING and k-PRECOLOURING EXTENSION are NP-complete if $k \geq 3$ and $d \geq k + 1$ and polynomial-time solvable otherwise.

Proof. We first consider (i). If $k \geq 3$ and $d \geq 3$, we use Theorem 4. If $k \leq 2$ or $d \leq 2$, we use Theorems 1 or 9, respectively.

We now consider (ii). We start with the hardness cases and so let $k \geq 3$ and $d \geq k + 1$.

First consider k-PRECOLOURING EXTENSION. Theorem 2 implies that 3-COLOURING is NP-complete for graphs of maximum degree at most d for all $d \geq 4$. The $k = 3$ case follows immediately from this result. Suppose $k \geq 4$ and $d \geq k + 1$. Consider a graph G of maximum degree 4. For each vertex v, we add $k - 3$ new vertices $x_1^v, \ldots, x_{k-3}^v$ and edges $vx_1^v, \ldots, vx_{k-3}^v$. Let G' be the resulting graph. Note that G' has maximum degree at most $4 + k - 3 = k + 1 \leq d$. We define a precolouring c on the newly added vertices by assigning colour $i + 3$ to each x_i^v. Then G' has a k-colouring extending c if and only if G has a 3-colouring.

Now consider k-REGULAR LIST COLOURING. The $k = 3$ case follows immediately from Theorem 2. Suppose $k \geq 4$ and $d \geq k + 1$. Consider a graph G of maximum degree 4. We define the list $L(v) = \{1, \ldots, k\}$ for each vertex $v \in V(G)$. For each vertex v, we add $k - 3$ new vertices $x_1^v, \ldots, x_{k-3}^v$ and edges $vx_1^v, \ldots, vx_{k-3}^v$. We define the list $L(x_i^v) = \{i, k+1, k+2, \ldots, 2k-1\}$ for each x_i^v. For each vertex x_i^v, we also add k new vertices $w_1(x_i^v), \ldots, w_k(x_i^v)$ and edges such that $x_i^v, w_1(x_i^v), \ldots, w_k(x_i^v)$ form a clique (on $k + 1$ vertices). We define the list $L(w_j(x_i^v)) = \{k + 1, \ldots, 2k\}$ for each $w_j(x_i^v)$. Let G' be the resulting graph. Note that G' has maximum degree at most $k + 1$ and that the resulting list assignment L is a k-regular list assignment of G'. Then G' has a k-colouring respecting L if and only if G has a 3-colouring.

We continue with the polynomial-time solvable cases. If $k \leq 2$, the result follows from Theorem 1. Suppose that $k \geq 3$ and $d \leq k$. Then the result for k-REGULAR LIST COLOURING follows from Theorems 9 and 10 and the result for k-PRECOLOURING EXTENSION follows from Theorem 12.   □

Note that Corollary 8 does not contain a dichotomy for k-COLOURING restricted to graphs of maximum degree at most d. A full classification of this problem is open, but a number of results are known. Molloy and Reed [17] classified the complexity for all pairs (k, d) for sufficiently large d. Emden-Weinert et al. [8] proved that k-COLOURING is NP-complete for graphs of maximum degree at most $k + \lceil \sqrt{k} \rceil - 1$.

2   The Proof of Theorem 5

We need an additional result (proof omitted).

Theorem 13. *For every integer $p \geq 3$, 3-LIST COLOURING is NP-complete for planar graphs of girth at least p that have a list assignment in which each list is one of $\{1, 2\}$, $\{1, 3\}$, $\{2, 3\}$, $\{1, 2, 3\}$.*

We are now ready to prove Theorem 5, which we restate below.

Theorem 5 (Restated). *Let $\mathcal{H}$ be a finite set of 2-connected planar graphs. Then* 4-REGULAR LIST COLOURING *is* NP-*complete for planar $\mathcal{H}$-subgraph-free graphs if there exists a planar $\mathcal{H}$-subgraph-free graph that is not 4-choosable.*

Proof. The problem is readily seen to be in NP. Let F be a planar $\mathcal{H}$-subgraph-free graph with a 4-regular list assignment L such that F has no colouring respecting L. We may assume that F is minimal (with respect to the subgraph relation). In particular, this means that F is connected. Let r be the length of a longest cycle in any graph of $\mathcal{H}$. We reduce from the problem of 3-LIST COLOURING restricted to planar graphs of girth at least $r + 1$ in which each vertex has list $\{1,2\}$, $\{1,3\}$, $\{2,3\}$ or $\{1,2,3\}$. This problem is NP-complete by Theorem 13. Let a graph G and list assignment L_G be an instance of this problem. We will construct a planar $\mathcal{H}$-subgraph-free graph G' with a 4-regular list assignment L' such that G has a colouring that respects L_G if and only if G' has a colouring that respects L'.

If every pair of adjacent vertices in F has the same list, then the problem of finding a colouring that respects L is just the problem of finding a 4-colouring which, by the Four Colour Theorem [2], we know is possible. Thus we may assume that, on the contrary, there is an edge $e = uv$ such that $|L(u) \cap L(v)| \leq 3$. Let $F' = F - e$. Then, by minimality, F' has at least one colouring respecting L, and moreover, for any colouring of F' that respects L, u and v are coloured alike (otherwise we would have a colouring of F that respects L). Let T be the set of possible colours that can be used on u and v in colourings of F' that respect L and let $t = |T|$. As $T \subseteq L(u) \cap L(v)$, we have $1 \leq t \leq 3$. Up to renaming the colours in L, we can build copies of F' with 4-regular list assignments such that

(i) the set T is any given list of colours of size t, and
(ii) the vertex corresponding to u has any given list of 4 colours containing T.

We will implicitly make use of this several times in the remainder of the proof.

We say that a vertex w in G is a *bivertex* or *trivertex* if $|L_G(w)|$ is 2 or 3, respectively. We construct a planar $\mathcal{H}$-subgraph-free graph G' and 4-regular list assignment L' as follows.

First suppose that $t = 1$. For each bivertex w in G, we do as follows. We add two copies of F' to G, which we label $F_1(w)$ and $F_2(w)$. The vertex in $F_i(w)$ corresponding to u is labelled u_i^w for $i \in \{1,2\}$ and we set $U^w = \{u_1^w, u_2^w\}$. We add the edges wu_1^w and wu_2^w. We give list assignments to the vertices of $F_1(w)$ and $F_2(w)$ such that $T = \{4\}$ for F_1 and $T = \{5\}$ for F_2. We let $L'(w) = L_G(w) \cup \{4,5\}$. For each trivertex w in G, we do as follows. We add one copy of F' to G, which we label $F_1(w)$. The vertex in $F_1(w)$ corresponding to u is labelled u_1^w and we set $U^w = \{u_1^w\}$. We add the edge wu_1^w. We give list assignments to vertices of $F_1(w)$ such that $T = \{4\}$ for F_1. We let $L'(w) = L_G(w) \cup \{4\}$. This completes the construction of G' and L' when $t = 1$.

Now suppose that $t = 2$. Let $s = r$ if r is even and $s = r + 1$ if r is odd (so s is even in both cases). For each bivertex w in G, we do as follows.

We add a copy of F' to G, which we label $F_1(w)$, and identify the vertex in $F_1(w)$ corresponding to u with w. We give list assignments to vertices of $F_1(w)$ such that $T = L_G(w)$ and $L'(w) = L_G(w) \cup \{4,5\}$. For each trivertex w in G, we do as follows. We add s copies of F' to G which we label $F_i(w)$, $1 \leq i \leq s$. The vertex in $F_i(w)$ corresponding to u is labelled u_i^w. Let $U^w = \{u_i^w \mid 1 \leq i \leq s\}$. Add edges such that the union of w and U^w induces a cycle on $s+1$ vertices. For all $1 \leq i \leq s$, we give list assignments to vertices of $F_i(w)$ such that $T = \{4,5\}$. We let $L'(w) = \{1,2,3,4\}$. This completes the construction of G' and L' when $t = 2$.

Now suppose that $t = 3$. For each bivertex w in G, we do as follows. We add two copies of F' to G which we label $F_1(w)$ and $F_2(w)$, such that for $i \in \{1,2\}$, the vertex in $F_i(w)$ corresponding to u is identified with w. We give list assignments to vertices of $F_1(w)$ and $F_2(w)$ such that $T = L_G(w) \cup \{4\}$ for $F_1(w)$, $T = L_G(w) \cup \{5\}$ for $F_2(w)$ and $L'(w) = L_G(w) \cup \{4,5\}$. For each trivertex w in G, we do as follows. We add a copy of F' to G which we label $F_1(w)$, such that the vertex in $F_1(w)$ corresponding to u is identified with w. We give list assignments to the vertices of $F_1(w)$ such that $T = \{1,2,3\}$ and $L'(w) = \{1,2,3,4\}$. This completes the construction of G' and L' when $t = 3$.

Note that G' is planar. Suppose that there is a subgraph H in G' that is isomorphic to a graph of $\mathcal{H}$. Since F is $\mathcal{H}$-subgraph-free, and since F' is obtained from F by removing one edge, F' is also $\mathcal{H}$-subgraph-free. Therefore for all w, H is not fully contained in any $F_i(w)$. Since H is 2-connected and since for all w only one vertex of any $F_i(w)$ has a neighbour outside of $F_i(w)$, we find that H has at most one vertex in each $F_i(w)$. In particular, H cannot contain any vertex of any $F_i(w)$ in which the vertex corresponding to u has been attached to w (as opposed to being identified with w); this includes the case when the union of w and U^w induces a cycle on $s + 1$ vertices. Hence we have found that H is a subgraph of G, which contradicts the fact that G has girth at least $r + 1$. Therefore G' is $\mathcal{H}$-subgraph-free.

Note that in any colouring of G' that respects L', each copy of F' must be coloured such that the vertices corresponding to u and v have the same colour, which must be one of the colours from the corresponding set T. If $t = 1$ and w is a trivertex, this means that the unique neighbour of w in U^w must be coloured with colour 4, so w cannot be coloured with colour 4. Similarly, if $t = 1$ and w is a bivertex or $t = 2$ and w is a trivertex then the two neighbours of w in U^w must be coloured with colours 4 and 5, so w cannot be coloured with colours 4 or 5. If $t = 2$ and w is a bivertex or $t = 3$ and w is a trivertex then w belongs to a copy of F' with $T = L_G(w)$, so w cannot have colour 4 or 5. If $t = 3$ and w is a bivertex then w belongs to two copies of F', one with $T = L_G(w) \cup \{4\}$ and one with $T = L_G(w) \cup \{5\}$. Therefore, w must be coloured with a colour from the intersection of these two sets, that is it must be coloured with a colour from $L_G(w)$. Therefore none of the vertices of G can be coloured 4 or 5. Thus the problem of finding a colouring of G' that respects L' is equivalent to the problem of finding a colouring of G that respects L_G. This completes the proof. $\qquad\square$

3   Conclusions

As well as filling the complexity gaps of a number of colouring problems for graphs with bounded maximum degree, we have given several dichotomies for the k-REGULAR LIST COLOURING problem restricted to subclasses of planar graphs. In particular we showed NP-hardness of the cases $k = 3$ and $k = 4$ restricted to planar $\mathcal{H}$-subgraph-free graphs for several sets $\mathcal{H}$ of 2-connected planar graphs. Our method implies that for such sets $\mathcal{H}$ it suffices to find a counterexample to 3-choosability or to 4-choosability, respectively. It is natural to ask whether we can determine the complexity of 3-REGULAR LIST COLOURING and 4-REGULAR LIST COLOURING for any class of planar $\mathcal{H}$-subgraph-free graphs. However, we point out that even when restricting $\mathcal{H}$ to be a finite set of 2-connected planar graphs, this would be very hard (and beyond the scope of this paper) as it would require solving several long-standing conjectures in the literature. For example, when $\mathcal{H} = \{C_4, C_5, C_6\}$, Montassier [18] conjectured that every planar $\mathcal{H}$-subgraph-free graph is 3-choosable.

A drawback of our method is that we need the set of graphs $\mathcal{H}$ to be 2-connected. If we forbid a set $\mathcal{H}$ of graphs that are not 2-connected, the distinction between polynomial-time solvable and NP-complete cases is not clear, and both cases may occur even if we forbid only one graph.

Acknowledgements. We thank Steven Kelk for helpful comments on an earlier version of this paper.

References

1. Alon, N., Tarsi, M.: Colorings and orientations of graphs. Combinatorica **12**, 125–134 (1992)
2. Appel, K., Haken, W.: Every planar map is four colorable, Contemporary Mathematics 89, AMS Bookstore (1989)
3. Brooks, R.L.: On colouring the nodes of a network. Math. Proc. Camb. Philos. Soc. **37**, 194–197 (1941)
4. Chen, M., Montassier, M., Raspaud, A.: Some structural properties of planar graphs and their applications to 3-choosability. Discrete Math. **312**, 362–373 (2012)
5. Chlebík, M., Chlebíková, J.: Hard coloring problems in low degree planar bipartite graphs. Discrete Appl. Math. **154**, 1960–1965 (2006)
6. Chudnovsky, M.: Coloring graphs with forbidden induced subgraphs. Proc. ICM **IV**, 291–302 (2014)
7. Dvořák, Z., Lidický, B., Škrekovski, R.: Planar graphs without 3-, 7-, and 8-cycles are 3-choosable. Discrete Math. **309**, 5899–5904 (2009)
8. Emden-Weinert, T., Hougardy, S., Kreuter, B.: Uniquely colourable graphs and the hardness of colouring graphs of large girth. Comb. Probab. Comput. **7**, 375–386 (1998)
9. Erdős, P., Rubin, A.L., Taylor, H.: Choosability in graphs. In: Proceedings of the West Coast Conference on Combinatorics, Graph Theory and Computing (Humboldt State Univ., Arcata, Calif., 1979), Congress. Numer., XXVI, pp. 125–157. Winnipeg, Man., Utilitas Math. (1980)

10. Garey, M.R., Johnson, D.S., Stockmeyer, L.J.: Some simplified NP-complete graph problems. In: Proceedings of STOC, pp. 47–63 (1974)
11. Golovach, P.A., Heggernes, P., van 't Hof, P., Paulusma, D.: Choosability on H-free graphs. Inform. Process. Lett. **113**, 107–110 (2013)
12. Golovach, P.A., Johnson, M., Paulusma, D., Song, J.: A survey on the computational complexity of colouring graphs with forbidden subgraphs, Manuscript, arXiv: 1407.1482v4 (2014)
13. Gutner, S.: The complexity of planar graph choosability. Discrete Math. **159**, 119–130 (1996)
14. Kratochvíl, J.: Precoloring extension with fixed color bound. Acta Mathematica Universitatis Comenianae **62**, 139–153 (1993)
15. Kratochvíl, J., Tuza, Z.: Algorithmic complexity of list colourings. Discrete Appl. Math. **50**, 297–302 (1994)
16. Lam, P.C.B., Xu, B., Liu, J.: The 4-choosability of plane graphs without 4-cycles. J. Comb. Theory Ser. B **76**, 117–126 (1999)
17. Molloy, M., Reed, B.: Colouring graphs when the number of colours is almost the maximum degree. J. Comb. Theory Ser. B **109**, 134–195 (2014)
18. Montassier, M.: A note on the not 3-choosability of some families of planar graphs. Inform. Process. Lett. **99**, 68–71 (2006)
19. Montassier, M., Raspaud, A., Wang, W.: Bordeaux 3-color conjecture and 3-choosability. Discrete Math. **306**, 573–579 (2006)
20. Thomassen, C.: Every planar graph is 5-choosable. J. Comb. Theory Ser. B **62**, 180–181 (1994)
21. Thomassen, C.: 3-List-coloring planar graphs of girth 5. J. Comb. Theory Ser. B **64**, 101–107 (1995)
22. Vizing, V.G.: Coloring the vertices of a graph in prescribed colors. In: Diskret. Analiz., no. 29, Metody Diskret. Anal. v. Teorii Kodov i Shem, vol. 101, pp. 3–10 (1976)
23. Vizing, V.G.: Vertex colorings with given colors. Diskret. Analiz. **29**, 3–10 (1976)
24. Voigt, M.: List colourings of planar graphs. Discrete Math. **120**, 215–219 (1993)
25. Voigt, M.: A not 3-choosable planar graph without 3-cycles. Discrete Math. **146**, 325–328 (1995)
26. Voigt, M.: A non-3-choosable planar graph without cycles of length 4 and 5. Discrete Math. **307**, 1013–1015 (2007)
27. Wang, Y., Lu, H., Chen, M.: Planar graphs without cycles of length 4, 5, 8, or 9 are 3-choosable. Discrete Math. **310**, 147–158 (2010)
28. Wang, Y., Lu, H., Chen, M.: Planar graphs without cycles of length 4, 7, 8, or 9 are 3-choosable. Discrete Appl. Math. **159**, 232–239 (2011)
29. Wang, D.-Q., Wen, Y.-P., Wang, K.-L.: A smaller planar graph without 4-, 5-cycles and intersecting triangles that is not 3-choosable. Inform. Process. Lett. **108**, 87–89 (2008)

Combinatorial Properties of Full-Flag Johnson Graphs

Irving Dai[1](✉)

Department of Mathematics, Princeton University, Princeton, NJ 08544, USA
idai@math.princeton.edu

Abstract. The Johnson graphs $J(n, k)$ are a well-known family of combinatorial graphs whose applications and generalizations have been studied extensively in the literature. In this paper, we present a new variant of the family of Johnson graphs, the Full-Flag Johnson graphs, and discuss their combinatorial properties. We show that the Full-Flag Johnson graphs are Cayley graphs on S_n generated by certain well-known classes of permutations and that they are in fact generalizations of permutahedra. Our main result will be to establish a tight $\Theta(n^2/k^2)$ bound for the diameter of the Full-Flag Johnson graph $FJ(n, k)$.

Keywords: Johnson graph · Permutahedron · Irreducible permutation · Cayley graph · Diameter

1 Introduction

For positive integers n and k with $k < n$, the Johnson graph $J(n, k)$ is an undirected graph whose vertex set is given by the collection of all k-element subsets of $\{1, 2, \ldots, n\}$. Two vertices are adjacent if and only if their intersection has size $k - 1$. For example, two vertices in $J(4, 3)$ are $u = \{1, 2, 3\}$ and $v = \{1, 3, 4\}$; u and v are adjacent since $|u \cap v| = 2$. The Johnson graphs are known to be Hamiltonian [1] and their spectra are given by the Eberlein polynomials [14]. Different generalizations of Johnson graphs and various related families have been studied by several authors (see e.g. [2,6]).

The permutahedra are another well-known family of combinatorial graphs. For a positive integer n, the permutahedron of order n has vertex set consisting of all permutations of $(1, 2, \ldots, n)$. Two vertices are adjacent if and only if they are of the form $(u_1, u_2, \ldots, u_i, u_{i+1}, \ldots, u_n)$ and $(u_1, u_2, \ldots, u_{i+1}, u_i, \ldots, u_n)$, respectively. That is, u and v are adjacent if they differ by a permutation which transposes two consecutive elements. (We call such transpositions "neighboring transpositions"). Permutahedra appear frequently in geometric combinatorics and are Hamiltonian by the Steinhaus-Johnson-Trotter algorithm [10]. Similar combinatorial families of graphs (in particular, associahedra) and their generalizations appear widely in algebra and discrete mathematics (see e.g. [5,17]).

In this paper, we present and discuss some combinatorial properties of a new variant of the set of Johnson graphs, the Full-Flag Johnson graphs. Roughly

© Springer International Publishing Switzerland 2016
Z. Lipták and W.F. Smyth (Eds.): IWOCA 2015, LNCS 9538, pp. 112–123, 2016.
DOI: 10.1007/978-3-319-29516-9_10

speaking, the Full-Flag Johnson graphs are constructed by imposing (an index-shifted version of) the Johnson graph adjacency condition on the collection of all full-flags of $\{1, 2, \ldots, n\}$. We show that Full-Flag Johnson graphs are Cayley graphs on the symmetric group S_n generated by certain classes of permutations, and that they are in fact generalizations of permutahedra.

Our main result will be to derive a tight $\Theta(n^2/k^2)$ bound for the diameter of the Full-Flag Johnson graph $FJ(n, k)$. Several authors have studied bounds for the diameters of generalized Johnson graphs and associahedra; see for example Bautista-Santiago et al. on the diameter of generalized Johnson graphs [18] and Manneville and Pilaud on the graph-theoretic properties of graph associahedra [15]. In our case, we take a rather different approach to bounding the diameter of $FJ(n, k)$ by translating the question into the *minimum-length generator problem*: given a permutation $\sigma \in S_n$ and fixed subset S of S_n, how many applications of elements of S suffice to sort σ?

As might be expected, in general this problem is difficult (NP-hard [8]). Bounds for specific instances of S have been studied widely in the literature; for example: sorting by reversals, cyclic transpositions [9], block transpositions [3], bounded-block transpositions [11], and so on. It turns out that for the class of permutations at hand, utilizing a parallel sorting algorithm of Baudet and Stevenson [4] allows us to easily obtain an upper bound on the number of generators needed and thus the diameter of $FJ(n, k)$.

Many of the introductory results and lemmas were proven jointly with Michael Greenberg, Noah Schoem, and Matt Tanzer at the Program in Mathematics for Young Scientists (Boston University, Summer 2010). The author would like to thank Paul Gunnells for formulating and proposing that project, out of which this paper eventually grew. The author would also like to thank Ho-Kwok Dai for various helpful conversations and ideas during the course of writing this paper.

2   Definitions and Examples

Let n be a positive integer. Denote by $[n]$ and (n) the unordered and ordered sets $\{1, 2, \ldots, n\}$ and $(1, 2, \ldots, n)$, respectively. A *full-flag of subsets* of $[n]$ is a sequence of nested subsets $U = (U_1, U_2, \ldots, U_n)$ of $[n]$ such that $|U_i| = i$ for all $i \in [n]$ and $U_i \subsetneq U_{i+1}$ for all $i \in [n-1]$. For example, one full-flag of subsets of $\{1, 2, 3, 4\}$ is $(\{3\}, \{3, 1\}, \{3, 1, 2\}, \{3, 1, 2, 4\})$. For a non-negative integer k such that $k < n$, the *Full-Flag Johnson graph* $FJ(n, k)$ has vertex set given by the collection of all possible full-flags of $[n]$. Two vertices $U = (U_1, U_2, \ldots, U_n)$ and $V = (V_1, V_2, \ldots, V_n)$ are adjacent in $FJ(n, k)$ if and only if $U_i \neq V_i$ for exactly k integers $i \in [n]$. If we view U and V as collections of subsets of $[n]$, then U and V are adjacent if and only if $|U \cap V| = n - k$.

The following equivalent definition of $FJ(n, k)$ simplifies the vertex set at the expense of complicating the adjacencies. Let $U = (U_1, U_2, \ldots, U_n)$ be any vertex in $FJ(n, k)$. For each $1 < i \leq n$, the difference $U_i - U_{i-1}$ is a singleton element which we denote by u_i. Letting u_1 be the singleton element of U_1, we may identify

U uniquely with the sequence $u = (u_1, u_2, \ldots, u_n)$. It is clear that u must be a permutation of (n) and that $U_i = \{u_1, u_2, \ldots, u_i\}$. Since every permutation of (n) corresponds to a full-flag of subsets of $[n]$ in this manner, we may view the vertex set of $FJ(n, k)$ as the collection of all permutations of (n). Two vertices $u = (u_1, u_2, \ldots, u_n)$ and $v = (v_1, v_2, \ldots, v_n)$ are adjacent if and only if there exist exactly k positive integers $i \in [n]$ such that $\{u_1, u_2, \ldots, u_i\} \neq \{v_1, v_2, \ldots, v_i\}$.

Example 1. Let $u = (1, 2, 3, 4, 5)$ and $v = (2, 1, 3, 5, 4)$ be two vertices in $FJ(5, 2)$. Then u and v are adjacent since $\{u_1, u_2, \ldots, u_i\} \neq \{v_1, v_2, \ldots, v_i\}$ for exactly two values of $i \in [5]$ ($i = 1$ and $i = 4$).

Given any permutation $u = (u_1, u_2, \ldots, u_n)$ of (n), we use $u(i)$ to denote the set $\{u_1, u_2, \ldots, u_i\}$. Occasionally, we will abuse this notation slightly and allow i to be 0, with the understanding that $u(0)$ is the empty set. Some algebraic rules for these sets are immediately evident. For instance, since the elements of the sequence u are distinct, for all x and y such that $0 \leq x < y \leq n$, we have $u(y) - u(x) = \{u_{x+1}, u_{x+2}, \ldots, u_{y-1}, u_y\}$. Thus, if $u(x) = v(x)$ and $u(y) = v(y)$, then clearly $u(y) - u(x) = v(y) - v(x)$.

It is easily seen that the Full-Flag Johnson graph $FJ(n, 0)$ is the isolated graph, in which each vertex is adjacent only to itself. Less trivially, we have:

Lemma 1. *The graph $FJ(n, 1)$ is the permutahedron of order n, in which two vertices are adjacent if and only if they are related by a neighboring transposition.*

Proof. Let n be any positive integer. Two vertices u and v are adjacent in $FJ(n, 1)$ if and only if $u(i) \neq v(i)$ for exactly one $i \in [n]$. In particular, $u(x) = v(x)$ for each positive integer $x < i$, which implies $u_1 = v_1$, $u_2 = v_2, \ldots, u_{i-1} = v_{i-1}$. Since $u(i-1) = v(i-1)$ but $u(i) \neq v(i)$, we have $u_i \neq v_i$. Hence the first $i - 1$ elements of the sequence u are equal to the first $i - 1$ elements of the sequence v, and the ith elements of u and v differ.

Now, it cannot be that $i = n$, since $u(n) = v(n) = \{1, 2, \ldots, n\}$. Thus $i < n$, and $u(i+1) = v(i+1)$. Since $u(i-1) = v(i-1)$, we then have $\{u_i, u_{i+1}\} = \{v_i, v_{i+1}\}$. But $u_i \neq v_i$, so it must be that $v_i = u_{i+1}$ and $v_{i+1} = u_i$.

Finally, $u(x) = v(x)$ for all $x > i$, which implies $u_{i+2} = v_{i+2}, u_{i+3} = v_{i+3}, \ldots, u_n = v_n$. Hence u and v are of the form

$$u = (u_1, u_2, \ldots, u_i, u_{i+1}, \ldots, u_n) \text{ and}$$
$$v = (u_1, u_2, \ldots, u_{i+1}, u_i, \ldots, u_n),$$

respectively; that is, they are related by a neighboring transposition. Conversely, it is evident that two vertices related by a neighboring transposition are adjacent in $FJ(n, 1)$. □

We close this section with a rough generalization of Lemma 1 that may help give some intuition for Full-Flag Johnson graphs in the general case. A much more precise combinatorial description of $FJ(n, k)$ will be formulated presently, but the following lemma will prove useful in subsequent sections.

Let u and v be two permutations of (n). We say that u and v are related by a k-*neighboring permutation* if they are of the form:

$$u = (u_1, u_2, \ldots, u_{i-1}, u_i, u_{i+1}, \ldots, u_{i+k-1}, u_{i+k}, \ldots u_n) \text{ and}$$
$$v = (u_1, u_2, \ldots, u_{i-1}, u_i', u_{i+1}', \ldots, u_{i+k-1}', u_{i+k}, \ldots u_n),$$

where $(u_i', u_{i+1}', \ldots, u_{i+k-1}')$ is any permutation of $(u_i, u_{i+1}, \ldots, u_{i+k-1})$. This generalizes the notion of a neighboring transposition, which (in this terminology) is a 2-neighboring permutation.

Lemma 2. *Let u and v be two permutations of (n) related by a $(k+1)$-neighboring permutation. Then u and v are connected by a path of no more than two edges in $FJ(n,k)$.*

Proof. Let u and v be two permutations of (n) related by a $(k+1)$-neighboring transposition, so that

$$u = (u_1, u_2, \ldots, u_i, u_{i+1}, \ldots, u_{i+k}, \ldots u_n) \text{ and}$$
$$v = (u_1, u_2, \ldots, u_i', u_{i+1}', \ldots, u_{i+k}', \ldots u_n),$$

where $(u_i', u_{i+1}', \ldots, u_{i+k}')$ is a permutation of $(u_i, u_{i+1}, \ldots, u_{i+k})$. If $u_i = u_{i+k}'$, we claim that u and v are already adjacent in $FJ(n,k)$. Indeed, suppose that this is the case. Then it is easily checked that $u(x) = v(x)$ for all $x < i$ and $x \geq i + k$. On the other hand, $u(x) \neq v(x)$ for all $i \leq x < i + k$, since $u_i \in u(x)$ but $u_i = u_{i+k}' \notin v(x)$ for these x. Hence $u(x) \neq v(x)$ for exactly k indices x, showing that u and v are adjacent in $FJ(n,k)$.

Thus we may assume $u_i \neq u_{i+k}'$. Choose any permutation of $(u_i, u_{i+1}, \ldots, u_{i+k})$ that begins with u_{i+k}' and ends with u_i. Let this permutation be denoted by $(w_i, w_{i+1}, \ldots, w_{i+k})$. Let w be the vertex constructed by replacing the subsequence $(u_i, u_{i+1}, \ldots, u_{i+k})$ in u with the sequence $(w_i, w_{i+1}, \ldots, w_{i+k})$; i.e.,

$$w = (u_1, u_2, \ldots, w_i = u_{i+k}', w_{i+1}, \ldots, w_{i+k} = u_i, \ldots, u_n).$$

Then the argument of the above paragraph shows that u is adjacent to w and w is adjacent to v. This completes the proof. □

Lemma 2 immediately allows us to prove:

Lemma 3. *All non-trivial Full-Flag Johnson graphs are connected.*

Proof. Let $FJ(n,k)$ be a non-trivial Full-Flag Johnson graph. Given two arbitrary permutations u and v, we may form a sequence of permutations beginning with u and ending with v such that each pair of consecutive permutations is related by a neighboring transposition. Thus, to show that u and v are connected by a path in $FJ(n,k)$, it suffices to show that any two permutations related by a neighboring transposition are connected. So long as $k \geq 1$, every 2-neighboring permutation is also a $(k+1)$-neighboring permutation (where the permutation of $k+1$ elements in question fixes $k-1$ of them). Applying Lemma 2 completes the proof. □

3   Combinatorial Interpretation

Let σ be a permutation in S_n. Recall that σ is said to be *reducible* if there exists an index $i \in [n-1]$ such that $\sigma([i]) = [i]$. If no such index exists, then we say that σ is *irreducible*. In general, to every permutation σ we may associate the unique finest partition $[n] = I_1 \cup \cdots \cup I_m$ of $[n]$ such that $\sigma(I_x) = I_x$ for each $x \in [m]$. If this partition is of cardinality m, then we call σ an *m-reducible permutation*. It is easily seen that the maximality of the partition (with respect to subdivision) is equivalent to σ acting on each I_x as an irreducible permutation. For further details, see e.g. [13].

Example 2. The permutation $\sigma = (2,1,3)$ is 2-reducible, since $\sigma(\{1,2\}) = \{2,1\}$ and $\sigma(\{3\}) = \{3\}$. The permutation $\sigma = (2,3,1)$ is irreducible. Note that every irreducible permutation is said to be "1-reducible".

The class of m-reducible permutations was introduced by Comtet [7], who gave an enumeration of the number of m-reducible permutations in S_n using generating functions. Irreducible permutations in particular are well-studied combinatorial objects and have appeared extensively in combinatorics as well as occasionally in ergodic theory [12] and number theory [16]. The appearance of these permutations is suggestive of further interesting structure in $FJ(n,k)$.

Lemma 4. *Two vertices in $FJ(n,k)$ are adjacent if and only if one is an $(n-k)$-reducible permutation of the other.*

Proof. Let $u = (u_1, \ldots u_n)$ and $v = (v_1, \ldots, v_n)$ be two permutations of (n). Define the set of indices

$$I = \{i \in [n] \mid u(i) = v(i)\},$$

and enumerate the elements of I in ascending order. Consider any two successive integers in this enumeration, say x and y. Since $u(x) = v(x)$ and $u(y) = v(y)$, we see that the subsequence $(v_{x+1}, v_{x+2}, \ldots, v_y)$ of v is a permutation of the subsequence $(u_{x+1}, u_{x+2}, \ldots, u_y)$ of u. We claim that these two subsequences are in fact irreducible permutations of each other. Indeed, it is evident that otherwise there would be some index z with $x < z < y$ such that $u(z) = v(z)$, contradicting the fact that x and y are successive elements in our ordering of I.

With this in mind, let our enumeration of I be given by $i_1, i_2, \ldots, i_m$. Note that since $u(n) = v(n)$, we must have $i_m = n$. Setting $i_0 = 0$, we partition the index set $\{1, 2, \ldots, n\}$ into m intervals as follows:

$$\{1, 2, \ldots, n\} = \bigcup_{r=0}^{m-1} \{i_r + 1, i_r + 2, \ldots, i_{r+1}\}.$$

Then the subsequences of u and v corresponding to any one partition are irreducible permutations of each other. That is, for each $r = 0, 1, \ldots, m-1$, the subsequence $(v_{i_r+1}, v_{i_r+2}, \ldots, v_{i_{r+1}})$ of v is an irreducible permutation of the subsequence $(u_{i_r+1}, u_{i_r+2}, \ldots, u_{i_{r+1}})$ of u. Hence v is an m-reducible permutation of u, where m is the cardinality of I.

If u and v are adjacent in $FJ(n,k)$ then the cardinality of I is by definition $n - k$. Hence v is an $(n-k)$-reducible permutation of u. Conversely, it is easily seen that if u and v are $(n-k)$-reducible permutations of each other, then they are adjacent in $FJ(n,k)$. This completes the proof. □

We formalize our work so far in:

Theorem 1. *The full-flag Johnson graph $FJ(n,k)$ is the Cayley graph on S_n generated by the class of $(n-k)$-reducible permutations.*

It is easily checked that the class of n-reducible permutations in S_n consists precisely of the identity, and the class of $(n-1)$-reducible permutations consists of all neighboring transpositions. Hence both Lemma 1 and the trivial case $k = 0$ follow immediately from Theorem 1.

Example 3. Consider the Full-Flag Johnson graph $FJ(4,2)$. The 2-reducible permutations of S_4 are:

$$\{(123), (213), (13), (234), (324), (24), (12)(34)\}.$$

(Here we are using the standard cycle notation). We thus see that $FJ(4,2)$ is a Cayley graph of regularity seven; see Fig. 1 on the following page.

4 Lower Bound for the Diameter

In this section, we investigate the diameter of $FJ(n,k)$ as a function of the input parameters n and k. We first consider the extremal cases $k = 1$ and $k = n - 1$, for which exact expressions are easily derived. Indeed, it is well-known that the permutahedron of order n has a diameter of $\binom{n}{2}$ for all $n \geq 2$. For $k = n - 1$, we have the following result:

Lemma 5. *The diameter of $FJ(n, n-1)$ is 2 for all $n \geq 3$.*

Proof. Since any two permutations of (n) are clearly related by an n-neighboring permutation, Lemma 2 shows that every two vertices of $FJ(n, n-1)$ are connected by a path of at most two edges. Moreover, if $n \geq 3$, it is possible to find two vertices in $FJ(n, n-1)$ that are not actually adjacent. (Choose for example two distinct permutations that both begin with the same element). This completes the proof. □

The interesting cases are thus when k is a small (but appreciable) fraction of n. We begin by bounding the diameter of $FJ(n,k)$ from below. Given a permutation $\sigma \in S_n$, recall that the disorder $f(\sigma)$ of σ is defined to be the number of inversions in σ; that is, the number of index pairs (i,j) such that $i < j$ but $\sigma(i) > \sigma(j)$. The following lemma estimates how much the value of f can change when traversing an edge in $FJ(n,k)$:

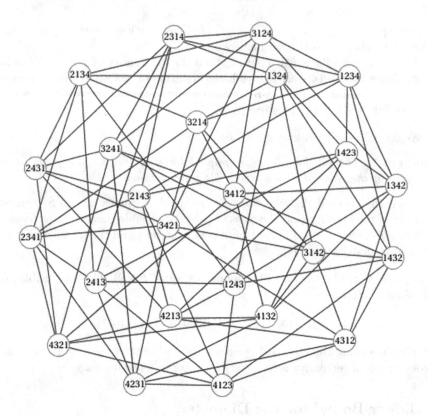

Fig. 1. $FJ(4, 2)$.

Lemma 6. *Let $\sigma \in S_n$ be an m-reducible permutation. Then any application of σ changes the disorder by no more than $\binom{n-m+1}{2}$.*

Proof. Let $\sigma \in S_n$ be an m-reducible permutation and consider the maximal partition $[n] = I_1 \cup \cdots \cup I_m$ of $[n]$ such that $\sigma(I_x) = I_x$ for each $x \in [m]$. Since any permutation of k contiguous elements changes the disorder by at most $\binom{k}{2}$, it is easily checked that σ changes the disorder by no more than

$$|\Delta f| \le \binom{|I_1|}{2} + \cdots + \binom{|I_m|}{2}.$$

Here, $\binom{|I_x|}{2} = |I_x| \cdot (|I_x| - 1)/2 = 0$ if $|I_x| = 1$. We claim that the right-hand side of this inequality is bounded above by $\binom{n-m+1}{2}$. Indeed, using the fact that $n = |I_1| + \cdots + |I_m|$, we have:

$$\binom{n-m+1}{2} = \frac{1}{2}(n-m+1)(n-m) = \frac{1}{2}\left(\sum_{x=1}^{m}(|I_x|-1)+1\right)\left(\sum_{x=1}^{m}(|I_x|-1)\right)$$

$$= \frac{1}{2}\left[\left(\sum_{x=1}^{m}(|I_x|-1)\right)^2 + \left(\sum_{x=1}^{m}(|I_x|-1)\right)\right]$$

$$\geq \frac{1}{2}\left[\left(\sum_{x=1}^{m}(|I_x|-1)^2\right) + \left(\sum_{x=1}^{m}(|I_x|-1)\right)\right]$$

$$= \frac{1}{2}\sum_{x=1}^{m}|I_x|\cdot(|I_x|-1) = \sum_{x=1}^{m}\binom{|I_x|}{2}.$$

$\square$

This immediately yields:

Theorem 2. *The diameter of every non-trivial Full-Flag Johnson graph $FJ(n,k)$ is bounded below by $\lceil\binom{n}{2}/\binom{k+1}{2}\rceil$.*

Proof. Let $u = (n, n-1, \ldots, 1)$ and $v = (1, 2, \ldots n)$, so that $f(u) = \binom{n}{2}$ and $f(v) = 0$. By Theorem 1, two vertices are adjacent in $FJ(n,k)$ if and only if they are related by an $(n-k)$-reducible permutation. But by Lemma 6, any $(n-k)$-reducible permutation changes the disorder by at most $\binom{k+1}{2}$. Hence every path from u to v must have at least $\lceil\binom{n}{2}/\binom{k+1}{2}\rceil$ edges. $\square$

5   Upper Bound for the Diameter

In this section, we derive an $O(n^2/k^2)$ upper bound for the diameter of $FJ(n,k)$. Somewhat surprisingly, we will not need to exploit the full adjacency structure of $FJ(n,k)$ to establish this result. Instead, recall that according to Lemma 2, any two vertices related by a $(k+1)$-neighboring permutation are at most two edges apart in $FJ(n,k)$. For the purposes of big-O notation, it is thus clear that at the cost of dropping a factor of two, we may bound the diameter of $FJ(n,k)$ by studying the following sorting problem:

Let σ be an arbitrary permutation of $(1, 2, \ldots, n)$. How many applications of $(k+1)$-neighboring permutations does it take to sort σ?

For instance, in the case $k = 1$, no more than $n(n-1)/2$ neighboring transpositions are needed. In general, if we can show that an arbitrary $\sigma \in S_n$ can be sorted by applying d or fewer $(k+1)$-neighboring permutations, then we obtain an $O(d)$-bound for on the diameter of $FJ(n,k)$.

As an initial attempt, we begin by considering a selection-sort-like algorithm that successively moves the elements 1, 2, and so on, to the front of σ via $(k+1)$-neighboring permutations. Since each application of a $(k+1)$-neighboring permutation allows us to shift the index of a given element by up to $k+1$, it is evident that each element of σ may be moved to its appropriate sorted index using no more than $\lceil n/(k+1) \rceil$ $(k+1)$-neighboring permutations. Hence we obtain a rough upper bound of $O(n^2/k)$ for the diameter of $FJ(n,k)$.

In order to obtain a tighter bound, we will utilize an algorithm of Baudet and Stevenson originally designed for efficient parallel sorting [4]. In what follows, we assume for the sake of the exposition that n is a positive integer multiple of $k + 1$ and that $k + 1 = 2m$ is even. It will be easily seen that the algorithm may be adapted to the general case with only minor modifications.

Let σ_0 be an arbitrary permutation of S_n. We perform the following sort on σ_0 by proceeding in stages, where at the ith stage ($i \geq 0$) we derive a new permutation σ_{i+1} from the previous permutation σ_i. Set $d = n/m$. For $0 \leq j < d$, denote by b_j the jth block of indices $\{jm + 1, jm + 2, \ldots, jm + m\}$, so that $\{1, 2, \ldots, n\}$ is partitioned into size-m blocks as $b_0 \cup b_1 \cup \ldots \cup b_{d-1}$. Then our sort proceeds as follows:

1. If i is even, individually sort the subsequences of σ_i of index $b_0 \cup b_1, b_2 \cup b_3, b_4 \cup b_5, \ldots$, each from least to greatest.
2. If i is odd, individually sort the subsequences of σ_i of index $b_1 \cup b_2, b_3 \cup b_4, b_5 \cup b_6, \ldots$, each from least to greatest.

Example 4. We illustrate this process with the following example. Let $n = 12$ and $k = 3$, so that $m = 2$ and $d = 6$. Suppose that we start with the initial permutation:

$$\sigma_0 = \left(\boxed{5 \quad 2 \quad 11 \quad 8 \mid 7 \quad 3 \quad 12 \quad 1 \mid 4 \quad 10 \quad 6 \quad 9} \right).$$

Here, we have boxed the subsequences of σ_0 that will be sorted during stage zero: the three subsequences of index $b_0 \cup b_1 = \{1, 2, 3, 4\}$, $b_2 \cup b_3 = \{5, 6, 7, 8\}$, and $b_4 \cup b_5 = \{9, 10, 11, 12\}$. The successive stages of our sort (displayed at the beginning of each stage) are then given by:

$$\sigma_1 = \left(\; 2 \quad 5 \; \boxed{8 \quad 11 \quad 1 \quad 3 \mid 7 \quad 12 \quad 4 \quad 6} \; 9 \quad 10 \; \right)$$

$$\sigma_2 = \left(\boxed{2 \quad 5 \quad 1 \quad 3 \mid 8 \quad 11 \quad 4 \quad 6 \mid 7 \quad 12 \quad 9 \quad 10} \right)$$

$$\sigma_3 = \left(\; 1 \quad 2 \; \boxed{3 \quad 5 \quad 4 \quad 6 \mid 8 \quad 11 \quad 7 \quad 9} \; 10 \quad 12 \; \right)$$

$$\sigma_4 = \left(\boxed{1 \quad 2 \quad 3 \quad 4 \mid 5 \quad 6 \quad 7 \quad 8 \mid 9 \quad 11 \quad 10 \quad 12} \right)$$

$$\sigma_5 = \left(\; 1 \quad 2 \; \boxed{3 \quad 4 \quad 5 \quad 6 \mid 7 \quad 8 \quad 9 \quad 10} \; 11 \quad 12 \; \right).$$

Note that our sort has completed by stage $i = 5$.

Roughly speaking, the idea behind this strategy is to gradually increase the general order of the entire sequence, rather than concentrating on one particular element at a time (as in selection sort). The difficulty is then to determine the running time of the sort. It turns out that using Knuth's 0–1 sorting principle, it is not too difficult to show that the sort terminates in at most d stages [4]. Here we give a straightforward alternative proof of the same linear bound (although our constant is slightly worse).

Our proof depends on the following technical definition and lemma. Fix any stage i and element $x \in \{1, 2, \ldots, n\}$. For any index block b_s, we say that x "dominates" b_s at stage i if x is greater than every element of σ_i with index in b_s. We denote this condition by $x > \sigma_i(b_s)$. Now suppose that the index of x in σ_i lies in b_j. We say that x is "good" at stage i if it dominates at least one of

b_{j-2} or b_{j-1}, with the convention that if one or both of these do not exist, we automatically say that x is good.

Example 5. In σ_0 above, the elements $5, 2, 11$, and 8 are automatically good. The element 7 is good since it dominates b_0, and the element 12 is good since it dominates both b_1 and b_2. The element 10 is good, since it dominates b_2. No other elements are good.

Example 6. In σ_1 above, it can be checked that all the elements of the previous example are still good, even though some of them now appear in different blocks. (This is not a coincidence). In addition, the element 9 is now good in σ_1, since it dominates b_4. No other elements are good.

Lemma 7. *Suppose that x is good at stage i. Then x is good at all subsequent stages.*

Proof. Assume that x is good at stage i. Let the index of x in σ_i lie in b_j; for convenience, we will assume that $j > 1$. We divide into cases based on whether j is the same parity as the stage number i.

Case 1. Suppose that j is the same parity as i, so that σ_{i+1} is derived from σ_i by sorting σ_i as follows:

$$\sigma_i = \left(\quad \cdots \quad \boxed{\; b_{j-2} \quad b_{j-1} \;|\; b_j \quad b_{j+1} \;} \quad \cdots \quad \right).$$

Since x dominates at least one of b_{j-2} and b_{j-1}, x is greater than (at least) k elements of $\sigma_i(b_{j-2} \cup b_{j-1})$. Hence x is certainly greater than the *least* k elements of $\sigma_i(b_{j-2} \cup b_{j-1})$, which implies $x > \sigma_{i+1}(b_{j-2})$. If the index of x in σ_{i+1} still lies in b_j, then this shows that x is good at stage $i + 1$. But if the index of x in σ_{i+1} is in b_{j+1}, then by construction we must have $x > \sigma_{i+1}(b_j)$, so again we see that x is good at stage $i + 1$.

Case 2. Suppose that j and i are of opposite parity, so that σ_{i+1} is derived from σ_i by sorting σ_i as follows:

$$\sigma_i = \left(\quad \cdots \quad \boxed{\; b_{j-3} \quad b_{j-2} \;|\; b_{j-1} \quad b_j \;} \quad \cdots \quad \right).$$

If x dominates b_{j-1}, then the index of x in σ_{i+1} is in b_j and $x > \sigma_{i+1}(b_{j-1})$. Hence we may assume that x dominates b_{j-2}. As in the previous case, this implies that $x > \sigma_{i+1}(b_{j-3})$. If the index of x in σ_{i+1} is in b_{j-1}, then this shows x is good at stage $i + 1$. Otherwise, the index of x in σ_{i+1} is in b_j and $x > \sigma_{i+1}(b_{j-1})$ by construction.

The above reasoning easily extends to the exceptional cases $j = 0$ and $j = 1$, showing that if x is good at stage i, then x is good at stage $i + 1$. This completes the proof. □

The following claim shows why Lemma 7 is of interest.

Lemma 8. *Fix any element $x \in \{1, 2, \ldots, n\}$. Then there exists at least one stage i, $0 \le i \le d$, such that x is good at stage i.*

Proof. Let $x \in \{1, 2, \ldots, n\}$ and let the index of x in σ_0 lie in b_j. Suppose that j is odd. Then at each stage, x moves to the block immediately to its left until it reaches the zeroth block or else dominates the block to its left. Since there are d blocks, this shows that x must be good at stage i for some $0 \leq i < d$. If j is even, then at the zeroth stage x can either move one block to the right or stay in the same block. In the former case, x is then good, while in the latter case the previous reasoning subsequently applies. □

Combining Lemmas 7 and 8, we see that every element is good by stage d. We use this to prove the main theorem of the section.

Theorem 3. *Let n be a positive integer multiple of $k + 1 = 2m$ and let σ_0 be an arbitrary permutation of S_n. Set $d = n/m$, as above. Performing a diffusion sort on σ_0 terminates in no more than $3d$ stages.*

Proof. We claim that for each j such that $0 \leq j < d$, the elements $b_j = \{jm + 1, jm + 2, \ldots, jm + m\}$ are in correct sorted position by stage $d + 2j$ (and all stages thereafter). This will show that diffusion sort terminates within $3d$ stages. We prove the claim by strong induction on j.

It is clear that the elements $b_0 = \{1, 2, \ldots, m\}$ will be in sorted position by stage d, since at every stage (except possibly the first) each one of these elements moves one block to the left until reaching the zeroth block. This establishes the base case. Thus, assume that at stage $d + 2j$, the elements of $b_0 \cup b_1 \cup \cdots \cup b_j$ are all in sorted position. Now, each element of b_{j+1} is good at stage $d + 2j$. The inductive hypothesis then implies that the index of each $x \in b_{j+1}$ in σ_{d+2j} must be in either b_{j+1} or b_{j+2}, since this is the only way for x to be good. It is then evident that within two more stages, every element of b_{j+1} will be in sorted position. This establishes the inductive step and completes the proof. □

We now formulate the desired diameter bound:

Theorem 4. *The diameter of $FJ(n, k)$ is bounded above by $O(n^2/k^2)$.*

Proof. Each stage of diffusion sort consists of the application of either $n/(k + 1)$ or $n/(k + 1) - 1$ $(k + 1)$-neighboring permutations. Since $d = 2n/(k + 1)$, we thus see that any permutation σ may be sorted by using no more than $6n^2/(k + 1)^2$ $(k + 1)$-neighboring permutations. Technically, our proof is only valid in the case that n is divisible by $k + 1$ and $k + 1$ is even, but we may easily adapt diffusion sort to the general case as follows. If $k + 1$ is odd, then we replace $k + 1$ by k. This substitution is valid since every k-neighboring permutation is a $(k + 1)$-neighboring permutation and the replacement clearly does not affect the big-O running time of the sort. If n is not divisible by $k + 1$, the sort is likewise easily adapted by replacing $n/(k + 1)$ with $\lceil n/(k + 1) \rceil$ wherever appropriate. □

Combining Theorems 2 and 4, we have the main result:

Theorem 5. *The diameter of $FJ(n, k)$ is $\Theta(n^2/k^2)$.*

References

1. Alspach, B.: Johnson graphs are Hamilton-connected. Ars Math. Contemporanea **6**, 21–23 (2013)
2. Araujo, G., Dumitrescu, A., Hurtado, F., Noy, M., Urrutia, J.: On the chromatic number of some geometric type Kneser graphs. Comput. Geom. Theor. Appl. **32**, 59–69 (2005)
3. Bafna, V., Pevzner, A.: Sorting by transpositions. SIAM J. **2**, 224–240 (1998)
4. Baudet, G., Stevenson, D.: Optimal sorting algorithms for parallel computers. IEEE Trans. Comput. **27**, 84–87 (1978)
5. Carr, M., Devadoss, S.: Coxeter complexes and graph-associahedra. Topology Appl. **153**, 2155–2168 (2006)
6. Chen, B., Lih, K.: Hamiltonian uniform subset graphs. J. Comb. Theor. Ser. B **42**, 257–263 (1987)
7. Comtet, L.: Advanced Combinatorics. D. Reidel Publishing Co., Dordrecht (1974)
8. Even, S., Goldreich, O.: The minimum-length generator sequence problem is NP-hard. J. Algorithms **2**, 311–313 (1981)
9. Jerrum, M.: The complexity of finding minimum-length generator sequences. Theor. Comp. Sci. **36**, 265–289 (1985)
10. Johnson, S.: Generation of permutations by adjacent transposition. Math. Comput. **17**, 282–285 (1963)
11. Heath, L., Vergara, J.: Sorting by bounded block moves. Discrete Appl. Math. **88**, 181–206 (1998)
12. Keane, M.: Interval exchange transformations. Math. Z. **141**, 25–31 (1975)
13. Klazar, M.: Irreducible and connected permutations. Inst. Teoretické Informatiky (ITI) Ser. **122** (2003)
14. Krebs, M., Shaheen, A.: On the spectra of Johnson graphs. Electron. J. Linear Algebra **17**, 154–167 (2008)
15. Manneville, T., Pilaud, V.: Graph properties of graph asociahedra. Preprint (2014). arxiv:1409.8114
16. Panaitopol, L.: A formula for $\pi(x)$ applied to a result of Koninck-Ivić. Nieuw. Arch. Wiskd **1**, 55–56 (2000)
17. Postnikov, A.: Permutohedra, associahedra and beyond. Int. Math. Res. Not. **6**, 1026–1106 (2009)
18. Bautista-Santiago, C., Cano, J., Fabila-Monroy, R., Flores-Peñaloza, D., González-Aguilar, H., Lara, D., Sarmiento, E., Urrutia, J.: On the connectedness and diameter of a geometric Johnson graph. Discrete Math. and Theor. Computer Sci. **15**, 21–30 (2013)

List Colouring and Partial List Colouring of Graphs On-line

Martin Derka[⊠], Alejandro López-Ortiz, and Daniela Maftuleac

D. Cheriton School of Computer Science, University of Waterloo,
Waterloo, ON N2L 3G1, Canada
{mderka,alopez-o,dmaftule}@uwaterloo.ca

Abstract. In this paper, we investigate the problem of graph list colouring in the on-line setting. We provide several results on paintability of graphs in the model introduced by Schauz [18] and Zhu [25]. We prove that the on-line version of Ohba's conjecture is true for the class of planar graphs. We show that the conjecture for partial list colouring on-line holds for several graph classes, namely claw-free graphs, maximal planar graphs, series-parallel graphs, and chordal graphs.

1 Introduction

We consider the study of graph colouring in an on-line streaming manner. This is both an approach of practical interest when dealing with large graphs such as social networks as well as the subject of independent theoretical study. As we know, storing and analysing a large social graph in main memory of a single computer is not always possible. If the graph exceeds the capacity of the main memory, it has to be swapped to external memory, which is an expensive overhead. Additionally social networks commonly make their graphs accessible only via a streaming API. If an external application aims to analyse the entire graph, it has to employ a local neighbourhood discovery protocol similar to web crawlers. In such a case, the application incurs costs associated with accessing a vertex of the graph in the form of the network communication necessary for issuing the API call. Both of the aforementioned problems make the traditional off-line analysis of social network graphs challenging and sometimes even infeasible. Thus, there is renewed interest in analysing graphs on-line.

Aside from their practical application, on-line graph algorithms have also been a rich source of theoretical problems with, for example, the celebrated theoretical results of Schauz [18] and Zhu [25].

In this paper, we investigate the graph list colouring problem in the on-line setting. In list colouring, the vertices of a graph are pre-assigned lists of colours, and the task is to properly colour the graph so that every vertex receives a colour from its list. In what follows, we study the problem from a theoretical point of view and resolve several open questions on this subject.

M. Derka—The first author was supported by Vanier CGS.

Z. Lipták and W.F. Smyth (Eds.): IWOCA 2015, LNCS 9538, pp. 124–135, 2016.
DOI: 10.1007/978-3-319-29516-9_11

We also study the partial list colouring on-line which is a "best-effort" variant of graph list colouring. In this setting, we are given a limited universe of colours which does not suffice to colour the entire graph, and the aim is to colour as many vertices as possible.

2 Definitions and Previous Work

The graphs considered in this paper are simple and undirected. We follow the standard terminology of graph theory (cf. for instance [4]).

Let G be a graph. A *list assignment* is a function $L : V(G) \to 2^{\mathbb{N}}$ which assigns every vertex of G a list of admissible colours. A proper colouring $c :$ $V(G) \to \mathbb{N}$ is called an *L-colouring* of G if it assigns every vertex v a colour $c(v)$ from its list $L(v)$. The *choosability number* of G, denoted by $\mathrm{ch}(G)$, is the minimum number k such that G has an L-colouring whenever L assigns every vertex a list of size at least k. For any $k \geq \mathrm{ch}(G)$, graph G is called *k-choosable*. The choosability number of a graph is also called the *list-chromatic number*.

The (off-line) list colouring problem—to decide whether a graph has an L-colouring—was introduced by Vizing in 1976 [22]. The choosability of graphs was investigated by Erdös, Rubin and Taylor [8] and later by many others. If L assigns every vertex the same list of colours, the instance of the list colouring problem becomes an instance of the "standard" vertex colouring problem. Thus,

$$\chi(G) \leq \mathrm{ch}(G)$$

and the problem is NP-complete. Voigt [23] showed in 1993 that the choosability number $\mathrm{ch}(G)$ can be strictly larger than the chromatic number $\chi(G)$ even for planar graphs.

The list colouring problem was brought to the on-line setting independently by Schauz [18] and Zhu [25] in 2009. Both the authors formulated the problem as a game of two players. In this paper, we follow the terminology of Schauz [18].

The game is played by two players called Mr. Paint and Mrs. Correct on a known graph G. In each round, the first player, Mr. Paint, takes a new colour c and colours some (at least one) uncoloured vertices. The colour c cannot be used again. There are no restrictions on the colouring of Mr. Paint—he can colour two adjacent vertices with the same colour. The other player, Mrs. Correct, attempts to correct Mr. Paint's mistakes. For this purpose, she has a finite number of so-called *erasers* assigned to every vertex. She can use an eraser to remove the colour c from any subset of vertices which were coloured by Mr. Paint in this round. An eraser can be used only once. By doing so, the number of erasers available for the given vertex decreases. The game ends when the entire graph is properly coloured in which case Mrs. Correct wins, or when Mrs. Correct cannot correct the colouring because she ran out of erasers for some vertex. In such a case, Mr. Paint wins.

If L is an assignment of number of erasers to the vertices of G (for brevity, we call it just an *assignment of erasers*) and Mrs. Correct has a winning strategy leading to a proper colouring of G, the graph is called *L-paintable*. If $\ell \in \mathbb{N}$ is a number of erasers that need to be assigned to every vertex of G for Mrs. Correct to always have a winning strategy, the graph is called $(\ell + 1)$-paintable. The minimum such number $(\ell + 1)$ is the *paintability number* of a graph, and with respect to a graph G, it is denoted by $\mathrm{ch}^{\mathrm{OL}}(G)$.

Note that if Mr. Paint writes down all the colours suggested for each vertex into a list and we ask who has the winning strategy, we get an instance of off-line list colouring. Both Schauz [18] and Zhu [25] noted that if G is not k-choosable, it cannot be k-paintable. So, the choosability number provides a lower bound on the paintability number. We get that

$$\chi(G) \leq \mathrm{ch}(G) \leq \mathrm{ch}^{\mathrm{OL}}(G).$$

Schauz [18] provided an example of a graph and an assignment of erasers L where Mr. Paint has a winning strategy, i.e., the graph is not L-paintable, however it has an (off-line) list colouring for any list assignment with lists of the respective sizes (see [5, Appendix A]). Zhu [25] proved that the complete bipartite graphs $K_{6,q}$ for $q \geq 9$ are not 3-paintable, however both $K_{6,9}$ and $K_{6,10}$ are 3-choosable. Thus, there are graphs with choosability strictly smaller than paintability.

In 1994, Thomassen [21] showed that all planar graphs are 5-choosable. Schauz [18] adapted this technique to the on-line list colouring model to show that every planar graph G is 5-paintable. In the same paper, Schauz also noted that "*ℓ-paintability is stronger than the ℓ-list-colourability (ℓ-choosability), but not by much. Although [...] there is a gap between these two notions, most theorems about list colourability hold for paintability as well.*"

In [17], Ohba investigated the classes of graphs where the choosability number equals the chromatic number. He showed that if a graph is sufficiently dense, namely, if $|V(G)| \leq \chi(G) + \sqrt{2\chi(G)}$, then $\chi(G) = \mathrm{ch}(G)$. As a strengthening of the result, he conjectured[1] that if G is a graph with $|V(G)| \leq 2\chi(G) + 1$ then $\chi(G) = \mathrm{ch}(G)$.

Kim et al. [13] studied Ohba's conjecture for multipartite graphs in the on-line setting. They pointed out that, unlike the off-line case, graphs $K_{2\star(k-1),3}$ (the complete multipartite graphs with $k - 1$ parts of size 2 and one part of size 3) are not chromatic choosable on-line and thus adjusted the inequality:

Conjecture 1 (Ohba's On-line Conjecture [13]). *Let G be a graph with $|V(G)| \leq 2\chi(G)$. Then, $\chi(G) = \mathrm{ch}^{\mathrm{OL}}(G)$.*

A step towards proving Conjecture 1 was made by Kozik, Micek and Zhu [15], who showed that it holds for the graphs with independence number of at most 3. Furthermore, they proved that the conjecture holds for graphs with $|V(G)| \leq \chi(G) + \sqrt{\chi(G)}$.

Lemma 2 (Kozik, Micek, Zhu [15]). *Conjecture 1 is true for any graph G with independence number $\alpha(G) \leq 3$.*

---

[1] Ohba's conjecture was proved by Noel, Reed and Wu [16] in 2014.

Additionally, there are various other results concerning the choosability and paintability of specific graph classes, see e.g. [11, 13–15].

Our Contribution. The main result of this paper is a proof of Conjecture 1 for the class of planar graphs (Sect. 4). We also prove several results about paintability of classes of sparse graphs (cf. Sect. 3). In Sect. 5, we provide an introduction to and investigate the partial list colouring problem in the on-line setting. We show that the conjecture for on-line partial list colouring of graphs holds for several graph classes, namely claw-free graphs, maximal planar graphs, series-parallel graphs, and chordal graphs.

3 Classical Model

In this section, we focus on the "classical" game-theoretic model of list colouring introduced by Schauz [18]. We investigate and extend results about paintability of graphs with a small number of edges. In order to do so, we work with the recursive definition of the on-line list colouring problem: The game starts on a graph G with assignment of erasers ℓ. Once the players finish a round, i.e., Mr. Paint colours a set of vertices V_P and Mrs. Correct erases the colours from some of them, denote her move by $V_C \subseteq V_P$, the vertices in $V_P \setminus V_C$ that remain coloured can be removed from the graph—those vertices are properly coloured and Mr. Paint will never use the same colour again. So, the game proceeds on a graph $G' = G[(V(G) \setminus V_P) \cup V_C]$ with one less eraser for the vertices in V_C. See [5, Appendix B] for a formal definition.

Let us begin with the following observation of [2] (proof is provided in [5, Appendix B]):

Proposition 3 (Carraher et al. [2]). *If G is a graph and ℓ an assignment of erasers to its vertices, the following holds for the game model of on-line list colouring: (a) If G is ℓ-paintable, every subgraph H of G is ℓ-paintable; (b) If ℓ assigns every vertex v of degree k at least k erasers, G is ℓ-paintable if and only if $G - v$ is ℓ-paintable.*

The following theorem is an easy consequence of Proposition 3 (see [5, Appendix B] for the proof).

Theorem 4. *Graphs with degeneracy $k \geq 0$ are $(k + 1)$-paintable.*

Series-parallel graphs are graphs with two distinguished vertices s and t called *source* and *sink*. The class itself is defined inductively as follows: (1) an edge (s, t) is a series-parallel graph; and (2) any graph G that can be obtained from two series-parallel graphs by a series or parallel composition on theirs sources and sinks is a series-parallel graph. An *Apollonian network* is either a triangle or a planar graph which can be obtained from a triangle by repeatedly inserting a vertex of degree 3 into an interior triangular face.

Theorem 4 states upper bounds for paintability of some graph classes (including series-parallel graphs and Apollonian networks) summarized by the following corollary:

Corollary 5. *(a) Forests are 2-paintable. (b) Outer planar graphs are 3-paintable. (c) Series-parallel graphs are 3-paintable. (d) Apollonian networks are 4-paintable. (e) k-regular graphs are $(k+1)$-paintable. (f) Planar graphs are 5-paintable by inductive argument in [18]. By degeneracy, they are trivially 6-paintable.*

It is easy to see that the class of series parallel graphs is a subclass of planar graphs and thus, they are 5-paintable. It is well-known that series-parallel graphs are 2-degenerate, so they are even 3-paintable. We wish to offer an alternate inductive argument which implies this statement, but works with the direct definition of series-parallel graphs. The significance of this result is given by the following: while most of the techniques for list colouring off-line can be transferred to the on-line setting, it is *not always possible* to do so for inductive arguments. As pointed out by an anonymous referee, it is not clear whether this strategy of Mrs. Correct for series-parallel graphs differs from the strategy implied by degeneracy. This is an intriguing point which, in our opinion, further increases the interest in this result. Inspired by Thomassen's proof of 5-choosability of planar graphs [21], and also by the proof of their 5-paintability by Schauz [18], we prove a slightly stronger claim (the proof is provided in [5, Appendix B]):

Theorem 6. *Let $G = (V, E)$ be a series-parallel graph with source s and sink t, and ℓ an assignment of erasers such that $\ell(s) = 0$, $\ell(t) = 1$ and $\ell(v) = 2$ for any other vertex. Graph G is ℓ-paintable.*

4   Ohba's On-line Conjecture

In this section, we prove the on-line version of Ohba's conjecture (cf. Conjecture 1) for planar graphs (see Theorem 8). We begin with the following preliminary lemma:

Lemma 7. *Let G be a graph and ℓ be an assignment of erasers such that Mr. Paint has a winning strategy. The winning strategy can be pursued by always selecting a set P such that $G[P]$ is connected.*

Proof. Let P_1, P_2 be two subsequent moves of Mr. Paint such that there are no edges between the vertices in P_1 and P_2, and let C_1, C_2 be arbitrary respective moves of Mrs. Correct. Denote the graph obtained by playing the moves by H. Observe that the graph obtained by playing moves $P_1 \cup P_2$ and $C_1 \cup C_2$ by Mr. Paint and Mrs. Correct is equal to H.

So, let P' be a move in Mr. Paint's winning strategy such that $G[P']$ is disconnected. Let H be a maximal connected component of $G[P']$. Select $P_1 := V(H)$ and $P_2 := P' \setminus P_1$, and replace the move P' with two subsequent moves P_1, P_2 (in this order). Assume that Mrs. Correct has a winning strategy by moves C_1, C_2 after this modification. By the observation above, Mrs. Correct's response $C_1 \cup C_2$ is a winning response to Mr. Paint's move P'. This is a contradiction. Repeat this argument to produce a winning strategy of Mr. Paint such that he always colours a connected induced subgraph of G. □

Before diving into the proof of Theorem 8, we encourage the reader to recall Proposition 3 from Sect. 3 which we frequently utilize.

Theorem 8. *Let G be a planar graph with $|V(G)| \leq 2\chi(G)$. Then, $\chi(G) = ch^{OL}(G)$.*

Proof. Let G be a connected planar graph with $|V(G)| \leq 2\chi(G)$. Recall that if G has independence number of at most 3, the statement holds by Lemma 2. By the Four Colour Theorem, $\chi(G) \leq 4$, so we proceed in four cases based on $\chi(G)$.

Case 1: $\chi(G) = 1$. If the chromatic number is 1, the graph has no edges. Thus, Mrs. Correct does not need any erasers, and the graph is 1-paintable.

Case 2: $\chi(G) = 2$. If the chromatic number is 2, the graph is bipartite. One should consider planar graphs of size up to 4 vertices. In fact, using Proposition 3, it is sufficient to prove the claim for the complete bipartite graphs on 4 vertices. There are two[2] possible distributions of vertices into the two partitions, so there are precisely two such graphs: $K_{1,3}$ and $K_{2,2} = C_4$. Graph $K_{1,3}$ is a tree, so it is 2-paintable (cf. Corollary 5).

For $C_4 = (v_1, v_2, v_3, v_4)$, Zhu [25] proved that even cycles are 2-paintable. We include the argument for this special case for the sake of completeness of our proof. We consider three cases of Mr. Paint's first move. If he colours one vertex only, no erasers are used up and the game continues on a tree. So, Mrs. Correct has a winning strategy. If Mr. Paint colours three or more vertices, it follows that two of them are not adjacent to each other. Let v_1 and v_3 be these two vertices. Mrs. Correct leaves v_1 and v_3 coloured and uses erasers for the rest. Then the game continues on a graph with two isolated vertices where no erasers are needed. So, the only option of Mr. Paint is to initially colour two adjacent vertices, say v_1 and v_2. Mrs. Correct uses an eraser for one of them, say v_2. The game continues on a path (v_2, v_3, v_4) where v_2 has no erasers, and the remaining vertices have each one eraser available. One can easily see that Mrs. Correct wins the game here too.

Case 3: $\chi(G) = 3$. Applying Proposition 3, it is sufficient to show that only graphs with 6 vertices and chromatic number 3 need to be considered. Graphs with no odd cycles are bipartite, and are thus 2-chromatic. We divide the case into two subcases: when G contains a cycle of length 3 and 5.

If G contains a cycle of length 5, any independent set contains at most two vertices of this cycle. Together with the last vertex, the independence number of G is at most 3 and thus, the claim holds by Lemma 2.

If G contains a cycle C of length 3, any independent set contains at most one vertex of this cycle. As $|V(G)| = 6$, the independence number is at most 4, in which case the vertices not in C, call them u, v, w, must be part of the independent set. Assume that this is the case (otherwise the claim again holds). Observe that u, v, w are connected to at most two vertices of C otherwise they cannot be all members of the independent set. Hence, their degrees are at most two, and G is 2-degenerate. Hence, G is 3-paintable by Theorem 4.

---

[2] Note that $K_{0,4}$ is not connected, so it is both 1-chromatic and 1-paintable.

Case 4: $\chi(G) = 4$. As $\chi(G)$ cannot be more than 4, if one can prove the claim for triangulated graphs on 8 vertices, it holds for all the 4-chromatic planar graphs on 8 vertices by Proposition 3.

Let v be a vertex in G. As G is triangulated, the neighbours $N(v)$ of v form a cycle C. The subgraph C together with v and its attachments to the vertices in C is called a *wheel*. Vertex v is called a *hub* of this wheel and C is its *rim*. In order to show that the independence number of G is at most 3, we analyse G based on its wheels.

Observe that G, being a planar triangulation on 8 vertices, contains precisely 12 triangular faces and 18 edges. The maximum size of an independent set is at most 4 as every triangular face can contribute at most one vertex. A wheel of size k contains $2(k-1)$ edges. Furthermore, the maximum independence number of such a wheel is $\lfloor \frac{k-1}{2} \rfloor$ and it cannot include the hub (the only independent set which includes the hub has size 1 as it cannot include any other vertex).

The sum of vertex degrees in G is $18 \cdot 2 = 36$. Hence, G must contain a vertex of degree at least 5. If G contains a vertex of degree 5, the wheel around this vertex has size 6, the rim is a 5-cycle and the wheel has 10 edges. Also, G has two vertices u, v that do not belong to this wheel. In order to obtain an independent set of size 4, both u and v have to be added into an independent set S of size 2 found in the wheel. To fill the remaining 8 edges into G, at least one of u, v must have degree at least 4. Thus, it has to be attached to one of the vertices that are already in S. Hence, the independence number of G is at most 3.

If G contains a vertex v of degree 6, in order to construct an independent set of size 4, one has to find an independent set S of size 3 in the wheel around v and fill in the additional vertex u that is not part of the wheel. Refer to Fig. 1. Without loss of generality, one can select vertices a, c, e into S. As u needs to belong to S as well, it cannot be connected to either of a, c or e. As G is triangulated, it must contain edges (f, b), (b, d), (d, f) that enclose the wheels around a, c, e respectively. Then u must be connected to b, d, f. One can notice that u, a, c, e, v, b, d, f is a 3-degenerate ordering, which immediately implies 4-paintability by Theorem 4.

If G contains a vertex of degree 7, all the vertices in the graph form a wheel of size 8 around this vertex. Hence, the maximum independent set has size 3. So, in all subcases of Case 4, graph G has independence at most 3 and the claim holds by Lemma 2, or it is the graph depicted in Fig. 1, which is 4-paintable by degeneracy.   □

5   Partial List Colouring On-line

Let G be a graph with n vertices and with chromatic number $\chi(G) = s$. We know that G can be properly coloured with s colours. However, we can still attempt to colour such a graph with less than s colours. A natural question in such a situation is what portion of the graph can be properly coloured. We know that if the number of available colours is $t \leq s$, then one can properly colour at least $\frac{tn}{s}$

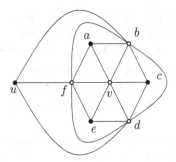

Fig. 1. An illustration for case 4 of the proof of Theorem 8. The graph has independent set of size 4, but it is 3-degenerate, so 4-paintable.

of vertices in G—we can use t colours to colour the t largest classes of a proper s-colouring of G.

A similar question was asked for list colouring by Albertson, Grossman and Hass in [1]: Given a graph G with choosability s and an arbitrary list assignment ℓ_t which assigns every vertex a list of least t colours, is it possible to colour at least $\frac{tn}{s}$ vertices in G? They conjectured [1, Conjecture 1] that the answer to this question is positive. Their conjecture is still open, although, it has been verified for some special classes of graphs.

This concept was brought to the on-line setting by Wong and Zhu in [24]. Formally, let G be an s-paintable graph, and let ℓ_t be an assignment of erasers to the vertices of G such that every vertex receives $t < s$ erasers. Let Mr. Paint and Mrs. Correct play the *partial list colouring game* as follows: In round i, Mr. Paint chooses a nonempty subset V_i of vertices in $V(G) \backslash \bigcup_{j=1}^{i-1} C_j$. Mrs. Correct chooses an independent set $C_i \subseteq V_i$ of vertices that will retain the colour i, and uses an eraser (if available) for each vertex in the rest of V_i. If Mrs. Correct chose to erase a colour from a vertex, but an eraser is not available, the vertex is marked as *finished*. The game proceeds on a subgraph of G obtained by removing vertices in C_i and the finished vertices. A vertex v is called *coloured* if it belongs to C_j for some j, and it is called *uncoloured* otherwise. The game ends when every uncoloured vertex in G is finished. The goal of Mrs. Correct is to maximize the number of coloured vertices at the end of the game. This number is denoted by λ_t^{OL}, where $t-1$ is the number of erasers available. The goal of Mr. Paint is to minimize this number.

Wong and Zhu in [24] stated an on-line version of the aforementioned partial list colouring conjecture:

Conjecture 9 (Wong, Zhu [24]). If G is an s-paintable graph and $t < s$, then $\lambda_t^{OL}(G) \geq \frac{tn}{s}$.

In this section, we build on the work of [12] and prove Conjecture 9 for several graph classes, namely claw-free graphs, maximal planar graphs, series-parallel graphs and chordal graphs. We also investigate the relationship between on-line partial list colouring and the treewidth of a graph.

Claw-free graphs are graphs that do not have $K_{1,3}$ as an induced subgraph. This class of graphs includes line graphs, complements of triangle-free graphs or comparability graphs. See [9] for a comprehensive survey on claw-free graphs.

Theorem 10. *Let G be a claw-free graph that has online list colouring with s colours, and let $t < s$. If ℓ_t is an assignment of t erasers to every vertex and the players play a partial list colouring game, then at least $\frac{tn}{s}$ vertices in the graph can be properly coloured.*

Proof. Let ℓ_t be an assignment of t erasers to vertices of G. Let Mrs. Correct play according to a strategy which maximizes the number of vertices coloured in each round. Recall that V_i is the set of vertices that Mr. Paint colours in round i, and C_i is the set of vertices that Mrs. Correct leaves coloured in round i (i.e., she does not use an eraser for those vertices). Let p be the number of rounds played in the game. Denote the set of vertices that are finished but uncoloured at the end of the game by Q. Since G is s-paintable, Q can be partitioned into $s - t$ independent sets, call them $Q_1, \ldots, Q_{s-t}$, such that the sets $C_1, \ldots, C_p, Q_1, \ldots, Q_{s-t}$ are the colour classes of a proper colouring of G.

If $|Q_1 \cup \ldots \cup Q_{s-t}| \leq \frac{(s-t)n}{s}$, then the theorem is proved. So, assume for contradiction that $|Q_1 \cup \ldots \cup Q_{s-t}| > \frac{(s-t)n}{s}$. From here, we get that

$$\sum_{i=1}^{p} |C_i| \leq \frac{tn}{s}.$$

Furthermore, there must be a set of vertices Q_σ with $\sigma \in \{1, \ldots, s - t\}$ with size above average, i.e., $|Q_\sigma| > \frac{n}{s}$. Since Mrs. Correct had t erasers available for every vertex in Q_σ and used them all up, we know that

$$\sum_{i=1}^{p} |Q_\sigma \cap V_i| > \frac{tn}{s}.$$

Putting the inequalities together, we conclude that there is some j with $|C_j| < |Q_\sigma \cap V_j|$. For brevity, we will denote such a $Q_\sigma \cap V_j$ by Z.

We now claim that there is a vertex in C_j which has at least two neighbours in Z. First, notice that Z cannot contain vertices that have no neighbours in C_j. Such vertices could have been included in C_j already — Mrs. Correct plays optimally. So, every vertex in Z has a neighbour in C_j. But Z is strictly bigger than C_j. Hence, at least two vertices in Z, call them z_1, z_2, must have a common neighbour in C_j. Observe that vertices in Z cannot have more than two neighbours in C_j, because C_j is an independent set and G would have a claw. The same holds vice versa for the vertices in C_j.

Now, find a minimal set $X \subseteq Z$ such that the neighbourhood of X in C_j, denoted by $N(X) \cap C_j$, is smaller than X. We know that such a set X exists, because at the very least, it contains z_1, z_2 which we found previously.

So, Mrs. Correct could have chosen to keep X coloured with colour j and leave the vertices from $N(X) \cap C_j$ uncoloured instead. This would increase the size of the properly coloured portion of the graph in that round, which is a contradiction with the assumed strategy. $\square$

5.1   Treewidth and Partial List Colouring On-line

In this section, we investigate the relationship between treewidth and partial list colouring on-line. From this relationship, we derive that Conjecture 9 is true for the class of series-parallel graphs.

Lemma 11. *Let $\mathcal{G}$ be a hereditary graph family. If for every graph $G \in \mathcal{G}$ we have $\mathrm{ch}^{OL}(G) = \chi(G) = s$, then for every $0 < t < s, \lambda_t^{OL}(G) \geq \frac{tn}{s}$.*

Proof. Let G be a graph in $\mathcal{G}$ and $s = \mathrm{ch}^{OL}(G) = \chi(G)$. Fix an arbitrary $0 < t < s$. Mrs. Correct will be using a proper (off-line) s-colouring of G to guide her strategy. So, let C_s be such a colouring. Find subgraph $G' \subseteq G$ formed by the vertices in the t largest colour classes in C_s. The size of G' is $\geq \frac{tn}{s}$. Graph G' has a t-colouring (we constructed it using such a colouring) and since it is in $\mathcal{G}$, we also have that $t = \mathrm{ch}^{OL}(G') = \chi(G')$. Hence, Mrs. Correct can play so that G' is painted using her $t - 1$ available erasers, and this is sufficient. $\square$

We say that a graph is *chordal* if every cycle of length more than 3 has a chord. In other words, the graph does not contain any induced cycle greater than 3. It is well-known that for a chordal graph G, we have $\chi(G) = \mathrm{ch}(G) = \omega(G)$, where $\omega(G)$ is the size of the largest clique in G. In order to proceed, we need to show that for chordal graphs, $\chi(G) = \mathrm{ch}^{OL}(G) = \omega(G)$. We will do so using so-called *perfect elimination order*—a special ordering of vertices in a chordal graph. Let us define this concept first.

Definition 12. *A perfect elimination order of a graph G is a vertex ordering $v_1, \ldots, v_n$ such that for every vertex v_i, the set of neighbours of v_i with index smaller than i, that is $\{v_j \in N(v_i) \mid j < i\}$, induces a clique in G.*

Lemma 13 (Fulkerson, Gross [10]). *Graph G is chordal if and only if it has a perfect elimination order.*

It is easy to see that a perfect elimination ordering of a graph G is a certificate that G is $(\omega(G) - 1)$-degenerate. Hence, by Theorem 4, G must be $\omega(G)$-paintable:

Theorem 14. *If G is a chordal graph, then $\mathrm{ch}^{OL}(G) = \omega(G)$.*

Proof. Let G be a chordal graph and ℓ an assignment of erasers which gives every vertex $\omega(G) - 1$ erasers. Mrs. Correct will precompute a perfect elimination order $v_1, \ldots, v_n$ of G and pursue the following strategy: if Mr. Paint suggests a set which contains vertices $v_i, v_j, i < j$ connected by an edge, Mrs. Correct uses an eraser for v_j (the vertex with bigger index). Let v_{i*} be the vertex which uses the most erasers. For sure, it uses at most $deg^-(v_{i*})$ which is the number of neighbours of v_{i*} preceding v_{i*} in the perfect elimination order. Thus, $\mathrm{ch}^{OL}(G) \leq deg^-(v_{i*}) + 1$. Since all such neighbours of v_{i*} form a clique, and v_{i*} is attached to all of them, we know that $\omega(G) \geq deg^-(v_{i*}) + 1$. However, it is well-known that $\chi(G) \geq \omega(G)$ and hence, also $\mathrm{ch}^{OL}(G) \geq \omega(G)$. Putting the inequalities together, we get that $\mathrm{ch}^{OL}(G) = \omega(G)$. $\square$

Corollary 15. *Let G be a chordal graph with $ch^{OL}(G) = s$. Then for every positive t less than s, we have $\lambda_t^{OL}(G) \geq \frac{tn}{s}$.*

Proof. Follows from Theorem 14, Lemma 11 and $\omega(G) = \chi(G)$ for chordal graphs.   □

Let G be a graph and H be a chordal graph that contains G as a subgraph (such a graph can be obtained by inserting edges into G) such that the clique number of H is minimal among all such graphs. It is well known that treewidth of G, denoted by $tw(G)$, is $tw(G) = \omega(H) - 1$ and the degeneracy of G is at most $tw(G)$.

Theorem 16. *Let G be a graph and $ch^{OL}(G) = s$. Then for every positive $t < s$, we have $\lambda_t^{OL}(G) \geq \frac{tn}{tw(G)+1}$.*

Proof. Let G' be a chordal graph obtained by adding edges to G such that the size of a largest clique in G' is $tw(G) + 1$. Since G' is chordal, we know that $tw(G) + 1 = \omega(G') = \chi(G') = ch^{OL}(G')$. From Corollary 15, we get that $\lambda_t^{OL}(G') \geq \frac{tn}{tw(G)+1}$. As G is a subgraph of G', the strategy of Mrs. Correct for G' is valid on G as well, so we get that $\lambda_t^{OL}(G) \geq \frac{tn}{tw(G)+1}$.   □

Note that this also implies that Conjecture 9 is true for the series-parallel graphs (Corollary 17) as their treewidth is at most 2 and paintability at most 3.

Corollary 17. *Let G be a series-parallel graph with $ch^{OL}(G) = s$. Then for every positive t less than s, we have $\lambda_t^{OL}(G) \geq \frac{tn}{s}$.*

6   Conclusions

We extended the previous results about paintability of planar graphs to some specific graph classes. We provided an inductive argument for 3-paintability of series-parallel graphs, and proved the on-line version of Ohba's conjecture for planar graphs. For future work, we would like to suggest extending the following two theorems to the on-line setting:

Any planar triangle-free graph without 4-cycles adjacent to 4- and 5-cycles is 3-choosable. Any graph that can be drawn with at most two crossings is 5-choosable. The proofs of both the theorems are inductive and we believe that extension to the on-line setting is possible.

Finally, while we advanced the set of graphs in which Conjecture 9 is known to hold, it remains to be fully resolved for arbitrary graphs.

References

1. Albertson, M., Grossman, S., Haas, R.: Partial list colorings. Discrete Math. **214**(1), 235–240 (2000)
2. Carraher, J., Loeb, S., Mahoney, T., Puleo, G.J., Tsai, M.-T., West, D.B.: Three topics in online list coloring. J. Comb. **3**, 1–10 (2013)

3. Cormen, T.H., Leiserson, C.E., Rivest, R.L., Stein, C.: Introduction to Algorithms, pp. 73–90, Sections 4.3 (The master method) and 4.4 (Proof of the master theorem), 2nd edition. MIT Press and McGraw-Hill (2001)
4. Diestel, R.: Graph Theory, 3rd edn. Springer, Berlin (2005)
5. Derka, M., López-Ortiz, A., Maftuleac, D.: List Colouring Big Graphs On-Line (2015). http://arxiv.org/abs/1502.02557
6. Dvořák, Z., Lidický, B., Škrekovski, R.: 3-Choosability of triangle-free planar graphs with constraints on 4-cycles. SIAM J. Discr. Math. **24**(3), 934–945 (2010)
7. Dvořák, Z., Lidický, B., Škrekovski, R.: Graphs with two crossings are 5-choosable. SIAM J. Discrete Math. **25**(4), 1746–1753 (2011)
8. Erdös, P., Rubin, A.L., Taylor, H.: Choosability in graphs. Proc. West Coast Conf. Comb. Graph Theor. Comput. Arcata, Congressus Numerantium **26**, 125–157 (1979)
9. Faudree, R., Flandrin, E., Ryjáček, Z.: Claw-free graphs - a survey. Discrete Math. **164**, 87–147 (1997)
10. Fulkerson, D.R., Gross, O.A.: Incidence matrices and interval graphs. Pacific J. Math. **15**, 835–855 (1965)
11. Huang, P., Wong, T., Zhu, X.: Application of polynomial method to on-line colouring of graphs. European J. Comb. **35**(5), 872–883 (2011)
12. Janssen, J., Mathew, R., Rajendraprasad, D.: Partial list colouring of certain graphs, manuscript, 11 March 2014. arxiv:1403.2587v1
13. Kim, S.-J., Kwon, Y., Liu, D.D., Zhu, X.: On-line list colouring of complete multipartite graphs. Electron J. Comb. **19**, 13 (2012). Paper #P41
14. Kierstead, H.A.: On the choosability of complete multipartite graphs with part size three. Discrete Math. **211**, 255–259 (2000)
15. Kozik, J., Micek, P., Zhu, X.: Towards an on-line version of Ohba's conjecture. European J. Comb. **36**, 110–121 (2014)
16. Noel, J., Reed, B., Wu, H.: A Proof of a Conjecture of Ohba (2014, to appear). arxiv:1211.1999v2
17. Ohba, K.: On chromatic-choosable graphs. J. Graph Theory **40**(2), 130–135 (2002)
18. Schauz, U.: Mr. Paint and Mrs. Correct. Electron J. Comb. **16**(1), #R77 (2009)
19. Schauz, U.: Flexible color lists in Alon and Tarsi's theorem, and time scheduling with unreliable participants. Electron J. Comb. **17**, 18 (2010). #R13
20. Schauz, U., Zhu, X., Mahoney, T.: REGS in Combinatorics. University of Illionis (2011) http://www.math.uiuc.edu/~west/regs/paint.htm
21. Thomassen, K.: Every planar graph is 5-choosable. J. Comb. Theory, Series B **62**(1), 180–181 (1994)
22. Vizing, V.G.: Colouring the vertices of a graph with prescribed colours. Metody Diskret, Anal. v Teorii Kodov i Shem **29**, 3–10 (1976). (in Russian)
23. Voigt, M.: List colourings of planar graphs. Discrete Math. **120**, 215–219 (1993)
24. Wong, T., Zhu, X.: Partial online list coloring of graphs. J. Graph Theory **74**(3), 359–367 (2013)
25. Zhu, X.: On-line list colouring of graphs. Electron. J. Comb. **16**(1), #R127 (2009)

About Ungatherability of Oblivious and Asynchronous Robots on Anonymous Rings

Gabriele Di Stefano[1], Pietro Montanari[2], and Alfredo Navarra[2(✉)]

[1] Dipartimento di Ingegneria e Scienze dell'Informazione e Matematica,
Università degli Studi dell'Aquila, Via G. Gronchi 18, 67100 L'Aquila, Italy
gabriele.distefano@univaq.it
[2] Dipartimento di Matematica e Informatica, Università degli Studi di Perugia,
Via Vanvitelli, 1, 06123 Perugia, Italy
pietro.montanari@studenti.unipg.it, alfredo.navarra@unipg.it

Abstract. We investigate on gathering of identical, memoryless, and mobile robots placed on the nodes of anonymous graphs. According to the well-known *Look-Compute-Move* model, robots operate in asynchronous cycles. In one cycle, a robot takes a snapshot of the current configuration (Look), decides whether to stay idle or to move to one of its neighbors (Compute), and in the latter case makes the computed move (Move). Cycles are performed asynchronously for each robot. The gathering problem asks for a strategy that brings all robots to a common node.

Several papers have been investigating the problem for various settings on ring graphs due its combinatorial relevance. However, none of the provided solutions can cope with the case of four robots, the only case still open on ring graphs, even though it is conjectured that the gathering is possible. We consider the specific cases of four robots placed on a ring of seven and nine nodes. We present an exhaustive proof about the impossibility of designing a strategy that solves the gathering in the considered setting. The proof makes use of both theoretical and computer-assisted approaches. Despite the specific cases considered, the relevance of the provided proof is twofold. On the one hand, it disproves the conjecture posed by previous works. On the other hand, it provides a new approach and new insights to the gathering problem on rings.

1 Introduction

We study one of the basic problems concerning self-organization of mobile entities, known in the literature as the *gathering* problem. In particular, we consider oblivious (memoryless) robots initially located on different nodes of an anonymous ring that have to gather at a common node (not determined in advance) and there remain. Neither nodes nor edges are labeled. Initially, some of the nodes of the ring are occupied by the robots and there is at most one robot in

Work supported by the Italian Ministry of Education, University, and Research: PRIN 2010N5K7EB "ARS TechnoMedia" and PRIN 2012C4E3KT "AMANDA", and by the National Group for Scientific Computation (GNCS-INdAM).

Z. Lipták and W.F. Smyth (Eds.): IWOCA 2015, LNCS 9538, pp. 136–147, 2016.
DOI: 10.1007/978-3-319-29516-9_12

each node. Robots operate in Look-Compute-Move cycles. In each cycle, a robot takes a snapshot of the current global configuration (Look), then, based on the perceived configuration, takes a decision to stay idle or to move to one of its adjacent nodes (Compute), and in the latter case it makes an instantaneous move to this neighbor (Move). Cycles are performed asynchronously for each robot. This means that the time between Look, Compute, and Move operations is finite but unbounded, and it is decided by the adversary for each robot. Hence, robots may move based on significantly outdated perceptions. The only constraint is that moves are instantaneous, and hence any robot performing a Look operation sees all other robots at nodes of the ring and not on edges. Robots are all identical, anonymous, and execute the same deterministic algorithm. They cannot leave any marks at visited nodes, nor send messages to other robots. This model is referred in the literature also as the *CORDA* model [9,16]. However, it is assumed that a robot has the ability to perceive during the Look operation whether there is one or more robots located at a given node of the ring, but not the exact number. This capability of robots is important and well-studied in the literature under the name of *multiplicity detection* (see e.g. [11] for a discussion), as a node with more than one robot located on it is called *multiplicity*. By definition, *initial* configurations do not contain multiplicities.

1.1 Related Works

The problem of let mobile entities meet on graphs or open spaces has been extensively studied in the last decades. When only two robots are involved, the problem is usually referred to as the *rendezvous* problem [3,4,7,15]. Under the Look-Compute-Move model, the rendezvous problem has been proved to be unsolvable on rings [13], hence instances with more than two robots have been investigated. The relevance of the ring topology is motivated by its completely symmetric structure. It means that algorithms for rings are more difficult to devise as they cannot exploit any topological structure, assuming that all nodes look the same. In the literature, different types of configurations have required different approaches. In particular periodicity and symmetry arguments have been investigated for rings. A configuration is called *periodic* if it is invariable under non-trivial (i.e., non-complete) rotation. A configuration is called *symmetric* if the ring admits a geometrical *axis of symmetry* that reflects single robots into single robots, multiplicities into multiplicities, and empty nodes into empty nodes. A symmetric configuration admits a *node-edge symmetry* if the axis passes through one node and one edge (see, e.g. configurations (i)–(iii) in Fig. 1); an *edge-edge symmetry* if the axis passes through (the middles of) two edges; a *node-node symmetry* if the axis passes through two nodes; a *robot-on-axis symmetry* if there is at least one node on the axis occupied by a robot.

In [13], it is proved that the gathering is not feasible if the configuration is periodic, or symmetric of type edge-edge, or contains only two multiplicities, or if the multiplicity detection capability is removed. Then all configurations with an odd number of robots, and all the asymmetric configurations with an even number of robots have been solved by different algorithms. In [12], the problem was solved in the symmetric cases with an even number of robots greater than 18. This left open the cases of symmetric configurations of types node-node or node-edge with even number of robots between 4 and 18. The case of 4 robots has been addressed in [10,14]. In [10], symmetric configurations of type robot-on-axis with $2k$ robots, $k \geq 2$, have been addressed. Moreover, in [12] it has been observed that configurations of 4 robots on a five node ring are ungatherable. The case of 6 robots has been solved in [6]. Finally, in [5], a unified approach dealing with all the gatherable cases has been designed. Besides the new techniques, the algorithm also exploits some of the previous results. In particular, the resolution of symmetric configurations with only 4 and 6 robots is delegated to the previous algorithms from [6,14], respectively. However, as we will show, the algorithm proposed in [14] cannot cope with all the symmetric cases of 4 robots.

1.2 Our Results

Although all the configurations with 4 robots on rings with more than five nodes have been claimed to be gatherable as long as the initial configurations are asymmetric and aperiodic (from [13]), or symmetric of type node-edge, node-node, or robot-on-axis (from [14]), we revise the very special case of 4 robots on a seven and nine nodes ring. We then show there cannot exist any strategy allowing the gathering in such cases. The obtained result points out a twofold aspect. On the one hand, the obtained result disproves some claimed conjectures of previous works concerning gatherable configurations of 4 robots. In particular, the algorithm proposed in [14] cannot cope with all the symmetric cases of 4 robots on odd rings. On the other hand, the new approach exploited to prove the impossibility result provides useful advances in the study of the gathering task. It is worth remarking that configurations with few robots imply more difficulties in designing suitable gathering strategies as the movement of a robot easily incurs in making the current configuration symmetric or even periodic. Our main result is constituted by the following theorem:

Theorem 1. *Four robots on a seven or nine nodes ring are ungatherable.*

Indeed, the above theorem proved by means of theoretical and computer-assisted analysis reveals sufficient hints for a more general claim. Consider the intervals of free nodes between two nodes occupied by robots (an interval could be empty if the robots are adjacent). Let $SP4$ be the set of symmetric initial configurations with four robots on odd rings where the maximal odd interval of free nodes cut by the axis is bigger than the even one (see, e.g., configurations (i) and (ii) of Fig. 1, and configurations (i)–(iv) of Fig. 7).

The initial configuration of four robots on a five nodes ring belongs to $SP4$ and it can be proved to be ungatherable by an exhaustive proof on the possible moves that can be performed. Indeed, specific configurations in $SP4$ could be gatherable but requiring suitable strategies difficult to be generalized. The main difficulty faced when dealing with configurations in $SP4$ comes from the fact that among the two intervals cut by the axis, the odd one is bigger than the even one. In [8], it has been proved that the middle node of the odd interval is the only possible candidate to finalize the gathering. Hence, when robots move towards such a node , it may happen that only one of the two symmetric robots allowed to move makes the movement. The subsequent configuration contains now two intervals of even size corresponding to those intervals originally cut by the axis of symmetry. Possibly, they can be of the same size, hence inducing a different symmetry with respect to the original one.

Proving that initial configurations with four robots on a seven nodes ring are ungatherable is challenging as exploring exhaustively all the possible moves becomes computationally intractable. In fact, we exploit both theoretical and computer-assisted analysis to obtain the proof of Theorem 1.

2   Definitions and Notation

We consider anonymous rings without orientation consisting of either seven or nine nodes. Initially, four nodes of the ring are occupied by single robots. For instance, all possible initial configurations for a ring of seven nodes are shown in Fig. 1. During a Look operation, a robot perceives the relative locations on the ring of multiplicities and single robots. We remind that a multiplicity occurs when more than one robot occupy the same node.

The current configuration of the system can be described in terms of the view of a robot r that is performing the Look operation at the current moment. It is the sequence of robots, multiplicities and empty nodes seen by r starting from its position and proceeding towards an arbitrary direction. It follows that, given a configuration, a robot recognizes its own position in the ring if the configuration is asymmetric. In case of symmetry, the robot has two possible choices for its position. For instance, referring to Fig. 1, robots denoted x and x' are indistinguishable. In initial symmetric configuration we will denote x and x' the two robots closest to the node on the axis, whereas the other two nodes will be denoted y and y'.

In a symmetric configuration, the axis of symmetry passes through one node. A move of a robot r towards such a node is denoted by $r{\uparrow}$, while $r{\downarrow}$ denotes the move towards the opposite direction. In the asymmetric configuration (iv) in Fig. 1, the robots are referred to as a, b, c, and d. The move of a robot r in the direction of a robot r' is denoted by $r \rightarrow r'$.

The *strategy* of a gathering algorithm specifies for each configuration which is the robot that has to move and the direction of the movement. Note that in case of symmetric configuration or multiplicities a single robot can not be identified as well as a single direction can not be defined.

3   Four Robots on a Seven Nodes Ring

Consider four robots on a seven nodes ring, the only possible initial configurations are shown in Fig. 1. The first three configurations are symmetric, while the last one is asymmetric.

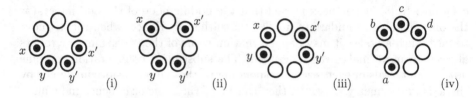

Fig. 1. The four possible initial configurations of four robots over a seven nodes ring. Edges between consecutive nodes are not drawn.

Being in an asynchronous system, it is clearly impossible to design a gathering algorithm that forces more than one robot to move from the same configuration. In fact, this would rely on the assumption that such robots have started their Look-Compute-Move cycles from the same configuration. Moreover, if an algorithm relies on the movement of more than one robot from a same configuration, the adversary can always force a robot to wake up after the movement of another robot, i.e. form different configurations.

By the above discussion, the next two lemmata show that for each initial configuration there exists only one possible move that any gathering strategy can allow.

Lemma 1. *Let C be a symmetric initial configuration among (i), (ii), and (iii), a strategy to solve the gathering task can allow only the $x\uparrow$ move.*

Proof. Considering Fig. 1, let C coincide with (i). The move $x\downarrow$ may produce two multiplicities as well as $y\uparrow$, and from [13] it follows that such configurations are ungatherable. In fact, in a configuration composed of just two multiplicities, robots might behave exactly like in a configuration where there are only two single robots. By allowing $y\downarrow$, configuration (i) can be obtained infinitely many times as y and y' may move simultaneously and exchange their positions. Hence, only $x\uparrow$ remains. If C coincides with (ii), then $x\downarrow$ would output configuration (i) from which we have shown that only $x\uparrow$ is allowed, hence the configuration would cycle between (i) and (ii), infinitely many times. Move $y\uparrow$ could generate configuration (ii) itself if only one robot moves. Move $y\downarrow$ again generates (ii) infinitely many times. Again, only $x\uparrow$ remains. If C coincides with (iii), then $x\downarrow$ may produce two multiplicities as well as $y\uparrow$. Move $y\downarrow$ would outputs configuration (ii) from which we have shown that only $x\uparrow$ is allowed, hence the configuration would cycle between (ii) and (iii) if a single robot moves, infinitely many times. It follows that $x\uparrow$ is the only possible move from each symmetric initial configuration. □

Lemma 2. *Let C be the asymmetric initial configuration (iv), a strategy to solve the gathering task can allow only the move $d \to c$.*

Proof. Let C be the asymmetric configuration (iv) shown in Fig. 1. If $a \to b$ is allowed, then configuration (i) is created, from which again configuration C can occur, by Lemma 1. If $a \to d$ is allowed, then again configuration (iv) is obtained with node a that should move backwards to the original position. If $b \to a$ is allowed, then configuration (iii) is created, from which again configuration (iv) can occur by applying the move of Lemma 1. If $b \to c$ is allowed, then the sequence of configurations shown below might occur. By the arrow on top of some robots we denote the decision made by the corresponding robot during its Compute operation to move towards the indicated direction. We refer to such a move as a *pending* move that will be performed during the Move operation, eventually.

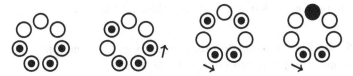

The above sequence of configurations shows that starting from (i), by Lemma 1, both x and x' can start their Look-Compte-Move cycle and move the configuration to (iv) with a pending move. Then, by hypothesis, $b \to c$ is applied. While $b \to c$ remains pending, the previous pending move is performed, leading the configuration to (iii) but with a pending move. Finally, again $x\uparrow$ is applied leading the configuration to have one multiplicity (represented by the full-black node) and a pending move. Since the last configuration might lead to two multiplicities, the considered move $b \to c$ cannot be allowed.

If $c \to b$ is allowed, then the sequence of configurations shown below might occur, leading to a configuration with two multiplicities.

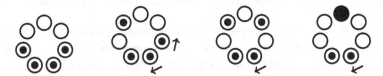

If $c \to d$ is allowed, then the cycling sequence of configurations shown below might occur.

If $d \to a$ is allowed, then the cycling sequence of configurations shown below might occur.

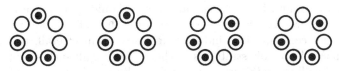

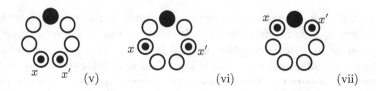

Fig. 2. The three possible symmetric configurations with one multiplicity.

Then the only move left is $d \to c$, and the claim follows.   □

Lemma 3. *Let C be a symmetric configuration with one multiplicity, a strategy to solve the gathering task can allow only the $x\uparrow$ move.*

Proof. Let C be configuration (v) shown in Fig. 2. If x and x' move toward each other, the same configuration might be obtained. If robots composing the multiplicity are allowed to move, then configuration (ii) might be obtained it the two robots move simultaneously in opposite directions (note that no algorithm can move the two robots in the same direction due to the symmetry of the configuration). From (ii), by Lemma 2, again configuration (v) can be obtained.

From configuration (vi), if single robots are allowed to move toward each other, then configuration (v) can be obtained. By the above discussion, the two single robots should move back, and configuration (vi) would be again obtained. If robots composing the multiplicity are allowed to move, then configuration (iii) might be obtained. From (iii), by Lemma 2, again configuration (vi) can be obtained.

From configuration (vii), if single robots are allowed to move away from the multiplicity, then configuration (vi) can be obtained. By the above discussion, the two single robots should move back, and configuration (vii) would be again obtained. If robots composing the multiplicity are allowed to move, then a configuration with two multiplicity can be obtained.

In any case, the only move left concerns single robots to move toward the multiplicity.   □

3.1   Further Simplifications

All configurations with two multiplicities must not be reached since they are ungatherable, hence any strategy does not need to specify a move for such configurations. Similarly, any configuration with a multiplicity made of four robots is final, hence no moves must be specified. All remaining configurations are reported in Fig. 3, and for each one, a strategy should specify one move. From (viii) to (xiii) there are six possible moves for each configuration to be tested, while the other configurations induce four moves each. Overall, there are still $6^6 * 4^3 = 2985984$ possible strategies to check. Testing all such strategies might be computationally prohibitive, so we need to eliminate some possibilities.

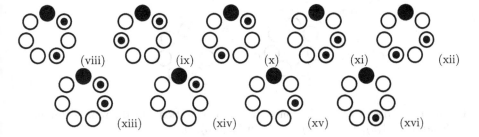

Fig. 3. All the remaining configurations that must be managed by a strategy.

For instance, from (viii), we cannot move the multiplicity toward left since if only one robot composing the multiplicity moves, then configuration (iv) is created and by Lemma 2 again configuration (viii) is obtained. From (ix) (from (xii), resp.), the move of the single robot closest to the multiplicity toward right would generate configuration (vi) (configuration (v), resp.) and by Lemma 3 again configuration (ix) (configuration (xii), resp.) can be obtained. From (x), moving the multiplicity to the right may generate the same configuration if only one robot moves. From (xi) and (xiii), we can avoid moving one single robot toward the other one as it would generate two multiplicities. From (xii), if the multiplicity is moved to the left, the same configuration is obtained. From (xiv), moving the multiplicity toward the single robot may produce two multiplicities, while the opposite move may generate (vii), and by Lemma 3, again configuration (xiv) can be obtained. From (xvi), moving the single robot or the multiplicity toward left may generate the same configuration.

Finally, since configuration (i) can generate any other initial configuration by applying the move of Lemma 1, when testing a strategy it can be discarded if it generates (i) since this implies a cyclic sequence of configurations. Then from (ix) and from (xiii), the move of the multiplicity toward left can be avoided. Similarly, the move of the multiplicity toward right in (xi) can be eliminated.

By removing all such moves, there remain 57600 possible strategies. In the next section, we make use of an automatic generator to check whether there is at least one strategy that allows gathering.

3.2   Computer-Assisted Results

We made use of the functional language OCaml [1]. Each strategy is represented by a string of 9 digits, corresponding to the nine configurations from (viii) to (xvi). The i-th digit j represents the j-th move associated with the i-th configuration. For instance, the moves associated with (xiv) are in order: the single robot moves away from the multiplicity or it moves toward the multiplicity. As shown in the previous section no further moves can be associated with configuration (xiv). It follows that in each string representing a strategy, the 7-th digit (corresponding to configuration (xiv)) ranges from 1 to 2. So the maximum string is 545343242 since there are 5 moves for (viii), 4 for (ix), and so forth.

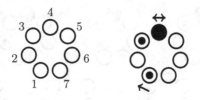

Fig. 4. The order of representation for each configuration and the representation of configuration $[(1,9);(0,0);(1,0);(2,14);(0,0);(0,0);(0,0)]$.

Each configuration is represented by a string of 7 pairs of integers. The first integer of each pair represents the number of robots lying in the corresponding node. The second integer represents the possible pending moves that corresponding robots might implement. As shown in Fig. 4, the correspondence of pairs with nodes is given by the clockwise order, starting from the bottom left node. For implementing pending moves, a robot is associated with 9 if its pending move is clockwise with respect to the current representation, with 5 if the pending move is anti-clockwise. Combinations of such numbers determine different pending moves provided by robots lying at the same node. The configuration shown in the figure admits a multiplicity with two robots that implement two opposite pending moves, and in fact, they are represented by the pair $(2, 14)$.

For each strategy we explore the graph of configurations that can be obtained by starting from (i). A strategy is successful if all the branches lead to the final configuration, that is the generated graph is a tree, and each leaf is the final configuration. Contrary, we stop a test if a cycle is found or if a configuration with two multiplicities (and without pending moves) is generated. We then report in an output file the sequence of configurations that makes the tested strategy fail. For instance, strategy 422232221 corresponds to set of moves depicted in Fig. 5. This may induce the sequence of configurations depicted in Fig. 6 that ends up in a cycle given by configurations (7)–(9).

By exhaustively exploring all the 57600 strategies left, our computations show that there exists no successful strategy, that is, the first part of Theorem 1

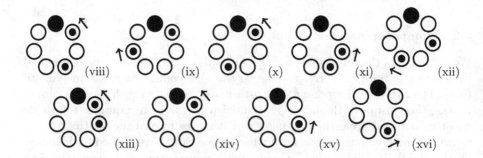

Fig. 5. Strategy 422232221. Arrows represent the defined moves.

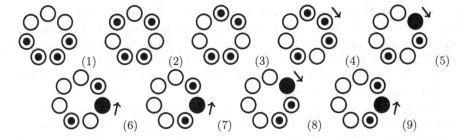

Fig. 6. Cyclic sequence of configurations generated by strategy 422232221.

is proven. Interested readers can found the implementation as well as the output of our computations in [2].

4 Four Robots on a Nine Nodes Ring

When considering four robots on a nine nodes ring, it can be easily verified there are 10 possible initial configurations out of which 6 are symmetric. Among those 6 configurations, 4 belong to $SP4$, see the first four configurations of Fig. 7. Recall from [8] that the only node where gathering can be potentially finalized in configurations belonging to $SP4$ is the middle one of the odd interval of free nodes cut by the axis. Moreover, it is worth noting that in order to gather a configuration belonging to $SP4$ it is necessary to reach a configuration not belonging to $SP4$ (like configuration (v) in Fig. 7). We now prove the second part of Theorem 1 by defining a specific behavior of the adversary. Starting from one configuration in $SP4$, whatever a gathering algorithm specifies to move, the adversary allows synchronous moves as long as the configuration remains in $SP4$, otherwise the adversary allows only one robot to move.

Similarly to the case shown for the seven nodes ring, from (i) only $x\uparrow$ can be allowed, hence reaching (ii). From (ii) only two moves can be allowed by any

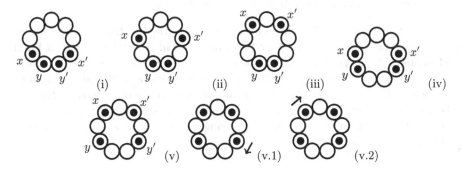

Fig. 7. The four possible initial configurations belonging to $SP4$ of four robots over a nine nodes ring and a configuration not in $SP4$ with two possible pending moves

gathering algorithm, that is $x\uparrow$ and $y\uparrow$ that lead to (iii) and (iv), respectively. From (iv), the only way to exit class $SP4$ is by $x\uparrow$, in which case the adversary makes only one robot move, hence leading to (iii). From (iii), the only ways to exit $SP4$ are by $x\uparrow$ or $y\uparrow$. By making move only one robot, in the former case (iv) is again obtained while in the latter case (v) is obtained with a possible pending move, as shown in (v.1).

From (v), by considering $x\uparrow$, the adversary can bring the configuration to (iii) starting from (v.1) and performing both the move of the pending robot and $x\uparrow$. Moves $x\downarrow$ and $y\downarrow$ bring to (iv) and (iii), respectively. So, it remains only $y\uparrow$. The adversary brings the configuration to (v.2) by starting from (v.1), performing the pending move and leaving pending $y\uparrow$ which now corresponds to $x\uparrow$. From (v.2), by performing both the pending moves $x\uparrow$ and $y\uparrow$, again (ii) is obtained.

We conclude there exists no strategy to exit class $SP4$.

5   Concluding Remarks

We have shown that the gathering of four robots on a seven or nine nodes ring is not feasible. Apart from its own interest as combinatorial problem, the obtained results disprove some previous works. It is worth nothing that some initial configurations are gatherable according to [13] if considered alone within the set of aperiodic and asymmetric configurations. Indeed, our results state the impossibility for gathering in general, i.e., when no assumptions are made on initial configurations. Moreover, configurations belonging to the set $SP4$ do not allow any gathering strategy even when considered the only possible initial configurations. Unfortunately, similar arguments applied for the case of nine nodes ring do not extend for rings of different size (including seven). Gathering configurations in $SP4$ with more than nine nodes still constitute an open problem. Actually, we have tried some strategies by extending the simulator described in Sect. 3.2 to gather configurations with eleven nodes but without succeeding so far. In fact, the conducted computer-assisted approaches seem to reveal the following statement:

Conjecture 1. Configurations in $SP4$ are ungatherable.

The intuition comes by observing that when starting from configurations where the four robots lay one after the other (like in configuration (i) of Fig. 1 and configuration (i) of Fig. 7), we know that by [8] the only node ν where gathering can be potentially finalized is the one opposite to the middle robots. To reach such a node, robots should be moved toward the other pole of the ring where ν lies. If they all move then it is possible to create a configuration similar to the initial one near to ν. If only two symmetric robots move, then orthogonal symmetries can easily occur.

References

1. http://caml.inria.fr/ocaml/
2. https://onedrive.live.com/?cid=7DEC52E233F33396&
   id=7DEC52E233F333961172
3. Alpern, S.: The rendezvous search problem. SIAM J. Control Optim. **33**, 673–683 (1995)
4. Czyzowicz, J., Labourel, A., Pelc, A.: How to meet asynchronously (almost) every-where. In: Proceedings of the 21st Annual ACM-SIAM Symposium on Discrete Algorithms (SODA), pp. 22–30 (2010)
5. D'Angelo, G., Di Stefano, G., Navarra, A.: Gathering on rings under the look-compute-move model. Distrib. Comput. **27**(4), 255–285 (2014)
6. D'Angelo, G., Di Stefano, G., Navarra, A.: Gathering six oblivious robots on anony-mous symmetric rings. J. Discrete Algorithms **26**, 16–27 (2014)
7. Dessmark, A., Fraigniaud, P., Kowalski, D., Pelc, A.: Deterministic rendezvous in graphs. Algorithmica **46**, 69–96 (2006)
8. Di Stefano, G., Navarra, A.: Optimal gathering of oblivious robots in anonymous graphs. In: Moscibroda, T., Rescigno, A.A. (eds.) SIROCCO 2013. LNCS, vol. 8179, pp. 213–224. Springer, Heidelberg (2013)
9. Flocchini, P., Prencipe, G., Santoro, N., Widmayer, P.: Hard tasks for weak robots: the role of common knowledge in pattern formation by autonomous mobile robots. In: Aggarwal, A.K., Pandu Rangan, C. (eds.) ISAAC 1999. LNCS, vol. 1741, pp. 93–102. Springer, Heidelberg (1999)
10. Haba, K., Izumi, T., Katayama, Y., Inuzuka, N., Wada, K.:On gathering problem in a ring for 2n autonomous mobile robots. Technical report COMP2008-30, IEICE, Japan (2008)
11. Izumi, T., Izumi, T., Kamei, S., Ooshita, F.: Randomized gathering of mobile robots with local-multiplicity detection. In: Guerraoui, R., Petit, F. (eds.) SSS 2009. LNCS, vol. 5873, pp. 384–398. Springer, Heidelberg (2009)
12. Klasing, R., Kosowski, A., Navarra, A.: Taking advantage of symmetries: gathering of many asynchronous oblivious robots on a ring. Theor. Comput. Sci. **411**, 3235–3246 (2010)
13. Klasing, R., Markou, E., Pelc, A.: Gathering asynchronous oblivious mobile robots in a ring. Theor. Comput. Sci. **390**, 27–39 (2008)
14. Koren, M.: Gathering small number of mobile asynchronous robots on ring. Zesz. Nauk. Wydzialu ETI Politech. Gdanskiej. Technol. Informacyjne **18**, 325–331 (2010)
15. Lim, W., Steve, A.: Minimax rendezvous on the line. SIAM J. Control Optim. **34**, 1650–1665 (1996)
16. Prencipe, G.: Impossibility of gathering by a set of autonomous mobile robots. Theor. Comput. Sci. **384**, 222–231 (2007)

Dynamic Subtrees Queries Revisited:
The Depth First Tour Tree

Gabriele Farina[1,2] and Luigi Laura[1,3]([⊠])

[1] Italian Association for Informatics and Automatic Calculus (AICA), Milano, Italy
gabriele2.farina@mail.polimi.it, laura@dis.uniroma1.it
[2] Polytechnic University of Milan, Milano, Italy
[3] "Sapienza" Università di Roma, Rome, Italy

Abstract. In the *dynamic tree problem* the goal is the maintenance of an arbitrary n-vertex forest, where the trees are subject to joining and splitting by, respectively, adding and removing edges. Depending on the application, information can be associated to nodes or edges (or both), and queries might require to combine values in path or (sub)trees. In this paper we present a novel data structure, called the *Depth First Tour Tree*, based on a linearization of a DFS visit of the tree. Despite the simplicity of the approach, similar to the ET-Trees (based on a Euler Tour), our data structure is able to answer queries related to both paths and (sub)trees. In particular, focusing on subtree computations, we show how to customize the data structure in order to answer queries for a concrete application: keeping track of the biconnectivity measures, including the impact of the removal of articulation points, of a dynamic undirected graph.

1 Introduction

In the *dynamic tree problem* the goal is the maintenance of an arbitrary n-vertex forest, where the trees are subject to joining and splitting by, respectively, adding and removing edges. Depending on the application, information can be associated to nodes or edges (or both), and queries might require to combine values in path or (sub)trees.

The dynamic tree problem has several applications, ranging from network flows [12,21,22], one of the original motivations, to other graph algorithms including connectivity [13], biconnectivity [8], and minimum spanning trees [8,13], and other combinatorial problems [15,16]. With such a wealth of applications, it is not surprising the fact that there are several approaches to solve (at least partially) the dynamic tree problem using $\mathcal{O}(\log n)$ time per operation: ST-trees [20,21], ET-trees [13,22], topology trees [8–10], top trees [3,4,23], RC-trees [1,2], and Mergeable Trees [11] that build up on the ST-tree and, as the name suggests, support also the *merge* operation. All these approaches map a generic tree into a balanced one, and can be divided into three main categories: *path decomposition* (ST-trees, Mergeable Trees), *tree contraction* (topology trees, top trees, RC-trees), and *linearization* (ET-trees); refer to the dissertation of

© Springer International Publishing Switzerland 2016
Z. Lipták and W.F. Smyth (Eds.): IWOCA 2015, LNCS 9538, pp. 148–160, 2016.
DOI: 10.1007/978-3-319-29516-9_13

Werneck [25] and the experimental comparison of Tarjan and Werneck [24] for a more complete picture about techniques and applications.

Approach. In this paper we present a novel data structure, called the *Depth First Tour Tree* (DFT-TREE), to solve the dynamic tree problem; the DFT-TREE, as the ET-Tree, is based on a linearization: as the name suggests, we linearize the tree following a DFS visit of it (see Fig. 1, where is shown for comparison also the Euler Tour). The main consequence of this approach is that the whole subtree of a node is stored contiguously, thus allowing us fast operations on the subtree, as we will detail in the rest of the paper. As we can see from Fig. 1, for example, the subtree of node 4 is contiguous in the DFT-TREE, whilst node 4 itself appears twice in its own subtree in the corresponding ET-Tree. DFT-TREE data structure can be easily implemented on top of any Balanced Binary Search Tree (BBST), such as Splay Trees [21] and Red-Black Trees [6].

The idea of linearizing the tree according to its DFS visit and maintaining the linearization in an efficient data structure is not new in the literature. Indeed, the very idea was exploited in other works, most notably [14,17,18], in the context of succinct trees. However, given the additional constraint of succinctness, the focus of these works is inherently different, and the set of supported queries is weaker and less oriented to data-processing operations.

The DFT-TREE supports all the operations shown in Table 1, that are divided in three groups: (i) structural operations, i.e. the ones that alter the structure of the tree, (ii) structural queries, and (iii) operations related to the values stored in the vertices; as we can see, it supports all the traditional dynamic tree operations together with others, such as LCA and CONDENSE, that are not completely standard and, thus, not supported by all the data structures; CONDENSE, in particular, allows to use the DFT-TREE to implement the Block Forest structure, following the exact algorithm of Westbrook and Tarjan [27].

Furthermore, the DFT-TREE supports three non standard *generic* operations, to be customized depending on the applications, that are:

- COMBINE(v), that aggregates values in the path between vertex v and the root of the tree;
- REDUCE-CHILDREN(v), that aggregates values of the children of v;
- REDUCE-CHILD-SUBTREES(v), that aggregates values in the subtrees rooted in the children of v.

These *generic* functions are, probably, the most interesting aspect of DFT-TREES.

Contribution. We propose a novel data structure, combining the simplicity of the Euler Tour trees with the expressiveness of the Depth First visit of a tree. We believe that the contribution of our approach is twofold:

- the resulting data structure is simple, using only elementary concepts, and thus is easy to understand, analyze and implement;
- we give a *unified framework* for treating a vast class of data aggregation tasks on subtrees.

While our data structure is able to support basic operations on paths, it is primarily designed to aggregate data on subtrees, an operation which is usually non-trivial with other data structures.

Unlike ST-trees, topology trees and RC-trees, DFT-TREES do not require the underlying forest to have vertices with bounded (constant) degree in order to efficiently cope with subtree queries. Degree restrictions can be avoided by *ternarizing* the input forest but, as observed in [26], "this introduces a host of special cases" and complicates the data structure. In the special case of ST-trees, some work has been done [19] to support queries on subtrees for a restricted set of operations (for example, giving the minimum element of a given subtree) without the need for ternarization, but the resulting data structure is still very complicated, both to analyze and implement. The same task can be performed extremely easily with DFT-TREES.

Furthermore, DFT-TREES can naturally aggregate on all the children subtrees of a node v in parallel without having to pay a cost proportional to the degree of v itself: for example, as we will see, given a node v it takes $\mathcal{O}(\log n)$, independently from the degree of v, to answer the child of v whose subtree is the largest. This is an interesting feature that distinguishes our data structure, and can be useful for practical problems, as we will demonstrate in the final sections of this paper.

The extreme flexibility of use of the structure comes at the cost of its structural rigidity. In particular, while all other structural operations require logarithmic time in the forest size, the EVERT operation requires a cost proportional to the depth of the node being everted. However, when either the number of eversions is small compared to the total number of queries performed, or the costs of the eversion is amortized, the cost of EVERT can be regarded as being $\mathcal{O}(\log n)$ like all the other structural operations. This is the case in all the applications we present here and in the extended paper[1].

Applications. In order to explain the versatility of the approach, we show how to customize the above functions for a concrete application:

- Given a streaming graph, for which we maintain all the biconnected properties using the mentioned approach of Westbrook and Tarjan, we can also compute the *impact* of an articulation point u, introduced in the context of the Autonomous Systems (AS) graph, as a measure of the resiliency of the network. The impact of u is defined as the number of vertices that gets disconnected from the main connected components after the removal of u. This application requires the determination of the subtree of a node having maximum size.

Refer to the extended paper for further applications:

- The *betweenness centrality* of a vertex v in a tree. This requires to count the sum of the squares of the sizes inside subtrees.
- The *closeness centrality* of a vertex v in a tree. This requires the sum of the distances to every node in the subtree and in the tree above v.

---

[1] The extended paper can be found at http://arxiv.org/abs/1502.05292.

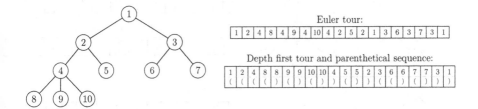

Fig. 1. An example of Euler Tour, Depth First Tour and *parenthetical sequence* of a tree (introduced in Sect. 3).

Table 1. DFT-TREE operations on an n vertex forest. The complexity values reported are amortized complexity if we implement the DFT-TREE with Splay Trees [21] and worst-case complexity if we use Red-Black Trees [6].

Operation	Complexity	Description
LINK(u, v)	$\mathcal{O}(\log n)$	Makes the root of the tree containing vertex v a child of vertex u
CUT(v)	$\mathcal{O}(\log n)$	Deletes the edge connecting v to its parent, splitting the tree. If v is the root of the tree, nothing happens
CONDENSE(v)	$\mathcal{O}(\log n)$	Deletes vertex v; its children become children of the parent of v. If vertex v is the root, the number of connected components of the forest increases by $d - 1$, with d being the degree of v
ERASE(v)	$\mathcal{O}(\log n)$	Deletes vertex v and all its adjacent edges
EVERT(v)	$\mathcal{O}(d \log n)^a$	Re-roots the tree containing vertex v at vertex v
ROOT(v)	$\mathcal{O}(\log n)$	Returns the root of the tree containing node v
SAME-TREE(u, v)	$\mathcal{O}(\log n)$	Tests if nodes u and v belong to the same tree
IS-DESCENDANT(u, v)	$\mathcal{O}(\log n)$	Answers whether node u is a descendant of v
PARENT(v)	$\mathcal{O}(\log n)$	Returns the parent of node v
ANCESTOR(v, k)	$\mathcal{O}(\log n)$	Returns the ancestor of node v at depth $d_v - k$, where d_v represents the depth of v, if existent
LCA(u, v)	$\mathcal{O}(\log n)$	Returns the lowest common ancestor of nodes u and v (if they belong to the same tree)
DEGREE(v)	$\mathcal{O}(\log n)$	Returns the degree of node v
LIST-CHILDREN(v)	$\mathcal{O}(\delta \log n)^b$	Returns a list containing the children of vertex v
CHANGE-VAL(v, x)	$\mathcal{O}(\log n)$	Assigns val$(v) = x$
REDUCE-CHILDREN	$\mathcal{O}(\log n)^c$	*See description in the text, Sect. 4*
REDUCE-CHILD-SUBTREES	$\mathcal{O}(\log n)^c$	*See description in the text, Sect. 4*
COMBINE	$\mathcal{O}(\log n)^c$	*See description in the text, Sect. 4*

[a] Where d is the depth of the node involved. We note that the EVERT operation is slow in the worst case, but it is possible to amortize it by always everting the smallest tree.

[b] Where δ is the degree of the node passed as argument to DEGREE.

[c] Assuming that the operations (denoted with $\oplus$ and $\otimes$) in REDUCE-CHILDREN, REDUCE-CHILD-SUBTREES and COMBINE take constant time when called with two nodes.

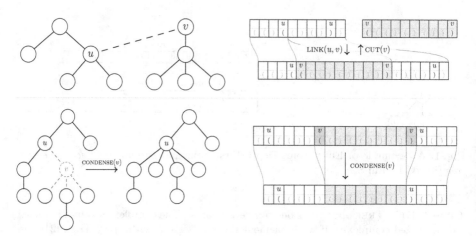

Fig. 2. Effect of the LINK, CUT and CONDENSE operations.

2 Preliminaries

We assume the reader is familiar with basic concepts of graph theory (see, e.g., [7]). We recall that, in an undirected graph G, a *connected component* is a maximal set of vertices $V' \subseteq V$ such that, given $u, v \in V'$, there is at least one path between u and v in G; an *articulation point* is a vertex $v \in V$ such that its removal from the graph G increases the number of connected components of G; similarly a *bridge* is an edge $e \in E$ such that its removal from the graph G increases the number of connected components of G. A *biconnected component* is a maximal set of vertices $V'' \subseteq V$ such that after the removal of any $v \in V''$, the remaining graph V''/v is connected. Following [5], the *impact* of an articulation point is the number of vertices that get disconnected from the largest connected component when v is removed from the graph.

3 Depth First Tour Trees

In this section we describe the main idea of the DFT-TREES, which builds up on the *Depth First Visit* of the tree and its linearization into an array; for the sake of the exposition we will populate this array with (opening and closing) parentheses that will be denoted as the *parenthetical sequence* of the tree. The other key ingredient of the DFT-TREES is a summary defined over the parenthetical sequence: in the underlying BBST the node corresponding to vertex v is augmented with both the information about v and the summary of its subtree (in the BBST). The depth first visit of a tree is constructed by recursively visiting nodes in a depth-first fashion. When a node is entered for the first time, it is appended to the back of depth first tour, along with a tag indicating it was a newly-opened node (called an *open-node*); when all its children have been visited, we push back the node again before returning the call, this time with tag

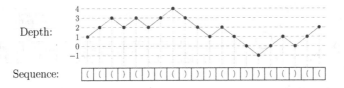

Fig. 3. Depth of a sequence of parentheses. In this case, the summary of the sequence is the pair $(-1, 3)$. The summary of the first four parentheses is $(0, 2)$.

indicating this is a fully explored node (called a *close-node*). Since every node is appended to the list exactly twice, the size of the depth first tour of a tree of size n is $2n$.

Figure 1 shows the depth first tour of an example tree of size 10, together with its linearization: an array that contains its parenthetical sequence; the Euler Tour of the same tree is shown for comparison: note that in an Euler Tour a node can appear several time; the size of an Euler Tour is $1 + 2m = 2n - 1$, since an Euler Tour begins with a node and then, for each edge of the tree, both its endpoints are added exactly once, when entering the node. In Fig. 2 we can see the effects of the LINK, CUT and CONDENSE operations on the tree and the corresponding parenthetical sequence.

Definition 1 (depth of a parenthesis). We define the *depth* of a parenthesis in a sequence of parentheses as the difference between the number of open parentheses and the number of closed parentheses in the prefix of the given sequence ending in that parenthesis.

The sequence of the depths of the parentheses coincides with the prefix sums of the sequence obtained by replacing every open parenthesis with a 1 and every closed parenthesis with a -1.

Definition 2 (summary of a sequence of parentheses). We define the *summary* of a sequence of parentheses as the pair of integers (a, b), where a is the minimum between 0 and the minimum depth of the parentheses of the sequence, and b is equal to the difference between the depth of the last parenthesis and a.

In the following, we refer to the first value of the summary as to the *down-value*, and to the second as to the *up-value*. Note that the down-value of a summary is always non-positive, while the up-value is always non-negative. In Fig. 3 we show a graphical representation of the depth of the parentheses in the sequence: for example, the summary of the whole sequence is the pair $(-1, 3)$, whilst the summary of the first four parentheses is $(0, 2)$. It should be clear that the summary of the sequence made of just one open parenthesis is $(0, 1)$, while the summary of the sequence made of just one closed parenthesis is $(-1, 0)$.

The following lemmas hold for any sequence of parentheses:

Lemma 1. *The down-values of the prefixes, taken in order, of any sequence of parentheses form a monotonically decreasing sequence of integers.*

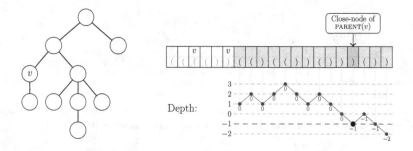

Fig. 4. Characterization of the parent of node v, as stated in Lemma 5. The values under the small dots represent the down-values of the prefixes.

Lemma 2. *A sequence of parentheses is balanced if, and only if, its summary is equal to $(0,0)$. Any prefix of a balanced parenthetical sequence has down-value 0.*

Lemma 3. *Let S_1, S_2 be two sequences of parenthesis having summary (a_1, b_1) and (a_2, b_2) respectively. The summary of the sequence $S_1 + S_2$ obtained by concatenating S_1 and S_2 is the pair $(a_1, b_1) \boxplus (a_2, b_2)$, where the sum between summaries is defined as:*

$$(a_1, b_1) \boxplus (a_2, b_2) = \begin{cases} (a_1, b_1 + a_2 + b_2) & \text{if } b_1 + a_2 \geq 0 \\ (a_1 + b_1 + a_2, b_2) & \text{otherwise.} \end{cases}$$

Lemma 4. *The sum of two summaries defined above is an associative operation.*

As a consequence of Lemma 4, as we mentioned before, we can store in each vertex of the BBST the sum of the summaries of all the vertices in its subtree. We proceed with the following lemma:

Lemma 5. *Let `close-v` be the close-node associated with the non-root node v. The close-node associated with the parent of v is the first (leftmost) node u after `close-v` reaching depth -1 relative to `close-v`.*

Lemma 5, together with the associativity of $\boxplus$ and the monotonicity of the down values of the prefixes of any (sub)sequence of parentheses (Lemma 1), gives us an efficient way to locate the parent of any non-root node: we simply binary search the smallest prefix having a negative down-value, inside the suffix of the parenthetical sequence starting after `close-v`. Refer to Fig. 4 for a visual insight. Similar properties hold for lca and ancestor: for example, for the k-th ancestor we can (binary) search the first node reaching relative depth $-k$ with respect to `close-v`, after `close-v`.

4   Subtree (and Path) Operations

In this section we detail the subtree and path operations. As we mentioned before, we assume that each node v has an associated value (note that values can

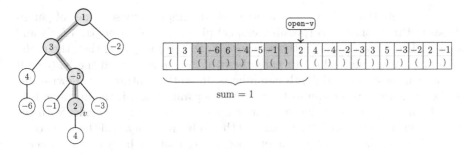

Fig. 5. Visual insight for Lemma 6. The numbers written in the nodes of the tree on the left represent the values assigned to the vertices.

be generic objects, not only numbers), denoted by val(v). We have the following three generic operations on a node that operate, respectively, on its children, on its subtree, and on the path from the node to the root:

- REDUCE-CHILDREN($v, \oplus$): Computes the value of val(c_1) $\oplus \cdots \oplus$ val(c_d), where $c_1, \ldots, c_d$ are the children of node v, and $\oplus$ is an associative operation (not necessarily invertible).
- REDUCE-CHILD-SUBTREES($v, \oplus, \otimes$): Computes the value of $\Sigma(c_1) \otimes \Sigma(c_2) \otimes \Sigma(c_3) \otimes \cdots \otimes \Sigma(c_d)$ where $c_1, \ldots, c_d$ are the children of node v, $\oplus$ and $\otimes$ are associative operations (not necessarily invertible), and $\Sigma(x) = $ val(x_1) $\oplus \cdots \oplus$ val(x_m) is some information about the subtree rooted at x and containing nodes $x_1, \ldots, x_m$.
- COMBINE($v, \odot$): Computes the value of val(v_1) $\odot \cdots \odot$ val(v_m), where $v = v_1, v_2, \ldots, v_m$ are the nodes in the path from v to the root of the tree, and $\odot$ is an associative and invertible operation.

Differently from all other arguments, the operations denoted with $\oplus$, $\otimes$ and $\odot$ used in the three operations above have to be known in advance, so that the DFT-TREE knows what partial evaluations it should memoize in the nodes.

Among the three operations, COMBINE is the most straightforward, implementation-wise. The idea is to assign a value to both the open-nodes and close-nodes of the DFT-TREE: we assign the value of the vertex val(v) to the open-node of v, and the opposite value $-$val(v), i.e. the inverse of val(v) with respect to operation $\odot$, to the corresponding close-node. We can thus state the following lemma, depicted in Fig. 5 for the case $\odot$ is the traditional sum operator '$+$':

Lemma 6. *Let* open-v *be the open node associated with the tree node v. The value of* COMBINE($v, \odot$) *is equal to the $\odot$-combination of the values of the nodes in the prefix of the* DFT-TREE *ending in* open-v.

In order to implement REDUCE-CHILDREN and REDUCE-CHILD-SUBTREE, we need to extend the summary of a sequence of parentheses.

Let us note that it is possible to uniquely decompose any sequence of parentheses in three contiguous (possibly empty) pieces, namely a *prefix*, a *body* and a *suffix*. If the down-value of the sequence is (strictly) negative, then the prefix ends in leftmost minimal-depth parenthesis of the sequence, and the body ends in the rightmost minimal-depth parenthesis. If, on the contrary, the down-value of the sequence is 0, we can distinguish two separate cases: if the up-value is 0, then both the prefix and the suffix are empty, and the body coincides with the whole sequence; else, both the prefix and the body are empty, and the suffix coincides with the whole sequence. In any case, notice that the body of a sequence is a balanced subsequence, made of zero or more *subtrees*. As an example, consider these five sequences:

-)()((): the prefix is), the body is () and the suffix is ((
-)()): the prefix is)()), both body and suffix are empty
-))(: the prefix is)), the body is empty and the suffix is (
- ((): both the prefix and the body are empty, and the suffix is ((
- (()()): both the prefix and the suffix are empty, while the body is (()())

We use this property, i.e. the unique decomposition of a sequence of parentheses, in the two summaries, used respectively by REDUCE-CHILDREN and REDUCE-CHILD-SUBTREE to incrementally aggregate information about subtrees. Below we report the simpler one, used in REDUCE-CHILDREN:

Definition 3 (rc-summary). An *rc-summary* of a sequence of parentheses is a tuple having these fields:

- *prefix-depth*, the depth of the minimal-depth parenthesis
- *body-combination*, the ⊕-combination of the values of the nodes associated with the subtrees of the body of the sequence.
- *suffix-depth*, the difference between the depth of the last parenthesis and the depth of any minimal-depth parenthesis.
- *suffix-info*, the value associated with the first node of the suffix, if any.

The similar *rcs-summary*, used in REDUCE-CHILD-SUBTREE, is reported in the extended paper. These two summaries, to be stored as usual in the nodes of the underlying BBST, and the three generic functions above can be used to implement several functions, and below we report few examples.

Functions implemented using REDUCE-CHILDREN. We can use REDUCE-CHILDREN to implement:

- CHILDREN-SUM(v): Finds the sum of the values of the children of node v. This is equivalent to REDUCE-CHILDREN(v, +).
- CHILDREN-MAX(v): Finds the maximal value among those of the children of node v. This is equivalent to REDUCE-CHILDREN(v, max).

Note that, if we set val(x) = 1 for every vertex in the forest, DEGREE(v) can be derived as well from REDUCE-CHILDREN(v, +).

Functions implemented using REDUCE-CHILD-SUBTREES. In the case of REDUCE-CHILD-SUBTREES we can implement:

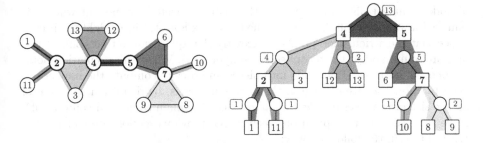

Fig. 6. A graph (left) and its Block Forest [27] (right).

- SUBTREE-SUM(v): Finds the sum of the values of the nodes in the subtree of node v, and is equivalent to val(v) + REDUCE-CHILD-SUBTREES($v, +, +$).
- SUBTREE-SIZE(v): Finds how many nodes are there in the subtree of node v, and is equivalent to SUBTREE-SUM(v) when val(x) = 1 for every node x of the forest.
- SUBTREE-MAX(v):
  Finds the maximal value among those of the nodes in the subtree of node v, and is equivalent to max(val(v), REDUCE-CHILD-SUBTREES(v, max, max).
- MAXSUM-CHILD(v): Finds the maximal value of SUBTREE-SUM among the children of node v. This is equivalent to REDUCE-CHILD-SUBTREES($v, +$, max)).

Functions implemented using COMBINE. A simple example of COMBINE is DEPTH(v), which finds the depth of node v, i.e. the distance from v to the root of the tree v belongs to. Indeed, this is equivalent to COMBINE($v, +$), assuming val(x) = 1 for every node x of the forest. We can implement DISTANCE(u, v), i.e. the distance in the tree between u and v, by computing DEPTH(u) + DEPTH(v) − 2 · DEPTH(LCA(u, v)). If we want to compute the distances in a weighted tree (i.e., we have weights on the edges), the same idea holds; since we store the information in the nodes, we store the weight of an edge connecting a child node to the parent node inside the child node.

5 Application: Biconnectivity Properties and Impact of Articulation Points

The DFT-TREE can be used to maintain all the (bi)connectivity properties of a streaming graph, following the same approach proposed by Westbrook and Tarjan [27]: as we mentioned before, it is sufficient to observe that the DFT-TREE supports all the operations needed by the algorithm of Westbrook and Tarjan to maintain the Block Forest (shown in Fig. 6), including CONDENSE that, as we mentioned before, is not a standard operation in the case of the *dynamic tree* problem. Indeed, it is possible to maintain connected and biconnected components, and bridges and articulation points of a streaming graph.

We now show how to answer queries on the *impact* of an articulation point. We recall, from [5], that the impact of an articulation point v is the number of nodes that get disconnected from the main connected component when v is removed from the graph. Looking at the the Block Forest, Fig. 6 (right), it is easy to see that the articulation points are exactly the square nodes that connect two or more round nodes (the biconnected components). When an articulation point is removed, its Block Tree splits into pieces: in order to compute the impact, we need to know the size of each of them: the impact is, by definition, the sum of all the size of the trees except the largest one (the main connected component). If we refer the subtree operations seen in the previous section, we can use the DFT-TREE in the following way:

- The value in each round node in the tree is 0 (they corresponds to biconnected components), and 1 in each square node (corresponding to real nodes in the graph).
- The size of the Block Tree can be computed by finding the root of the tree, using ROOT and then computing its SUBTREE-SIZE.
- The size of the maximum subtree of v can be computed using MAXSUM-CHILD.

It is easy to see that, with the operations described above, we can compute the impact of a node, and thus we can state the following result.

Lemma 7. *Using a DFT-TREE, it is possible to answer* impact *queries of a vertex in time* $\mathcal{O}(\log n)$.

6   Conclusion and Future Works

In this paper we presented a novel data structure, the *Depth First Tour Tree*. This structure is based on a linearization of a DFS visit of the tree, similarly to the ET-Trees (based on a Euler Tour). The structure is simple and easy to implement; it provides a framework for a large class of data aggregation tasks – especially on subtrees, a task that is usually non-trivial with other data structures. Furthermore, DFT-TREES can naturally aggregate on all the children subtrees of a node v in parallel without having to pay a cost proportional to the degree of v itself: as we already mentioned, given a node v it takes $\mathcal{O}(\log n)$, independently from the degree of v, to answer the child of v whose subtree is the largest. This flexibility, related to subtree queries, is paid by the EVERT operation, that requires a cost proportional to the depth of the node being everted. However, as discussed, when either the number of eversions is small compared to the total number of queries performed, or the costs of the eversion is amortized, the cost of EVERT can be regarded as being $\mathcal{O}(\log n)$ like all the other structural operations. This is the case in the illustrated application.

In the future, we plan to experimentally assess the performance of our data structure, and compare it with the existing alternatives, following the approach of [24]. We believe that the simplicity of our approach, when compared e.g. to the work of [19] in the context of the SUBTREE-MAX operation, is likely to deliver faster and more readable code in practice.

References

1. Acar, U.A., Blelloch, G.E., Harper, R., Vittes, J.L., Woo, S.L.M.: Dynamizing static algorithms, with applications to dynamic trees and history independence. In: SODA (2004)
2. Acar, U.A., Blelloch, G.E., Vittes, J.L: An experimental analysis of change propagation in dynamic trees. In: ALENEX (2005)
3. Alstrup, S., Holm, J., de Lichtenberg, K., Thorup, M.: Minimizing diameters of dynamic trees. In: ICALpP (1997)
4. Alstrup, S., Holm, J., Lichtenberg, K.D., Thorup, M.: Maintaining information in fully dynamic trees with top trees. ACM Trans. Algorithms 1(2), 243–264 (2005)
5. Ausiello, G., Firmani, D. and Laura, L.: Real-time analysis of critical nodes in network cores. In: IWCMC (2012)
6. Thomas, H.C., Charles, E.L., Ronald, L.R., Clifford, S.: Introduction to Algorithms, 3rd edn. MIT Press, Cambridge (2009)
7. Diestel, R.: Graph Theory, 4th edn. Springer, Heidelberg (2010)
8. Frederickson, G.N.: Data structures for on-line updating of minimum spanning trees, with applications. SIAM J. Comput. 14, 781–798 (1985)
9. Frederickson, G.N.: Ambivalent data structures for dynamic 2-edge-connectivity and k smallest spanning trees. SIAM J. Comput. 26(2), 484–538 (1997)
10. Frederickson, G.N.: A data structure for dynamically maintaining rooted trees. J. Algorithms 24(1), 37–65 (1997)
11. Georgiadis, L., Kaplan, H., Shafrir, N., Tarjan, R.E., Werneck, R.F.: Data structures for mergeable trees. ACM Trans. Algorithms 7(2), 14 (2011)
12. Goldberg, A.V., Grigoriadis, M.D., Tarjan, R.E.: Use of dynamic trees in a network simplex algorithm for the maximum flow problem. Math. Program 50(3), 277–290 (1991)
13. Henzinger, M.R., King, V.: Randomized fully dynamic graph algorithms with polylogarithmic time per operation. J. ACM 46(4), 502–516 (1999)
14. Joannou, S., Raman, R.: Dynamizing succinct tree representations. In: Klasing, R. (ed.) SEA 2012. LNCS, vol. 7276, pp. 224–235. Springer, Heidelberg (2012)
15. Kaplan, H., Molad, E., Tarjan, R.E.: Dynamic rectangular intersection with priorities. In: STOC (2003)
16. Langerman, S.: On the shooter location problem. In: CCCG (2000)
17. Munro, J.I., Raman, V.: Succinct representation of balanced parentheses and static trees. SIAM J. Comput. 31(3), 762–776 (2001)
18. Navarro, G., Sadakane, K.: Fully functional static and dynamic succinct trees. ACM Trans. Algorithms 10(3), 16 (2014)
19. Radzik, T.: Implementation of dynamic trees with in-subtree operations. ACM J. Exp. Algorithms 3, 9 (1998)
20. Sleator, D.D., Tarjan, R.E.: A data structure for dynamic trees. J. Comput. Syst. Sci. 26(3), 362–391 (1983)
21. Sleator, D.D., Tarjan, R.E.: Self-adjusting binary search trees. J. ACM 32(3), 652–686 (1985)
22. Tarjan, R.E.: Dynamic trees as search trees via Euler tours, applied to the network simplex algorithm. Math. Program. 78(2), 169–177 (1997)
23. Tarjan, R.E., Werneck, R.F.: Self-adjusting top trees. In: SODA (2005)
24. Tarjan, R.E., Werneck, R.F.: Dynamic trees in practice. ACM J. Exp. Algorithmics 14, 5 (2009)

25. Werneck, R.F.: Design and analisys of data structures for dynamic trees. Ph.D thesis (2006)
26. Werneck, R.F.: Dynamic trees. In: Kao, M.-Y. (ed.) Encyclopedia of Algorithms. Springer, Heidelberg (2008)
27. Westbrook, J., Tarjan, R.E.: Maintaining bridge-connected and biconnected components on-line. Algorithmica **7**(1–6), 433–464 (1992)

Schröder Partitions and Schröder Tableaux

Luca Ferrari[1]([⊠])

Dipartimento di Matematica e Informatica "U. Dini",
Università degli Studi di Firenze, Viale Morgagni 65, 50134 Florence, Italy
luca.ferrari@unifi.it

Abstract. We introduce the notions of *Schröder shape* and of *Schröder tableau*, which provide some kind of analogs of the classical notions of Young shape and Young tableau. We investigate some properties of the partial order given by containment of Schröder shapes. Then we propose an algorithm which is the natural analog of the well known RS correspondence for Young tableaux, and we characterize those permutations whose insertion tableaux have some special shapes. We end our paper with a few suggestions for possible further work.

1 Introduction

Given a positive integer n, a *partition* of n is a finite sequence of positive integers $\lambda = (\lambda_1, \lambda_2, \ldots, \lambda_r)$ such that $\lambda_1 \geq \lambda_2 \geq \cdots \geq \lambda_r$ and $n = \lambda_1 + \lambda_2 + \cdots + \lambda_r$. When λ is a partition of n we also write $\lambda \vdash n$. A graphical way of representing partitions is given by Young shapes. The *Young shape* of the above partition $\lambda \vdash n$ consists of r left-justified rows having $\lambda_1, \ldots, \lambda_r$ boxes (also called cells) stacked in decreasing order of length. The set of all Young shapes can be endowed with a poset structure by containment (of top-left justified shapes). Such a poset turns out to be in fact a lattice, called the *Young lattice*. A *standard Young tableau* with n cells is a Young shape whose cells are filled in with positive integers from 1 to n in such a way that entries in each row and each column are (strictly) increasing.

Young tableaux are among the most investigated combinatorial objects. The widespread interest in Young tableaux is certainly due both to their intrinsic combinatorial beauty (which is witnessed by several surprising facts concerning, for instance, their enumeration, such as the hook length formula and the RSK algorithm) and to their usefulness in several algebraic contexts, typically in the representation theory of groups and related matters (such as Schur functions and the Littlewood-Richardson rule).

Apart from their classical definition, there are several alternative ways to introduce Young tableaux. In the present paper we are interested in the possibility of defining standard Young tableaux in terms of a certain lattice structure

---

L. Ferrari—Partially supported by INdAM - GNCS 2015 project "Problemi di consistenza, unicitá e ricostruzione per grafi ed ipergrafi" and by MIUR PRIN 2010–2011 grant "Automi e Linguaggi Formali: Aspetti Matematici e Applicativi", code H41J12000190001.

Z. Lipták and W.F. Smyth (Eds.): IWOCA 2015, LNCS 9538, pp. 161–172, 2016.
DOI: 10.1007/978-3-319-29516-9_14

on Dyck paths. The main advantage of this point of view lies in the possibility of giving an analogous definition in a modified setting, in which Dyck paths are replaced by some other class of lattice paths. Here we will try to see what happens if we replace Dyck paths with Schröder paths, just scratching the surface of a theory that, in our opinion, deserves to be better studied.

Given a Cartesian coordinate system, a *Dyck path* is a lattice path starting from the origin, ending on the x-axis, never falling below the x-axis and using only two kinds of steps, $u(p) = (1, 1)$ and $d(own) = (1, -1)$. A Dyck path can be encoded by a word w on the alphabet $\{u, d\}$ such that in every prefix of w the number of u's is greater than or equal to the number of d's and the total number of u and d in w is the same (the resulting language is called *Dyck language* and its words *Dyck words*). The *length* of a Dyck path is the length of the associated Dyck word (which is necessarily an even number).

Consider the set $\mathbf{D}_n$ of all Dyck paths of length $2n$; it can be endowed with a very natural poset structure, by declaring $P \leq Q$ whenever P lies weakly below Q in the usual two-dimensional drawing of Dyck paths (for any $P, Q \in \mathbf{D}_n$). This partial order actually induces a distributive lattice structure on $\mathbf{D}_n$, to be denoted $\mathcal{D}_n$ and called *Dyck lattice of order n*. This can be shown both in a direct way, using the combinatorics of lattice paths (see [FP]), and as a consequence of the fact that $\mathcal{D}_n$ is order-isomorphic to (the dual of) the Young lattice of the staircase partition $(n - 1, n - 2, \ldots, 2, 1)$ (that is the principal down-set generated by such a staircase partition in the Young lattice). Referring to the latter approach, any $P \in \mathbf{D}_n$ uniquely determines a Young shape, which can be obtained by taking the region included between P and the maximum path of $\mathcal{D}_n$, then slicing it into square cells using diagonal lines of slope 1 and -1 passing through all points having integer coordinates, and finally rotating the sheet of paper by 45° anticlockwise (see Fig. 1).

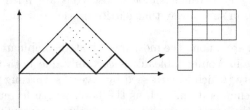

Fig. 1. A Dyck path and the associated Young shape.

It is well known that there is a bijection between standard Young tableaux of a given shape and saturated chains in the Young lattice starting from the empty shape and ending with that shape. Translating this fact on Dyck lattices, we can thus state that standard Young tableaux of a given shape are in bijection with saturated chains (inside a Dyck lattice of suitable order) starting from the Dyck path associated with that shape and ending with the maximum of the lattice.

This suggests us to try to find an analog of this fact in which Dyck paths are replaced by other types of paths. As already mentioned, the case treated in the present paper is that of Schröder paths.

In Sect. 2 we introduce the notion of Schröder shape and study some properties of the poset of Schröder shapes (in some sense analogous to those of the Young lattice). In Sect. 3 we introduce the notion of Schröder tableau and we define an algorithm which, given a permutation, produces a pair of Schröder tableaux having the same Schröder shape; this is made in analogy with the classical RS algorithm. In particular, we will address the problem of determining which permutations are mapped into the same Schröder insertion tableau, and we solve it for a few special shapes. Finally, we devote Sect. 4 to the presentation of some directions of further research.

2   The Poset of Schröder Partitions

A *Schröder shape* is a set of triangular cells in the plane obtained from a Young shape by drawing the NE-SW diagonal of each of its (square) cells, and possibly adding at the end of some rows one more triangular cell, provided that, in a group of rows having equal length, only the first (topmost) one can have an added triangle. The number of cells of a Schröder shape is called the *order* of that shape. An example of a Schröder shape is illustrated in Fig. 2.

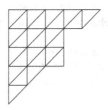

Fig. 2. A Schröder shape of order 25.

A Schröder shape has triangular cells of two distinct types, which will be referred to as *lower triangular cells* and *upper triangular cells*. In particular, rows having an odd number of cells necessarily terminate with an upper triangular cell. A Schröder shape determines a unique integer partition, whose parts are the number of cells in the rows of the shape. For instance, the partition associated with the shape in Fig. 2 is $(9, 6, 6, 3, 1)$. As a consequence of the definition of a Schröder shape, it is clear that not every partition can be represented using a Schröder shape. More precisely, we have the following result, whose proof is completely trivial and so it is left to the reader.

Proposition 1. *An integer partition can be represented with a Schröder shape if and only if its odd parts are* simple *(i.e. have multiplicity 1).*

Those integer partitions which can be represented with a suitable Schröder shape will be called *Schröder partitions*. The set of all Schröder partitions will be denoted **Sch**, and the set of Schröder partitions of order n with **Sch**$_n$. From now on we will frequently refer to Schröder shapes and to Schröder partitions interchangeably, when no confusion is likely to arise.

From the enumerative point of view, the number of Schröder partitions is known, and is recorded in [Sl] as sequence A006950. In particular, the generating function of Schröder partitions is given by

$$\prod_{k>0} \frac{1 + x^{2k-1}}{1 - x^{2k}}.$$

There are several combinatorial interpretations for the resulting sequence, however an appropriate reference for the present one (in terms of Schröder partitions) appears to be [D]. In that paper the author proves a far more general result, concerning partitions such that the multiplicity of each odd part is in a prescribed set and the multiplicity of each even part is unrestricted.

It is interesting to notice that this sequence is also relevant from an algebraic point of view. Indeed it coincides with the sequence of numbers of nilpotent conjugacy classes in the Lie algebras $o(n)$ of skew-symmetric $n \times n$ matrices. This suggests that Schröder partitions have a role in representation theory that certainly deserves to be better investigated.

Here we propose a refined enumerative result, namely we describe a simple recurrence for the number of Schröder partitions of n into k parts.

Proposition 2. *Denote with $s_{n,k}$ the number of Schröder partitions of n into k parts and with $s'_{n,k}$ the number of Schröder partitions of n into k parts having smallest part different from 1. Then, for all $n \geq k \geq 1$:*

(i) $s_{n,k} = s'_{n,k} + s'_{n-1,k-1}$;
(ii) $s'_{n,k} = s'_{n-2,k-1} + s'_{n-2k-1,k-1} + s'_{n-2k,k}$.

Proof. We immediately observe that the set of Schröder partitions of n into k parts whose smallest part is equal to 1 is in bijection with the set of Schröder partitions of n into $k - 1$ parts whose smallest part is different from 1. This gives at once the formula in (i).

Concerning (ii), given a Schröder partition λ of n into k parts with no part equal to 1, we distinguish two cases. If λ has at least one part equal to 2, then removing it leaves us with a Schröder partition of $n - 2$ into $k - 1$ parts, still having no part equal to 1. Otherwise, removing the first two columns of λ returns a Schröder partition of $n - 2k$ into k parts, possibly having one part equal to 1. From here, using (i), we immediately obtain (ii). ∎

Though the formalism of Schröder shapes seems not to add relevant information on the enumerative combinatorics of Schröder partitions, it suggests at least an interesting family of maps on integer partitions, which turns out to define a family of involutions if suitably restricted. Consider the family of maps $(c_n)_{n \in \mathbf{N}}$ defined on the set of all integer partitions as follows: given a partition λ and a positive integer n, $c_n(\lambda)$ is the integer partition $\mu = (\mu_1, \mu_2, \ldots, \mu_k)$ (of the same size as λ) whose i-th part μ_i is given by the sum of the n columns of (the Young shape of) λ from the $((i-1)n+1)$-th one to the (in)-th one. So, for instance, $c_3((7,6,6,6,4,3,3,1)) = (22,13,1)$. Since each of the above maps preserves the size of a partition, it is clearly an endofunction when restricted to the set of all integer partitions of size n. Notice that c_1 is the well-known conjugation map (which exchanges rows with columns in a Young shape). In spite of the fact that c_1 is an involution (on the set of all partitions), it is easy to see that all the other c_n's are not involutions. However, it is possible to characterize the set of those partitions for which c_n^2 acts as the identity map.

Proposition 3. *Given $n \in \mathbf{N}$ and an integer partition λ (whose i-th part will be denoted λ_i, as usual), we have that $c_n^2(\lambda) = \lambda$ if and only if the following two conditions hold:*

- *if $\lambda_i \not\equiv 0 \pmod{n}$, then λ_i is simple;*
- *there is at most one part λ_i of λ such that $\lambda_i < n$.*

Proof. For any given λ, suppose that there exists one part of $c_n(\lambda)$ which is $\not\equiv 0 \pmod{n}$; denoting with μ' the first of them, this means that λ has a set of n consecutive columns whose sum is equal to μ'. Since $\mu' \not\equiv 0 \pmod{n}$, this implies that such n columns are not all equal. Now, since in a Young shape columns are in decreasing order of length, it is impossible that the successive n columns of λ sum up to μ', hence μ' is simple. We have thus proved that the first of the two conditions in the above statement holds for every partition in the image of c_n. This is enough to conclude that, if $c_n^2(\lambda) = \lambda$, then necessarily the same condition holds for λ (which lies indeed in the image of c_n). Moreover, if λ has at least two parts $< n$, then certainly $c_n^2(\lambda) \neq \lambda$, since each part of a partition in the image of c_n has length at least n, except at most for its smallest part.

Conversely, observe that we can represent every partition λ by means of a Young-like shape, which is obtained from the usual Young shape of λ by simply grouping together the cells of each row n by n. In this way we obtain a shape (call it $\tilde{\lambda}$) in which each cell is a horizontal rectangle made of n cells of the original Young shape, except at most the last cell of each row, which is a horizontal rectangle having *at most* n cells. Now observe that $c_n(\lambda)$ can be obtained by exchanging the rows and the columns of $\tilde{\lambda}$ and then breaking the horizontal rectangles of the resulting shape into n square cells. This construction is illustrated below for the partition $\lambda = (9,7,6,6,6,4,3,3,2)$ and $n = 3$: cells with the same label have to be grouped together, and the resulting partition $c_3(\lambda) = (26,16,4)$ is depicted on the right.

It is now obvious that, performing twice this operation, one gets back to the original partition λ, that is $c_n^2(\lambda) = \lambda$, as desired. ∎

a	a	a	b	b	b	c	c	c
a	a	a	b	b	b	c		
a	a	a	b	b	b			
a	a	a	b	b	b			
a	a	a	b	b	b			
a	a	a	b					
a	a	a						
a	a	a						
a	a							

a	a	a	a	a	a	a	a	a	a	a	a	a	a	a	a	a	a	a	a	a	a	a	a
b	b	b	b	b	b	b	b	b	b	b	b	b	b	b	b								
c	c	c	c																				

As already mentioned, as a special case of the above proposition we have that the set of all integer partitions is the set of fixed points of the map c_1^2 (where c_1 is the conjugation map). Another consequence is recorded in the following corollary, which shows the role of Schröder partitions in this context.

Corollary 1. *The set of Schröder partitions is the set of fixed points of the map c_2^2.*

Proof. Just observe that, setting $n = 2$ in the previous proposition, the first condition tells that odd parts have to be simple, whereas the second condition becomes a special case of the first. ∎

The set **Sch** of all Schröder shapes can be naturally endowed with a poset structure, by declaring $\lambda \leq \mu$ whenever the set of cells of the shape λ is a subset of the set of cells of the shape μ, provided that we draw the two shapes in such a way that their top left cells coincide. This is equivalently (and perhaps more formally) expressed in terms of Schröder partitions: if $\lambda = (\lambda_1, \ldots, \lambda_h)$ and $\mu = (\mu_1, \ldots, \mu_k)$, then $\lambda \leq \mu$ when $h \leq k$ and, for all $i \leq h$, $\lambda_i \leq \mu_i$. Therefore the poset $\mathscr{S}$ of Schröder shapes is actually a subposet of the Young lattice. However, it seems not at all a trivial one; notice, in particular, that an interval of the Young lattice whose endpoints are Schröder partitions does not contain only Schröder partitions (apart from very simple cases). In general, it appears to be very hard (if not impossible) to infer nontrivial properties of the Schröder poset from properties of the Young lattice.

One of the most fundamental properties the Schröder poset shares with the Young lattice is recorded in the next theorem.

Theorem 1. *The Schröder poset $\mathscr{S}$ is a distributive lattice.*

Proof. Since every sublattice of a distributive lattice is distributive, it will be enough to show that $\mathscr{S}$ is a sublattice of the Young lattice.

Given two Schröder partitions λ and μ, their join in the Young lattice is the partition $\lambda \vee \mu$ whose i-th part is the maximum between λ_i and μ_i, for all i. We will now show that $\lambda \vee \mu$ is a Schröder partition.

Suppose that $(\lambda \vee \mu)_i$ is an odd part of $\lambda \vee \mu$. Moreover, suppose w.l.o.g. that $(\lambda \vee \mu)_i = \lambda_i$ (which means that $\lambda_i \geq \mu_i$). Now $\lambda_i > \lambda_{i+1}$ (since we are supposing that λ_i is odd). Moreover, if μ_i is even, then $\lambda_i > \mu_i \geq \mu_{i+1}$ (once again since λ_i is odd), whereas if μ_i is odd, then $\lambda_i \geq \mu_i > \mu_{i+1}$ (since μ is a Schröder partition). Thus, in all cases $(\lambda \vee \mu)_i = \lambda_i > \lambda_{i+1}, \mu_{i+1}$, hence $(\lambda \vee \mu)_i > (\lambda \vee \mu)_{i+1}$. Using a similar argument it is possible to show that $(\lambda \vee \mu)_i < (\lambda \vee \mu)_{i-1}$. We have thus shown that $(\lambda \vee \mu)$ has simple odd parts, i.e. it is a Schröder partition.

Using a completely similar argument one can also show that the meet of two Schröder partitions in the Young lattice is again a Schröder partition, thus completing the proof.   ∎

Remark. Notice that the above theorem can be generalized as follows. For a given function $f : \mathbf{N} \to \mathbf{N} \cup \{\infty\}$, consider the set of integer partitions in which part i appears at most $f(i)$ times. Such set is a sublattice of Young lattice (with partwise join and meet). This generalization, though interesting, will play no role in the present paper.

3   An RSK-like Algorithm for Schröder Tableaux

From the algorithmic point of view, the main application of Young tableaux is in the context of the RSK algorithm. This algorithm, named after Robinson, Schensted and Knuth, takes as input a word (on the alphabet of positive integers) of length n and produces in output two semistandard Young tableaux with n cells having the same shape. For what concerns us, we will deal with a special case of the RSK algorithm, often referred to as *Robinson-Schensted correspondence* (briefly, RS correspondence), in which the input is a permutation of length n and the output is given by a pair of standard Young tablueaux. A brief description of such an algorithm is given below (Algorithm 1, where $\pi = \pi_1 \pi_2 \cdots \pi_n$ is a generic permutation of length n).

The RSK algorithm is extensively described in the literature. For instance, the interested reader can find a modern and elegant presentation of it in [Be]. Among other things, one of the most beautiful properties of the RS correspondence is that it establishes a bijection between permutations of length n and pairs of standard Young tableaux with n cells having the same shape. This fact bears important enumerative consequences, as well as strictly algebraic ones. For a given permutation π, the tableaux of the pair(P, Q) returned by the RS algorithm are usually referred to as the *insertion tableau* (the tableau P) and the *recording tableau* (the tableau Q). As a consequence, we have the following nice result, which can again be found in [Be].

Theorem 2. *Denote with f^λ the number of standard Young tableaux of shape λ. Then we have:*
$$n! = \sum_{\lambda \vdash n} (f^\lambda)^2.$$

A *standard Schröder tableau* (from now on, simply *Schröder tableau*) with n cells is a Schröder shape whose cells are filled in with positive integers from 1 to n in such a way that entries in each row and each column are (strictly) increasing.

Algorithm 1. RS(π)

$P := \boxed{\pi_1}$;

$Q := \boxed{1}$;

for k from 2 to n **do**

    $\alpha := \pi_k$;

    for $i \geq 1$ **do**

        if α is bigger than all elements in the i-th row of P **then**

            append a cell with π_k inside at the end of the i-th row of P;

            append the cell $\boxed{k}$ at the end of the i-th row of Q;

            break;

        else

            write α in the cell of the i-th row containing the smallest element β bigger than α;

            $\alpha := \beta$;

        end

    end

end

We propose here a natural analog of the RS algorithm for Schröder tableaux. The main difference (which is due to the specific underlying shape of a Schröder tableaux) lies in the fact that there are two distinct ways of managing the insertion of a new element in the tableau, depending on whether the cell it should be inserted in is an upper triangle or a lower triangle. As a consequence, our algorithm does not establish a bijection between permutations and pairs of Schröder tabealux; nevertheless, due to the strict analogy with the RS correspondence, we believe that it is very likely to have interesting combinatorial properties. A description of our algorithm is given below (Algorithm 2, where π is as in Algorithm 1).

Example. Consider the permutation $\pi = 465193287$. The pair (P, Q) of Schröder tableaux produced by applying the algorithm Sch to π is:

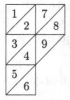   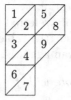

In this section we aim at starting the investigation of the combinatorial properties of this RS-analog. More specifically, we will address the following problem: given a Schröder shape P, can we characterize those permutations having a Schröder tableau of shape P as their insertion tableau? How many of them are there? This problem seems to be quite difficult in its full generality; here we will deal with very few simple cases, for which we can provide complete answers.

Algorithm 2. Sch(π)

$P :=$ the 1-cell Schröder tableau with π_1 written in the cell;
$Q :=$ the 1-cell Schröder tableau with 1 written in the cell;
for k from 2 to n **do**

    $\alpha := \pi_k$;
    for $i \geq 1$ **do**

        if α is bigger than all elements in the i-th row of P **then**
            append a cell (either an upper or a lower triangle) with π_k inside at
            the end of the i-th row of P;
            append a cell (either an upper or a lower triangle) with k inside at
            the end of the i-th row of Q;
            break;

        else
            let A be the cell of the i-th row containing the smallest element
            bigger than α;
            if A is an upper triangle **then**
                $\beta :=$ content of the lower triangle immediately below A;
                move the content of A to the lower triangle immediately below
                A;
                write α in A;
                $\alpha := \beta$;
            else
                $\beta :=$ content of A;
                write α in A;
                $\alpha := \beta$;
            end
        end
    end
end

3.1   Permutations with Given Schröder Insertion Shape: Some Cases

The first case we investigate is that of a Schröder shape consisting of a single row (which can terminate either with an upper or a lower triangle). To state our result we first need to recall a classical definition.

Given a permutation $\pi = \pi_1 \cdots \pi_n$, we say that π_i is a *left-to-right maximum* (or, briefly, *LR maximum*) whenever $\pi_i = \max(\pi_1, \ldots, \pi_i)$.

Proposition 4. *Let* $\pi = \pi_1 \cdots \pi_n$ *be a permutation of length* n. *The Schröder insertion tableau of* π *has a single row if and only if, for all* $i \geq n$:

1. *if* i *is odd, then* π_i *is a LR maximum of* π;
2. *if* i *is even, then* π_i *is a LR maximum of the permutation obtained from* π *by removing* π_{i-1} *(and suitably renaming the remaining elements).*

Proof. Suppose we are inserting π_i in the insertion tableau P, which is assumed to consist of a single row. If i is odd, then the last cell of P is a lower triangle;

in order not to create new rows, π_i has necessarily to be a LR maximum. On the other hand, if i is even, then the last cell of P is an upper triangle; in this case, π_i can be inserted in P in two ways: either π_i is a LR maximum, and so it is simply appended at the end of the unique row of P, or π_i is greater than all previous elements of π but π_{i-1}, hence π_i is inserted in the cell containing π_{i-1} (which is the last cell of the unique row of P, and so it is an upper triangle) and a new cell (a lower triangle) containing π_{i-1} is added at the end of the unique row of P.

Conversely, it is easy (and so left to the reader) to check that a permutation satisfying conditions 1 and 2 in the statement of the present proposition must have a Schröder insertion tableau consisting of a single row.   ∎

The permutations π of length n whose Schröder insertion tableau have a single row can therefore be simply characterized as follows: for all i, $\{\pi_{2i+1}, \pi_{2i+2}\} = \{2i + 1, 2i + 2\}$. As a consequence of this fact, a formula for the number of such permutations follows immediately.

Proposition 5. *The set of permutations of length n whose Schröder insertion tableau consists of a single row has cardinality $2^{\lfloor \frac{n}{2} \rfloor}$.*

The second case we consider is the natural counterpart of the previous one, that is Schröder shapes having a single column. Despite the similarities with the previous case, it turns out that the set of permutations having Schröder insertion tableau of this form can be nicely described in terms of *pattern avoidance*.

Given two permutations σ and $\tau = \tau_1 \cdots \tau_n$ (of length k and n respectively, with $k \leq n$), we say that there is an *occurrence* of σ in τ when there exists indices $i_1 < i_2 < \cdots < i_k$ such that $\tau_{i_1} \tau_{i_2} \cdots \tau_{i_k}$ is order isomorphic to σ. When there is an occurrence of σ in τ, we also say that τ *contains the pattern* σ. When τ does not contain σ, we say that τ *avoids the pattern* σ. The set of all permutations of length n avoiding a given pattern σ is denoted with $Av_n(\sigma)$. Some useful references for the combinatorics of patterns in permutations are [Bo,K], whereas similar notions of patterns in set partitions and in compositions and words are studied in [M,HM], respectively.

Proposition 6. *Let $\pi = \pi_1 \cdots \pi_n$ be a permutation of length n. The Schröder insertion tableau of π has a single column if and only if $\pi \in Av_n(123, 213)$.*

Proof. An argument similar to that of the preceding proposition shows that the Schröder insertion tableau of π has a single column if and only if, for all $i \leq n$, $\pi_i < \min(\{\pi_1, \ldots, \pi_{i-1}\} \setminus \min\{\pi_1, \ldots, \pi_{i-1}\})$ (i.e., π_i is smaller than the second minimum of set of all previous elements). Thus π can be factored into subpermutations (made of consecutive elements of π), say $\pi = \tilde{\pi}_1 \cdots \tilde{\pi}_r$, in such a way that each factor $\tilde{\pi}_i$ is isomorphic to a permutation of the form $1t(t-1) \cdots 32$ (for some t) and each element of $\tilde{\pi}_i$ is greater than each element of $\tilde{\pi}_{i+1}$ (for all i). In the language of permutation patterns, this is usually expressed by saying that π is a *skew sum* of permutations of the form $1t(t - 1) \cdots 32$. It is now a known fact (see, for instance, [AA]) that such permutations are precisely those avoiding the two patterns 123 and 213.   ∎

Many classes of permutations avoiding a given set of patterns have been enumerated. The above one is among them, see [SiSc].

Proposition 7. *The set of permutations of length n whose Schröder insertion tableau consists of a single column has cardinality 2^{n-1}.*

We close this section by simply stating (without proof) one more case, which is, in some sense, a generalization of both the cases described above. Namely, we consider the case of what can be called *Schröder hooks*, that is Schröder shapes having at most one row and one column with more than one cell.

Again, we need to recall a classical definition, and also to give a new one. A *shuffle* of two permutations σ and τ (having length n and m, respectively) is a permutation of length $n + m$ having two disjoint subpermutations (not made in general by adjacent elements of π) isomorphic to σ and τ. Moreover, if the subpermutations of σ and τ formed by the first k elements are isomorphic, a *k-rooted shuffle* of σ and τ is a permutation obtained by concatenating the permutation formed by the first k elements of σ (or τ) (with elements suitably renamed) with a shuffle of the subpermutations formed by the remaining elements of σ and τ. For instance, a shuffle of 25143 and $\overline{4132}$ is given by $4\overline{7}92\overline{1}8\overline{536}$, and a 3-rooted shuffle of 253$\underline{461}$ and 254$\overline{13}$ is given by 3756$\overline{1}$82$\overline{4}$.

Proposition 8. *Let $\pi = \pi_1 \cdots \pi_n$ be a permutation of length n. The Schröder insertion tableau of π is a Schröder hook if and only if π is a 2-rooted shuffle of two permutations having a single row Schröder insertion tableau and a single column Schröder insertion tableau, respectively.*

4   Further Work

The algebraic and combinatorial properties of the distributive lattice $\mathscr{S}$ of Schröder shapes needs to be further investigated. In particular, the analogies with differential posets should be much deepened.

We have just started the characterization and enumeration of permutations having a given Schröder insertion tableau. Many more shapes should be investigated. Moreover, we still have to understand the role of the recording tableau.

Can we find a nice closed formula for the number of Schröder tableaux of a given shape? In the case of Young tableaux there is a famous *hook formula*, which however seems to be unlikely in our case, since we have numerical evidence that, for certain shapes, this number has large prime factors.

An alternative presentation of Schröder tableaux is as Young shapes whose cells are filled in with pairs of distinct integers. This description shows that Schröder tableaux could be somehow related to *interval orders*.

The analogies between Young tableaux and Schröder tableaux should be investigated more, especially from a purely algebraic point of view. Combinatorial objects related to Young tableaux, such as Schur functions and the plactic monoid, as well as algorithmic and algebraic constructions, such as Schützenberger's *jeu de taquin*, the Littlewood-Richardson rule and the Schubert calculus on Grassmannians and flag varieties, could have some interesting counterparts in the context of Schröder tableaux.

Acknowledgement. We are very grateful to the anonymous referees for an extremely careful reading and many useful suggestions, which led to a significant improvement of the presentation.

References

[AA] Albert, M., Atkinson, M.D.: Pattern classes and priority queues. Pure Math. Appl. (PU.M.A.) **23**, 161–177 (2012)

[Be] Bergeron, F.: Algebraic Combinatorics and Coinvariant Spaces. CMS Treatise in Mathematics. A K Peters/CRC Press, Natick (2009)

[Bo] Bóna, M.: Combinatorics of Permutations. Discrete Mathematics and its Applications. CRC Press, Taylor & Francis Group, Boca Raton (2012)

[D] Drake, B.: Limits of areas under lattice paths. Discrete Math. **309**, 3936–3953 (2009)

[FP] Ferrari, L., Pinzani, R.: Lattices of lattice paths. J. Statist. Plann. Infer. **135**, 77–92 (2005)

[HM] Heubach, S., Mansour, T.: Combinatorics of Compositions and Words. Chapman & Hall/CRC, Taylor & Francis Group, Boca Raton (2009)

[K] Kitaev, S.: Patterns in Permutations and Words. EATCS Monographs in Theoretical Computer Science. Springer, Berlin (2011)

[M] Mansour, T.: Combinatorics of Set Partitions. Chapman & Hall/CRC, Taylor & Francis Group, Boca Raton (2012)

[SiSc] Simion, R., Schmidt, F.W.: Restricted permutations. Eur. J. Comb. **6**, 383–406 (1985)

[Sl] Sloane, N.J.A.: The On-Line Encyclopedia of Integer Sequences. http://oeis.org

[St] Stanley, R.P.: Differential posets. J. Amer. Math. Soc. **1**, 919–961 (1988)

Algorithmic Aspects of the S-Labeling Problem

Guillaume Fertin[1], Irena Rusu[1], and Stéphane Vialette[2]($\boxtimes$)

[1] LINA UMR CNRS 6241, Université de Nantes, Nantes, France
{guillaume.fertin,irena.rusu}@univ-nantes.fr
[2] LIGM CNRS UMR 8049, Université Paris-Est, Champs-sur-Marne, France
vialette@univ-mlv.fr

Abstract. Given a graph $G = (V, E)$ of order n and maximum degree Δ, the **NP**-complete S-LABELING problem consists in finding a labeling of G, *i.e.* a bijective mapping $\phi : V \rightarrow \{1, 2 \ldots n\}$, such that $\mathrm{SL}_\phi(G) = \sum_{\{u,v\} \in E} \min\{\phi(u), \phi(v)\}$ is minimized. A preliminary study of the S-LABELING problem has been undertaken in [9]; here, we prolongate this study, and focus more specifically on algorithmic results concerning the problem. We first give intrinsic properties of optimal labelings, which will prove useful for our algorithmic study. We then show that the S-LABELING problem is polynomial-time solvable for (sets of) caterpillars. We also provide upper and lower bounds on $\mathrm{SL}_\phi(G)$, that in turn allow us to determine polynomial-time approximation algorithms for different classes of graphs such as regular graphs, connected graphs and forests, but also for general graphs. Concerning exact algorithms, we show that the problem is solvable in $O^*(1.44225^{n\Delta})$ time, and that deciding whether there exist a labeling ϕ of G such that $\mathrm{SL}_\phi(G) \leq |E| + k$ is solvable in $O^*(2^{2\sqrt{k}} (2\sqrt{k})!)$.

1 Introduction

A *labeling* of a graph G is an assignment of distinct integers to its vertices, in such a way that a certain objective function is optimized. In the literature, a graph labeling of G is also called a *linear arrangement*, a *linear layout* or a *numbering* of the vertices of G. A large amount of relevant combinatorial problems in different areas can be rephrased as graph labeling problems, which have been widely investigated during the last decades. These include numerical analysis [17], VLSI circuit design [4], network reliability [15], computational biology [16], scheduling [1], parallel processing [18], etc. However, for most objective functions, the derived graph labeling problem turns out to be **NP**-complete. Popular graph labeling problems arising in the above-mentioned applications include BANDWIDTH [10], MINIMUM LINEAR ARRANGEMENT [13], CUTWIDTH [2] and VERTEX SEPARATION [19]. There exist several surveys that deal with different aspects of graph labeling problems [3,5,8].

In this paper, we are interested in the S-LABELING problem [9,22], which, given a graph $G = (V, E)$, asks for a graph labeling $\phi : V \rightarrow \{1, 2 \ldots |V|\}$ such that the total sum $\mathrm{SL}_\phi(G) = \sum_{\{u,v\} \in E} \min\{\phi(u), \phi(v)\}$ is minimized, and we write $\mathrm{SL}(G)$ for the minimum $\mathrm{SL}_\phi(G)$ over all possible labelings of G. Note

© Springer International Publishing Switzerland 2016
Z. Lipták and W.F. Smyth (Eds.): IWOCA 2015, LNCS 9538, pp. 173–184, 2016.
DOI: 10.1007/978-3-319-29516-9_15

that, in this problem (and thus, all along this paper), we do not require G to be connected. However, we can always assume G has no isolated vertices. Indeed, if this was the case, let V_I be the set of isolated vertices in G. Since in any labeling ϕ of G, no vertex from V_I contributes to $\mathrm{SL}_\phi(G)$, we can always label the vertices with the highest possible labels from V_I, or equivalently, change G into $G' = (V \backslash V_I, E)$.

The S-LABELING problem has been proved to be **NP**-complete for planar at most cubic graphs [22], but its complexity is unknown for trees, forests, and more generally, for bipartite graphs. The problem has also been studied in [9], where $SL(G)$ is determined for classical classes of graphs (paths, cycles, complete (bipartite) graphs). Upper and lower bounds are also given for general graphs, and approximation ratios are given for trees, regular graphs, and general graphs. The S-LABELING problem has also been proved to be polynomial-time solvable for split graphs [9].

This paper can be seen as a (late) follow-up of [9], mainly focused on algorithmic aspects of the S-LABELING problem, and is organized as follows. Section 2 introduces terminology used throughout this paper. Section 3 presents essential properties of optimal labelings. In Sect. 4, we prove that the S-LABELING problem is polynomial-time solvable for (sets of) caterpillars. In Sect. 5, we investigate polynomial-time approximation and exponential-time algorithms. Due to space constraints, most proofs are omitted and deferred to the full version of this paper.

2   Notations

A *graph* is an ordered pair $G = (V, E)$ comprising a set V of vertices together with a set E of edges, which are 2-element subsets of V. The *order n* (resp. *size m*) of G is its number of vertices (resp. edges). The degree of a vertex $u \in V$, denoted $d_G(u)$ (or $d(u)$ if clear from the context), is the number of edges that are incident to u. We write $\Delta(G)$ (or Δ, if clear from the context) for the maximum degree of G. A *vertex cover* of $G = (V, E)$ is a subset of vertices $V' \subseteq V$ such that each edge $e \in E$ is incident to at least one vertex of V'. The size of a minimum cardinality vertex cover of G is denoted $\tau(G)$ (or τ, if clear from the context). An *independent set* of $G = (V, E)$ is a subset of vertices $V' \subseteq V$, no two of which are adjacent. A *tree* is a graph in which any two vertices are connected by exactly one simple path, and a *caterpillar* is a tree in which all the vertices are within distance 1 of a central path (equivalently, caterpillars are the trees in which there exists a path that contains every node of degree two or more). A *forest* is a collection of trees.

Let $G = (V, E)$ be a graph of order n. A *labeling* of G is a bijective mapping $\phi : V \to \{1, 2, \ldots, n\}$, and we denote by $\Phi(G)$ the set of all labelings of G. The *S-labeling number* of G with respect to some labeling $\phi \in \Phi(G)$, denoted $\mathrm{SL}_\phi(G)$, is defined to be $\mathrm{SL}_\phi(G) = \sum_{\{u,v\} \in E} \min\{\phi(u), \phi(v)\}$. To abbreviate notations, we write $\mathrm{SL}(G) = \min\{\mathrm{SL}_\phi(G) : \phi \in \Phi(G)\}$ and we let $\Phi_{\mathbf{opt}}(G) \subseteq \Phi(G)$ stand for the set of all optimal labelings (*i.e.*, $\Phi_{\mathbf{opt}}(G) = \{\phi \in \Phi(G) : \mathrm{SL}_\phi(G) = \mathrm{SL}(G)\}$). Let $\phi \in \Phi(G)$ be a labeling of G.

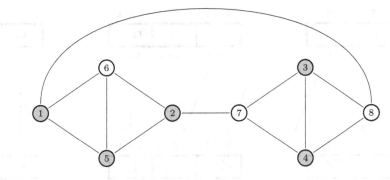

Fig. 1. A graph G together with a labeling ϕ. Vertices in V_ϕ^+ are in grey. We have $c_\phi(\phi^{-1}(1)) = c_\phi(\phi^{-1}(2)) = c_\phi(\phi^{-1}(3)) = 3$, $c_\phi(\phi^{-1}(4)) = 2$, $c_\phi(\phi^{-1}(5)) = 1$, and $c_\phi(\phi^{-1}(6)) = c_\phi(\phi^{-1}(7)) = c_\phi(\phi^{-1}(8)) = 0$, yielding $\mathrm{SL}_\phi(G) = (3 \times 1) + (3 \times 2) + (3 \times 3) + (2 \times 4) + (1 \times 5) + (0 \times 6) + (0 \times 7) + (0 \times 8) = 31$.

For any vertex $u \in V$, the *contribution* $c_\phi(u)$ of u to the S-labeling number SL_ϕ is the integer defined as $c_\phi(u) = |\{v \in V : \{u,v\} \in E \text{ and } \phi(u) < \phi(v)\}|$. We will be mostly interested, in the rest of the paper, in those vertices with non-zero contribution. To this aim, we define $V_\phi^+ = \{u \in V : c_\phi(u) > 0\}$. An example illustrating these notions is given in Fig. 1.

3   Properties of Optimal Labelings

Towards investigating computational issues, it is critical to support a deeper understanding of the structure of optimal labelings. This is the purpose of the current section.

Property 1. For any graph $G = (V, E)$ and any labeling $\phi \in \Phi(G)$: (a) V_ϕ^+ is a vertex cover of G, (b) $\sum_{u \in V_\phi^+} c_\phi(u) = |E|$, and (c) $\mathrm{SL}_\phi(G) = \sum_{u \in V_\phi^+} c_\phi(u)\,\phi(u)$.

Lemma 1. *[9] Let $G = (V, E)$ be a graph, and $\phi \in \Phi_{\mathbf{opt}}(G)$. For any two vertices $u, v \in V$, if $\phi(u) < \phi(v)$ then $c_\phi(u) \geq c_\phi(v)$.*

Lemma 2. *Let $G = (V, E)$ be a graph, and $\phi \in \Phi_{\mathbf{opt}}(G)$. For any distinct vertices $u, v \in V$, if $c_\phi(u) = c_\phi(v)$ then $\{u, v\} \notin E$.*

We now turn to proving that, in any optimal labeling ϕ of G, the vertex ranked first by ϕ is a maximum degree vertex.

Lemma 3. *Let $G = (V, E)$ be a graph, let $\phi \in \Phi_{\mathbf{opt}}(G)$, and let $u \in V$ be the vertex satisfying $\phi(u) = 1$. Then $c_\phi(u) = \Delta(G)$.*

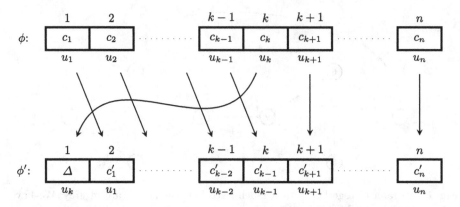

Proof. Let $G = (V, E)$ be a graph, and $\phi \in \Phi_{\mathbf{opt}}(G)$ be an optimal labeling of G. Let $k = \min\{\phi(u) : u \in V \wedge c_\phi(u) = \Delta(G)\}$. We show that $k = 1$ thereby proving the lemma. Write Δ for $\Delta(G)$. For convenience, for $1 \leq i \leq n$, let u_i and c_i stand for $\phi^{-1}(i)$ and $c_\phi(\phi^{-1}(i)) = c_\phi(u_i)$, respectively.

Suppose, aiming at a contradiction, that $k > 1$. Consider the labeling ϕ' of G defined as follows: $\phi'(u_k) = 1$, $\phi'(u_i) = \phi(u_i) + 1$ for any $1 \leq i \leq k - 1$ and $\phi'(u_i) = \phi(u_i)$ for any $k + 1 \leq i \leq n$. For convenience, for $1 \leq i \leq n$, $i \neq k$, let c_i' stand for $c_{\phi'}(u_i)$ (see above figure). We thus have $\mathrm{SL}_\phi(G) = \sum_{i=1}^{n} i c_i$ and $\mathrm{SL}_{\phi'}(G) = \Delta + \sum_{i=1}^{k-1}(i+1)c_i' + \sum_{i=k+1}^{n} i c_i$. Now, if we let D stand for $\mathrm{SL}_\phi(G) - \mathrm{SL}_{\phi'}(G)$, we obtain $D = \sum_{i=1}^{n} i c_i - \Delta - \sum_{i=1}^{k-1}(i+1)c_i' - \sum_{i=k+1}^{n} i c_i = \sum_{i=1}^{k} i c_i - \Delta - \sum_{i=1}^{k-1}(i+1)c_i'$ (Eq. 1). We need the following inequality.

Claim 1. $\sum_{i=1}^{k-1}(i+1)c_i' \leq \sum_{i=1}^{\Delta-c_k}(i+1)(c_i - 1) + \sum_{i=\Delta-c_k+1}^{k-1}(i+1)c_i$.

Proof. By construction, $c_i - 1 \leq c_i' \leq c_i$ for $1 \leq i \leq k - 1$. Moreover, since exactly $\Delta - c_k$ vertices of $\{u_1, u_2, \ldots, u_{k-1}\}$ are connected with vertex u_k, then there exist $S \subseteq \{1, 2, \ldots, k - 1\}$ of size $\Delta - c_k$ such that $c_i' = c_i - 1$ if and only if $i \in S$. If we let $\overline{S}$ stand for $\{1, 2, \ldots, k - 1\}\backslash S$, then $\sum_{i=1}^{k-1}(i+1)c_i' = \sum_{i \in S}(i+1)(c_i - 1) + \sum_{i \in \overline{S}}(i+1)c_i = \sum_{i=1}^{k-1}(i+1)c_i - \sum_{i \in S}(i+1)$. This latter sum is certainly maximized for $S = \{1, 2, \ldots \Delta - c_k\}$, and hence $\sum_{i=1}^{k-1}(i+1)c_i' \leq \sum_{i=1}^{\Delta-c_k}(i+1)(c_i - 1) + \sum_{i=\Delta-c_k+1}^{k-1}(i+1)c_i$. $\square$

Combining Claim 1 with (Eq. 1) yields $D \geq \sum_{i=1}^{k} i c_i - \Delta - \sum_{i=1}^{\Delta-c_k}(i+1)$ $(c_i-1) - \sum_{i=\Delta-c_k+1}^{k-1}(i+1)c_i = \sum_{i=1}^{k} i c_i - \Delta - \sum_{i=1}^{k-1} i c_i - \sum_{i=1}^{k-1} c_i + \sum_{i=1}^{\Delta-c_k}(i+1)$ $= k c_k - \Delta - \sum_{i=1}^{k-1} c_i + \sum_{i=1}^{\Delta-c_k}(i+1) = k c_k - \Delta - \sum_{i=1}^{k-1} c_i + \frac{(\Delta-c_k+3)(\Delta-c_k)}{2}$.

But $c_i \leq \Delta - 1$ for $1 \leq i \leq k - 1$ (this follows from $k > 1$ and Lemma 1), and hence $\sum_{i=1}^{k-1} c_i \leq (k - 1)(\Delta - 1)$. Then it follows that $D \geq k c_k - \Delta - (k - 1)(\Delta - 1) + \frac{(\Delta-c_k+3)(\Delta-c_k)}{2} = (k - 1) - k(\Delta - c_k) + \frac{(\Delta-c_k+3)(\Delta-c_k)}{2} = (k - 1) + \frac{(\Delta-c_k)(\Delta-c_k-2k+3)}{2}$. Combining the above with $\Delta - c_k \geq 1$ yields $D \geq (k - 1) + \frac{4-2k}{2} \geq 1$, and hence $\mathrm{SL}_\phi(G) > \mathrm{SL}_{\phi'}(G)$. This is the desired contradiction since ϕ is an optimal labeling of G. $\square$

Lemma 4. *Let $G = (V, E)$ be a graph, $\phi \in \Phi_{\mathbf{opt}}(G)$, and let ϕ' be the labeling obtained from ϕ by swapping the labels of any two vertices $u, v \in V$ such that $c_\phi(u) = c_\phi(v)$. Then $\phi' \in \Phi_{\mathbf{opt}}(G)$.*

Lemma 5. *Let $G = (V, E)$ be a graph, $\phi \in \Phi_{\mathbf{opt}}(G)$, and let $u, v \in V$ such that $\{u, v\} \in E$ and $c_\phi(u) = c_\phi(v) - 1$. Then an optimal labeling ϕ' with $\phi'(u) = \phi'(v) - 1$ may be obtained from ϕ by a series of label swaps involving only vertices $z \in V$ with $c_\phi(z) \in \{c_\phi(u), c_\phi(v)\}$ and $\phi(v) < \phi(z) < \phi(u)$.*

Coming back to V_ϕ^+, and as discussed in [9], in the light of Property 1(a), it would be tempting to claim that $\tau(G) = |V_\phi^+|$ for any –or at least one– optimal labeling $\phi \in \Phi_{\mathbf{opt}}(G)$. Unfortunately, this is not true: there exist a graph G for which any optimal labeling ϕ satisfies $\frac{|V_\phi^+|}{\tau(G)} = \frac{5}{4}$ [9]. This raises the following question: for any graph G, does there exist a labeling $\phi \in \Phi_{\mathbf{opt}}(G)$ such that $\frac{|V_\phi^+|}{\tau(G)} = O(1)$? This question remains open. However, we have the following result, which improves Lemma 1.3 from [9] by a factor $\sqrt{2}$.

Lemma 6. *For any graph G of maximum degree Δ, there exist an optimal labeling $\phi \in \Phi_{\mathbf{opt}}(G)$ for which $\frac{|V_\phi^+|}{\tau(G)} < \sqrt{\Delta}$.*

4 A Polynomial-Time Algorithm for Sets of Caterpillars

In this section, we prove that the S-LABELING problem is in **P** when the input instance is a (set of) caterpillar(s) (see Proposition 1). This result can be seen as a step towards understanding the complexity of the problem in trees and forests, which remains unknown. Recall that a caterpillar is a tree $G = (V, E)$ for which all the vertices are within distance 1 of a dominating path. It is easy to see that every longest path P of G has the property that, for all $v \in V$, either v belongs to P or v is a leaf adjacent to a vertex of P, which is not an endpoint of P. We call *linear representation* of G, denoted LR(G), any drawing of G in which the vertices of some longest path P of G lie along an horizontal line according to their order on the path. We call *set of caterpillars* any vertex-disjoint set of caterpillars. A set of caterpillars is seen as a graph $G = (V, E)$ whose connected components $C_1, C_2 \dots C_p$ are caterpillars. The linear representation of G, still denoted LR(G), is in this case the sequence LR(C_1), LR(C_2) ... LR(C_p) of the linear representations of its connected components, successively on the same horizontal line. Recall that G being a set of caterpillars on n vertices, G contains $O(n)$ edges.

Lemma 7. *Let G be a set of caterpillars, and u be its leftmost vertex in LR(G) such that $d_G(u) = \Delta$. There is an optimal labeling $\phi_{\mathbf{opt}}$ of G such that $\phi_{\mathbf{opt}}(u) = 1$.*

Algorithm 1. Greedy computation of a labeling of a set of caterpillars.

Algorithm: GreedyCat

Data: A set of caterpillars $G = (V, E)$

Result: A labeling ϕ of G

$i \leftarrow 1$

while $G \neq \emptyset$ **do**

    Compute $\mathrm{LR}(G)$

    Let u be the leftmost vertex in $\mathrm{LR}(G)$ with $d_G(u) = \Delta(G)$

    $\phi(u) \leftarrow i$

    $i \leftarrow i + 1$

    $G \leftarrow G - u$

return (ϕ)

Proof. Notice that if we prove the existence of an optimal labeling $\phi \in \Phi_{\mathbf{opt}}(G)$ such that $c_\phi(u) = \Delta$, then we are done. Indeed, by Lemma 3, we know that the vertex z with label 1 has contribution Δ, and by Lemma 4, we know that swapping the labels of u and z yields an optimal labeling $\phi_{\mathbf{opt}}$ such that $\phi_{\mathbf{opt}}(u) = 1$.

We now show that there is an optimal labeling ϕ of G such that $c_\phi(u) = \Delta$. Let ϕ' be an optimal labeling of G such that every leaf v of G that is adjacent to u (if any) satisfies $\phi'(v) > \phi'(u)$ (say this is a *regular* leaf). To prove that such a labeling exists, assume by contradiction that this is not the case, and let ϕ' be chosen so as to minimize the number of leaves with $\phi'(v) < \phi'(u)$. Denote v_0 the leaf with this property and such that $\phi'(v_0)$ is maximum. By Lemma 1, we know that $c_{\phi'}(u) \leq c_{\phi'}(v_0)$, with $c_{\phi'}(v_0) = 1$. We thus necessarily have $c_{\phi'}(u) = 0$ and $c_{\phi'}(v_0) = 1$ by Lemma 2, $\{u, v_0\}$ being an edge in G. In this case, Lemma 5 ensures that some labels of ϕ' may be swapped to ensure $\phi'(v_0) > \phi'(u)$, and that these swaps do not affect regular leaves. But this contradicts the initial choice of ϕ'. As a consequence, all the leaves adjacent to u are regular, and $c_{\phi'(u)} \geq \Delta - 2$ since at most two neighbors of u are not leaves.

Now, let v (respectively w) be the right (respectively left) neighbor of u, if it exists. Then, since u is the leftmost vertex in $\mathrm{LR}(G)$ with degree Δ, we have $d_G(w) < \Delta$ and $d_G(v) \leq \Delta$. We transform ϕ' into an optimal labeling ϕ such that $\phi(u) < \phi(w)$ and $\phi(u) < \phi(v)$, in two steps. First, if $\phi'(u) < \phi'(w)$, then we define $\phi'' = \phi'$. Otherwise, we have $\phi'(u) > \phi'(w)$, and since u and w are adjacent, we deduce that $c_{\phi'}(u) < c_{\phi'}(w)$ by Lemmas 1 and 2. The remarks that $c_{\phi'}(u) \geq \Delta - 2$ and $c_{\phi'}(w) \leq \Delta - 1$ further imply that we necessarily have $c_{\phi'}(u) = \Delta - 2$ and $c_{\phi'}(w) = \Delta - 1$, thus allowing us to apply Lemma 5 in order to deduce the existence of an optimal labeling ϕ'' with $\phi''(u) < \phi''(w)$. Second, if $\phi''(u) < \phi''(v)$, then we define $\phi = \phi''$. Otherwise, since u and v are adjacent, we deduce that $c_{\phi''}(u) < c_{\phi''}(v)$ by Lemmas 1 and 2. The remarks that $c_{\phi''}(u) \geq \Delta - 1$ (we know that $\phi''(w) > \phi''(u)$) and $c_{\phi''}(v) \leq \Delta$ further imply that we necessarily have $c_{\phi''}(u) = \Delta - 1$ and $c_{\phi''}(v) = \Delta$, thus allowing us to apply Lemma 5 in order to deduce the existence of an optimal labeling ϕ with $\phi(u) < \phi(v)$. Notice that the label changes performed according to Lemma 5 do not affect the label of w. Then, $\phi(u) < \phi(x)$ for all the neighbors x of u, and thus $c_\phi(u) = \Delta$. $\square$

Proposition 1. *Algorithm* **GreedyCat** *optimally solves the S-LABELING problem for any set of caterpillars of order n, in $O(n \log n)$ time.*

Proof. By contradiction, assume the labeling ϕ obtained by **GreedyCat** is not an optimal labeling. Let $v_1, v_2 \ldots v_n$ be the vertices of G ordered such that $\phi(v_i) = i$. Now, consider an optimal labeling $\phi_{\mathbf{opt}}$ which maximizes the value k with the property that $\phi_{\mathbf{opt}}(v_i) = \phi(v_i)$ $(= i)$ for all i, $1 \leq i \leq k$. Let G_{k+1} be the subgraph induced by $\{v_{k+1} \ldots v_n\}$ in G.

According to **GreedyCat**, $v_1 \ldots v_n$ are labeled successively in this order, and when v_{k+1} is labeled, it has maximum degree in G_{k+1} and is the leftmost in $\mathrm{LR}(G)$ with this property. It is easy to note that a labeling ϕ_0 of G with $\phi_0(v_i) = i$ for all i, $1 \leq i \leq k$, is optimal for G if and only if the labeling ϕ_1 of G_{k+1}, defined by $\phi_1(v_i) = \phi_0(v_i) - k$ for $k + 1 \leq i \leq n$, is optimal for G_{k+1}. This is due to the constant cost of the edges $\{v_i, v_j\}$ with $i \leq k$ or $j \leq k$ within $\mathrm{SL}(G)$. By Lemma 7, there is an optimal labeling ϕ_1 of G_{k+1} such that $\phi_1(v_{k+1}) = 1$. The resulting labeling $\phi_0(v_i) = \phi_1(v_i) + k$ for $k + 1 \leq i \leq n$, extended to G by $\phi_0(v_i) = i$ for $1 \leq i \leq k$, is then optimal and satisfies $\phi_0(v_i) = \phi(v_i)$ $(= i)$ for all i, $1 \leq i \leq k + 1$. This contradicts the choice of $\phi_{\mathbf{opt}}$.

Concerning the complexity, the initial computation of $\mathrm{LR}(G)$ is done in $O(n)$ time by using traversals starting with a 1-degree vertex. The vertices occurring on the horizontal line of the representation are then renumbered in increasing order from left to right, whereas the leaves receive the remaining numbers. Then, for each $d = 0, 1, 2 \ldots \Delta$, we store all vertices of degree d in a balanced BST denoted $B[d]$. We furthermore need another balanced BST to store all the degrees d for which $B[d]$ is non-empty. It is easy to see that initializing all the BSTs is done in global time of $O(n \log n)$ by $n + \Delta$ insertions taking $O(\log n)$ time each. The update of the data structures we use involves, at each step, only the neighbors of the vertex u removed from G. There are $O(d_G(u))$ such neighbors, and the operations take no more than $O(\log n)$ time for each neighbor, which yields the required complexity.   □

5   Algorithmic Issues

This section contains two parts: the first one is devoted to approximating the S-LABELING problem, whereas the second part focuses on exact (*i.e.* exponential-time algorithms.

Approximating the S-labeling Problem. We begin by giving accurate lower and upper bounds of optimal labelings. For completeness, in Lemma 8, we recall the lower bound given in [9].

Lemma 8 [9]. *For any graph G of order n, size m and maximum degree Δ,* $\mathrm{SL}(G) \geq \left(m - \frac{\Delta}{2} \left\lfloor \frac{m}{\Delta} \right\rfloor\right) \left(\left\lfloor \frac{m}{\Delta} \right\rfloor + 1\right).$

Another general lower bound is given in Lemma 9 below.

Lemma 9. *For any graph G of size m, let ϕ be any labeling of G, and $|V_\phi^+| = k$. Then* $\mathrm{SL}_\phi(G) \geq \frac{k^2-k+2m}{2}$.

Proof. Let ϕ be any labeling of G, and let $|V_\phi^+| = k$. Since every vertex in V_ϕ^+ has a strictly positive contribution (by definition), summing these over the k vertices in V_ϕ^+ adds up to $\frac{k(k+1)}{2}$, and covers k edges. For the remaining $m - k$ edges in G, the best scenario is that they all are incident to the vertex having label 1, which adds up to $m - k$. Hence the result.   □

We now turn to giving upper bounds for optimal labelings. The first two upper bounds are achieved by the following generic greedy algorithm, that we call **GreedyGen**: while G has edges, take a vertex u of maximum degree in G, assign the smallest available label to u (starting from 1), and remove u and its incident edges from G. Notice that a randomized algorithm achieving the same general upper bound as in Lemma 10 was given in [9]. Here, we improve this previous result by providing a *deterministic* algorithm, achieving the same performances for general graphs, and improving them for acyclic graphs.

Lemma 10. *For any graph G of order n and size m, Algorithm* **GreedyGen** *computes, in $O(m \log n)$ time, a labeling ϕ such that* $\mathrm{SL}_\phi(G) \leq \frac{m(n+1)}{3}$. *Furthermore, if G is acyclic with $\Delta \geq 3$, ϕ satisfies* $\mathrm{SL}_\phi(G) \leq \frac{m(n+1)}{4}$.

As a side remark, we observe that there exist classes of graphs for which each of the previously given upper and lower bounds are reached. Indeed, for any even n, $\mathrm{SL}(C_n) = \frac{n^2+2n}{4} = \frac{\Delta\lfloor\frac{m}{\Delta}\rfloor(\lfloor\frac{m}{\Delta}\rfloor+1)}{2}$; for any $k \leq m$, $\mathrm{SL}(K_{1,m-k+1} \cup (k-1)K_2) = \frac{k^2-k+2m}{2}$; $\mathrm{SL}(K_n) = \frac{1}{6}n(n^2-1) = \frac{m(n+1)}{3}$; and, finally, for any odd n, $\mathrm{SL}(P_n) = \frac{m(n+1)}{4}$. Another general upper bound can be obtained by starting from a vertex cover V' of G, and applying a greedy strategy, similar to the one of Algorithm **GreedyGen**, but only taking into account vertices in V'. Since all vertices in $V \backslash V'$ have a zero contribution, they can be arbitrarily labeled from $|V'| + 1$ to n.

Lemma 11. *For any graph G of order n and size m, let V' be a vertex cover of G, with $|V'| = p$. A labeling $\phi_{V'}$ satisfying* $\mathrm{SL}_{\phi_{V'}}(G) \leq \frac{1}{2}m(p + 1)$ *can be computed in $O(m \log p)$ time.*

We are now ready to state our results concerning approximation algorithms, which essentially rely on using the proper upper and lower bounds among the ones presented above. Some of these results improve or generalize the ones from [9], others are new, and all are deterministic (as opposed to some results from [9]). If G is a graph of size m and order n, we denote by $d = \frac{2m}{n}$ its average degree. The first general result we have is the following.

Proposition 2. *For any graph,* **GreedyGen** *is a $\frac{4\Delta}{3d}$-approximation algorithm for the S-LABELING problem.*

Since we can always assume that G has no isolated vertices, we always have $d \geq 1$, and thus for any graph, we now that a $\frac{4\Delta}{3}$ approximation algorithm exists. Concerning regular graphs, setting $d = \Delta$ in Proposition 2 yields the following corollary.

Corollary 1. *For any regular graph,* **GreedyGen** *is a $\frac{4}{3}$-approximation algorithm for the S-*LABELING* problem.*

Now, if we restrict ourselves to specific classes of graphs, we are able to obtain better approximation ratios. Let $\mathcal{F}_{\leq C}$ (resp. $\mathcal{F}_{\geq C}$) be the set of forests having at most (resp. at least) C connected components.

Proposition 3. *For any graph G,* **GreedyGen** *is an approximation algorithm having the following ratios: (a) $\frac{\Delta}{2}$ if $G \in \mathcal{F}_{\leq \Delta - 1}$, provided $\Delta \geq 3$; (b) Δ if $G \in \mathcal{F}_{\geq \Delta}$; (c) $\frac{2\Delta}{3}$ if G is connected.*

We now give one more approximation algorithm, in relation to the MINIMUM VERTEX COVER problem. Let $\mathcal{G}_{\mathcal{VC}}^{\alpha}$ denote the class of graphs for which the cardinality of a minimum vertex cover can be approximated within ratio α. The following result concerns graphs belonging to that class.

Proposition 4. *For any graph G in $\mathcal{G}_{\mathcal{VC}}^{\alpha}$, there exist an $\alpha\Delta$-approximation algorithm for the S-*LABELING* problem.*

In particular, for all graphs for which a minimum vertex cover can be computed in polynomial-time, the S-LABELING problem can be approximated within ratio Δ.

Exact Algorithms. After having discussed approximation algorithms, we now turn to exact algorithms. We first concentrate on exponential-time algorithms.

Proposition 5. *For any graph G of order n and maximum degree Δ, the S-*LABELING* problem is solvable in time $O^*(1.44225^{n\Delta})$, and in time $O^*(1.41422^{n\Delta})$ if G is a tree.*

Let G be a graph and $\phi \in \Phi_{\mathbf{opt}}(G)$ be an optimal labeling of G. For every positive integer $c \neq 0$, let $V_\phi^c = \{u : u \in V \wedge c_\phi(u) = c\}$, and, for any non-empty V_ϕ^c, let $l_{min}^c = \min\{\phi(u) : u \in V_\phi^c\}$ be the smallest label given by ϕ among the vertices in G having contribution equal to c by ϕ. Because of Lemma 2, we have the following corollary.

Corollary 2. *For any graph G, let $\phi \in \Phi_{\mathbf{opt}}(G)$ be an optimal labeling of G. Then V_ϕ^c is a maximal independent set on the vertices of maximum degree in the induced graph $G[\{v : \phi(v) \geq l_{min}^c\}]$.*

Proof (of Proposition 5). We use a Δ-bounded search tree based algorithm. The root of the search tree is labeled by $(G, 0, \emptyset)$. We explore the tree as follows. For a node (H, i, L), we consider all maximal independent sets on the maximum degree vertices of H. For each maximal independent set $V' = \{u_1, u_2, \ldots, u_k\}$, we get a child labeled $(H', i+k, L')$, where $H' = H \setminus V'$ and $L' = L \cup \{(u_j, i+j) : 1 \le j \le k\}$. A leaf in the search tree is labeled (H, i, L) for some edgeless graph H and it evaluates to $\sum_{\{u,v\} \in E} \min\{\phi(u), \phi(v)\}$, where $\phi \in \Phi(G)$ is obtained from L in the natural way, i.e., for every $u \in V$, $\phi(u) = i$ if $(u, i) \in L$. The optimal labeling is the minimum evaluation of a leaf of the search tree.

Correctness of the algorithm follows from Corollary 2. What is left is to prove that the search tree has depth bounded by $\Delta(G)$. This is again a consequence of Corollary 2, since the maximum degree of the graphs strictly decreases during the exploration of the tree. The time complexity now follows from the following result from [14]: any graph of order n contains at most 1.44225^n maximal independent sets, and these maximal independent sets can be enumerated in $O^*(1.44225^n)$ time. Besides, if G is a tree, we know by [23] that the number of maximal independent sets G is upper bounded by $O((\sqrt{2})^n)$ (more precisely, the number is $2^{n/2-1} + 1$ if n is even, and $2^{(n-1)/2}$ is n is odd). □

As a side remark concerning Corollary 2, note that it would be tempting to push ahead and replace *maximal* by *maximum* in this corollary. However, that is pushing the argument too far, even for trees. Indeed, consider the tree given Fig. 2 with 3 maximum degree vertices. For one, starting from the unique maximum independent set $\{u, w\}$ on degree $\Delta = 3$ vertices yields a labeling with total sum 34. For another, starting from the other maximal independent set on degree $\Delta = 3$ vertices $\{v\}$ yields a labeling with total sum 31.

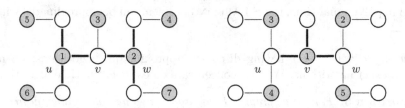

Fig. 2. A graph G with three maximum degree vertices u, v and w, each of degree $\Delta = 3$. (Left) An S-labeling ϕ_1 of G that starts with the unique maximum independent set on maximum degree vertices of cardinality 2 (namely, $\{u, w\}$), optimally completed with labels 3 to 8, yielding $\mathrm{SL}_{\phi_1}(G) = 34$. (Right) An S-labeling ϕ_2 of G that starts with the unique maximal independent set on maximum degree vertices of cardinality 1 (namely, $\{v\}$), optimally completed with labels 3 to 8, yielding $\mathrm{SL}_{\phi_2}(G) = 31 < \mathrm{SL}_{\phi_1}(G)$.

Now, note that for any graph G of order n and size m, $\mathrm{SL}(G) = \Omega(m)$. But since we can always assume G has no isolated vertices, $n \le 2m$ and thus $\mathrm{SL}(G) = \Omega(n)$. Therefore, the S-LABELING problem is fixed-parameter tractable [21] for any graph G, when the parameter is the size of the solution $\mathrm{SL}(G)$. However, the above result relies on a parameter whose value can be considered as too high. Thus, we focus on the following problem: Given a graph $G = (V, E)$ of size $|E| = m$ and a positive integer k, is there a labeling $\phi \in \Phi(G)$ such that

$SL_\phi(G) \leq m + k$? We need the following result that gives a lower bound for $SL_\phi(G)$ in terms of $|V_\phi^+|$.

Lemma 12. *For any graph G of size m, let $\phi \in \Phi(G)$ be a labeling of G such that $SL_\phi(G) \leq m + k$ for some positive integer k. Then $|V_\phi^+| \leq 2\sqrt{k}$.*

Proof. Let $G = (V, E)$ be a graph and $\phi \in \Phi(G)$ be a labeling of G such that $SL_\phi(G) \leq |E| + k$ for some positive integer k. By Lemma 9, we know that $SL_\phi(G) \geq \frac{|V_\phi^+|^2 - |V_\phi^+| + 2m}{2}$, which yields $|V_\phi^+|^2 - |V_\phi^+| \leq 2k$. Solving this second degree inequality leads to $|V_\phi^+| \leq x$, where $x = \frac{1 + \sqrt{8k+1}}{2}$. A simple computation then shows that $x \leq 2\sqrt{k}$ for any positive integer k, which proves the result. $\square$

Proposition 6. *For any graph G of size m and any positive integer k, it can be decided in $O^*(2^{2\sqrt{k}} (2\sqrt{k})!)$ time whether $SL(G) \leq m + k$.*

In the above proposition, it is worth noticing that $k \geq \frac{m(n-4)}{4}$ for regular graphs (see Lemma 8), and hence Proposition 6 does not give fixed-parameter tractability for the above-guarantee parameterization variant [11,20]. However, supposing G is of order n, we now determine how Proposition 6 compares to the brute-force $O^*(n!)$ time algorithm. First, we have $SL(G) \leq mn$, therefore $SL(G) \leq m+k$ implies $k \leq mn - m = m(n-1) \leq \frac{n(n-1)^2}{2}$, and hence $k = O(n^3)$. We now need the following lemma.

Lemma 13. *For positive x and n, if $2^x x! = n!$ then $x = \Theta(n)$.*

Substituting x by $2\sqrt{k}$ in Lemma 13 yields $k = \Theta(n^2)$. Since $2^x x!$ is an increasing function, we conclude that Proposition 6 improves on the brute-force $O^*(n!)$ time algorithm for any k, up to $\Theta(n^2)$.

6   Conclusion

We would like to end this paper with several open problems. First, what is the complexity of the S-LABELING problem for trees, forests, or more generally bipartite graphs? Second, does there exist a PTAS for the S-LABELING problem, for any graph G, or at least for specific classes of graphs? The same question holds for constant approximation ratios.

References

1. Adolphson, D.: Single machine job sequencing with precedence constraints. SIAM J. Comput. **6**, 40–54 (1977)
2. Adolphson, D., Hu, T.C.: Optimal linear ordering. SIAM J. Appl. Math. **25**(3), 403–423 (1973)
3. Bezrukov, S.: Edge isoperimetric problems on graphs. In: Lovasz, L., Gyárfás, A., Katona, B.O.H., Recski, A., Szekely, L. (eds.) Graph Theory and Combinatorial Biology, pp. 157–197. Janos Bolyai Mathematical Society, Budapest (1999)

4. Bhatt, S.N., Leighton, F.T.: A framework for solving VLSI graph layout problems. J. Comput. Syst. Sci. **28**, 300–343 (1984)

5. Chung, F.R.K.: Labeling of graphs. In: Wilson, R.J., Beineke, L.W. (eds.) Selected Topics in Graph Theory, vol. 3, pp. 151–168. Academic Press, London (1988)

6. Corless, R.M., Gonnet, G.H., Hare, D.E.G., Jeffrey, D.J., Knuth, D.E.: On the Lambert W function. Adv. Comput. Math. **5**, 329–359 (1996)

7. Cormen, T.H., Leiserson, C.E., Rivest, R.L., Stein, C.: Introduction to Algorithms, 3rd edn. MIT Press, Cambridge (2009)

8. Dìaz, J., Petit, J., Serna, M.: A survey on graph layout problems. J. ACM Comput. Surv. (CSUR) **34**(3), 313–356 (2002)

9. Fertin, G., Vialette, S.: On the S-labeling problem. In: Proceedings of the 5th European Conference on Combinatorics, Graph Theory and Applications (EuroComb). Electronic Notes on Discrete Mathematics, vol. 34, pp. 273–277 (2009)

10. Golub, G.H., Van Loan, C.F.: Matrix Computations, 3rd edn. Johns Hopkins University Press, Baltimore, London (1996)

11. Gutin, G., Rafiey, A., Szeider, S., Yeo, A.: The linear arrangement problem parameterized above guaranteed value. Theor. Comput. Syst. **41**(3), 521–538 (2007). Erratum-ibid [12]

12. Gutin, G., Rafiey, A., Szeider, S., Yeo, A.: Corrigendum. the linear arrangement problem parameterized above guaranteed value. Theor. Comput. Syst. **53**(4), 690–691 (2013)

13. Harper, L.H.: Optimal numberings and isoperimetric problems on graphs. SIAM J. **12**(1), 131–135 (1964)

14. Johnson, D.S., Yannakakis, M., Papadimitriou, C.H.: On generating all maximal independent sets. Inf. Process. Lett. **27**, 119–123 (1988)

15. Karger, D.R.: A randomized fully polynomial approximation scheme for all terminal network reliability problem. In: 27th ACM Symposium on Theory of Computing (STOC), pp. 11–17 (1995)

16. Karp, R.M.: Mapping the genome: some combinatorial problem arising in molecular biology. In: 24th ACM Symposium on Theory of Computing (STOC), pp. 278–285 (1993)

17. Kitaev, S.: Iterative Methods for Sparse Linear Systems, 2nd edn. Society for Industrial and Applied Mathematics, Philadelphia (2003)

18. Leighton, F.T.: Introduction to Parallel Algorithms and Architectures: Arrays, Trees, Hypercubes. Morgan Kaufmann, San Mateo (1993)

19. Lipton, R.J., Tarjan, R.E.: A separator theorem for planar graphs. SIAM J. Appl. Math. **36**(2), 177–189 (1979)

20. Mahajan, M., Raman, V.: Parameterizing above guaranteed values: MaxSat and MaxCut. J. Algorithms **31**(2), 335–354 (1999)

21. Niedermeier, R.: Invitation to Fixed Parameter Algorithms. Lecture Series in Mathematics and Its Applications. Oxford University Press, Oxford (2006)

22. Vialette, S.: Packing of $(0, 1)$-matrices. Theor. Inform. Appl. RAIRO **40**(4), 519–536 (2006)

23. Wilf, H.S.: On the number of maximal independent sets in a tree. SIAM J. Algebraic Discrete Methods **7**(1), 125–130 (1986)

Contagious Sets in Dense Graphs

Daniel Freund[1], Matthias Poloczek[2]([✉]), and Daniel Reichman[3]

[1] Center for Applied Mathematics, Cornell University, Ithaca, NY, USA
[2] School of Operations Research and Information Engineering,
Cornell University, Ithaca, NY, USA
{df365,poloczek}@cornell.edu
[3] Department of Computer Science, Cornell University, Ithaca, NY, USA
daniel.reichman@gmail.com

Abstract. We study the activation process in undirected graphs known as bootstrap percolation: a vertex is active either if it belongs to a set of initially activated vertices or if at some point it had at least r active neighbors, for a threshold r that is identical for all vertices. A contagious set is a vertex set whose activation results with the entire graph being active. Let $m(G, r)$ be the size of a smallest contagious set in a graph G. We examine density conditions that ensure that a given n-vertex graph $G = (V, E)$ has a small contagious set. With respect to the minimum degree, we prove that if G has minimum degree $n/2$ then $m(G, 2) = 2$. We also provide tight upper bounds on the number of rounds until all nodes are active.

For $n \geq k \geq r$, we denote by $M(n, k, r)$ the maximum number of edges in an n-vertex graph G satisfying $m(G, r) > k$. We determine the precise value of $M(n, k, 2)$ and $M(n, k, k)$ assuming that n is sufficiently large compared to k.

Keywords: Bootstrap percolation · Target set selection · Extremal graph theory

Dynamic processes that model the diffusion of information through a social network have received considerable interest recently: they capture for instance "word of mouth" effects occurring in viral marketing, where the information is only revealed to a small group of persons initially, who subsequently share it with their friends and so. Similarly, we can think of cascading effects in finance, where an institute might default if a certain number of business partners fail (cp. [2,9,12,13] and the references therein).

In this article we study the *r-neighbor bootstrap percolation* process. Here we are given an undirected graph $G = (V, E)$ and an integer $r \geq 1$. Every vertex is

D. Freund—Supported in part by U.S. Army Research Office grant W911NF-14-1-0477.

M. Poloczek—Supported by the Alexander von Humboldt Foundation within the Feodor Lynen program, and in part by NSF grant CCF-1115256.

D. Reichman—Supported in part by NSF grants IIS-0911036 and CCF-1214844, AFOSR grant FA9550-08-1-0266, and ARO grant W911NF-14-1-0017.

© Springer International Publishing Switzerland 2016
Z. Lipták and W.F. Smyth (Eds.): IWOCA 2015, LNCS 9538, pp. 185–196, 2016.
DOI: 10.1007/978-3-319-29516-9_16

either *active* or *inactive*. We say a set A of vertices is active if all vertices in A are active. The vertices that are active initially are called *seeds*, and the set of seeds is denoted by A_0. If vertices become active thereafter we also refer to them as *infected*. A contagious process evolves in discrete rounds. The set of active vertices in round $i > 0$ is

$$A_i = A_{i-1} \cup \{v : |N(v) \cap A_{i-1}| \geq r\},$$

where $N(v)$ is the set of neighbors of v. That is, a vertex becomes active irrevocably in a given round if it has at least r active neighbors. We refer to r as the *threshold*. Let $\langle A_0 \rangle$ be the set of nodes that will eventually become infected if we activate A_0.

Definition 1. *Given $G = (V, E)$, a set $A_0 \subseteq V$ is called* contagious *if $\langle A_0 \rangle = V$. In words, activating A_0 results in the infection of the entire vertex set. The size of the smallest contagious set is denoted by $m(G, r)$. For a contagious set A_0, the number of rounds until total infection is the smallest integer t with $A_t = V$.*

The term bootstrap percolation is used sometimes to model the case where the seeds are chosen independently at random. In this work we use this term also with respect to the deterministic selection of a contagious set. Bootstrap percolation was first studied by statistical physicists [8]. Since then, this model has found applications in many fields. Furthermore, various questions related to bootstrap percolation have been examined for a large variety of graphs including hypercubes [4], grids [5,6], several models of random graphs [3,7,11], and expanders [10].

A natural question is to determine for a given integer k, what combinatorial properties of graphs ensure that the minimum size of a contagious set is at most k. Such a characterization seems difficult even for $k = 2$ (and $r = 2$). Indeed the family of all graphs with a contagious set of size two include, for example, cliques, bipartite cliques (with both sides larger than one), and binomial random graphs with edge probability $p \geq n^{-1/2+\epsilon}$ [11].

Previous works have examined the connection between $m(G, k)$ and the degree sequence of G [1,16]. Here we continue this line of investigation and study two basic (and interrelated) graph parameters: the minimum degree and edge cardinality. More concretely, our goal is to determine what conditions on these parameters imply that $m(G, r) = k$ where k is small compared to the number of vertices in G, and $r \leq k$. We study the cases that $r = k$ or $r = 2$.

How large does the minimum degree have to be in order to guarantee a contagious set of size two, if all thresholds are two? We prove that $n/2$ suffices, where n is the number of vertices. A graph with this property is called Dirac graph. Note that this condition on the minimum degree is the best possible: if the minimum degree is $n/2 - 1$, then G might be disconnected implying that $m(G, 2) > 2$ (provided that G has at least three vertices). We also prove the existence of a contagious set of size two for Ore graphs. Ore graphs are a generalization of Dirac graphs, where each pair of nonadjacent vertices u, v obeys

$\deg(u) + \deg(v) \geq n$. Similarly to Dirac graphs, they have been studied in the context of Hamiltonicity.

Contagious sets may vary in the number of rounds they require in order to infect the whole graph (e.g., [15]). For Dirac graphs which are not isomorphic to two cliques of equal size connected by a perfect matching, we are able to derive upper bounds on the number of rounds required to infect the whole of G. More specifically, we show that for such graphs, *all* subsets of three nodes are contagious, and that any such subset will infect the whole graph in at most three rounds. Observe that it is easy to determine the family of all contagious sets in the Dirac graph consisting of two cliques connected by a perfect matching (as well as the number of rounds until all nodes are infected).

A classic question in graph theory is to determine the minimum number of edges in an n-vertex graph G that ensures that G possesses a monotone graph property. Here, we examine extremal questions related to the existence of small contagious sets.

Definition 2. *Given integers $n \geq k \geq r$, we denote by $M(n, k, r)$ the maximum number of edges in an n vertex graph G, where G satisfies $m(G, r) > k$.*

First we study $M(n, k, k)$: a necessary condition for a graph $G = (V, E)$ of $n > k$ vertices to satisfy $m(G, k) = k$ is that G is *connected*. Here we show that the minimum number of edges that guarantees connectivity is also sufficient to ensure $m(G, k) = k$ for $n \geq 2k + 2$, i.e. we have $M(n, k, k) = \binom{n-1}{2}$. Next we consider the case where $r = 2$. For $k \ll n$, we prove that $M(n, k, 2) = \binom{n-k+1}{2} + \lfloor \frac{k+1}{2} \rfloor - 1$ holds.

Preliminaries. All graphs are undirected. Given a graph $G = (V, E)$, we will always assume it has n vertices. The degree of a node $v \in V$ is denoted by $\deg(v)$. The set of all neighbors of a vertex v are denoted by $N(v)$. For a set $S \subseteq V$ we shorthand $\overline{S} := V \backslash S$. Given two disjoint sets A, B of vertices, $E(A, B)$ is the number of edges with one endpoint in A and one endpoint in B, and $E(A)$ is the number of edges with both endpoints in A.

1   Contagious Sets in Dirac and Ore Graphs

We focus in this section exclusively on the case $r = 2$. Recall that an n-vertex graph is a Dirac graph if every vertex in the graph is of degree at least $n/2$. For an Ore graph $G = (V, E)$ we have that $u, v \in V$ with $(u, v) \notin E$ implies $\deg(u) + \deg(v) \geq n$.

1.1   The Existence of Small Contagious Sets

The upper bound in [1, 16] shows that in a Dirac graph there *exists* a contagious set of size three. Here we prove that in Dirac graphs there is in fact a contagious set of size two.

The following family of Dirac graphs will be of particular interest: for n even, let DC_n be the undirected graph composed of two disjoint cliques of size $\frac{n}{2}$ connected by a perfect matching. Note that a set of two nodes is not contagious in DC_n only if both nodes are either in the same clique or connected by the perfect matching.

We begin with the following simple lemma.

Lemma 1. *Let $G = (V, E)$ be a Dirac graph and assume that every node has threshold two. If more than $\frac{n}{2}$ nodes are active, then in the next round all remaining nodes become infected.*

Proof. Let $A \subset V$ with $|A| > \frac{n}{2}$ denote a set of active, resp. infected nodes. Then every node in $\overline{A}$ can have at most $|\overline{A}| - 1 \leq \lceil \frac{n}{2} \rceil - 2$ neighbors outside A. Thus, it must have two neighbors in A, since its degree is at least $\lceil \frac{n}{2} \rceil$.   □

Lemma 2. *Let $G = (V, E)$ be a Dirac graph that is not DC_n for some $n \geq 2$. Then every set of vertices of size three is contagious.*

Proof. Let A be a set of active vertices of size k, where $2 < k < n/2$ holds, and note that then every vertex in A has at least $n/2 - k + 1$ neighbors in $\overline{A}$. Thus, A has at least $k \cdot (n/2 - k + 1)$ edges with one endpoint in A and the other one in $\overline{A}$.

On the other hand, there are only $n - k$ nodes outside A; in particular, if $k \cdot (n/2 - k + 1) > n - k$ holds, then there must be a node in $\overline{A}$ that has two neighbors in A and hence will be infected in the next round. The equation $k \cdot (n/2 - k + 1) = n - k$ has exactly two roots: $k = 2$ and $k = \frac{n}{2}$. Hence every set A satisfying $3 \leq |A| < n/2$ necessarily infects a vertex in $\overline{A}$. In addition, if a set B has more than $n/2$ nodes, then by Lemma 1 it will infect every vertex in $\overline{B}$.

Hence if a set C of size at least 3 does not infect the whole of G, then it will necessarily infect a set D of size $n/2$ eventually: as long as $|C| < n/2$, one node in $\overline{C}$ is infected in the subsequent round. The only way D does not infect an additional vertex is that it is connected by a perfect matching to $\overline{D}$. In this case by the degree condition both D and $\overline{D}$ are cliques. This proves the lemma.   □

Theorem 1. *Every Dirac graph has a contagious set of size two.*

Proof. If the graph is a clique, we can activate two arbitrary vertices. Otherwise, the degree constraints guarantee that any two non-adjacent, activated vertices will infect a third vertex. Unless the graph is DC_n, with $n \geq 2$, Lemma 2 then applies. In case of DC_n with $n \geq 2$, in the first round the two nodes that are adjacent to the seeds will be infected, and in the second round all remaining nodes.   □

We wish to generalize Theorem 1 to Ore graphs. However, some new ideas are required, as Ore graphs may not share properties of Dirac graphs used in the proof of Theorem 1. For example, Lemma 2 does not extend to Ore graphs. In fact, there exist n-vertex Ore graphs such that there is a selection of up to $\lfloor \frac{n}{2} \rfloor$ nodes that do not form a contagious set.

Example 1. We construct the graph as follows: the set $S = \{v_1, v_2, \ldots, v_c\}$ forms a clique. The remaining $n - c$ nodes also form a clique, and are partitioned into c disjoint groups $G_1, G_2, \ldots, G_c$. We let $c \leq \lfloor \frac{n}{2} \rfloor$, thus every G_i is non-empty. Every node in G_i is adjacent to v_i but not to any other node in S. Hence S is not a contagious set. Moreover, note that for any pair $(u, v) \in S \times \overline{S}$ we have

$$\deg(u) + \deg(v) = (c - 1 + 1) + (n - c - 1 + 1) = n,$$

hence we have constructed an Ore graph. Here it is crucial to note that pairs of nodes within S (and in $\overline{S}$ resp.) are adjacent and hence their degrees are not required to sum up to n in a pairwise manner. Notice that for $c = \frac{n}{2}$, the constructed graph is DC_n. □

Now we show the following.

Theorem 2. *Every Ore graph $G = (V, E)$ has a contagious set of size two.*

Proof. For Dirac graphs any three nodes form a contagious set, but we have seen that this statement is not valid for Ore graphs. However, activating three arbitrarily selected nodes with degree $\frac{n}{2}$ each will infect at least half of the nodes, as we show in Lemma 4.

Interestingly, such an active set of size three can be obtained by activating two nodes only: according to Lemma 3 there are two nodes u, v with degree at least $\frac{n}{2}$, such that both are adjacent to a third node w of degree at least $\frac{n}{2}$ as well. Then activating u and v will infect w and subsequently at least half of the nodes.

Thereafter, the infection will reach all nodes unless the graph is isomorphic to DC_n. This is proven in Lemma 5. On the other hand, if the graph is isomorphic to DC_n, Theorem 1 implies that $m(G, 2) \leq 2$.

Lemma 3. *In an Ore graph there exists a vertex w of degree at least $\frac{n}{2}$ that is adjacent to at least two vertices u, v with $\deg(u), \deg(v) \geq n/2$.*

Proof. Let S be the set of vertices with degree at least $\lceil \frac{n}{2} \rceil$. We want to show that there exists a vertex in S with two neighbors in S.

First we show that S must have size at least $\lfloor \frac{n}{2} \rfloor$: if there is a vertex $x \notin S$, then x has at most $\lceil \frac{n}{2} \rceil - 1$ neighbors, denoted by $N(x)$. All vertices that do not belong to $x \cup N(x)$ must belong to S (in order to satisfy the degree constraint for non-adjacent nodes); note that there are at least $\lfloor \frac{n}{2} \rfloor$ such nodes outside $\{x\} \cup N(x)$.

If there is no vertex in S with two neighbors in S, then $E(S, \overline{S}) \geq (\lceil \frac{n}{2} \rceil - 1) \cdot \lfloor \frac{n}{2} \rfloor$ as $|S| \geq \lfloor \frac{n}{2} \rfloor$ and every vertex in S has at least $\lceil \frac{n}{2} \rceil - 1$ neighbors outside S. Observe that

$$\sum_{v \in \overline{S}} \deg(v) \leq |\overline{S}| \cdot \left(\left\lceil \frac{n}{2} \right\rceil - 1 \right).$$

Thus, $|E(\overline{S})|$ is bounded above by the difference of this product on the RHS and the lower bound on $|E(S, \overline{S})|$:

$$|\overline{S}| \cdot \left(\left\lceil \frac{n}{2} \right\rceil - 1 \right) - \left(\left\lceil \frac{n}{2} \right\rceil - 1 \right) \cdot \left\lfloor \frac{n}{2} \right\rfloor = \left(\left\lceil \frac{n}{2} \right\rceil - 1 \right) \cdot \left(|\overline{S}| - \left\lfloor \frac{n}{2} \right\rfloor \right). \tag{1}$$

Recall that we showed $|\overline{S}| \leq \left\lceil \frac{n}{2} \right\rceil$. Thus, the bound given in Eq. (1) is nonnegative only if $|\overline{S}| \in \left\{ \left\lfloor \frac{n}{2} \right\rfloor, \left\lceil \frac{n}{2} \right\rceil \right\}$, and hence the upper bound equals $\left\lceil \frac{n}{2} \right\rceil - 1$ or 0. But $\overline{S}$ has to be a clique by choice of S and the degree requirement of Ore graphs. Therefore, the number of edges inside $\overline{S}$ must be $\binom{|\overline{S}|}{2} = \binom{\lceil \frac{n}{2} \rceil}{2}$, which contradicts the upper bound of $\left\lceil \frac{n}{2} \right\rceil - 1$ or 0 on the number of edges inside $\overline{S}$ if $n > 4$ holds.

For $n \in \{3, 4\}$ we recall that every Ore graph has a Hamiltonian cycle [14]; observe that the statement of the lemma follows immediately in this case $\qquad \square$

Thus, once we activate u, v, the node w will become infected and then eventually half of the nodes.

Lemma 4. *The activation of three vertices with degrees at least $\frac{n}{2}$ each will infect at least half of the nodes in an Ore graph.*

Proof. Let A_0 consist of three vertices of degree at least $\frac{n}{2}$. Let $A := \langle A_0 \rangle$, i.e. the set of nodes that will eventually be active if we activate A_0. Observe that $A_0 \subseteq A$ holds by definition.

Assume for the sake of contradiction that $|A| < \frac{n}{2}$ and recall that $\overline{A} := V \backslash A$ is the set of nodes that do not become active. We claim that each of the vertices in $A \backslash A_0$ must have at least one neighbor in $\overline{A}$: vertices in $\overline{A}$ have at most one neighbor in A and thus degree at most $|\overline{A}|$ each. If $a \in A$ and $b \in \overline{A}$ are non-adjacent, we have that

$$\deg(a) + \deg(b) \leq |\overline{A}| + |A| - 1 + |N(a) \cap \overline{A}|.$$

As this quantity has to be at least n in an Ore graph and $|A| + |\overline{A}| = n$ holds, $N(a) \cap \overline{A}$ must be non-empty.

Each of the nodes in A_0 has degree at least $\frac{n}{2}$ by assumption of the lemma, and hence each of them has at least $\left(\frac{n}{2} - (|A| - 1) \right)$ neighbors in $\overline{A}$. Recall that the other $|A| - 3$ vertices in A must have at least one neighbor in $\overline{A}$ each. But since each node in $\overline{A}$ can have at most one neighbor in A, otherwise it would be infected, we get

$$\begin{aligned} |\overline{A}| &\geq 3 \cdot \left(\frac{n}{2} - (|A| - 1) \right) + (|A| - 3) \\ &= 3 \cdot \frac{n}{2} - 3 \cdot |A| + 3 + |A| - 3 \\ &= \frac{n}{2} + n - 2 \cdot |A|. \end{aligned}$$

Thus, we have $|\overline{A}| + 2 \cdot |A| = n + |A| \geq n + \frac{n}{2}$ and the desired contradiction $|A| \geq \frac{n}{2}$ follows. $\qquad \square$

Next, we show

Lemma 5. *Consider an n-vertex Ore graph that is not equal to DC_n. Then any set of three vertices with degree at least $\frac{n}{2}$ each is a contagious set.*

Proof. Pick any three vertices with degree at least $\frac{n}{2}$ as seed and let A be the set of eventually infected vertices. By Lemma 4 we have $|\overline{A}| \leq |A|$. Again every $b \in \overline{A}$ is adjacent to at most one node in A, otherwise b would be infected, and hence we have $\deg(b) \leq |\overline{A}|$. Then every node in A that is non-adjacent to some $b \in \overline{A}$ must have degree at least $n - |\overline{A}|$ to meet the degree requirement of Ore graphs.

It follows that every vertex $a \in A$ must have at least one neighbor in $\overline{A}$: if a is adjacent to all vertices in $\overline{A}$, the claim holds. If a is non-adjacent to at least one, then we have already shown that a has degree at least $n - |\overline{A}| = |A|$. Since node a can have only $|A| - 1$ neighbors in A, a must have at least one within $\overline{A}$.

No vertex in A can have more than one neighbor in $\overline{A}$, since this would imply the existence of a vertex in $\overline{A}$ with two active neighbors, as $|\overline{A}| \leq |A|$; but this would contradict the choice of $\overline{A}$. Thus, each vertex in A has exactly one neighbor in $\overline{A}$ and we have that $|A| = |\overline{A}| = n/2$. Notice that A and $\overline{A}$ must both be cliques as otherwise two non-adjacent vertices in A (resp., $\overline{A}$) would have degree less than $\frac{n}{2}$ each and thus their degrees add up to less than n contradicting the property of an Ore graph. But then the graph is isomorphic to DC_n. $\qquad\square$

This concludes the proof of Theorem 2. $\qquad\square$

1.2   The Speed of Spreading in Dirac Graphs

In the case of DC_n, it is easy to see that any contagious set actually infects the entire graph in just two rounds. We will now prove a tight bound on the speed of spreading in arbitrary Dirac graphs.

Theorem 3. *Let G be a Dirac graph that does not coincide with DC_n for any $n \geq 2$, and let A_0 be an arbitrary selection of three vertices. Then the activation of A_0 will infect the whole vertex set within three rounds.*

Proof. Recall from Lemma 1 that all nodes will be infected at the end of the subsequent round if $|A_i| > \frac{n}{2}$ holds for any round i. Moreover, if $|A_1| = \frac{n}{2}$ holds, then the whole graph will be active after the third round, since the graph is not DC_n. In particular, any contagious set will infect one new vertex in the first round, hence $|A_1| \geq 4$ because of $|A_0| = 3$; thus, we may assume $\frac{n}{2} > 4$, or equivalently $n \geq 9$, and

$$|A_1| \leq \left\lceil \frac{n}{2} \right\rceil - 1 \tag{2}$$

from now on.

Since each node in A_1 has at most $|A_1| - 1$ neighbors in A_1, it has at least $\left\lceil \frac{n}{2} \right\rceil - |A_1| + 1$ edges to nodes outside A_1. Moreover, since each node outside A_1 is adjacent to at most one node in A_0, we observe that the neighborhood of A_0 in $\overline{A}_1$, denoted by N, has size $|N| \geq |A_0| \cdot \left(\left\lceil \frac{n}{2} \right\rceil - |A_1| + 1 \right) =$

$3 \cdot \left(\left\lceil \frac{n}{2} \right\rceil - |A_1| + 1 \right)$. Finally, consider the $|A_1| - 3$ nodes of $A_1 \backslash A_0$ and note that there are at least $(|A_1| - 3) \cdot \left(\left\lceil \frac{n}{2} \right\rceil - |A_1| + 1 \right)$ edges between $A_1 \backslash A_0$ and $\overline{A}_1$; we denote the set of these edges by F.

There are $n - |A_1|$ nodes outside A_1. For the sake of contradiction, let us assume now that there are at most $\left\lfloor \frac{n}{2} \right\rfloor - |A_1|$ nodes in $\overline{A}_1$ that have more than one neighbor in A_1. Then the number of edges between $\overline{A}_1$ and $A_1 \backslash A_0$ is at most

$$(n - |A_1| - |N|) + (|A_1| - 3) \cdot \left(\left\lfloor \frac{n}{2} \right\rfloor - |A_1| \right). \tag{3}$$

To see this upper bound, first observe that at most $n - |A_1| - |N|$ edges of F can be placed so that every node in $\overline{A}_1$ is adjacent to exactly one node in A_1. The second summand follows from the observation that each node in $\overline{A}_1$ is incident with at most $|A_1| - 3$ edges of F.

But this yields the desired contradiction, since we know that there are at least $(|A_1| - 3) \cdot \left(\left\lceil \frac{n}{2} \right\rceil - |A_1| + 1 \right)$ edges in F. Subtracting the upper bound on number of edges given by Eq. (3), that was implied by the assumption that there are at most $\left\lfloor \frac{n}{2} \right\rfloor - |A_1|$ nodes in $\overline{A}_1$ with more than one neighbor in A_1, we obtain

$$(|A_1| - 3) \cdot \left(\left\lceil \frac{n}{2} \right\rceil - |A_1| + 1 \right) - (n - |A_1| - |N|) - (|A_1| - 3) \cdot \left(\left\lfloor \frac{n}{2} \right\rfloor - |A_1| \right)$$
$$\geq |A_1| - 3 + |N| - n + |A_1|$$
$$\geq 2 \cdot |A_1| - 3 + 3 \cdot \left(\left\lceil \frac{n}{2} \right\rceil - |A_1| + 1 \right) - n$$
$$\geq 3 \cdot \left\lceil \frac{n}{2} \right\rceil - |A_1| - n \geq \left\lceil \frac{n}{2} \right\rceil - |A_1| \geq 1,$$

where the last inequality is implied by Eq. (2). Thus, in $\overline{A}_1$ there are at least $\left\lfloor \frac{n}{2} \right\rfloor + 1 - |A_1|$ nodes with at least two neighbors in A_1, and all remaining nodes will be infected in the third round according to Lemma 1. □

Corollary 1. *Any contagious set of size two in a Dirac graph infects the entire graph within at most four rounds.*

Proof. Any contagious set satisfies $|A_1| > |A_0| = 2$ after the first round; then Theorem 3 gives that the whole vertex set will be infected after three additional rounds. □

We show that the bounds given in Theorem 3 and Corollary 1 are tight.

Example 2. Consider the following graph on the vertex set $V = \{v_1, \ldots, v_8\}$: let v_1, v_2, v_3 be a clique, v_4 and v_5 be adjacent to v_1, while v_7 and v_8 are adjacent to v_2. Moreover, let v_3 be adjacent to v_4 and v_6. v_4 and v_5 are adjacent to each other as well as to v_7 and v_8. v_7 and v_8 are adjacent to each other and to v_6. v_6 is also adjacent to v_5. Every vertex has degree at least four. Thus, if v_1 and v_2 are activated, it takes four rounds for the entire graph to be infected. Moreover, if we activate v_1, v_2, and v_3 then it takes three rounds for the entire graph to be infected. □

2   Extremal Number of Edges

What is the maximum number of edges in a graph with n nodes such that there is no contagious set of size k assuming that all nodes have threshold k? We provide the following tight bound for the case $n \geq 2k + 2$.

Theorem 4. *Let $k \geq 1$. For $n \geq 2k + 2$ we have $M(n, k, k) = \binom{n-1}{2}$.*

Proof. To see $M(n, k, k) \geq \binom{n-1}{2}$, note that a clique of $n - 1$ nodes plus an isolated node is a disconnected graph with n nodes and $\binom{n-1}{2}$ edges. However, no disconnected graph can have a contagious set of size $k < n$ when the thresholds are k. In the sequel we show $M(n, k, k) \leq \binom{n-1}{2}$, i.e. every graph on n nodes with at least $\binom{n-1}{2} + 1$ edges has a contagious set of size k if all thresholds are k.

If a set $S \subseteq V$ with $|S| = k$ is not contagious for all thresholds equal to k, then there is a set T with $S \subseteq T$ such that each node in $\overline{T}$ has at most $k - 1$ neighbors in T; only then the infection does not spread outside of T.

The gist is that there are at least

$$|\overline{T}| \cdot (|T| - (k-1)) = (n - |T|) \cdot (|T| - (k-1))$$

pairs of nodes in $T \times \overline{T}$ that are not adjacent. In particular, we claim that if $|T| \in \{k+1, k+2, ..., n-3, n-2\}$ then the number of non-adjacent node pairs is larger than $n - 2$. However, at most $n - 2$ pairs of nodes are not adjacent in a graph with n nodes and at least $\binom{n-1}{2} + 1$ edges, since $\binom{n-1}{2} + 1 = \binom{n}{2} - (n-2)$ holds. Thus, if the number of active vertices is at least $k + 1$ and at most $n - 2$, then in the subsequent round at least one more node is infected newly, and therefore the infection does not stop until at least $n - 1$ nodes are active.

Now we prove the claim. First observe that $f(|T|) = (n - |T|) \cdot (|T| - (k-1))$ is a quadratic function in $|T|$ and has roots $|T| = n$ and $|T| = k - 1$. In the former case, the process has already infected all nodes, and the latter case cannot occur, since $S \subseteq T$ and $|S| = k$. Next we show that $f(|T|)$ is larger than $n - 2$ for values of $|T| \in \{k+1, ..., n-2\}$. Recall that we assumed $n \geq 2k + 2$, and observe that the number of non-adjacent pairs in $T \times \overline{T}$ is minimized for any fixed k by setting $n = 2k + 2$. Therefore the number of such pairs is at least $(2k + 2 - |T|) \cdot (|T| - k + 1)$. On the one hand, if $|T| = k + 1$ holds, then their number is $(k+1) \cdot 2 = 2k + 2 = n$. On the other hand, for $|T| = n - 2 = 2k$ their number is $2 \cdot (k + 1) = n$ again. Thus, the claim holds for both values of $|T|$, and furthermore for all choices of $|T|$ in between, since $f''(|T|) = -2$ and hence f is concave.

Thus, we focus on $|T| \in \{k, n - 1\}$ in the sequel. First we show how to select A_0 with $|A_0| = k$ such that $|A_1| \geq k + 1$ holds. If the graph does not contain any node of degree less than k, we pick any node v and choose A_0 to contain k neighbors of v. Then $v \in A_1$ holds and hence $|A_1| \geq k + 1$.

Now assume there is a node u with degree $d < k$. Note that any node of degree smaller k is non-adjacent to at least $n - 1 - \frac{n-2}{2} = \frac{n}{2}$ nodes, where we use $\frac{n-2}{2} \geq k$. Hence there can be at most one such node because there are at most $n - 2$ non-adjacent pairs of nodes in the graph.

Let G' be the graph after removing u and its d incident edges. Note that the degree of each node in G' was reduced by at most one due to the removal of u, therefore all degrees in G' are at least $k - 1$. Hence we pick any node w that was adjacent to u (recall that the graph is connected) and choose A_0 to contain u and $k - 1$ neighbors of w. Then in the first round w will be infected, thus we have $|A_1| \geq k + 1$.

We have already shown that if there are at least $k + 1$ active nodes, then the process does not stop until there are $n - 1$ active nodes. Assume that the process does not infect the last node w. Then w has degree less than k; but in this case w was selected for A_0. Thus, the process cannot stop at $n - 1$ nodes. $\square$

Example 3. Note that the statement of Theorem 4 does not hold for arbitrary $k \in \{1, \ldots, n-1\}$ if n is fixed. For a clique on n vertices we pick a perfect matching M and delete the edges of M. Let $k = n-1$. Each vertex has degree $n-2$, hence there is no contagious set of size $n-1$. $\square$

For the case that all nodes have threshold two we give in Theorem 5 a tight bound on the number of edges that guarantees the existence of a contagious set of size $k \ll n$.

Theorem 5. *For all $k \geq 2$ there exists $n_k \in \mathbb{N}$, such that for all $n \geq n_k$,*

$$M(n, k, 2) = \binom{n - k + 1}{2} + \left\lfloor \frac{k - 1}{2} \right\rfloor.$$

Due to space restrictions the proof is deferred to the full version of the paper.

We did not attempt to find the exact $k(n)$ for which Theorem 5 holds. It should be noted that certain restrictions on k have to be imposed, as is shown the following example.

Example 4. We construct a family of graphs to demonstrate that

$$M(n, k, 2) = \binom{n - k + 1}{2} + \left\lfloor \frac{k - 1}{2} \right\rfloor$$

does not hold for arbitrary k and n. Consider for $k \geq \frac{2n+2}{3}$ a clique on $n - k$ vertices together with a star on k vertices. All $k - 1$ leaves of the star must be contained in a contagious set and so do two vertices from the clique, so there is no contagious set of size k. However, the number of edges is

$$\binom{n - k}{2} + (k - 1) = \binom{n - k + 1}{2} + \frac{k}{2} + \frac{3k}{2} - (n + 1)$$

$$\geq \binom{n - k + 1}{2} + \frac{k}{2} > \binom{n - k + 1}{2} + \left\lfloor \frac{k - 1}{2} \right\rfloor.$$

$\square$

3   Conclusion

We have examined conditions on the minimum degree and the average degree of undirected graphs ensuring the existence of contagious sets of size $k \geq 2$.

We focused primarily on the case where all thresholds equal two and showed tight bounds on the number of rounds it takes to infect all vertices for graphs whose minimum degree guarantees the existence of a contagious set of size two.

There are several questions that arise from this work. One is to determine the value of $M(n, k, r)$ for all $n \geq k \geq r$. Another question is to find how large the minimum degree of G needs to be in order to ensure that $m(G, r) = k$, where $k \geq r > 2$. Finally, it might be of interest to discover additional graph properties implying $m(G, 2) = 2$.

Acknowledgments. The authors would like to thank the reviewers for their valuable comments that helped improving the presentation of the paper significantly.

References

1. Ackerman, E., Ben-Zwi, O., Wolfovitz, G.: Combinatorial model and bounds for target set selection. Theor. Comput. Sci. **411**, 4017–4022 (2010)
2. Amini, H., Cont, R. Minca, A.: Resilience to contagion in financial networks. Math. Finance (2013)
3. Amini, H., Fountoulakis, N.: What I tell you three times is true: bootstrap percolation in small worlds. In: Goldberg, P.W. (ed.) WINE 2012. LNCS, vol. 7695, pp. 462–474. Springer, Heidelberg (2012)
4. Balogh, J., Bollobás, B.: Bootstrap percolation on the hypercube. Prob. Theor. Relat. Fields **134**, 624–648 (2006)
5. Balogh, J., Bollobás, B., Duminil-Copin, H., Morris, R.: The sharp threshold for bootstrap percolation in all dimensions. Trans. Am. Math. Soc. **364**, 2667–2701 (2012)
6. Balogh, J., Pete, G.: Random disease on the square grid. Random Struct. Algorithms **13**, 409–422 (1998)
7. Balogh, J., Pittel, B.: Bootstrap percolation on the random regular graph. Random Struct. Algorithms **30**, 257–286 (2007)
8. Chalupa, J., Leath, P.L., Reich, G.R.: Bootstrap percolation on a Bethe lattice. J. Phys. C Solid State Phys. **12**, L31 (1979)
9. Chen, N.: On the approximability of influence in social networks. SIAM J. Discrete Math. **23**, 1400–1415 (2009)
10. Coja-Oghlan, A., Feige, U., Krivelevich, M., Reichman, D.: Contagious sets in expanders. In: Proceedings of the 26th Symposium on Discrete Algorithms (SODA 2015), pp. 1953–1987 (2015)
11. Janson, S., Łuczak, T., Turova, T., Vallier, T.: Bootstrap percolation on the random graph $G_{n,p}$. Ann. Appl. Prob. **22**, 1989–2047 (2012)
12. Kempe, D., Kleinberg, J.M., Tardos, É.: Maximizing the spread of influence through a social network. Theor. Comput. **11**, 105–147 (2015)
13. Nichterlein, A., Niedermeier, R., Uhlmann, J., Weller, M.: On tractable cases of Target Set Selection. Soc. Netw. Anal. Min. **3**, 233–256 (2013)

14. Ore, O.: Note on Hamilton circuits. Am. Math. Monthly **67**, 55 (1960)
15. Przykucki, M.: Maximal percolation time in hypercubes under two-dimensional bootstrap percolation. Electron. J. Comb. **19**, 1–13 (2012)
16. Reichman, D.: New bounds for contagious sets. Discrete Math. **312**, 1812–1814 (2012)

Solving the Tree Containment Problem for Genetically Stable Networks in Quadratic Time

Philippe Gambette[1(✉)], Andreas D.M. Gunawan[2], Anthony Labarre[1],
Stéphane Vialette[1], and Louxin Zhang[2]

[1] Université Paris-Est, LIGM (UMR 8049), UPEM, CNRS, ESIEE, ENPC,
77454 Marne-la-Vallée, France
philippe.gambette@univ-mlv.fr
[2] Department of Mathematics,
National University of Singapore, Singapore 119076, Singapore

Abstract. A phylogenetic network is a rooted acyclic digraph whose leaves are labeled with a set of taxa. The tree containment problem is a fundamental problem arising from model validation in the study of phylogenetic networks. It asks to determine whether or not a given network displays a given phylogenetic tree over the same leaf set. It is known to be NP-complete in general. Whether or not it remains NP-complete for stable networks is an open problem. We make progress towards answering that question by presenting a quadratic time algorithm to solve the tree containment problem for a new class of networks that we call genetically stable networks, which include tree-child networks and comprise a subclass of stable networks.

1 Introduction

With thousands of genomes being fully sequenced, phylogenetic networks have been adopted to study "horizontal" processes that transfer genetic material from a living organism to another without descendant relation. These processes are a driving force in evolution which shapes the genome of a species [1,9].

A *rooted (phylogenetic) network* over a set X of taxa is a rooted acyclic digraph with a set of leaves (*i.e.*, vertices of outdegree 0) that are each labeled with a distinct taxon. Such a network represents the evolutionary history of the taxa in X, where the *tree nodes* (*i.e.*, nodes of indegree 1) represent speciation events. The nodes of indegree at least two are called *reticulations* and represent genetic material flow from several ancestral species into an "unrelated" species. A plethora of methods for reconstructing networks and related algorithmic issues have been extensively studied over the past two decades [4,5,8,10].

One of the ways of assessing the quality of a given phylogenetic network is to verify that it is consistent with previous biological knowledge about the species. Biologists therefore demand that the network display existing gene trees, and the corresponding algorithmic problem is known as the *tree containment* problem

© Springer International Publishing Switzerland 2016
Z. Lipták and W.F. Smyth (Eds.): IWOCA 2015, LNCS 9538, pp. 197–208, 2016.
DOI: 10.1007/978-3-319-29516-9_17

(or TC problem for short) [5], which is well-known to be NP-complete [6,7]. Great efforts have been devoted to identifying tractable subclasses of networks, such as binary galled trees [7], normal networks, binary tree-child networks, level-k networks [6], or nearly-stable networks [3]. One of the major open questions in this setting is the complexity of the TC problem on the so-called *stable networks*.

A node v in a network is *stable* if there exists a leaf such that every path from the root to the leaf passes through v. A network is *stable* (or *reticulation visible*) [5] if every reticulation is stable. Motivated by the study in [2], we make progress in this work towards determining the complexity of the TC problem on stable networks by presenting a quadratic-time algorithm for a new class that we call *genetically stable networks*. As we shall show, these networks comprise a subclass of stable, tree-sibling networks, including *tree-child* networks.

2   Concepts and Notions

2.1   Binary Networks

We focus in this paper on *binary* networks, *i.e.* networks whose root has indegree 0 and outdegree 2, whose internal nodes all have degree 3, and whose leaves all have indegree 1 and outdegree 0. An internal node in a network N is called a *tree node* if its indegree and outdegree are 1 and 2, respectively. It is called a *reticulation* (*node*) if its indegree and outdegree are 2 and 1, respectively. A node v is said to be *below* a node u if u is an ancestor of v, *i.e.* there is a directed path from u to v in N.

We also assume that in a binary network, there is a path from its root to every leaf and that a node can be of indegree 1 and outdegree 1. We also draw an open edge entering the root so that the root becomes a tree node with degree 3, as shown in Fig. 1. For a network or a subnetwork N, we use the following notation: $\rho(N)$ for its root, $\mathcal{L}(N)$ for its leaf set, $\mathcal{R}(N)$ for the set of reticulations, $\mathcal{T}(N)$ for the set of tree nodes, $\mathcal{V}(N)$ for its vertex set (*i.e.*, $\mathcal{R}(N) \cup \mathcal{T}(N) \cup \mathcal{L}(N) \cup \{\rho(N)\}$), $\mathcal{E}(N)$ for its edge set, $p(u)$ for the set of parents of $u \in \mathcal{R}(N)$ or the unique parent of u otherwise, children(u) for the set of children of $u \in \mathcal{T}(N)$ or the unique child of $u \in \mathcal{R}(N)$, and $\mathcal{P}_N(u,v)$ for the set of all paths from a node u to a node v in N.

A path P from u to v in a network is a *tree path* if every internal node of P, that is every node in $\mathcal{V}(P) \setminus \{u,v\}$, is a tree node. For a network N and an edge subset $E \subseteq \mathcal{E}(N)$, $N - E$ denotes the subnetwork with vertex set $\mathcal{V}(N)$ and edge set $\mathcal{E}(N) - E$. For a node subset $S \subset \mathcal{V}(N)$, $N - S$ denotes the subnetwork with vertex set $\mathcal{V}(N) - S$ and edge set $\{(u,v) \in \mathcal{E}(N) \mid u \notin S, v \notin S\}$. When E or S has only one element x, we simply write $N - x$. A leaf in the resulting network is a dummy leaf if it is not a leaf in the original network N.

2.2   The Tree Containment (TC) Problem

Let N be a binary network and T a binary tree over the same set of taxa. We say that N *displays* T if N contains a subtree T', obtained by removing an incoming edge for each reticulation in N, such that T can be obtained from T' by:

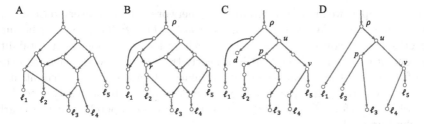

Fig. 1. (**A**) A nearly tree-child network. (**B**) A non-nearly tree-child network, in which the parents of r are not connected to any leaf by a tree path. (**C**) A subtree T' obtained by removing an incoming edge from each reticulation in the network in **B**. (**D**) A tree obtained from T' by contraction.

1. recursively removing dummy leaves (such as d in Fig. 1.C), and
2. contracting every path containing only nodes of degree 2 into a single edge (Fig. 1.B-D).

T' is then referred to as a *subdivision* of T in N. Given a binary network and a binary tree, the *tree containment (TC) problem* is to determine whether or not the network displays the tree [5]. This problem is known to be NP-complete [6,7], and a large part of the current research therefore focuses on finding tractable classes of binary networks that are as general as possible.

3   Genetically Stable Networks

Let N be a binary network and $u, v \in \mathcal{V}(N)$. Node u is *stable on node* v if every path from $\rho(N)$ to v passes through u. We denote by $\mathcal{PDL}_N(u)$ the set of leaves on which u is stable, and say that u is *stable* (or *visible*) if $\mathcal{PDL}_N(u) \neq \emptyset$. Network N is itself *stable* if every $r \in \mathcal{R}(N)$ is stable. The network in Fig. 1.A is stable, whereas the one in Fig. 1.B is not. The following result will be useful.

Proposition 1. *Let N be a binary network and $r \in \mathcal{R}(N)$ with $p(r) = \{u, v\}$.*

(a) *If $s \in \mathcal{V}(N)$ is a stable node, then* children(s) *contains a tree node.*
(b) *If N is stable, then both parents of each reticulation are tree nodes.*
(c) *For any descendant x of r, either u or v is not stable on x.*
(d) *If r and u are stable on the same leaf, then u is stable on v.*
(e) *If r is stable on $\ell \in \mathcal{L}(N)$ and v is stable on $\ell' \in \mathcal{L}(N)$ but not on ℓ, then u is not in a path from v to ℓ'. Additionally, there is no z in a path from v to ℓ' that is connected to u by a tree path.*

If a tree node is stable on a leaf ℓ, then its unique parent is also stable on ℓ, but the stability of a reticulation does not imply that of its parents. Cordue, Linz and Semple [2] recently introduced a class of stable networks that we call *nearly tree-child networks* and which satisfy the property that every reticulation has a parent connected to some leaf by a tree path (see Fig. 1.A for an example).

In this paper, we are interested in stable networks in which every reticulation has a stable parent. We coin the concept *genetic stability* (GS) to describe such networks, which conveys the idea that each reticulation inherits its stability from one of its parents. Note that in a nearly tree-child network, there is a tree path from one of the parents of every reticulation to a leaf, so that parent must be stable. On the other hand, a GS network may not be a nearly tree-child network[1]. Therefore, GS networks comprise a proper superclass of the nearly tree-child networks.

A network is *tree-sibling* if every reticulation has at least one sibling that is a tree node [5]. Interestingly, we also have the following fact.

Proposition 2. *Every GS network is tree-sibling.*

Our result on the complexity of the TC problem for binary GS networks therefore refines the complexity gap of the TC problem between the classes of binary tree-child networks, where it can be solved in polynomial time, and tree-sibling networks where it is NP-complete [6]. Furthermore, a study of the properties of networks simulated using the coalescent model with recombination shows that the percentage of simulated networks which are GS is significantly larger than that of tree-child networks (see http://phylnet.info/recophync/), thus making that new class significant in practice.

4   Solving the TC Problem for GS Networks

In this section, T denotes a binary tree and N is a genetically stable network on the same leaf set as T unless noted otherwise.

4.1   Overview of the Algorithm

A *cherry* is a subtree induced by two sibling leaves ℓ' and ℓ'' and their parent $\alpha_{\ell',\ell''}$, which we denote $\{\alpha_{\ell',\ell''}, \ell', \ell''\}$. It is easy to see that any tree can be transformed into a single node by repeatedly deleting the leaves of a cherry and their incident edges, since this operation turns their parent into a new leaf.

Our algorithm relies on the fact that for any cherry $\{\alpha_{\ell',\ell''}, \ell', \ell''\}$ in T, N displays T if and only if there exists a tree node $p \in \mathcal{T}(N)$ and two disjoint *specific paths* (defined later) $P' \in \mathcal{P}_N(p, \ell')$ and $P'' \in \mathcal{P}_N(p, \ell'')$ such that the modified network $N - [(\mathcal{V}(P') \cup \mathcal{V}(P'')) \setminus \{p\}]$ displays the modified tree $T - \{\ell', \ell''\}$, if we identify leaf p in the modified network with leaf $\alpha_{\ell',\ell''}$ (the parent node of the cherry in T) in the modified tree. Therefore, our algorithm is a recursive procedure which executes the following tasks at each recursive step:

S1: Select a cherry $\{\alpha_{\ell',\ell''}, \ell', \ell''\}$ in T, and determine the corresponding node p and paths P' and P''.

S2: If we fail to find such a node and such paths, N does not display T. Otherwise, recurse on $N - [(\mathcal{V}(P') \cup \mathcal{V}(P'')) \setminus \{p\}]$ and $T - \{\ell', \ell''\}$.

---

[1] See e.g. the network given at http://phylnet.info/isiphync/network.php?id=4.

4.2   Three Lemmas

The difficulty in implementing the proposed approach is that a network can display a tree through different subdivisions of the tree and the parent node and edges of a cherry may correspond to different tree nodes and paths in different subdivisions. Therefore, we first prove that the two paths corresponding to the edges of a cherry have special properties.

Lemma 1 (Cherry path). *Let N display T and $\{\alpha_{\ell',\ell''}, \ell', \ell''\}$ be a cherry in T. Then $\alpha_{\ell',\ell''}$ corresponds to a tree node p in each subdivision T' of T in N. Moreover, assume that P' and P'' are the paths in T' that correspond to edges $(\alpha_{\ell',\ell''}, \ell')$ and $(\alpha_{\ell',\ell''}, \ell'')$, respectively. Then the following properties hold:*

(1) The node p is not stable on any leaf $\ell \notin \{\ell', \ell''\}$.
(2) No vertex in $P' \setminus \{p\}$ is stable on a leaf other than ℓ'.
(3) No vertex in $P'' \setminus \{p\}$ is stable on a leaf other than ℓ''.

In the following discussion, we focus on paths P from an internal node x to a leaf ℓ having the following property:

$(\star)$ Each $u \in \mathcal{V}(P) \setminus \{x\}$ is either stable only on ℓ or not stable at all.

A path satisfying condition $(\star)$ is called a *specific path* (with respect to ℓ). We use $\mathcal{SP}_N(x, \ell)$ to denote the set of specific paths from x to $\ell \in \mathcal{L}(N)$. A path P from u to v is said to be *unstable specific* if no $x \in \mathcal{V}(P) \setminus \{u, v\}$ is stable, where u and v are non-leaf nodes. Note that in a GS network, an unstable specific path is a tree path, since every reticulation is stable. Finally, for a path P and $a, b \in \mathcal{V}(P)$, we use $P[a, b]$ to denote the subpath of P from a to b.

Lemma 2 (Cherry path uniqueness). *Let N be a GS network, $\ell_1, \ell_2 \in \mathcal{L}(N)$, and $a', a'' \in \mathcal{T}(N)$. If there exist two paths $P_1' \in \mathcal{SP}_N(a', \ell_1)$ and $P_2' \in \mathcal{SP}_N(a', \ell_2)$ such that $\mathcal{V}(P_1') \cap \mathcal{V}(P_2') = \{a'\}$ and two paths $P_1'' \in \mathcal{SP}_N(a'', \ell_1)$ and $P_2'' \in \mathcal{SP}_N(a'', \ell_2)$ such that $\mathcal{V}(P_1'') \cap \mathcal{V}(P_2'') = \{a''\}$, then:*
(1) Either a'' is a descendant of a' in $P_1' \cup P_2'$ or vice versa.
(2) If a'' is a descendant of a' in P_2' and u_1 is the highest common node in P_1' and P_1'' (Fig. 2.A), then one of the following facts holds:
   (a) $P_1'[a', u_1] = (a', u_1) \in \mathcal{E}(N)$, and $P_1''[a'', u_1]$ is unstable specific.
   (b) $P_1''[a'', u_1] = (a'', u_1) \in \mathcal{E}(N)$ and a'' is stable on ℓ_2.

Proof. (1) Assume the statement is false. Since both P_i' and P_i'' end at ℓ_i, they must intersect for $i = 1, 2$. Let u_i be the highest common node in P_i' and P_i'', $i = 1, 2$. Clearly, u_1 and u_2 are reticulations stable on ℓ_1 and ℓ_2, respectively; for each i, the only node common to $P_i'[a', u_i]$ and $P_i''[a'', u_i]$ is u_i (Fig. 2.B-D).

If $P_1'[a', u_1], P_1''[a'', u_1], P_2'[a', u_2]$, and $P_2''[a'', u_2]$ are all edges (Fig. 2.B), then a' and a'' are the parents of both u_1 and u_2. Since N is GS, either a' or a'' is stable. Clearly a' and a'' are not stable on ℓ_1 and ℓ_2, so stability should involve another leaf ℓ below u_1 or u_2; but this is not possible because there is always a path from a' (resp. a'') to ℓ avoiding a'' (resp. a'). Therefore, one of these four subpaths contains more than one edge. We assume without loss of generality that $P_1'[a', u_1]$ has more than one edge and v and w are the parents of u_1 in P_1' and P_1'', respectively, where $v \neq a'$. We consider two subcases.

1. If $w = a''$ (Fig. 2.C), a'' is clearly not stable on both ℓ_1 and ℓ_2. If a'' is stable on a leaf $\ell \notin \{\ell_1, \ell_2\}$, then ℓ cannot be a descendant of u_1, otherwise the path from a' to ℓ through u_1 would avoid a'', a contradiction. If ℓ is not a descendant of u_1 but is a descendant of the child z of a'' in P_2'', then every path from $\rho(N)$ to ℓ must contain the edge (a'', z) and z. This implies that z is stable on ℓ, contradicting that z is in the specific path P_2''. Therefore, a'' is not stable on any leaf. Since N is GS, v must be stable on a leaf ℓ_3. Since P_1' is a specific path and $v \neq a'$, $\ell_3 = \ell_1$. This implies that v is an ancestor of a'' and so is a', which contradicts the assumption.

2. If $w \neq a''$ (Fig. 2.D), either v or w is stable, because N is GS and they are the parents of u_1. Without loss of generality, we may assume w is stable. Since P_1'' is a specific path, w must be stable on ℓ_1. By Proposition 1.(d), w is stable on v, so w either is in $P_1'[a', v]$ or is an ancestor of a'. The former contradicts the fact that u_1 is the highest common node in P_1' and P_1'', whereas the latter implies that a' is a descendant of a'', which contradicts the assumption.

(2) Using the same notation as in (1) (Fig. 2.A), since N is GS, either v or w is stable. If v is stable on ℓ_1, by Proposition 1.(d), v is stable on w. So $v = a'$ and $P_1'[a', u_1]$ is just (v, u_1). Let x be in $P_1''[a'', u_1] - \{a''\}$. If x is stable, it must be stable on ℓ_1, since it is in P_1''. This contradicts the fact that there is a path from $\rho(N)$ to ℓ_1 through u_1 avoiding x. Therefore, $P_1''[a'', u_1]$ is unstable specific.

If v is stable on $\ell \neq \ell_1$, then $v = a'$. Otherwise, v would be an internal node of P_1', contradicting the fact that P_1' is in $\mathcal{SP}_N(a', \ell_1)$. If v is not stable, then w must be stable, there are two possible cases. If $w \neq a''$, it must be stable on ℓ_1, as it is in P_1''. Therefore, either w is in $P_1'[a', u_1]$, contradicting that u_1 is the highest common node in P_1' and P_1'', or w is an ancestor of a', contradicting that a' is an ancestor of a''. If $w = a''$, then it is stable on ℓ_2 and $P_1''[a'', u_1]$ is simply (w, u_1). □

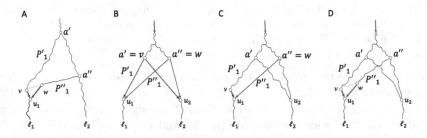

Fig. 2. Illustration of the different cases in the proof of Lemma 2.

Let α_{ℓ_1, ℓ_2} be the parent of ℓ_1 and ℓ_2 in T. Lemma 2.(1) implies that the set of nodes $\{a \mid \exists\ P_1 \in \mathcal{SP}_N(a, \ell_1), P_2 \in \mathcal{SP}_N(a, \ell_2) \text{ s.t. } \mathcal{V}(P_1) \cap \mathcal{V}(P_2) = \{a\}\}$ is totally ordered by the descendant relation, *i.e.* all its elements appear in a path from $\rho(N)$ to ℓ_1. So there is a unique tree node, say p, that is the lowest among all such nodes. Moreover, for any node a in the set, from which there are specific paths P_1 and P_2 going to ℓ_1 and ℓ_2 respectively, Lemma 2.(2) states that if p is a node in P_2, then the path from p to P_1 is an unstable specific path (and vice versa). The next lemma will utilize this property to show that there is a subtree T' of N that is a subdivision of T, in which p corresponds to α_{ℓ_1, ℓ_2}.

Let $t \in \mathcal{T}(N)$. For $\ell_1, \ell_2 \in \mathcal{L}(N)$ and two specific paths whose only common vertex is t, $P_1 \in \mathcal{SP}_N(t, \ell_1)$ and $P_2 \in \mathcal{SP}_N(t, \ell_2)$, we set $N(P_1, P_2)$ to be the subnetwork with vertex set $\mathcal{V}(N)$ and edge set $\mathcal{E}(N) - \{(x, y), (y, x) \mid x \in V(Q) \text{ and } y \notin V(Q)\} - \{(x, y), (y, x) \mid x \in V(P_1) \backslash \{t\} \text{ and } y \in V(P_2) \backslash \{t\}\}$ where $Q = (P_1 \cup P_2) \backslash \{t\}$. Note that $N(P_1, P_2)$ is the subnetwork obtained after removing all the edges not in the paths, but incident at some node in Q.

Lemma 3. (Choice of the lower path). *Let N be a GS network and ℓ_1 and ℓ_2 be two sibling leaves in T. Assume that $t \in \mathcal{T}(N)$ and $P_1 \in \mathcal{SP}_N(t, \ell_1)$ and $P_2 \in \mathcal{SP}_N(t, \ell_2)$ are two specific paths whose only common vertex is t such that $N(P_1, P_2)$ displays T. For any path P from u to v in which every $x \in V(P) \backslash \{u, v\}$ is not stable:*
(1) If $\mathcal{V}(P) \cap \mathcal{V}(P_j) = \{u\}$ and $\mathcal{V}(P) \cap \mathcal{V}(P_{j'}) = \{v\}$, where $\{j', j\} = \{1, 2\}$, T is also displayed in $N(P_j[u, \ell_j], P[u, v] \cup P_{j'}[v, \ell_{j'}])$.
(2) If $\mathcal{V}(P) \cap \mathcal{V}(P_{j'}) = \emptyset$ and $\mathcal{V}(P) \cap \mathcal{V}(P_j) = \{u, v\}$, where $\{j, j'\} = \{1, 2\}$, T is also displayed in $N(P_{j'}, P_j - P_j[u, v] + P[u, v])$.

Lemma 3.(1) implies that if N displays T, there is a subtree T' that is a subdivision of T, such that p corresponds to α_{ℓ_1, ℓ_2}. The next section includes an algorithm to find the node p.

4.3   The Algorithm

We use two lists at each node u to represent the input network N and the input tree T: the list parent(u) comprises the nodes from which u has an edge, and the list children(u) consists of nodes to which u has an edge.

We say that a node u is *reachable* from the network root if there is a path from the root to u. Using a breadth-first search, we can determine the sets of descendants for each vertex in $O(|\mathcal{E}(N)| + |\mathcal{V}(N)|)$ time. To determine the stability of a node u, one can compute the set $R_{not}(u)$ of leaves that are reachable from the root in $N - u$. Obviously, the set $\mathcal{PDL}_N(u)$ of nodes on which u is stable is $\mathcal{L}(N) - R_{not}(u)$, so u is stable if and only if $R_{not}(u) \neq \mathcal{L}(N)$. Therefore, we can determine whether or not a node is stable on a leaf in time $O(|\mathcal{E}(N)| + |\mathcal{V}(N)|)$.

We first find two sibling leaves l_1 and l_2 with parent α_{l_1, l_2}, which takes $O(|\mathcal{L}(T)|)$ time. We then extend a specific path starting at l_1 by moving a node up each time to find a $p \in \mathcal{T}(N)$ such that if N displays T, there is a subdivision of T in which p corresponds to α_{l_1, l_2}. Assume we arrive at a node w. If w is stable on a leaf $z \notin \{l_1, l_2\}$, then we conclude that N does not display T. If w is stable on l_2, or if there is a specific path from w to l_2, then w must be p if N displays T and we are done, so we continue our analysis by assuming otherwise.

If w is a tree node, we simply move up to its unique parent $p(w)$. If w is a reticulation, it is stable on l_1. Let $p(w) = \{u, v\}$. We have to chose either u or v

to move up using the stability property that w is only stable on l_1 and at least one of u and v is stable. By Proposition 1.(c), u and v cannot both be stable on l_1. If u is stable on l_1 and v is stable on l_2, by Proposition 1.(d), u must also be stable on l_2. Therefore, we just need to consider eight different conditions (Table 1) to choose u or v to move up.

Table 1. When w is stable on l_1, there are six combinations of its parents u and v for consideration. Here, $l(u, v) = u$ if u is a descendant of v, or v otherwise

Cond. S/N	Stability of u	Stability of v	Selection
C1	$\mathcal{PDL}_N(u) \backslash \{l_1, l_2\} \neq \emptyset$	$\mathcal{PDL}_N(v) \backslash \{l_1, l_2\} \neq \emptyset$	Neither
C2	$\mathcal{PDL}_N(u) \backslash \{l_1, l_2\} \neq \emptyset$	$\mathcal{PDL}_N(v) \subseteq \{l_1, l_2\}$	v
C3	u is stable on l_1 (and eventually on l_2)	v is not stable	v
C4	u is not stable	v is stable only on l_2	v
C5	$\mathcal{PDL}_N(u) \subseteq \{l_1, l_2\}$	$\mathcal{PDL}_N(v) \backslash \{l_1, l_2\} \neq \emptyset$	u
C6	u is not stable	v is stable on l_1 (and eventually on l_2)	u
C7	u is stable only on l_2	v is not stable	u
C8	u is stable on l_2	v is stable on l_2	$l(u, v)$

If condition C1 holds, w cannot be a node in the path corresponding to the edge (α_{l_1, l_2}, l_1) in any subdivision T' of T. This is because a leaf in either $\mathcal{PDL}_N(u) \setminus \{l_1, l_2\}$ or $\mathcal{PDL}_N(v) \setminus \{l_1, l_2\}$ will not appear in any T' that can be contracted into a tree in which l_1 and l_2 are siblings. Similarly, if C2 holds, u cannot be a node in the path corresponding to the edge (α_{l_1, l_2}, l_1) in a subdivision T' of T. Therefore, we select v. If C3 holds, since u and w are stable on l_1, by Proposition 1.(d), u is stable on v and we move to v if v is not stable. If C4 holds, by Proposition 1.(e), if u is below v, there is a reticulation r' such that there is a tree path from r' to u, r' is not above l_2, and r' is below v. This implies that r' is stable on a leaf other than l_1 and l_2, so we choose v. If u is not below v, then we also choose v because we need to choose the lower one. Conditions C5–C7 are symmetric to C2–C4 and so we select u to move up if they are true. If C8 holds, then, u is a descendant of v or vice versa. Clearly, we have to choose whichever is lower than the other. Algorithm 1 summarizes the whole procedure.

As we have seen, the property that each reticulation has a stable parent is crucial in enabling a correct choice at a reticulation stable on a leaf under consideration. A simple condition allows us to determine whether we have reached p while moving up from x: there is a unstable specific tree path from p to l_2 or to a reticulation stable on l_2, because there is a specific path from p to l_2. Thus, we obtain Algorithm 2 to solve the TC problem.

Algorithm 1. Move up one node to find p

Procedure *MoveUpInSpecificPath(w, l_1, l_2, P, N)*

  Input: node w, leaves l_1 and l_2 and path P in network N

  Output: **false** if N does not display T, **true** if no final decision can yet be made

1   **if** w *is a tree node* **then**

2      $P \leftarrow P \cup \{w\}$; $N \leftarrow N - (\text{parent}(w), w)$; $w \leftarrow \text{parent}(w)$;

     // Select a parent at a reticulation

3   **if** w *is a reticulation stable on l_1 with parents $\{u, v\}$* **then**

4   **if** *C1* **then**

5   **return false**;

6   **if** *C2 or C3 or C4 or (C8 and v is lower)* **then**

7        $P \leftarrow P \cup \{w\}$; $N \leftarrow N - (u, w)$; $w \leftarrow v$;

8   **if** *C5 or C6 or C7 or (C8 and u is lower)* **then**

9        $P \leftarrow P \cup \{w\}$; $N \leftarrow N - (v, w)$; $w \leftarrow u$;

10   **return true**;

11   **else**

12   **return false**;

Theorem 1. *Algorithm 2 solves the TC problem for GS networks in quadratic time.*

Proof. Assume the input network N displays the input tree T, and let $\mathcal{SD}_N(T)$ be the set of subdivisions of T in N. Let α_{ℓ_1,ℓ_2} be the parent of the sibling leaves ℓ_1 and ℓ_2 in T selected in line 4 of Algorithm 2. Recall that by Lemma 2, the set $\{a \mid \exists P_1 \in \mathcal{SP}(a, \ell_1), P_2 \in \mathcal{SP}(a, \ell_2) \text{ s.t. } \mathcal{V}(P_1) \cap \mathcal{V}(P_2) = \{a\}\}$ has a lowest element p. If N displays T, by Lemma 3, p must correspond to α_{ℓ_1,ℓ_2} in some subdivision of T in N. We now show that Algorithm 2 correctly finds p.

Let P_i be the path from p to ℓ_i in a subdivision $T' \in \mathcal{SD}_N(T)$ corresponding to the edge $(\{\alpha_{\ell_1,\ell_2}, \ell_1, \ell_2\}, \ell_i)$ in the cherry in T for $i = 1, 2$. By Lemma 1, P_1 and P_2 are specific paths. Let us prove that the first while-loop exits at $w_1 = p$. Assume t is the last vertex in P_1 at which the algorithm has moved off during the first while-loop before stopping at $w_1 = w \neq p$ (Fig. 3.A). So t is a reticulation with a parent v in P_1 and the other parent u to which the algorithm moved from t. Let P be the path consisting of all vertices visited by the algorithm after t.

Since t is a reticulation in P_1, it is stable on ℓ_1. By the definition of the moving up procedure MoveUpInSpecificPath, moving from t to u implies that C5, C6, C7 or C8 holds. C5 cannot be true, as v is in P_1 and cannot be stable on a leaf not equal to ℓ_1. If C8 holds, then $v = p$. u is not in P_2, otherwise u is lower than p and there are specific paths from u to ℓ_1, ℓ_2. If u is not in P_2, then it is above p since it is stable on ℓ_2, but then it is also above v, contradicting that we choose u. If C7 is valid, then the algorithm should stop at u, as u is stable on ℓ_2, implying $w = u$. This is impossible as w is not in P_2. If C6 is valid, then v is stable on ℓ_1 and u is not stable. By Proposition 1.(d), v is stable on u, which implies that v is an ancestor of w or vice versa.

1. If node v is an ancestor of w (Fig. 3.B), then P can be extended into a path $\overline{P}$ from v to u. Since v is stable on ℓ_1, there are no reticulations in $\overline{P}[v, w]$. Furthermore, no node in $\overline{P}[v, w]$ is stable on a leaf, since the first edge of $\overline{P}$ is not in T'. Otherwise, if a node y in $\overline{P}[v, w]$ is stable on ℓ, y is not in T', and then ℓ is not in T', contradiction. That the algorithm stopped at w implies that (i) w is a reticulation with both parents being stable on a leaf not in $\{\ell_1, \ell_2\}$, or (ii) there is an unstable specific path P' from w to a node s that is stable on ℓ_2.

   Case (i) is not true, because we have observed that the parent of w in $\overline{P}$ is not stable. If case (ii) is true, s must be in P_2. We have another pair of specific paths $P[w, t] \cup P_1[t, \ell_1]$ and $P'[w, s] \cup P_2[s, \ell_2]$, which is impossible because w is not in $P_1 \cup P_2$ (Lemma 2.(1)).

2. If node w is an ancestor of v (Fig. 3.C), then since v is stable on ℓ_1, the path P taken by the algorithm from u to w must go through v, contradicting the choice of t. Using an argument similar to the one presented above, we can show that the second while-loop stops at p correctly. After the execution of the two while loops, we have that $w_1 = w_2 = p$. By Lemma 3, the recursive call in Step 3 is correct.

This shows that if N displays T, our algorithm finds the lowest image of the parent of ℓ_1 and ℓ_2 together with specific paths P_1 and P_2 in a subdivision of T. By Lemma 3, N displays T if and only if $T - \ell_1 - \ell_2$ is displayed in $N - P_1 - P_2$. This concludes the proof of correctness of the algorithm.

Regarding the time complexity of the algorithm: note that each recursive step removes two sibling leaves from the input tree and that N has at most $|\mathcal{E}(N)| = O(|\mathcal{L}(N)|)$ nodes (see [3]). In different recursive steps, the nodes whose stability is examined are different, and the time spent on checking stability is at most $|\mathcal{V}(N)| \times O(|\mathcal{E}(N)| + |\mathcal{V}(N)|) = O(|\mathcal{L}(N)|^2)$. Before entering the next recursive step, the nodes that have been visited in the current step are removed. Therefore, the algorithm has quadratic time complexity.   □

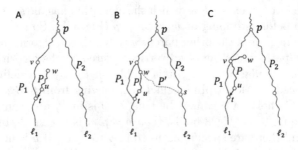

Fig. 3. Illustration for the proof of Theorem 1.

Algorithm 2. Deciding whether a given GS network displays a given tree.

Procedure *Tree-Display(N, T)*

     Input: a GS network N with information on stability, a tree T

     Output: **true** if N displays T, **false** otherwise

1   **if** T *is a single node* **then**

2 | **return true**;

3   Compute a cherry

4    $\{\alpha_{\ell_1,\ell_2}, \ell_1, \ell_2\}$ in T;

5    $w_1 \leftarrow \mathrm{parent}(\ell_1)$; $P_1 \leftarrow \{\ell_1, w_1\}$;   // Initialize to start with ℓ_1

      /* Move up to reach the lowest p corresponding to α_{ℓ_1,ℓ_2} in a

         subdivision of T */

6   **while** *no unstable specific path from* w_1 *to* ℓ_2 *or a node stable on* ℓ_2 **do**

7 | **if** *MoveUpInSpecificPath(w_1, ℓ_1, ℓ_2, P_1, N)* = **false then**

8 | | **return false**;

9    $w_2 \leftarrow \mathrm{parent}(\ell_2)$; $P_2 \leftarrow \{\ell_2, w_2\}$;   // Initialize to move up at ℓ_2

10   **while** $w_2 \neq w_1$ *and* w_2 *is below* w_1 **do**

11 | **if** *MoveUpInSpecificPath(w_2, ℓ_2, ℓ_1, P_2, N)* = **false then**

12 | | **return false**;

13   **if** $w_2 \neq w_1$ **then**

14 | **return false**;

15   **return** *Tree-Display($N - P_1 - P_2$, $T - \ell_1 - \ell_2$)*;

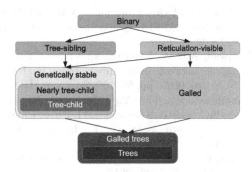

Fig. 4. Inclusion relationships between GS networks and other classes, represented by rectangles. A class that is drawn within another one is a subclass of the latter; an arrow points from a nested class cluster to another if classes in the former are all a superclass of the classes in the latter. A network is tree-child if every node in it has a child that is a tree node.

5 Conclusion

In the present work, we introduced the class of GS networks to study the TC problem. In [3], we developed a quadratic-time algorithm for nearly stable networks by iteratively selecting an edge entering a reticulation to delete in the end of a longest path in a nearly stable network. Here, using a different approach, we have proved that the TC problem can also be solved in quadratic time for GS networks.

A trivial $2^{|\mathcal{R}(N)|} \cdot \text{poly}(|\mathcal{L}(N)|)$ algorithm solves the TC problem as follows: for each reticulation, simply guess which entering edge to delete. However, the number of reticulations can be quite large e.g. in the case of bacterial genomes [9], and many gene families need to be examined. Therefore, our proposed algorithm with low time complexity is definitely valuable for model verification in comparative genomics.

Several problems remain open for future study. First, Fig. 4 summarizes the inclusion relationships between the network classes defined in this paper and other well-studied classes defined in [5]. Galled networks are a generalization of level-1 networks (also called galled trees), comprising a subclass of stable networks [5]. The complexity of the TC problem for galled networks is open.

Second, a natural generalisation of the TC problem is to decide whether a given network displays another given network. Is it possible to determine in polynomial time whether a given GS network displays another given one?

Acknowledgments. The project was financially supported by Merlion Programme 2013.

References

1. Chan, J.M., et al.: Topology of viral evolution. PNAS **110**, 18566–18571 (2013)
2. Cordue, P., Linz, S., Semple, C.: Phylogenetic networks that display a tree twice. Bulletin Math. Biol. **76**, 2664–2679 (2014)
3. Gambette, P., Gunawan, A.D.M., Labarre, A., Vialette, S., Zhang, L.: Locating a tree in a phylogenetic network in quadratic time. In: Przytycka, T.M. (ed.) RECOMB 2015. LNCS, vol. 9029, pp. 96–107. Springer, Heidelberg (2015)
4. Gusfield, D.: ReCombinatorics: The Algorithmics of Ancestral Recombination Graphs and Explicit Phylogenetic Networks. MIT Press, Cambridge (2014)
5. Huson, D.H., Rupp, R., Scornavacca, C.: Phylogenetic Networks. Cambridge University Press, Cambridge (2010)
6. van Iersel, L., Semple, C., Steel, M.: Locating a tree in a phylogenetic network. Inform. Proces. Lett. **110**, 1037–1043 (2010)
7. Kanj, I.A., Nakhleh, L., Than, C., Xia, G.: Seeing the trees and their branches in the network is hard. Theoret. Comput. Sci. **401**, 153–164 (2008)
8. Nakhleh, L.: Computational approaches to species phylogeny inference and gene tree reconciliation. Trends Ecol. Evol. **28**, 719–728 (2013)
9. Treangen, T.J., Rocha, E.P.: Horizontal transfer, not duplication, drives the expansion of protein families in prokaryotes. PLOS Genet. **7**, e1001284 (2011)
10. Wang, L., Zhang, K., Zhang, L.: Perfect phylogenetic networks with recombination. J. Comput. Biol. **8**, 69–78 (2001)

On the Complexity of Rainbow Coloring Problems

Eduard Eiben[1(✉)], Robert Ganian[1], and Juho Lauri[2]

[1] TU Wien, Vienna, Austria
eiben@ac.tuwien.ac.at, rganian@gmail.com
[2] Tampere University of Technology, Tampere, Finland
juho.lauri@tut.fi

Abstract. An edge-colored graph G is said to be *rainbow connected* if between each pair of vertices there exists a path which uses each color at most once. The *rainbow connection number*, denoted by $rc(G)$, is the minimum number of colors needed to make G rainbow connected. Along with its variants, which consider vertex colorings and/or so-called strong colorings, the rainbow connection number has been studied from both the algorithmic and graph-theoretic points of view.

In this paper we present a range of new results on the computational complexity of computing the four major variants of the rainbow connection number. In particular, we prove that the STRONG RAINBOW VERTEX COLORING problem is NP-complete even on graphs of diameter 3. We show that when the number of colors is fixed, then all of the considered problems can be solved in linear time on graphs of bounded treewidth. Moreover, we provide a linear-time algorithm which decides whether it is possible to obtain a rainbow coloring by saving a fixed number of colors from a trivial upper bound. Finally, we give a linear-time algorithm for computing the exact rainbow connection numbers for three variants of the problem on graphs of bounded vertex cover number.

1 Introduction

The concept of rainbow connectivity was introduced by Chartrand, Johns, McKeon, and Zhang in 2008 [7] as an interesting connectivity measure motivated by recent developments in the area of secure data transfer. Over the past years, this strengthened notion of connectivity has received a significant amount of attention in the research community. The applications of rainbow connectivity are discussed in detail for instance in the recent survey [21], and various bounds are also available in [8,22].

An edge-colored graph G is said to be *rainbow connected* if between each pair of vertices a, b there exists an $a - b$ path which uses each color at most once; such a path is called *rainbow*. The minimum number of colors needed to make G rainbow connected is called the *rainbow connection number* (rc), and the

The authors acknowledge support by the Austrian Science Fund (FWF, projects P26696 and W1255-N23).

© Springer International Publishing Switzerland 2016
Z. Lipták and W.F. Smyth (Eds.): IWOCA 2015, LNCS 9538, pp. 209–220, 2016.
DOI: 10.1007/978-3-319-29516-9_18

RAINBOW COLORING problem asks to decide if the rainbow connection number is upper-bounded by a number specified in the input. Precise definitions are given in Sect. 2.

The rainbow connection number and RAINBOW COLORING have been studied from both the algorithmic and graph-theoretic points of view. On one hand, the exact rainbow connection numbers are known for a variety of simple graph classes, such as wheel graphs [7], complete multipartite graphs [7], unit interval graphs [24], and threshold graphs [5]. On the other hand, RAINBOW COLORING is a notoriously hard problem. It was shown by Chakraborty et $al.$ [4] that already deciding if $rc(G) \leq 2$ is NP-complete, and Ananth et $al.$ [1] showed that for any $k > 2$ deciding $rc(G) \leq k$ is NP-complete. In fact, Chandran and Rajendraprasad [5] strengthened this result to hold for chordal graphs. In the same paper, the authors gave a linear time algorithm for rainbow coloring split graphs which form a subclass of chordal graphs with at most one more color than the optimum.

Later on, the inapproximability of the problem was investigated by Chandran and Rajendraprasad [6]. They proved that there is no polynomial time algorithm to rainbow color graphs with less than twice the minimum number of colors, unless $P = NP$.

Several variants of the notion of rainbow connectivity have also been considered. Indeed, a similar concept was introduced for vertex-colored graphs by Krivelevich and Yuster [18]. A vertex-colored graph H is $rainbow$ $vertex$ $connected$ if there is a path whose internal vertices have distinct colors between every pair of vertices, and this gives rise to the $rainbow$ $vertex$ $connection$ $number$ (rvc). The $strong$ $rainbow$ $connection$ $number$ (src) was introduced and investigated also by Chartrand et $al.$ [8]; an edge-colored graph G is said to be $strong$ $rainbow$ $connected$ if between each pair of vertices a, b there exists a shortest $a - b$ path which is rainbow. The combination of these two notions, $strong$ $rainbow$ $vertex$ $connectivity$ (srvc), was studied in a graph theoretic setting by Li et $al.$ [20].

Not surprisingly, the problems arising from the strong and vertex variants of rainbow connectivity are also hard. Chartrand et $al.$ showed that $rc(G) = 2$ if and only if $src(G) = 2$ [7], and hence deciding if $src(G) \leq k$ is NP-complete for $k = 2$. The problem remains NP-complete for $k > 2$ for bipartite graphs [1], and also for split graphs [17]. Furthermore, the strong rainbow connection number of an n-vertex bipartite graph cannot be approximated within a factor of $n^{1/2-\epsilon}$, where $\epsilon > 0$ unless $NP = ZPP$ [1], and the same holds for split graphs [17]. The computational aspects of the rainbow vertex connection numbers have received less attention in the literature. Through the work of Chen et $al.$ [10] and Chen et $al.$ [9], it is known that deciding if $rvc(G) \leq k$ is NP-complete for every $k \geq 2$. However, to the best of our knowledge, the complexity of deciding whether $srvc(G) \leq k$ (the k-SRVC problem) has not been previously considered.

In this paper, we present new positive and negative results for all four variants of the rainbow coloring problems discussed above.

– In Sect. 3, we prove that k-SRVC is NP-complete for every $k \geq 3$ even on graphs of diameter 3. Our reduction relies on an intermediate step which proves the NP-hardness of a more general problem, the k-SUBSET STRONG RAINBOW VERTEX COLORING problem. We also provide bounds for approximation algorithms (under established complexity assumptions), see Corollary 6.

– In Sect. 4, we show that all of the considered problems can be formulated in monadic second order (MSO) logic. In particular, this implies that for every fixed k, all of the considered problems can be solved in linear time on graphs of bounded treewidth, and the vertex variants can be solved in cubic time on graphs of bounded clique-width.

– In Sect. 5, we investigate the problem from a different perspective: we ask whether, given an n-vertex graph G and an integer k, it is possible to color G using k colors less than the known upper bound. Here we employ a win-win approach and show that this problem can be solved in time $\mathcal{O}(n)$ for any fixed k.

– In the final Sect. 6, we show that in the general case when k is not fixed, three of the considered problems admit linear-time algorithms on graphs of bounded vertex cover number. This is also achieved by exploiting a win-win approach, where we show that either k is bounded by a function of the vertex cover number and hence we can apply the result of Sect. 4, or k is sufficiently large which allows us to exploit the structure of the graph to precisely compute the connectivity number.

2 Preliminaries

2.1 Graphs and Rainbow Connectivity

We refer to [13] for standard graph-theoretic notions. We use $[i]$ to denote the set $\{1, 2, \ldots, i\}$. All graphs considered in this paper are simple and undirected. The *degree* of a vertex is the number of its incident edges, and a vertex is a *pendant* if it has degree 1. We will often use the shorthand ab for the edge $\{a, b\}$. For a vertex set X, we use $G[X]$ to denote the subgraph of G induced on X.

A *vertex coloring* of a graph $G = (V, E)$ is a mapping from V to $\mathbb{N}$, and similarly an *edge coloring* of G is a mapping from E to $\mathbb{N}$; in this context, we will often refer to the elements of $\mathbb{N}$ as colors. An $a - b$ *path* P *of length* p is a finite sequence of the form $(a = v_0, e_0, v_1, e_1, \ldots b = v_p)$, where $v_0, v_1, \ldots v_p$ are distinct vertices and $e_0, \ldots e_{p-1}$ are distinct edges and each edge e_j is incident to v_j and v_{j+1}. An $a - b$ path of length p is a *shortest path* if every $a - b$ path has length at least p. The *diameter* of a graph G is the length of its longest shortest path, denoted by $\mathrm{diam}(G)$. Given an edge (vertex) coloring α of G, a color $x \in \mathbb{N}$ *occurs* on a path P if there exists an edge (an internal vertex) z on P such that $\alpha(z) = x$.

A vertex or edge coloring of G is *rainbow* if between each pair of vertices a, b there exists an $a - b$ path P such that each color occurs at most once on P; in this case we say that G is *rainbow connected* or *rainbow colored*. We denote by $\mathrm{rc}(G)$ the minimum $i \in \mathbb{N}$ such that there exists a rainbow edge coloring

$\alpha : E \rightarrow [i]$. Similarly, $\mathrm{rvc}(G)$ denotes the minimum $i \in \mathbb{N}$ such that there exists a rainbow vertex coloring $\alpha : V \rightarrow [i]$. Furthermore, an edge or vertex coloring of G is a *strong rainbow coloring* if between each pair of vertices a, b there exists a shortest $a - b$ path P such that each color occurs at most once on P. We denote by $\mathrm{src}(G)$ $(\mathrm{srvc}(G))$ the minimum $i \in \mathbb{N}$ such that there exists a strong rainbow edge (vertex) coloring $\alpha : E \rightarrow [i]$ $(\alpha : V \rightarrow [i])$.

Let G and H be two graphs with n and n' vertices, respectively. The *corona* of G and H, denoted by $G \circ H$, is the disjoint union of G and n copies of H where the i-th vertex of G is connected by an edge to every vertex of the i-th copy of H. Clearly, the corona $G \circ H$ has $n(1 + n')$ vertices.

2.2   Problem Statements

Here we formally state the problems studied in this work.

RAINBOW k-COLORING (k-RC)
Instance: A connected undirected graph $G = (V, E)$.
Question: Is $\mathrm{rc}(G) \leq k$?

STRONG RAINBOW k-COLORING (k-SRC), RAINBOW VERTEX k-COLORING (k-RVC) and STRONG RAINBOW VERTEX k-COLORING (k-SRVC) are then defined analogously for $\mathrm{src}(G)$, $\mathrm{rvc}(G)$ and $\mathrm{srvc}(G)$, respectively. We also consider generalized versions of these problems, where k is given as part of the input.

RAINBOW COLORING (RC)
Instance: A connected undirected graph $G = (V, E)$, and a positive integer k.
Question: Is $\mathrm{rc}(G) \leq k$?

The problems SRC, RVC, and SRVC are also defined analogously. In Sect. 5 we consider the "saving" versions of the problem, which ask whether it is possible to improve upon the trivial upper bound for the number of colors.

SAVING k RAINBOW COLORS (k-SavingRC)
Instance: A connected undirected graph $G = (V, E)$.
Question: Is $\mathrm{rc}(G) \leq |E| - k$?
SAVING k RAINBOW VERTEX COLORS (k-SavingRVC)
Instance: A connected undirected graph $G = (V, E)$.
Question: Is $\mathrm{rvc}(G) \leq |V| - k$?

2.3   Structural Measures

Several of our results utilize certain structural measures of graphs. We will mostly be concerned with the *treewidth* and the *vertex cover number* of the input graph. Section 4 also mentions certain implications of our results for graphs of bounded *clique-width*, the definition of which can be found for instance in [12].

A *tree decomposition* of G is a pair $(T, \{X_i : i \in I\})$ where $X_i \subseteq V$, $i \in I$, and T is a tree with elements of I as nodes such that:

1. for each edge $uv \in E$, there is an $i \in I$ such that $\{u, v\} \subseteq X_i$, and
2. for each vertex $v \in V$, $T[\{i \in I \mid v \in X_i\}]$ is a (connected) tree with at least one node.

The *width* of a tree decomposition is $\max_{i \in I} |X_i| - 1$. The *treewidth* [23] of G is the minimum width taken over all tree decompositions of G and it is denoted by $\operatorname{tw}(G)$.

Fact 1 ([3]). *There exists an algorithm which, given a graph G and an integer p, runs in time $2^{p^{\mathcal{O}(1)}} \cdot (|V(G)| + |E(G)|)$, and either outputs a tree decomposition of G of width at most p or correctly determines that $\operatorname{tw}(G) > p$.*

A *vertex cover* of a graph $G = (V, E)$ is a set $X \subseteq V$ such that each edge in G has at least one endvertex in X. The cardinality of a minimum vertex cover in G is denoted as $\operatorname{vcn}(G)$. Given a vertex cover X, a *type* T is a subset of $V \setminus X$ such that any two vertices in T have the same neighborhood; observe that any graph contains at most $2^{|X|}$ many distinct types.

2.4   Monadic Second Order Logic

We assume that we have an infinite supply of individual variables, denoted by lowercase letters x, y, z, and an infinite supply of set variables, denoted by uppercase letters X, Y, Z. *Formulas* of MSO_2 logic are constructed from atomic formulas $I(x, y)$, $x \in X$, and $x = y$ using the connectives $\neg$ (negation), $\wedge$ (conjunction) and existential quantification $\exists x$ over individual variables as well as existential quantification $\exists X$ over set variables. Individual variables range over vertices and edges, and set variables range either over sets of vertices or over sets of edges. The atomic formula $I(x, y)$ expresses that vertex x is incident to edge y, $x = y$ expresses equality, and $x \in X$ expresses that x is in the set X.

MSO_1 logic is defined similarly as MSO_2 logic, with the following distinctions. Individual variables range only over vertices, and set variables only range over sets of vertices. The atomic formula $I(x, y)$ is replaced by $E(x, y)$, which expresses that vertex x is adjacent to vertex y.

Free and bound variables of a formula are defined in the usual way. A *sentence* is a formula without free variables. It is known that MSO_2 formulas can be checked efficiently as long as the graph has bounded tree-width.

Fact 2 ([11]). *Let ϕ be a fixed MSO_2 sentence and $p \in \mathbb{N}$ be a constant. Given an n-vertex graph G of treewidth at most p, it is possible to decide whether $G \models \phi$ in time $\mathcal{O}(n)$.*

Similarly, MSO_1 formulas can be checked efficiently as long as the graph has bounded *clique-width* [12] (or, equivalently, *rank-width* [15]). In particular, while the formula can be checked in linear time if a suitable rank- or clique-decomposition is provided, current algorithms for finding (or approximating) such a decomposition require cubic time.

Fact 3 ([12,15]). *Let ϕ be a fixed MSO_1 sentence and $p \in \mathbb{N}$ be a constant. Given an n-vertex graph G of clique-width at most p, it is possible to decide whether $G \models \phi$ in time $\mathcal{O}(n^3)$.*

3   Hardness of Strong Rainbow Vertex k-Coloring

It is easy to see that $\mathrm{srvc}(G) = 1$ if and only if $\mathrm{diam}(G) = 2$. We will prove that deciding if $\mathrm{srvc}(G) \leq k$ is NP-complete for every $k \geq 3$ already for graphs of diameter 3. This is done by first showing hardness of an intermediate problem, described below.

In the k-SUBSET STRONG RAINBOW VERTEX COLORING problem (k-SSRVC) we are given a graph G which is a corona of a complete graph and K_1, and a set P of pairs of pendants in G. The goal is to decide if the vertices of G can be colored with k colors such that each pair in P is connected by a vertex rainbow shortest path. We will first show this intermediate problem is NP-complete by reducing from the classical k-vertex coloring problem: given a graph G, decide if there is an assignment of k colors to the vertices of G such that adjacent vertices receive a different color. The k-vertex coloring problem is well-known to be NP-complete for every $k \geq 3$.

Lemma 4. *The k-SSRVC problem is NP-complete for every $k \geq 3$.*

Proof (Sketch). Let $G = (V, E)$ be an instance of the k-vertex coloring problem, where $k \geq 3$. We will construct an instance $\langle G', P \rangle$ of the k-SSRVC problem such that $\langle G', P \rangle$ is a YES-instance if and only if G is k-vertex colorable.

The graph $G' = (V', E')$ along with the set of pairs P are constructed as follows:

- $V' = V \cup \{p_v \mid v \in V\}$,
- $E' = \{uv \mid u, v \in V \wedge u \neq v\} \cup \{vp_v \mid v \in V\}$, and
- $P = \{\{p_u, p_v\} \mid uv \in E\}$.

Clearly, $G' = K_{|V|} \circ K_1$. This completes the construction of G'. Satisfying color assignments for V in G are necessarily satisfying color assignments for V in G', and vice versa. □

We are now ready to prove the following.

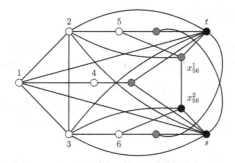

Fig. 1. The graph $K_3 \circ K_1$ transformed to a graph of diameter 3 with $P = \{\{4,5\},\{4,6\}\}$. The color c_1 is represented with grey, and the color c_2 with black. White vertices represent an unknown vertex coloring under which the pairs in P are strong rainbow vertex connected.

Theorem 5. *The problem k-SRVC is NP-complete for every integer $k \geq 3$, even when the input is restricted to graphs of diameter 3.*

Proof (Sketch). Let $k \geq 3$ and $\langle G = (V,E), P \rangle$ be an instance of the k-SSRVC problem. We will construct a graph $G' = (V', E')$ that is strong rainbow vertex colorable with k colors if and only if $\langle G = (V,E), P \rangle$ is a YES-instance of k-SSRVC.

Let V_1 denote the set of pendant vertices in G. For every vertex $v \in V_1$ we introduce a new vertex x_v. For every pair of pendant vertices $\{u,v\} \notin P$, we add two vertices x^1_{uv} and x^2_{uv}. We also add two new vertices s and t. In the following, we denote by k_v, where $v \in V_1$, the unique vertex that v is adjacent to in G. Formally, we construct a graph $G' = (V', E')$ such that:

- $V' = V \cup \{x_v \mid v \in V_1\} \cup \{x^1_{uv}, x^2_{uv} \mid \{u,v\} \in \binom{V_1}{2}\backslash P\} \cup \{s,t\}$,
- $E' = E \cup E_1 \cup E_2 \cup E_3 \cup E_4$,
- $E_1 = \{vx_v, sx_v, tx_v \mid v \in V_1\}$,
- $E_2 = \{ux^1_{uv}, x^1_{uv}x^2_{uv}, x^2_{uv}v \mid \{u,v\} \in \binom{V_1}{2}\backslash P\}$,
- $E_3 = \{sx^1_{uv}, tx^2_{uv}, k_ux^1_{uv}, k_vx^2_{uv} \mid \{u,v\} \in \binom{V_1}{2}\backslash P\}$, and
- $E_4 = \{sy, ty \mid y \in V\backslash V_1\}$.

This completes the construction of G' (see also Fig. 1). It is easy to verify that $\mathrm{diam}(G') = 3$. $\square$

It can be observed that the size of the above reduction does not depend on k, the number of colors. In fact, if the instance of the k-vertex coloring problem has n vertices, then the graph G' we build in Theorem 5 has no more than $\mathcal{O}(n^2)$ vertices. Furthermore, a strong rainbow vertex coloring of G' gives us a solution to the k-vertex coloring problem. Since the vertex coloring number of an n-vertex graph cannot be approximated within a factor of $n^{1-\epsilon}$ for any $\epsilon > 0$ unless $P = NP$ [26], we obtain the following corollary.

Corollary 6. *There is no polynomial time algorithm for approximating the strong rainbow vertex connection number of an n-vertex graph of bounded diameter within a factor of $n^{1/2-\epsilon}$ for any $\epsilon > 0$, unless $P = NP$.*

4   MSO Formulations

This section will present formulations of the k-coloring variants of rainbow connectivity in MSO logic, along with their algorithmic implications.

Lemma 7. *For every $k \in \mathbb{N}$ there exists a MSO_1 formula ϕ_k such that for every graph G, it holds that $G \models \phi$ iff G is a YES-instance of k-RVC. Similarly, for every $k \in \mathbb{N}$ there exists a MSO_2 formula ψ_k such that for every graph G, it holds that $G \models \psi$ iff G is a YES-instance of k-RC.*

Proof. In the case of k-RC, we wish to partition the edges of the graph $G = (V, E)$ into k color classes $C_1, \ldots, C_k$ such that each pair of vertices is connected by a rainbow path. Let us consider the following MSO_2 formula ψ_k.

$$\psi_k := \exists C_1, \ldots, C_k \subseteq E \Big(\forall e \in E \big(e \in C_1 \vee \cdots \vee e \in C_k \big) \Big)$$
$$\wedge \Big(\forall i, j \in [k], i \neq j : (C_i \cap C_j = \emptyset) \Big)$$
$$\wedge \Big(\forall u, v \in V \big((u \neq v) \implies \bigvee_{1 \leq i \leq k} \big(\exists e_1, \ldots, e_i \in E \big(\mathrm{Path}(u, v, e_1, \ldots, e_i) \big)$$
$$\wedge \mathrm{Rainbow}(e_1, \ldots, e_i) \big) \big) \Big) \Big),$$

Here, $\mathrm{Path}(u, v, e_1, \ldots, e_\ell)$ expresses that the edges $e_1, \ldots, e_\ell$ form a path between the vertices u and v. The predicate $\mathrm{Rainbow}(e_1, \ldots, e_\ell)$ expresses that the edges $e_1, \ldots, e_\ell$ are each in precisely one color class.

In the case of k-RVC, the MSO_1 formula ϕ_k is defined analogously, with the following distinctions:

1. instead of edges, we partition the vertices of G into color classes;
2. the predicate Pathspeaks of vertices instead of edges and uses the adjacency relation instead of the incidence relation; and
3. the predicate Rainbow tests the coloring of vertices instead of edges.   □

Using a similar approach, we obtain an analogous result for the strong variants of these problems.

Lemma 8. *For every $k \in \mathbb{N}$ there exists a MSO_1 formula ϕ_k such that for every graph G, it holds that $G \models \phi$ iff G is a YES-instance of k-SRVC. Similarly, for every $k \in \mathbb{N}$ there exists a MSO_2 formula ψ_k such that for every graph G, it holds that $G \models \psi$ iff G is a YES-instance of k-SRC.*

Theorem 9. *Let $p \in \mathbb{N}$ be fixed. Then the problems k-RC, k-SRC, k-RVC, k-SRVC can be solved in time $\mathcal{O}(n)$ on n-vertex graphs of treewidth at most p. Furthermore, k-RVC, k-SRVC can be solved in time $\mathcal{O}(n^3)$ on n-vertex graphs of clique-width at most p.*

Proof. The proof follows from Lemmas 7 and 8 in conjunction with Facts 2 and 3.   □

In the language of parameterized complexity [14], Theorem 9 implies that these problems are fixed-parameter tractable (FPT) parameterized by treewidth, and their vertex variants are FPT parameterized by clique-width.

5   The Complexity of Saving Colors

This section focuses on the saving versions of the rainbow coloring problems introduced in Subsect. 2.2, and specifically gives linear-time algorithms for k-SavingRC and k-SavingRVC. Our results make use of the following facts.

Fact 10 ([16]). *There is a* MSO_1 *predicate* VertexConnects *such that on a graph* $G = (V, E)$ VertexConnects(S, u, v) *is true iff* $S \subseteq V$ *is a set of vertices of* G *such that there is a path from* u *to* v *that lies entirely in* S.

The above is easily modified to give us the following.

Fact 11. *There is a* MSO_2 *predicate* EdgeConnects *such that on a graph* $G = (V, E)$ EdgeConnects(X, u, v) *is true iff* $X \subseteq E$ *is a set of edges of* G *such that there is path from* u *to* v *that lies entirely in* X.

Theorem 12. *For each* $k \in \mathbb{N}$, *the problem* k-SavingRC *can be solved in time* $\mathcal{O}(n)$ *on* n-*vertex graphs.*

Proof. Observe that by coloring each edge of a spanning tree of G with a distinct color we have that $\mathrm{rc}(G) \leq n - 1$. Thus, if $m \geq n + k$, we have a YES-instance of k-SavingRC. Otherwise, suppose $m < n + k$. Then G has a feedback edge set of size at most k, and hence G has treewidth at most k. We construct a MSO_2 formula ψ_k such that it holds that $G \models \psi_k$ is true iff G is a YES-instance of k-SavingRC. Using Fact 11, we construct ψ_k as follows:

$$\psi_k := \exists R_1, \ldots, R_k \subseteq E \Big(\forall i, j \in [k], i \neq j : (R_i \cap R_j = \emptyset) \Big)$$
$$\wedge \Big(\forall i \in [k] : \big(\exists e \in E(e \in R_i) \big) \Big) \wedge |R_1 \cup R_2 \cup \cdots \cup R_k| \geq 2k$$
$$\wedge \Big(\forall u, v \in V \big((u \neq v) \implies \big(\exists X \subseteq E \big(\mathrm{EdgeConnects}(X, u, v)$$
$$\wedge \forall e_1, e_2 \in X \big(\forall i \in [k] : (e_1 \in R_i \wedge e_2 \in R_i) \implies (e_1 = e_2) \big) \big) \big) \big) \Big).$$

In the above, the expression $|A| \geq 2k$ is shorthand for the existence of $2k$ pairwise-distinct edges in A, which can be expressed by a simple but lengthy MSO_2 expression. The formula ψ_k expresses that there exist k disjoint sets $R_1, \ldots, R_k$ of edges (each corresponding to a different color set with at least 1 edge) such that their union contains at least $2k$ edges, with the following property: there is a path using at most one edge from each set $R_1, \ldots, R_k$ between every pair of vertices. Formally, this property is stated as the existence of an edge-set X for each pair of vertices u, v such that the graph (V, X) contains an $u - v$ path that cannot repeat edges from any R_i.

The proof then follows by Fact 2.   □

To prove a similar result for k-SavingRVC, we will use the following result.

Fact 13 ([2]). *If the treewidth of a connected graph G is at least $2k^3$, then G has a spanning tree with at least k vertices with degree 1.*

Theorem 14. *For each $k \in \mathbb{N}$, the problem k-SavingRVC can be solved in time $\mathcal{O}(n)$ on n-vertex graphs.*

Proof. Using Fact 1, we will test if the treewidth of G is at least $2k^3$. If it is, then by Fact 13 the graph G has a spanning tree with at least k vertices of degree 1. Each of these k vertices can receive the same color, and we conclude we have a YES-instance. Otherwise, suppose the treewidth of G is less than $2k^3$, and we construct a MSO$_1$ formula ϕ_k such that it holds that $G \models \phi_k$ is true iff G is a YES-instance of k-SavingRVC. The construction is analogous to Theorem 12, but instead of EdgeConnects we use VertexConnects from Fact 10. The proof then follows by Fact 2.   □

6   Rainbow Coloring Graphs with Small Vertex Covers

In this section we turn our attention to the more general problem of determining whether the rainbow connection number is below a number specified in the input. Specifically, we show that RC, RVC, and SRVC admit linear time algorithms on graphs of bounded vertex cover number. In particular, this implies that RC, RVC, SRVC are FPT parameterized by $\mathrm{vcn}(G)$.

Lemma 15. *Let $G = (V, E)$ be a connected graph and $p = \mathrm{vcn}(G)$. Then $\mathrm{rvc}(G) \leq 2p$ and $\mathrm{srvc}(G) \leq p^2$.*

The following lemma will be useful in the proof of Lemma 17, a key component of our approach for dealing with RC on the considered graph classes. An *edge separator* is an edge e such that deleting e separates the connected component containing e into two connected components.

Lemma 16. *Let $G = (V, E)$ be a graph and X be a minimum vertex cover of G. Then there exist at most $2|X| - 2$ edge separators which are not incident to a pendant outside of X.*

For ease of presentation, we define the function β as $\beta(p) = 2p - 2 + p \cdot (p^2 + 2p \cdot 2^p)$. The next Lemma 17 will represent one part of our win-win strategy, as it allows us to precisely compute $\mathrm{rc}(G)$ when the number of edge separators is sufficiently large. We remark that an analogous claim does not hold for $\mathrm{src}(G)$ (regardless of the choice of β).

Lemma 17. *Let $G = (V, E)$ be a connected graph and $p = \mathrm{vcn}(G)$. Let z be the number of edge separators in G. If $z \geq \beta(p)$, then $\mathrm{rc}(G) = z$.*

Lemma 18. *Let $G = (V, E)$ be a graph with a vertex cover $X \subseteq V$ of cardinality p. Let z be the number of edge separators in G. If $z < \beta(p)$, then $\mathrm{rc}(G) \leq \beta(p) + p^2 + 2^p \cdot 2p$.*

Theorem 19. *Let $p \in \mathbb{N}$ be fixed. Then the problems* RC, RVC, SRVC *can be solved in time* $\mathcal{O}(n)$ *on n-vertex graphs of vertex cover number at most p.*

Proof. For RVC and SRVC, we first observe that if k (the queried upper bound on the number of colors) is greater than $2p$ and p^2, respectively, then the algorithm can immediately output YES by Lemma 15. Otherwise we use Theorem 9 and the fact that the vertex cover number is an upper bound on the treewidth to compute a solution in $\mathcal{O}(n)$ time.

For RC, it is well known that the total number of edge separators in G, say z, can be computed in linear time on graphs of bounded treewidth. If $z \geq \beta(p)$, then by Lemma 17 we can correctly output YES when $z \leq k$ and NO when $z > k$. On the other hand, if $z < \beta(p)$, then by Lemma 18 the value $rc(G)$ is upper-bounded by a function of p. We compare k and this upper bound on $rc(G)$; if k exceeds the upper bound on $rc(G)$, then we output YES, and otherwise we can use Theorem 9 along with the fact that the vertex cover number is an upper bound on the treewidth to compute a solution in $\mathcal{O}(n)$ time. $\square$

7   Concluding Notes

We presented new positive and negative results for the most prominent variants of rainbow coloring. We believe that the techniques presented above, and in particular the win-win approaches used in Sects. 5 and 6, can be of use also for other challenging connectivity problems.

It is worth noting that our results in Sect. 4 leave open the question of whether RAINBOW COLORING or its variants can be solved in (uniformly) polynomial time on graphs of bounded treewidth. Hardness results for related problems [19, 25] do not imply that finding an optimal coloring of a bounded-treewidth graph is hard, and it seems that new insights are needed to determine the complexity of these problems on graphs of bounded treewidth.

References

1. Ananth, P., Nasre, M., Sarpatwar, K.K.: Rainbow connectivity: hardness and tractability. In: FSTTCS 2011, pp. 241–251 (2011)
2. Bodlaender, H.: On linear time minor tests with depth-first search. J. Algorithms **14**(1), 1–23 (1993)
3. Bodlaender, H.L., Kloks, T.: Efficient and constructive algorithms for the path-width and treewidth of graphs. J. Algorithms **21**(2), 358–402 (1996)
4. Chakraborty, S., Fischer, E., Matsliah, A., Yuster, R.: Hardness and algorithms for rainbow connection. J. Comb. Optimizat. **21**(3), 330–347 (2009)
5. Chandran, L.S., Rajendraprasad, D.: Rainbow colouring of split and threshold graphs. In: Gudmundsson, J., Mestre, J., Viglas, T. (eds.) COCOON 2012. LNCS, vol. 7434, pp. 181–192. Springer, Heidelberg (2012)
6. Chandran, L.S., Rajendraprasad, D.: Inapproximability of rainbow colouring. In: FSTTCS 2013, pp. 153–162 (2013)

7. Chartrand, G., Johns, G., McKeon, K., Zhang, P.: Rainbow connection in graphs. Mathematica Bohemica **133**(1), 85–98 (2008)
8. Chartrand, G., Zhang, P.: Chromatic Graph Theory. CRC Press, Boca Raton (2008)
9. Chen, L., Li, X., Lian, H.: Further hardness results on the rainbow vertex-connection number of graphs. Theoret. Comput. Sci. **481**, 18–23 (2013)
10. Chen, L., Li, X., Shi, Y.: The complexity of determining the rainbow vertex-connection of a graph. Theoret. Comput. Sci. **412**(35), 4531–4535 (2011)
11. Courcelle, B.: The monadic second-order logic of graphs I. Recognizable sets of finite graphs. Inf. Comput. **85**, 12–75 (1990)
12. Courcelle, B., Makowsky, J.A., Rotics, U.: Linear time solvable optimization problems on graphs of bounded clique-width. Theory Comput. Syst. **33**(2), 125–150 (2000)
13. Diestel, R.: Graph Theory. Springer, Heidelberg (2010)
14. Downey, R.G., Fellows, M.R.: Fundamentals of Parameterized Complexity. Springer, London (2013)
15. Ganian, R., Hliněný, P.: On parse trees and Myhill-Nerode-type tools for handling graphs of bounded rank-width. Discrete App. Math. **158**(7), 851–867 (2010)
16. Gottlob, G., Lee, S.T.: A logical approach to multicut problems. Inf. Process. Lett. **103**(4), 136–141 (2007)
17. Keranen, M., Lauri, J.: Computing minimum rainbow, strong rainbow colorings of block graphs. Preprint available on arXiv:1405.6893 (2014)
18. Krivelevich, M., Yuster, R.: The rainbow connection of a graph is (at most) reciprocal to its minimum degree. J. Graph Theory **63**(3), 185–191 (2010)
19. Lauri, J.: Further hardness results for rainbow and strong rainbow connectivity. Discrete Appl. Math. (to appear)
20. Li, X., Mao, Y., Shi, Y.: The strong rainbow vertex-connection of graphs. Utilitas Mathematica **93**, 213–223 (2014)
21. Li, X., Shi, Y., Sun, Y.: Rainbow connections of graphs: a survey. Graphs Comb. **29**(1), 1–38 (2012)
22. Li, X., Sun, Y.: Rainbow Connections of Graphs. Springer, New York (2012)
23. Robertson, N., Seymour, P.: Graph minors. II. Algorithmic aspects of tree-width. J. Algorithms **7**(3), 309–322 (1986)
24. Chandran, L.S., Das, A., Rajendraprasad, D., Varma, N.M.: Rainbow connection number, connected dominating sets. J. Graph Theory **71**(2), 206–218 (2012)
25. Uchizawa, K., Aoki, T., Ito, T., Suzuki, A., Zhou, X.: On the rainbow connectivity of graphs: complexity and FPT algorithms. Algorithmica **67**(2), 161–179 (2013)
26. Zuckerman, D.: Linear degree extractors and the inapproximability of max clique and chromatic umber. In: STOC 2006, pp. 681–690 (2006)

How to Design Graphs with Low Forwarding Index and Limited Number of Edges

Frédéric Giroire[1]([⊠]), Stéphane Pérennes[1], and Issam Tahiri[2]

[1] CNRS, University of Nice Sophia Antipolis, I3S, UMR 7271, Inria, Coati,
06900 Sophia Antipolis, France
frederic.giroire@cnrs.fr
[2] Institut de Mathématiques, Université de Bordeaux, UMR 5251,
CNRS, Inria, 33405 Talence, France

Abstract. The *(edge) forwarding index* of a graph is the minimum, over all possible routings of all the demands, of the maximum load of an edge. This metric is of a great interest since it captures the notion of global congestion in a precise way: the lesser the forwarding-index, the lesser the congestion. In this paper, we study the following design question: Given a number e of edges and a number n of vertices, what is the least congested graph that we can construct? and what forwarding-index can we achieve? Our problem has some distant similarities with the well-known (Δ, D) problem, and we sometimes build upon results obtained on it. The goal of this paper is to study how to build graphs with low forwarding indices and to understand how the number of edges impacts the forwarding index. We answer here these questions for different families of graphs: *general graphs, graphs with bounded degree, sparse graphs with a small number of edges* by providing constructions, most of them asymptotically optimal. For instance, we provide an asymptotically optimal construction for $(n, n + k)$ cubic graphs - its forwarding index is $\sim \frac{n^2}{3k} \log_2(k)$. Our results allow to understand how the forwarding-index drops when edges are added to a graph and also to determine what is the *best (i.e. least congested) structure with e edges*. Doing so, we partially answer the practical problem that initially motivated our work: If an operator wants to power only e links of its network, in order to reduce the energy consumption (or wiring cost) of its networks, what should be those links and what performance can be expected?

Keywords: Graphs · Forwarding index · Routing · Design problem · Energy efficiency · Extremal graphs

---

This work has been partially supported by ANR project Stint under reference ANR-13-BS02-0007, ANR program Investments for the Future under reference ANR-11-LABX-0031-01, ANR VISE, CNRS-FUNCAP project GAIATO, the associated Inria team AlDyNet, the project ECOS-Sud Chile.

Z. Lipták and W.F. Smyth (Eds.): IWOCA 2015, LNCS 9538, pp. 221–234, 2016.
DOI: 10.1007/978-3-319-29516-9_19

1   Introduction

Given a graph $G = (V, E)$ with $n = |V|$ vertices, a *routing* R is a collection of paths connecting all the ordered pairs of vertices of G. A routing R induces on every edge e a *load* that is the number of paths going through e. The *edge-forwarding index* (or simply the *forwarding index*) $\pi(G, R)$ of G with respect to R is then the maximum number of paths in R passing through any edge of G. In other words, it corresponds to the maximum load of an edge of G when R is used. So $\pi(G, R)$ measures how congested is the routing R, hence-fore it is important to design routings minimizing this index. The forwarding index $\pi(G)$ of a connected graph G is the minimum $\pi(G, R)$ over all splittable (fractional) routings R's of G (We will also sometimes consider non-splittable (integral) routing and denote the minimum load $\pi_{\mathcal{I}}(G)$ in this case). By definition the forwarding index of a graph measures its intrinsic congestion, so it is a parameter as essential, and arguably more important than simpler parameters such as the diameter or the average distance.

Problem. In this paper, our goal is to provide for a given number of vertices n and for a given number of edges k graphs with the minimum forwarding indices, or at least graphs with low forwarding indices. For a given n, we will study how the number of edges of a graph impacts its forwarding index. Formally, we define the following design problem:

MIN CONGESTED (n, e)-GRAPH: *Given $n, e \in I\!N$, find a graph $(G = V, E)$ with $|V| = n$ vertices and $|E| = e$ edges such that $\pi(G = V, E)$ is minimum. We will denote this number $\pi^*(n, e)$ (when $e < n - 1$, note that $\pi^*(n, e) = \infty$).*

Here is an example. When restricted to the class of cubic graphs, the min congested $(8, 12)$-graph is the cube. Its forwarding index is equal to 8. The routing that achieves this load is the following: for each ordered pair of nodes (u, v), we connect u to v using all shortest paths from u to v. For instance there are 6 paths that connect the node $(0, 0, 0)$ to the node $(1, 1, 1)$. Each of those paths will hold a load of $1/6$ for this ordered pair. Since the cube is edge-transit if, this routing ensures that all edges will get the same load. Every node a is of distance 1 from 3 nodes, distance 2 from 3 nodes, and distance 3 from 1 node. Hence, the total load induced by order pairs that start with a is $1 \cdot 3 + 2 \cdot 3 + 3 \cdot 1 = 12$. Since there are 8 nodes in the cube, The total load on the graph is $12 \cdot 8 = 96$. Therefore the load on each edge is $96/12 = 8$. The optimality of this graph is proven in Sect. 5. For more examples, check Table 2.

Motivation. Our problem can be viewed as: for a given bound U on the forwarding index, find a spanner F of G with minimum number of edges such that $\pi(F) \leq U$ or reciprocally given a bound on the number of edges minimize $\pi(F)$.

First, to the best of our knowledge the problem of designing a (sub) graph with minimum forwarding index has not been studied when the main other constraint is the number of edges. Indeed, most of the results have been derived either for classical graphs and graphs families or have been considering other constraints, as example the bounded degree one. So even if a constraint such as the number of edges is both natural and of importance it has not been well studied so far. As example, one of our initial goal was to understand how the forwarding index drops from order n^2 for tree like graphs to order $n \log n$ for cubic graphs, and also to understand how adding a single edge can decrease significantly (or not) the forwarding index.

Second, the recent trend of "Energy Saving" has made our problem even more relevant in practice, especially for network operators willing to reduce the energy consumed by their networks. In fact, most of the network links consume a constant energy independently of the amount of traffic they are flowing. Therefore the only way to reduce the energy used by the network links is to turn some of them off, or more conveniently, put them on an idle mode. Outside the rush hours, several studies [1,2,4,5,7] show that a good choice of the links to turn off can lead to significant energy savings, while keeping the same communication quality. In the case where the throughputs from every node to every other node are of the same order, and where the capacities also lie in same small range, a good choice of those links amount to solve the problem of finding spanners of the network with low forwarding indices.

Related Work. The forwarding-index was introduced by Chung and Al in 87 [6], due to its importance this parameter has been studied quite extensively: on one side results have been given for different graph classes (e.g. random graphs [24], transitive and Cayley graphs [10,21] graphs with small numbers of vertices [3] and well-connected graphs [23]). On the other side deep relations with other expansion-related graph invariants have been established: Laplacian, Cheeger constant (see the survey [17]), Sparsest cut [12] and the "geometry of graphs" [13]. This notion has also been used to prove that some Markov chains mix fast using either canonical paths (routings) or "resistance" [20]. See the recent survey [25] for a global view on the known results. The problem is also known as the *maximum concurrent flow problem* and its dual was probably first introduced in [19] in which the authors also discussed the relation with the network throughput, in [22] a simple oblivious packet routing algorithm achieving network stability for any rate λ with $\lambda\pi < 1$ was provided. Some variants: load on arcs for digraphs ([14]) load on the vertices have also been studied.

The edge forwarding index is strongly related to distance properties of the graph. Indeed a usual naive lower bound on π (Average distance Bound) is:

$$\pi(G = V, E) \geq \frac{\sum_{(u,v)\in V^2} D(u, v)}{|E|} = \frac{\overline{D_G}|V|^2}{|E|} = 2|V|\frac{\overline{D_G}}{\overline{d_G}},$$

where $D(u, v), d(v), \overline{D_G}$ and $\overline{d_G}$ denote respectively the distance function, the degree function, the average distance in G and the average degree in G. This indicates that solving our design problem is strongly related to finding graphs with small average distance. The Degree-Diameter problem or (Δ, D)-DESIGN PROBLEM is about finding the graph with degree Δ and diameter D with the maximum number of vertices (or reciprocally it is about minimizing the diameter of a Δ-regular graph). It is quite a complex problem and it has been studied extensively (see [16] for a recent survey). Even after 30 years of steady efforts, generic constructions are still very far from being optimal. So, since good (n, e)-graphs should resemble (Δ, D) graphs, we may expect our problem to be complex. But we can also hope to be able to use results about the (Δ, D)-problem in our context.

Contributions and Plan of the Paper

- In Sect. 2, we consider our design problem for general graphs, that is when the only design constraint is the number of edges. We characterize the graphs with minimum forwarding index. When the number of edges is $k(n - k), k \in \mathbb{N}$, optimum graphs happen to have a simple structure since they are the complete bipartite graphs $K_{k,n-k}$. In between these values, the function $\pi^*(n, e)$ follows, rather surprisingly, a stepwise function (see Propositions 4 and 5).
- In Sect. 3, motivated by telecommunication networks, we study the case of bounded degree graphs. We provide almost optimal graphs for the different values of maximum degree Δ. We then focus on graphs with a small number of edges ($\Delta = 3$) as they correspond to the range of values for which the forwarding index greatly changes. We determine quite sharply how the minimum forwarding index behaves and evolves from $\Theta(n^2)$ to $\Theta(n \log n)$ when the number edges grows from $n - 1$ to $n + \frac{n}{2}$. We also develop a method that allow us simplify the design problem by considering the *graph skeleton*.
- We then examine the case $e = n + k$ with a fixed small $k \in \{1, 2, 3\}$ in Sect. 4. We determine the minimum forwarding index exactly for any n. This is possible because the main structure of the graph, that we called skeleton is finite, so we can explore all of them and use weight arguments in order to deal with a finite problem. Some of the results, as example Proposition 11, are strikingly counter intuitive.
- Last, in Sect. 5, we provide optimal cubic-graphs with *small number of vertices*, that is for $n \in [4, 22]$. Those graphs are not only interesting per se (and some structures again are surprising), but also because, as we shall see, their structure may be used as a *skeleton* to build good graphs with a few edges and arbitrarly size.

Due to the lack of space, all the proofs are omitted and can be found in a research report [9].

2   Minimally Congested Graphs

In this section, we study the design of minimally congested graphs for given numbers of vertices n and edges e. We first give a trivial lower bound of $\pi^*(n, e)$, the minimum forwarding index of a (n, e)-graph. We then provide families of minimally congested graphs reaching this bound for some couples of values (n, e), e.g. complete bipartite graphs $K_{i,n-i}$, complete k-partite graphs, or Kneser graphs, see Fig. 1. These graphs are edge-transitive and of diameter 2. In particular, we show that $K_{i,n-i}$ ($i \in \mathbb{N}, i \leq \lfloor n/2 \rfloor$) are minimally congested $(n, i(n - i))$-graphs with forwarding index $\pi^*(n, e) = 2(\frac{n(n-1)}{e} - 1)$. Last, we

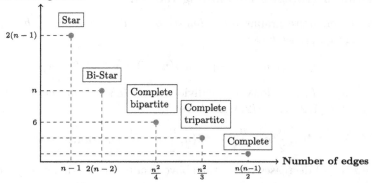

Fig. 1. Forwarding indices of minimally congested graphs with n vertices as a function of their number of edges.

study the behavior of $\pi^*(n, e)$ when e varies between two "perfect" cases, from $i(n - i)$ to $(i + 1)(n - (i + 1))$. Surprisingly, π^* follows a step-wise function in the sense of Propositions 4 and 5 and jumps suddenly from $\pi^*(n, i(n - i))$ to $\pi^*(n, (i + 1)(n - (i + 1)))$.

Proposition 1 (Lower bound on $\pi^*(n, e)$). *The forwarding index of an (n, e)-graph is lower bounded by:*

$$\pi^*(n, e) \geq \frac{2n(n - 1)}{e} - 2.$$

Proposition 2 (Optimal (n, e)-graph). *An (n, e)-graph that is edge-transitive and of diameter 2 is optimal. Its forwarding index is*

$$\frac{2n(n - 1)}{e} - 2.$$

Corollary 1 (Families of optimal graphs). *The following families of graphs are optimal:*

– *Complete bipartite graphs, giving:*

$$\pi^*(n, i(n - i)) = \frac{2n(n - 1)}{e} - 2, \qquad i \in I\!N, i \leq \lfloor n/2 \rfloor.$$

– *Turán graphs $T(n, r)$, for which r divides n (that is, complete multipartite regular graphs with r independent subsets of equal sizes), giving:*

$$\pi^*(n, \frac{n}{2}(n - \frac{n}{r})) = \frac{2n(n - 1)}{e} - 2, \qquad r \in I\!N, r \leq n.$$

– *Kneser graphs $KN_{\nu,\kappa}$ for which $\kappa \geq \nu/3$ (Kneser graphs of diameter 2), giving:*

$$\pi^*\left(\binom{\nu}{\kappa}, \frac{1}{2}\binom{\nu}{\kappa}\binom{\nu - k}{\kappa}\right) = \frac{2n(n - 1)}{e} - 2, \qquad \nu \in I\!N, \nu/3 \leq \kappa \leq \nu.$$

Proposition 3 (Integral Forwarding Index).

- Complete bipartite graphs *are (almost) optimal, in the sense that, for* $i \in \{1, 2, \ldots, \lfloor n/2 \rfloor\}$, *we have:*

$$\pi_{\mathcal{I}}^*(n, i(n-i)) \in \lceil \pi^*(n, i(n-i)) \rceil + \{0, 1, 2, 3, 4\}.$$

- Turán graphs $T(n, r)$, *for which* r *divides* n *are (almost) optimal, in the sense that, for* $i \in \{1, 2, \ldots, \lfloor n/2 \rfloor\}$, *we have:*

$$\pi_{\mathcal{I}}^*(n, \frac{n}{2}(n - \frac{n}{r})) = \pi^*(n, \frac{n}{2}(n - \frac{n}{r})) + \{0, 1, 2, 3, 4\}.$$

Since $\pi^*(n, e)$ decreases with e the above results implies that $\pi^*(n, e)$ evolves like $\Theta(\frac{2n^2}{e})$, but we don't know yet the precise behavior of $\pi^*(n, e)$ between two perfect cases (i.e. $e = i(i - k)$). As we shall prove this behavior is not a smooth linear decrease since it indeed proceeds with jumps occurring at values close to those perfect ones. First, we start studying the intermediary cases when e starts at $n - 1$ ($\pi^*(n, e) = 2(n - 1)$, optimal graph is a star) and grows to $e = 2(n - 2)$ ($\pi^*(n, e) = n - 2$. optimal graph is $K_{2, n-2}$). The next proposition shows that when e get larger than $n - 1$, first π^* does not decrease significantly and stays around $2(n-1)$ then it jumps abruptly down to $n-1$ when e get close to $2(n-2)$.

Proposition 4.

$$\forall e \in [n - 1, 2(n - 2) - o(n)] \quad \pi^*(n, e) = 2(n - 1) + o(n)$$
$$e = 2(n - 2) \qquad\qquad \pi^*(n, e) = (n - 1) + o(n)$$

The result can be extended to larger values of e ($e = n + k$ with $k = o(n)$), see Proposition 5.

Proposition 5. *For any fixed* $k \in I\!N$:

$$\forall e \in [kn, (k + 1)n - o(n)] \quad \pi^*(n, e) = \tfrac{2n}{k} + o(n).$$

3   Bounded Degree Graphs with Low Edge Forwarding Index

In the preceding section, we provided somewhat optimal families of graphs. This solves the question of minimally congested graphs in the general case. We now study graphs with a constraint on the degree (Δ will denote the maximum degree). The motivation comes from telecommunication & real interconnection networks for which the node degree is often small, see for example [8,18]. In this section, we consider first the general case for $\Delta \geq 3$ ($\Delta = 2$ is trivial) and we succeed in determining how the forwarding index drops from $\pi(n, e) = n^2/4$ to $\frac{2}{3}n \log_2 n$ when the average degree raises from 2 to 3 So, we focus on graphs with a *small number of edges*, namely graphs with average degree $\overline{\Delta} \in [2, 3[$, that is when $e \in [n, \frac{3}{2}n]$, and we study the transition of $\pi(n, e)$ from $\frac{n^2}{4}$ to $\Theta(n \log n)$ when the number of edges e raises from $n - 1$ to $\frac{3}{2}n$.

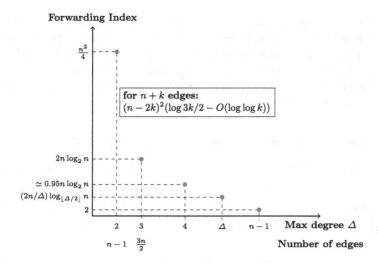

Fig. 2. Forwarding indices of minimaly congested graphs with n vertices as a function of their number of edges.

3.1   Graphs with Bounded Degree Δ: Some Remarks

For $\Delta = 3$, when $e = \frac{3n}{2}$, graphs such like the shuffle exchange provide deterministic generic constructions for which $\pi(G) \leq n \log_2 n$ (this is a folk result for people studying network throughput, one may see [25]). Since using the Moore bound (that bound claims by direct counting that the average distance in a Δ bounded degree graph is of order $\log_{\Delta-1}(n)$, see as example [16]) one can prove that $\pi^*(n, \frac{3n}{2}) \geq \frac{2}{3} n \log_2 n(1 + o(1))$ the lower and upper bounds matche up to factor of $\frac{2}{3}$. Moreover we shall prove that random cubic graphs are almost optimal since with high probability they are such that $\pi(G) = \frac{2}{3} n \log_2 n(1 + o(1))$. Moreover for larger values of Δ de Bruijn graphs and their variants provide Δ-regular graphs whose forwarding index is of the right order (see Fig. 2). So when the degree is bounded by Δ, the value of $\pi(n, \frac{\Delta}{2} n)$ is relatively well understood (see [6,11]), and structures close to the optimal are obtained using de Bruijn graphs or slight variants of it. Indeed, on the one hand, the Moore bound implies that:

$$\pi^*(n, \frac{\Delta}{2} n) \geq \frac{2}{\Delta} n \log_{\Delta-1} n(1 - o(1)).$$

On the other hand, for de Bruijn graphs, one has (see [6,11])

$$\pi(n, \frac{\Delta}{2} n) \leq \frac{2}{\Delta} n \log_{\lfloor \frac{\Delta}{2} \rfloor} n.$$

The argument that provides the above bound for the de Bruijn graph with degree $\Delta = 2d$ and d^n vertices, is quite simple since it exists in this graph an integral routing that is uniform on the edges and that connects each couple of vertices with a path of length exactly n. This length is only a constant factor larger than the minimum average distance predicted by the Moore bound, hence the ratio between the above upper and lower bound is at most 3 and decreases with Δ.

So our purpose is to understand what is happening between two well understood situations: $e = n - 1, \pi^*(n, e) = \frac{n^2}{4}$ and $e = \frac{3}{2}n, \pi^*(n, e) = \Theta(n \log n)$ that is when e evolves in $[n, \frac{3}{2}n]$, in other words we shall study the evolution of $\pi^*(n, e)$ when the number of edges e raises from $n - 1$ to $\frac{3}{2}n$.

3.2   A Lower Bound for the Case $e \in [n, \frac{3}{2}n]$ for $\Delta \le 3$

In this section, we provide a lower bound on the forwarding indices of graphs with $e \in [n, \frac{3}{2}n]$ and $\Delta \le 3$.

Proposition 6. *If G is a $(n, n + k)$ graph with $\Delta = 3$ then $\pi(G) \ge \frac{(n-2k)^2}{3k}(\log(3k/2) - O(\log \log(k))$.*

3.3   Construction of Minimaly-Congested Graph with Degree $\le \Delta$

Our construction simply reverts the previous operation and builds graphs with few extra edges from good skeletons.

Definition 1. Given a graph, we construct $Sub(G, \mathbf{W})$ as follows: we subdivide each edge ab by adding one node x_{ab} and we then attach a binary tree with weight $\mathbf{W}$ on x_{ab}.

Lemma 1. *Let G be a Δ-regular graph with x vertices, and let $H = sub(G, \mathbf{W})$ then $\pi(H) \le Max\left\{\pi(G)(\frac{\Delta}{2}\mathbf{W} + 1)^2 + \mathbf{W}(\frac{\Delta\mathbf{W}}{2} + 1)x, \mathbf{W}((\frac{\Delta\mathbf{W}}{2} + 1)x - \mathbf{W})\right\}$.*

To our surprise, we could not find the following result in the literature, moreover in the recent survey [11] the best bounds for cubic graphs were provided by shuffle exchange graphs, and more generally, for bounded degree graphs the best bounds known are derived using de Bruijn graphs. Those bounds are rather good since they differ from the lower bound only by a relatively small (always lesser than 2) constant factor. But indeed random regular graph are asymptotically optimal.

Proposition 7. *There exist cubic regular graphs such that $\pi(G) = \frac{2}{3}n \log_2(n)(1 + o(1))$, and Δ-regular graphs with $\pi(G) = \frac{2}{\Delta}n \log_{\Delta-1}(n)(1 + o(1))$.*

Remark 1. Note that the fair shortest path routing (in which each shortest path carries the same flow) is probably better and for small values of n it may even be significantly better, but we don't have currently a good method to evaluate its load and proving that so doing we get a better load. Probably the forwarding index of random cubic graph is $\frac{2}{3}n \log_2 n + \Theta(n)$, but we proved only a weaker result. Moreover the value of n for which our $(1 + o(1))$ becomes smaller than the $\frac{3}{2}$ are relatively high (order of 1000).

Proposition 8. *There exist $(n, e = n + k)$ cubic graphs such that $\pi(G) \le \frac{n^2}{3k} \log_2(k)(1 + o(1))$.*

4 Edge Forwarding Index of Cubic ($\Delta = 3$) Graphs with Few Extra Edges: $e = n + k$

When k is large, we provided in Sect. 3 asymptotically matching upper and lower bounds on the minimum congestion. This implies that $\pi^*(n, n + k)$ behaves like $\Theta(\frac{n^2}{k} \log \frac{n}{k})$ when both k and n are large. So, in order to get a complete picture of the situation, we still need to understand the case of $(n, n + k)$ graphs when k is fixed. In this section, we answer this question, that is we solve the MIN-CONGESTION DESIGN PROBLEM, for graphs with arbitrary n, but small values of k.

4.1 Method: The Skeleton Approach

From the results of Sect. 3, we know that $(n, n + k)$ graphs are constructed from a cubic skeleton on which are attached trees with size u. So, when k is small, we may enumerate all the possible skeletons (like we enumerated all the cubic graphs) and determine for each the best way to attach trees. Attaching trees means determining for each edge $e \in E$ the size $\alpha(e)$ of the tree that we attach in the edge. Hence, we want to find the best weight repartition $\alpha : E \to N$ that satisfies $\sum_{e \in E} \alpha(e) = n$ and $\forall e \in E, \alpha(e) \leq w_{max}$, where by best we mean with the smallest forwarding index. So, finding the best way to subdivide edges means solving a problem of the following flavor:

Definition 2. (Best Mass Repartition). Given a graph G and a maximum weight w_0 find a weight function $w : V \to \mathbb{R}^+$ with $\forall v \in V, w(V) = 1, w(v) \leq w_0$ such that $\pi(G, w)$ is minimum.

4.2 Optimal $(n, n - 1 + k)$ Cubic Graph for $k = 0, 1, 2, 3$

Results are listed below and corresponding constructions are given in Table 1.

When $k = 0$ and $e = n - 1$, the network is a tree with max degree $\Delta = 3$. The case of degree Δ trees is trivial since for such trees, considering the most balanced cut, we get: $\pi(T) \geq 2\Delta(\Delta - 1) \left(\frac{n}{\Delta}\right)^2$ and this value is attained using a balanced Δ-ary tree or a subdivided Δ-star with branches with equal size $\frac{n}{\Delta}$. So, for $\Delta = 3$. we have:

$$\pi^*(n, n - 1) = 2\Delta(\Delta - 1) \left(\frac{n}{\Delta}\right)^2 = 2\frac{(\Delta - 1)}{\Delta}n^2 = \frac{4}{3}n^2.$$

In this case, the first intuition is that the cycle C_n should be the optimal structure. Recall that $\pi(C_n) = \frac{n^2}{4}$ when n is even, and $\pi(C_n) = \frac{n-1}{2}\frac{n+1}{2}$ when n is odd (indeed $\pi(C_n) = \lceil \frac{n-1}{2} \rceil \lfloor \frac{n+1}{2} \rfloor$). The cycle is the only 2 connected structure but it is not the min-congested one since some graphs with bridges do have lesser congestion.

Table 1. Constructions of optimal graphs with n vertices and $n - 1 + k$ edges for different numbers of extra edges k.

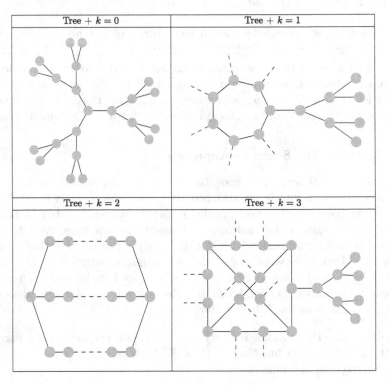

Proposition 9. $\pi^*(n, n) = \frac{12}{49}n^2$.

We provide a graph G_7 with $\pi(G) = Opt = \frac{12}{49}n^2$: we simply take the cycle C_7 and on each vertex we attach a tree (any tree can will do it) with $\frac{n}{7}$ nodes.

Proposition 10. $\pi^*(n, n + 1) = \frac{2}{9}n^2$.

A possible construction is then to use $\forall i \in \{1, 2, 3\}$ a path P_i of length $n/3$ for e_i, then one can cover all the request using 3 cycles of size $\frac{2n}{3}$ ($P_i \cup P_j, i \neq j$).

The next result is rather surprising since intuitively a uniform (or at least symmetric) subdivision of the K_4 should provide an optimal solution. But a phenomena similar to the one we already met in the case $k = 1$ (the C_7) happens again in a slightly more complex way.

Proposition 11. $\pi^*(n, n + 2) = \frac{20}{11^2}n^2$.

A graph reaching this bound is obtained by subdividing 5 edges of K_4 twice and one edge once, thus we add 11 new nodes. Then, we attach a tree with weight $\frac{n}{11}$ on each new node.

5 Graphs with a Small Number of Vertices ($\Delta = 3$)

We have seen in Sects. 3 and 4 the importance of having good skeletons to build graphs with low forwarding indices. In Table 2 on page 10, we present graphs with a small number of vertices which have the minimum possible forwarding indices. These graphs can serve as skeletons to build families of graphs with an arbitrary number of vertices. In some cases, optimality is easy to prove using:

- the Moore bound. In a cubic graph, and for a given vertex, the number of vertices that are at distance $0, 1, 2, 3, \ldots$, are respectively, at most $1, 3, 6, 12, \ldots$. When those bounds are reached for all the vertices of a cubic graph, the latter minimizes $\mathcal{L} = 2|V|\frac{\overline{D(G)}}{\overline{d(G)}}$ among all the graphs with the same size and

Table 2. Small cubic graphs with minimum edge forwarding index

$n = 4, \pi = \mathcal{L} = 2$ $\mathcal{L}$ given by a cut	$n = 6, \pi = \mathcal{L} = 4.66...$ $\mathcal{L}$ given by the Moore bound	$n = 8, \pi = \mathcal{L} = 8$ $\mathcal{L}$ given by a cut
$n = 10, \pi = \mathcal{L} = 10$ $\mathcal{L}$ given by a cut	$n = 12, \pi = \mathcal{L} = 14.26...$ $\mathcal{L}$ found by brute force	$n = 14, \pi = \mathcal{L} = 18$ $\mathcal{L}$ given by the Moore bound
$n = 16, \pi = \mathcal{L} = 22$ $\mathcal{L}$ found by brute force	$n = 18, \pi = \mathcal{L} = 26.66...$ $\mathcal{L}$ found by brute force	$n = 20, \pi = 30.84..., \mathcal{L} = 30$

with degree 3. When the graph is optimal for the Moore bound and is edge-transitive, its forwarding index is minimum. This is the case for $n = 6, 14$;
– cut arguments, for $n = 4, 8, 10$.

In other cases, ($n = 12, 16, 18$), the generic arguments fail to provide matching upper and lower bounds. We had to check all the possible cubic graphs [15].

5.1   Consequences for Unbounded n but a Few Edges

All those graphs can be used as skeletons, as example if one wishes to get a good $(n, 6)$ graph with $e = n + 6$ edges one can simply pick the Petersen graph as skeleton and apply Lemma 1. We use the uniform weight function $\mathbf{W} = \frac{n}{15}$ and using the generic routing of the lemma we get: $\pi(n, 5) \leq \pi(G) \leq 10 \left(\frac{n}{10}\right)^2 + 2\frac{n}{30} \times \frac{14n}{15} = \frac{n^2}{10} + \frac{14n^2}{225}$. This may be potentially improved by computing the exact forwarding index of the so defined weighted graph (that has only 15 vertices).

Solving the best mass repartition problem would allow us to go quite further, but currently we have no clue about what is the best repartition even for a small structure. It is certainly possible to repeat what we did for $0, 1, 2, 3$ extra edges, but the difficulty shall increase considerably each time we add one edge, finding a method that would scale more than considering cases by "hand" is certainly interesting.

6   Conclusion

In this paper, we provided a basic understanding of the interplay between the forwarding-index of a graph and its number of extra-edges. Our bounds are mostly asymptotically tight and explain as example how the transition happens between highly congested graphs (Trees, Paths, ...) to cubic regular graphs which have much lower congestion.

Some results, like the step-like behavior in Proposition 4 or irregular optimal structures, are also *fun*, since they are unexpected. Last, we believe that our work opens many questions:

– **Small cases**: In the case of a few extra-edges, we stopped at 3 extra edges (and even in those cases the proofs are not immediate). So, it may be interesting to go further and to understand if optimal graphs with k extra-edges are built using an optimal cubic graph with $\frac{k}{2}$ vertices (we determined such graphs till $k = 22$). As example: is the family of optimal graphs with 5 extra edges built using the Petersen and subdivising it properly? And, if so, how do we find the best subdivision (we saw the uniform subdivision is not always optimal).
– **Construction from skeletons**: Given a skeleton, we do not know how to affect weights in order to minimize the forwarding-index of the resulting graph. That problem can be expressed as a quadratic non convex problem and we conjecture that it is NP-Complete.

References

1. Araujo, J., Giroire, F., Liu, Y., Modrzejewski, R., Moulierac, J.: Energy efficient content distribution. In: IEEE International Conference on Communications (ICC 2013), pp. 4233–4238. IEEE (2013)
2. Baliga, J., Tucker, R., Ayre, R., Hinton, K.W., Sorin, W.: Energy consumption in IP networks. In: 34th European Conference on Optical Communication, ECOC 2008, p. 1 (2008)
3. Bouabdallah, A., Sotteau, D.: On the edge forwarding index problem for small graphs. Networks **23**(4), 249–255 (1993)
4. Restrepo, J. C. C., Gruber, C. G., Machuca, C. M.: Energy profile aware routing. In: Communications Workshops of IEEE International Conference on Communications (ICC), pp. 1–5 (2009)
5. Chiaraviglio, L., Mellia, M., Neri, F.: Energy-aware umts core network design. In: The 11th International Symposium on Wireless Personal Multimedia Communications (2008)
6. Chung, F.R.K., Coffman Jr., E.G., Reiman, M.I., Simon, B.: The forwarding index of communication networks. IEEE Trans. Inf. Theory **33**(2), 224–232 (1987)
7. Giroire, F., Mazauric, D., Moulierac, J., Onfroy, B.: Minimizing routing energy consumption: from theoretical to practical results. In: IEEE/ACM International Conference on Green Computing and Communications (GreenCom 2010), Hangzhou, China, p. 8 (2010)
8. Giroire, F., Perennes, S., Tahiri, I.: Grid spanners with low forwarding index for energy efficient networks. In: International Network Optimization Conference (INOC), Warsaw, Poland, May 2015
9. Giroire, F., Pérennes, S., Tahiri, I.: Graphs with optimal forwarding indices: what is the best throughput you can get with a given number of edges? Research Report RR-8752, INRIA Sophia Antipolis, INRIA, June 2015. https://hal.inria.fr/hal-01172725
10. Heydemann, M.C., Meyer, J.C., Sotteau, D.: On forwarding indices of networks. Discrete Appl. Math. **23**(2), 103–123 (1989)
11. Xu, J.-M., Xu, M.: The forwarding indices of graphs - a survey. xarchiv (2012)
12. Leighton, T., Rao, S.: Multicommodity max-flow min-cut theorems and their use in designing approximation algorithms. J. ACM **46**(6), 787–832 (1999)
13. Linial, N., London, E., Rabinovich, Y.: The geometry of graphs and some of its algorithmic applications. Combinatorica **15**, 577–591 (1994)
14. Manoussakis, Y., Tuza, Z.: The forwarding index of directed networks. Discrete Appl. Math. **68**(3), 279–291 (1996). http://www.sciencedirect.com/science/article/pii/0166218X9500072Y
15. Meringer, M.: Small cubic graphs, flinders Univ projet. http://www.flinders.edu.au/science_engineering/csem/research/programs/flinders-hamiltonian-cycle-project/graph-database.cfm
16. Miller, M., Širáň, J.: Moore graphs and beyond: a survey of the degree/diameter problem. Electron. J. Comb. **61**, 1–63 (2005)
17. Mohar, B.: Some applications of laplace eigenvalues of graphs. In: Hahn, G., Sabidussi, G. (eds.) Graph Symmetry. NATO ASI Series, vol. 497, pp. 225–275. Springer, Amsterdam (1997)
18. Orlowski, S., Wessäly, R., Pióro, M., Tomaszewski, A.: Sndlib 1.0—survivable network design library. Networks **55**(3), 276–286 (2010)

19. Shahrokhi, F., Matula, D.W.: The maximum concurrent flow problem. J. ACM **37**(2), 318–334 (1990). http://doi.acm.org/10.1145/77600.77620
20. Sinclair, A.: Improved bounds for mixing rates of markov chains and multicommodity flow. Comb. Probab. Comput. **1**, 351–370 (1992)
21. Solé, P.: Expanding and forwarding. Discrete Appl. Math. **58**(1), 67–78 (1995)
22. Tassiulas, L., Ephremides, A.: Stability properties of constrained queueing systems and scheduling policies for maximum throughput in multihop radio networks. IEEE Trans. Autom. Control **37**(12), 1936–1948 (1992)
23. de la Vega, W.F., Manoussakis, Y.: The forwarding index of communication networks with given connectivity. Discrete Appl. Math. **37–38**, 147–155 (1992)
24. de la Vega, F., Gordones, L.M.: The forwarding indices of random graphs. Random Struct. Algorithms **3**(1), 107–116 (1992)
25. Xu, J.M., Xu, M.: The forwarding indices of graphs - a survey. CoRR abs/1204.2604 (2012)

Enumeration and Maximum Number of Minimal Connected Vertex Covers in Graphs

Petr A. Golovach[1], Pinar Heggernes[1], and Dieter Kratsch[2(✉)]

[1] Department of Informatics, University of Bergen, Bergen, Norway
{petr.golovach,pinar.heggernes}@ii.uib.no
[2] Université de Lorraine, LITA, Metz, France
dieter.kratsch@univ-lorraine.fr

Abstract. CONNECTED VERTEX COVER is one of the classical problems of computer science, already mentioned in the monograph of Garey and Johnson [15]. Although the optimization and decision variants of finding connected vertex covers of minimum size or weight are well studied, surprisingly there is no work on the enumeration or maximum number of minimal connected vertex covers of a graph. In this paper we show that the maximum number of minimal connected vertex covers of a graph is $O(1.8668^n)$, and these can be enumerated in time $O(1.8668^n)$. For graphs of chordality at most 5, we are able to give a better upper bound, and for chordal graphs and distance-hereditary graphs we are able to give tight bounds on the maximum number of minimal connected vertex covers.

1 Introduction

The maximum number of minimal vertex covers that a graph on n vertices can have is equal to the maximum number of maximal independent sets, which is known to be $3^{n/3}$ by a celebrated result of Moon and Moser [23]. This result is easily extended to an algorithm that enumerates all the minimal vertex covers of a graph within a polynomial factor of the given bound. The bound is tight as a disjoint union of $n/3$ triangles has exactly $3^{n/3}$ minimal vertex covers. These results have been extremely useful in many algorithms, e.g., they were used by Lawler [21] to give an algorithm for graph coloring, which was the fastest algorithm for this purpose for decades. For special graph classes, better bounds have been obtained, e.g., the tight bound for triangle-free graphs is $2^{n/2}$, given by Hujtera and Tuza [19] with combinatorial arguments, and by Byskov [4] algorithmically. Also these results have been useful in several algorithms, e.g., for graph homomorphism [11]. Although connected vertex covers were defined and studied as early as vertex covers [15], interestingly the maximum number of minimal connected vertex covers in graphs or the enumeration of these have not been given attention.

The research leading to these results has received funding from the Research Council of Norway and the European Research Council under the European Union's Seventh Framework Programme (FP/2007–2013)/ERC Grant Agreement no. 267959.

Z. Lipták and W.F. Smyth (Eds.): IWOCA 2015, LNCS 9538, pp. 235–247, 2016.
DOI: 10.1007/978-3-319-29516-9_20

In this paper, we study exactly these questions, and we give an algorithm for enumerating all minimal connected vertex covers of a graph in time $O(1.8668^n)$. This also gives an upper bound on the number of such covers a graph can have. We provide a lower bound example, which is a graph that has $3^{(n-1)/3}$ minimal connected vertex covers, leaving a gap between these bounds on general graphs. We are able to narrow this gap for graphs of chordality at most 5, and almost close the gap for chordal graphs and distance-hereditary graphs. In particular, we show that the maximum number of minimal connected vertex covers in chordal graphs, graphs of chordality at most 5, and distance-hereditary graphs, respectively, is at most $3^{n/3}$, 1.6181^n, and $2 \cdot 3^{n/3}$. All our bounds are obtained by enumeration algorithms whose running times correspond to the given bounds up to polynomial factors.

We would like to emphasize that our motivation for the given bounds and enumeration algorithms is not for fast computation of connected vertex covers of minimum size. In fact, as we will see in the next section, such sets can be computed in time $O(1.7088^n)$ on general graphs. Furthermore, Escoffier et al. [9] have shown that this problem can be solved in polynomial time on chordal graphs. The problem of computing minimum connected vertex covers is indeed well studied with a large number of published results. These are nicely surveyed in the introduction given by Escoffier et al. [9].

Our motivation comes from the background given in the first paragraph, as well as the fact that the study of the maximum number of vertex subsets with given properties is a well established area in combinatorics and graph theory. More recently, exponential time enumeration algorithms for listing such vertex subsets in graphs have become increasingly popular and found many applications [13]. For most of these algorithms, an upper bound on the number of enumerated subsets follows from the running time of the algorithm. Examples of such recent results, both on general graphs and on some graph classes, concern the enumeration and maximum number of minimal dominating sets, minimal feedback vertex sets, minimal subset feedback vertex sets, minimal separators, maximal induced matchings, and potential maximal cliques [2,5,6,10,12,14,16,17].

2   Preliminaries

We consider finite undirected graphs without loops or multiple edges. For each of the graph problems considered in this paper, we let n denote the number of vertices and m the number of edges of the input graph. For a graph G and a subset $U \subseteq V(G)$ of vertices, we write $G[U]$ to denote the subgraph of G induced by U. We write $G - U$ to denote $G[V(G) \setminus U]$, and $G - u$ if $U = \{u\}$. A set $U \subseteq V(G)$ is *connected* if $G[U]$ is a connected graph. For a vertex v, we denote by $N_G(v)$ the *(open) neighborhood* of v, i.e., the set of vertices that are adjacent to v in G. The *closed neighborhood* is $N_G[v] = N_G(v) \cup \{v\}$. For a set of vertices $U \subseteq V(G)$, $N_G[U] = \cup_{v \in U} N_G[v]$ and $N_G(U) = N_G[U] \setminus U$. Two distinct $u, v \in V(G)$ are *false twins* if $N_G(u) = N_G(v)$. The *distance* $\text{dist}_G(u, v)$ between vertices u and v of G is the number of edges on a shortest (u, v)-path.

A path or cycle P is *induced* if it has no *chord*, i.e., there is no edge of G that joins any two vertices of P that are not adjacent in P. The *chordality*, $\text{chord}(G)$, of a graph G is the length of a longest induced cycle in G; if G has no cycles, then $\text{chord}(G) = 0$. A set of vertices is an *independent set* if there is no edge between any pair of these vertices, and it is a *clique* if all possible edges are present between pairs of these vertices. An independent set (clique) is *maximal* if no set properly containing it is an independent set (clique). A set of vertices $S \subset V(G)$ of a connected graph G is a *separator* if $G - S$ is disconnected. A vertex v is a *cut vertex* of a connected graph G if $\{v\}$ is a separator. For an edge $uv \in E(G)$, the *contraction* of uv is the operation that replaces u and v by a new vertex adjacent to $(N_G(u) \cup N_G(v)) \setminus \{u, v\}$. G/e denotes the graph obtained from G by contracting edge e. A graph G' is an *induced minor* of G if G' can be obtained from G by deleting vertices and contracting edges.

For a non-negative integer k, a graph G is *k-chordal* if $\text{chord}(G) \leq k$. A graph is *chordal* if it is 3-chordal. A graph is a *split* graph if its vertex set can be partitioned in an independent set and a clique. A graph is *cobipartite* if its vertex set can be partitioned into two cliques. A graph G is a *chordal bipartite* graph if G is a bipartite graph and $\text{chord}(G) \leq 4$. A graph G is *distance-hereditary* if for every connected induced subgraph H of G, $\text{dist}_H(u, v) = \text{dist}_G(u, v)$ for $u, v \in V(H)$. Each of the above mentioned graph classes can be recognized in polynomial (in most cases linear) time, and they are closed under taking induced subgraphs [3, 18]. See the monographs by Brandstädt et al. [3] and Golumbic [18] for more properties and characterizations of these classes and their inclusion relationships.

A set of vertices $U \subseteq V(G)$ is a *vertex cover* of G if for every $uv \in E(G)$, $u \in U$ or $v \in U$. A vertex cover U is *connected* if U is a connected set. A (connected) vertex cover U is *minimal* if no proper subset of U is a (connected) vertex cover. Observe that U is a minimal connected vertex cover of G if and only if for every vertex $u \in U$, either u is a cut vertex of $G[U]$ or there is an edge ux of G such that $x \notin U$. Hence given a vertex set $U \subseteq V(G)$, it can be decided in time $O(nm)$ whether U is a minimal connected vertex cover of G.

It is easy to see that U is a (minimal) vertex cover of G if and only if $V(G) \setminus U$ is a (maximal) independent set. The following upper bound for the number of maximal independent sets was obtained by Miller and Muller [22] and Moon and Moser [23].

Theorem 1 ([22, 23]). *The number of minimal vertex covers (maximal independent sets) of a graph is at most*

$$
\begin{cases}
3^{n/3} & \text{if } n \equiv 0 \ (\text{mod} 3), \\
4 \cdot 3^{(n-4)/3} & \text{if } n \equiv 1 \ (\text{mod} 3), \\
2 \cdot 3^{(n-2)/3} & \text{if } n \equiv 2 \ (\text{mod} 3).
\end{cases}
$$

Together with the fact that all maximal independent sets can be enumerated with polynomial delay (see, e.g., [20, 25]), this implies that all minimal vertex covers of a graph can be enumerated in time $O^*(3^{n/3})$, where the O^*-notation suppresses polynomial factors. Note that the same result can also be obtained by a branching algorithm (see, e.g. [13]).

The bounds of Theorem 1 are tight; a well known lower bound example is a graph consisting of $n/3$ disjoint triangles, which is a chordal distance-hereditary graph. By adding a vertex which is adjacent to every vertex of this graph, we can obtain a lower bound for the maximum number of minimal connected vertex covers of a graph.

Proposition 1. *There are chordal distance-hereditary graphs with at least $3^{(n-1)/3}$ minimal connected vertex covers.*

We do not know of any better lower bounds for the maximum number of minimal connected vertex covers on graphs in general. We will use the following simple observation to give upper bounds on the number of minimal connected vertex covers.

Observation 1. *Let S be a separator of a connected graph G. Then for every connected vertex cover U of G, $S \cap U \neq \emptyset$. In particular, if v is a cut vertex, then v belongs to every connected vertex cover.*

Recall that our motivation for enumerating the minimal connected vertex covers of a graph is not for the computation of a connected vertex cover of minimum size. In fact such a set can be computed in time $O(1.7088^n)$, using the following result of Cygan [7] about the parameterized complexity of the problem.

Theorem 2 ([7]). *It can be decided in time $O(2^k \cdot n^{O(1)})$ and in polynomial space, whether a graph has a connected vertex cover of size at most k.*

Combining the algorithm of Cygan [7] with brute force checking of all vertex subsets of size at most k, we obtain the following corollary.

Corollary 1. *It can be decided in time $O(1.7088^n)$ and in polynomial space, whether a graph has a connected vertex cover of size at most k.*

Our upper bounds will be given via enumeration algorithms that are recursive branching algorithms. For the analysis of the running time $T(n)$ of such an algorithm, we use standard terminology [13]. If at a step, the algorithm branches into t new subproblems, where the problem size decreases by $c_1, c_2, \ldots, c_t$ in each subproblem, respectively, we get the *branching vector* $(c_1, c_2, \ldots, c_t)$. In particular, a branching vector $(c_1, c_2, \ldots, c_t)$ results from the recurrence $T(n) \leq T(n-c_1) + T(n-c_2) + \ldots + T(n-c_t)$. In this case $T(n) = O^*(\alpha^n)$, where α is the unique positive real root of $x^n - x^{n-c_1} - \ldots - x^{n-c_t} = 0$ [13]. The number α is called the *branching number* of this branching vector. When different branching vectors are involved at different steps of an algorithm, the branching vector with the highest branching number gives an upper bound on $T(n)$. In the analysis of the running time of our branching algorithms the problem size is $|F|$, i.e. the number of free vertices of the instance.

3   General Graphs

Theorem 3. *The maximum number of minimal connected vertex covers of an arbitrary graphs is $O(1.8668^n)$, and these can be enumerated in time $O(1.8668^n)$.*

Proof. We give a branching algorithm that we call ENUMCVC(S, F), which takes as the input two disjoint sets $S, F \subseteq V(G)$ and outputs minimal connected vertex covers U of G such that $S \subseteq U \subseteq S \cup F$. We call ENUMCVC($\emptyset, V(G)$) to enumerate the minimal connected vertex covers of G. We say that $v \in V(G)$ is *free* if $v \in F$ and v is *selected* if $v \in S$. The algorithm branches on a subset of free vertices and either *selects* some of them to be included in a (potential) minimal connected vertex cover or *discards* some of them by forbidding them to be selected.

ENUMCVC(S, F)

1. If S is a minimal connected vertex cover then return S and stop.
2. If $F = \emptyset$, then stop.
3. If there are two adjacent free vertices $u, v \in F$, then branch as follows:
   - select u, i.e., set $S' = S \cup \{u\}$, $F' = F \backslash \{u\}$, and call ENUMCVC(S', F'),
   - discard u and select its neighbors, i.e., set $S' = S \cup N_G(u)$, $F' = F \backslash N_G[u]$, and call ENUMCVC(S', F').
4. If F is an independent set, then let s be the number of components of $G[S]$. Consider every non-empty set $X \subseteq F$ of size at most $s - 1$ and output $S \cup X$ if $S \cup X$ is a minimal connected vertex cover of G.

To argue that the algorithm is correct, consider a minimal connected vertex cover U of G such that $S \subseteq U \subseteq S \cup F$ and for every $v \in V(G) \backslash (S \cup F)$, $N_G(v) \subseteq S$. If $F = \emptyset$, then $U = S$ and the algorithm outputs U on Step 1. Assume inductively that ENUMCVC(S', F') outputs U for every pair of disjoint sets S', F' such that $S' \subseteq U \subseteq S' \cup F'$ and $|F'| < |F|$. Clearly, if S is a connected vertex cover of G, then $U = S$ by minimality and U is returned by the algorithm on Step 1. If S is not a connected vertex cover, then $F \cap U \neq \emptyset$ and the algorithm does not stop at Step 2. If there are two adjacent free vertices $u, v \in F$, then $u \in U$, or $u \notin U$ and $N_G(u) \subseteq U$, because at least one endpoint of every edge is in U. In the first case we have that $S' \subseteq U \subseteq S' \cup F'$, where $S' = S \cup \{u\}$ and $F' = F \backslash \{u\}$. In the second case, $S' \subseteq U \subseteq S' \cup F'$, where $S' = S \cup N_G(u)$ and $F' = F \backslash N_G[u]$. By induction, the algorithm outputs U when we call ENUMCVC(S', F'). Finally, if F is an independent set and S is not a connected vertex cover of G, then $U = S \cup X$ for $X \subseteq F$. Because F is independent and $N_G(v) \subseteq S$ for $v \in V(G) \backslash (S \cup F)$, S is a vertex cover of G, i.e., the vertices of X are included in U only to ensure the connectivity of $G[U]$. We have that each vertex of X is a cut vertex of $G[U]$. Since $G[S]$ has s components, $|X| \leq s - 1$. Therefore, the algorithm outputs U. To complete the correctness proof, it remains to notice that if $S = \emptyset$ and $F = V(G)$, then $S \subseteq U \subseteq S \cup F$ and $V(G) \backslash (S \cup F) = \emptyset$. Hence, ENUMCVC($\emptyset, V(G)$) outputs U. As U is an arbitrary minimal connected vertex cover, the algorithm outputs all minimal connected vertex covers.

We have that $\text{ENUMCVC}(\emptyset, V(G))$ lists all minimal connected vertex covers of G. To obtain the upper bound on the number of minimal connected vertex covers of G, we upper bound the number of leaves of the search tree produced by the algorithm.

Observe that by executing Steps 1–3, the algorithm either produces a leaf of the search tree or a node of the tree corresponding to a pair of sets S and F such that F is independent. We call a node corresponding to such S and F a *sub-leaf* or (S, F)-*sub-leaf*. Note that the children of a sub-leaf are leaves of the search tree produced by Step 4. The only branching rule (Step 3) has branching vector $(1, 2)$ since we remove at least one free vertex in the first branch and at least two, i.e., u and v, in the second one. This branching vector has branching number $\alpha \approx 1.61803$ [13]. Moreover, by executing Steps 1–3 the algorithm produces $O^*(\alpha^h)$ (S, F)-sub-leaves such that $h = n - |F|$.

Now we consider the (S, F)-sub-leaves of the search tree and Step 4. We have the following two cases.

Case 1. $h = n - |F| \geq n/3$. Then $|F| \leq 2n/3$. Clearly, there are at most 2^{n-h} sets $X \subseteq F$ of size at most $s - 1$ and such an (S, F)-sub-leaf has at most 2^{n-h} children. Since there are $O^*(\alpha^h)$ (S, F)-sub-leaves with $h = n - |F|$, the total number of children of these nodes is $O^*(\alpha^h \cdot 2^{n-h})$. Since, $h \geq n/3$ and $\alpha < 2$, the number of these children is $O^*(\alpha^{n/3} \cdot 2^{2n/3})$.

Case 2. $h = n - |F| < n/3$. Let $s \geq 2$ be the number of components of $G[S]$. We have that $s - 1 \leq h$. Let $\beta = h/n$. Then $h = \beta n$, $n - h = (1 - \beta)n$ and $h/(n - h) = \beta/(1 - \beta) \leq 1/2$. The number of non-empty sets $X \subseteq F$ such that $|X| \leq s - 1$ is

$$\binom{n - h}{1} + \ldots + \binom{n - h}{s - 1} \leq \binom{(1 - \beta)n}{1} + \ldots + \binom{(1 - \beta)n}{\beta n} \leq 2^{H(\beta/(1-\beta))(1-\beta)n},$$

where $H(x) = -x \log_2 x - (1 - x) \log_2(1 - x)$ is the entropy function (see, e.g., [13]). Let

$$f(\beta) = \alpha^\beta \cdot 2^{H(\beta/(1-\beta))(1-\beta)} = \left(\frac{1 + \sqrt{5}}{2}\right)^\beta \cdot \left(\frac{1 - \beta}{\beta}\right)^\beta \left(\frac{1 - \beta}{1 - 2\beta}\right)^{1-2\beta}.$$

The function $f(\beta)$ on the interval $(0, 1/3)$ has the maximum value[1] for $\beta^* = \frac{1}{2} - \frac{1}{2\sqrt{3 + 2\sqrt{5}}}$ and $f(\beta^*) \approx 1.86676$. Since the number of (S, F)-sub-leaves with $n - |F| = h$ is $O^*(\alpha^h)$, we obtain that the total number of children of these sub-leaves is $O^*(f(\beta^*)^n)$.   □

4   Graphs of Chordality at Most 5

The upper bound that we proved in the previous section leaves a gap between that bound and the best known lower bound given in Proposition 1. In this

---

[1] The computations have been done by computer.

section, we will narrow this gap for graphs of chordality at most 5. We will also close the gap for chordal graphs, i.e., graphs of chordality at most 3. We start with this latter class.

Theorem 4. *The maximum number of minimal connected vertex covers of a chordal graph is at most $3^{n/3}$, and these can be enumerated in time $O^*(3^{n/3})$.*

Proof. Let G be a chordal graph. If G has no edges, then the claim is trivial. Notice also that the removal of an isolated vertex does not influence connected vertex covers, and if G has two components with at least one edge each, G has no connected vertex cover. Hence, without loss of generality we can assume that G is a connected graph and $n \geq 2$.

Let S be the set of cut vertices of G and $G' = G - S$. We claim that $U \subseteq V(G)$ is a minimal connected vertex cover of G if and only if $S \subseteq U$ and $X = U \cap V(G')$ is a minimal vertex cover of G'.

Let X be a vertex cover of G'. We show that $U = S \cup X$ is a connected vertex cover of G. Because X is a vertex cover of G' and S covers the edges of $E(G) \backslash E(G')$, U is a vertex cover of G. To show that $G[U]$ is connected, assume for the sake of contradiction that it is not so. Let H_1 and H_2 be distinct components of $G[U]$ at minimum distance from each other. Let $P = v_0 \ldots v_k$ be a shortest path in G that joins a vertex of H_1 with a vertex of H_2. For $i \in \{1, \ldots, k\}$, $v_{i-1} \in U$ or $v_i \in U$, because U is a vertex cover of G. Since H_1 and H_2 are chosen to be components at minimum distance, $k = 2$. Since $v_1 \notin U$, v_1 is not a cut vertex of G. Therefore, $G - v_1$ has a shortest (v_0, v_2)-path $P' = u_0 \ldots u_s$. Because P' is an induced path and $v_0 v_2 \notin E(G)$, $v_1 u_i \in E(G)$ for $i \in \{1, \ldots, s-1\}$ by chordality. As $v_1 \notin U$, $u_i \in U$ for $i \in \{1, \ldots, u_{s-1}\}$. Therefore, $V(P') \subseteq U$ contradicting that H_1 and H_2 are distinct components of $G[U]$. Since U is a vertex cover and $G[U]$ is connected, U is a connected vertex cover of G.

Let U be a connected vertex cover of G. By Observation 1, $S \subseteq U$. As the vertices of S cover only the edges of $E(G) \backslash E(G')$, $X = U \backslash S$ is a vertex cover of G'.

We proved that $U \subseteq V(G)$ is a connected vertex cover of G if and only if $S \subseteq U$ and $X = U \cap V(G')$ is a vertex cover of G'. This implies that U is a minimal connected vertex cover of G if and only if $S \subseteq U$ and $X = U \cap V(G')$ is a minimal vertex cover of G'.

Since G' has at most $3^{n/3}$ minimal vertex covers by Theorem 1, G has at most $3^{n/3}$ minimal connected vertex covers. Because S can be found and G' can be constructed in polynomial time, the minimal connected vertex covers of G can be enumerated in time $O^*(3^{n/3})$ using the algorithms of e.g., [20,25] as mentioned in the preliminaries.   □

Proposition 1 shows that the upper bound is tight. Now we consider graphs of chordality at most 5. First a definition: a vertex in a graph is *weakly simplicial* if its neighborhood is an independent set and the neighborhoods of its neighbors form a chain under inclusion.

Lemma 1 ([24]). *A graph is chordal bipartite if and only if every induced subgraph of it has a weakly simplicial vertex. Furthermore, a nontrivial chordal bipartite graph has a weakly simplicial vertex in each partite set.*

Observation 2. *If e is an edge of a graph G, then* $\mathrm{chord}(G/e) \leq \mathrm{chord}(G)$.

Theorem 5. *The maximum number of minimal connected vertex covers of a graph of chordality at most 5 is at most 1.6181^n, and these can be enumerated in time $O(1.6181^n)$.*

Proof. We give a branching algorithm that we call ENUMCVC-CHORD(H, S, F), which takes as input an induced minor H of G and two disjoint sets $S, F \subseteq V(G)$ and outputs minimal connected vertex covers U of G such that $S \subseteq U \subseteq S \cup F$. We call ENUMCVC-CHORD$(G, \emptyset, V(G))$ to enumerate minimal connected vertex covers of G. As before, we say that $v \in V(G)$ is *free* if $v \in F$ and v is *selected* if $v \in S$. The algorithm branches on a set of free vertices and either selects some of them to be included in a (potential) minimal connected vertex cover or discards some of them by forbidding them to be selected.

ENUMCVC-CHORD(H, S, F)

1. If S is a minimal connected vertex cover then return S and stop. If S is a connected vertex cover but not minimal then stop.
2. If at least two distinct components of $G[S \cup F]$ contain vertices of S, then stop.
3. If there are two adjacent free vertices $u, v \in F$, then branch as follows:
   - select u, i.e., set $S' = S \cup \{u\}$, $F' = F \backslash \{u\}$, and call ENUMCVC-CHORD(H, S', F'),
   - discard u and select its neighbors, i.e., set $S' = S \cup N_G(u)$, $F' = F \backslash N_G[u]$, $H' = H - u$, and call ENUMCVC-CHORD(H', S', F').
4. If F is an independent set, then contract consecutively every edge $uv \in E(H)$ such that $u, v \notin F$ and denote by H' the obtained graph. Find a weakly simplicial vertex $u \in V(H') \backslash F$. For each $v \in N_{H'}(u)$, select v and discard $N_{H'}(u) \backslash \{v\}$, i.e., set $S' = S \cup \{v\}$, $F = F \backslash N_{H'}(u)$, $H'' = H' - (N_{H'}(u) \backslash \{v\})$, and call ENUMCVC-CHORD$(H'', S', F')$.

To show that the algorithm is correct, consider a minimal connected vertex cover U of G. Suppose that S and F are disjoint subsets of $V(G)$, and H is an induced minor of G such that

(i) $S \subseteq U \subseteq S \cup F$,
(ii) for every $v \in V(G) \backslash (S \cup F)$, $N_G(v) \subseteq S$, and
(iii) H is obtained from G by deleting vertices of $V(G) \backslash (S \cup F)$ and by contracting some edges uv such that $u, v \in S$.

If $F = \emptyset$, then $U = S$ and the algorithm outputs U on Step 1. Assume inductively that ENUMCVC-CHORD(H', S', F') outputs U for any disjoint S', F' and H' satisfying (i)–(iii) if $|F'| < |F|$.

Clearly, if S is a connected vertex cover of G, then $U = S$ by minimality, and U is returned by the algorithm on Step 1. Since $U \subseteq S \cup F$, and U is a connected set in G, all the vertices of S are in the same component of $G[S \cup F]$ and the algorithm does not stop at Step 2.

To argue the correctness of Step 3, suppose that there are two adjacent free vertices $u, v \in F$. Then $u \in U$, or $u \notin U$ and $N_G(u) \subseteq U$, because at least one endpoint of every edge is in U. In the first case we have that $S' \subseteq U \subseteq S' \cup F'$, where $S' = S \cup \{u\}$ and $F' = F \setminus \{u\}$, and then we call ENUMCVC-CHORD(H, S', F'). By induction, the algorithm outputs U. In the second case, $S' \subseteq U \subseteq S' \cup F'$, where $S' = S \cup N_G(u)$ and $F' = F \setminus N_G[u]$. Also $H' = H - u$; notice that H' is obtained from H by the deletion of a vertex of $V(G) \setminus (S' \cup F')$ and that all neighbors of u are in S'. Then we call ENUMCVC-CHORD(H', S', F'). Again, by induction, the algorithm outputs U.

To consider Step 4, suppose that F is an independent set and S is not a connected vertex cover of G. Observe that because of (ii), S is a vertex cover of G, and the vertices of $U \setminus S \subseteq F$ are used to ensure connectivity. Recall that the graph H' is obtained by contracting edges $uv \in E(H)$ such that $u, v \notin F$. This means that $V(H') \setminus F$ is an independent set, and we have that each vertex of $X = V(H') \setminus F$ is obtained by contracting a component of $G[S]$. For each $x \in X$, denote by $W_x \subseteq S$ the set of vertices of the component of $G[S]$ that is contracted to x. Because the algorithm did not stop at Step 1, S is not a connected vertex cover of G and $U \setminus S \neq \emptyset$. Hence, $F \neq \emptyset$. We have that H' is a bipartite graph such that X, F is the bipartition of $V(H')$. By Observation 2, $\mathrm{chord}(H') \leq \mathrm{chord}(G) \leq 5$. As H' is bipartite, $\mathrm{chord}(H') \leq 4$, i.e., H' is a chordal bipartite graph. Notice that because $G[S]$ is disconnected, $|X| \geq 2$. Because the algorithm did not stop at Step 2, all the vertices of S are in the same component of $G[S \cup F]$. Therefore, $d_{H'}(x) \geq 1$ for $x \in X$. By Lemma 1, there is a weakly simplicial vertex $u \in X$, and we have that $N_{H'}(u) \neq \emptyset$.

We show that $|N_{H'}(u) \cap U| = 1$. Because U is a connected set of G and $G[S]$ is disconnected, F has a vertex that is adjacent to a vertex of the component $G[W_u]$ of $G[S]$. Hence, $N_{H'}(u) \cap U \neq \emptyset$. Since u is a weakly simplicial vertex of H', the neighborhoods of the vertices of $N_{H'}(u)$ form a chain under inclusion. Let $v \in N_{H'}(u) \cap U$ be a vertex with the inclusion maximal neighborhood. Suppose that $(N_{H'}(u) \cap U) \setminus \{v\} \neq \emptyset$ and $w \in (N_{H'}(u) \cap U) \setminus \{v\}$. By the choice of v, $N_{H'}(w) \subseteq N_{H'}(v)$. Hence, if w is adjacent in G to a vertex of W_x for some $x \in F$, then v is also adjacent to a vertex of W_x. Because U is a connected set of G, we obtain that $U' = U \setminus \{w\}$ is also a connected set. Since $S \subseteq U'$, U' is a vertex cover of G, i.e., U' is a connected vertex cover of G, but this contradicts the minimality of U. Therefore, $N_{H'}(u) \cap U = \{v\}$.

Let v be the unique vertex of $N_{H'}(u) \cap U$. On Step 4 we branch on v. We set $S' = S \cup \{v\}$, $F = F \setminus N_{H'}(u)$, and $H'' = H' - (N_{H'}(u) \setminus \{v\})$, and we call ENUMCVC-CHORD(H'', S', F'). It remains to observe that the algorithm outputs U for this call by induction.

To complete the correctness proof, it remains to notice that if $S = \emptyset$ and $F = V(G)$, then $S \subseteq U \subseteq S \cup F$ and also $H = G$ is an induced minor of G. Clearly, (i)–(iii) are fulfilled for these S, F and H. Hence, ENUMCVC-CHORD$(G, \emptyset, V(G))$ outputs U. As U is an arbitrary minimal connected vertex cover, the algorithm outputs all minimal connected vertex covers.

To obtain the upper bound on the number of minimal connected vertex covers of G, it is sufficient to upper bound the number of leaves of the search tree produced by the algorithm. The algorithm ENUMCVC-CHORD(H, S, F) branches on Steps 3 and 4. Let $k = |F|$. In Step 3 the algorithm is called recursively for $|F'| = k - 1$ on the first branch and for $|F'| \leq k - 2$ on the second one. Due to the decrease of the number of free vertices, the branching vector is $(1, 2)$, whose branching number is $\alpha \approx 1.6181$. To analyze the branching in Step 4 let $t = d_{H'}(u)$. Then the algorithm has t branches and in each the new instance has $|F'| = k - t$ free vertices. Hence the branching vector is $(t, t, \ldots, t)$ with $t \geq 1$ entries, which is known to have the maximum branching number $3^{1/3} < 1.6181$ when $t = 3$. Hence the number $L(n)$ of the leaves of the search tree is bounded by the recurrence $L(n) \leq L(n - 1) + L(n - 2)$ with $L(1) = 1$. It can be easily shown by induction that $L(n) \leq 1.6181^n$.

Because each step of ENUMCVC-CHORD can be done in polynomial time, the bound on the number of leaves of the search tree immediately implies that the algorithm runs in time $O(1.6181^n)$. $\square$

5   Distance-Hereditary Graphs

Another graph class for which we are able to give a tight upper bound on the maximum number of minimal connected vertex covers, is the class of distance-hereditary graphs. First, we need some additional notations. Let G be a connected graph and $u \in V(G)$. We denote the levels of the breadth-first search (BFS) of G starting at u by $L_0(u), \ldots, L_{s(u)}(u)$. Hence for all $i \in \{0, \ldots, s(u)\}$, $L_i(u) = \{v \in V(G) \mid \text{dist}_G(u, v) = i\}$. Clearly, the number of levels in this decomposition is $s(u) + 1$. For $i \in \{1, \ldots, s(u)\}$, we denote by $\mathcal{G}_i(u)$ the set of components of $G[L_i(u) \cup \ldots \cup L_{s(u)}(u)]$, and $\mathcal{G}(u) = \cup_{i=1}^{s(u)} \mathcal{G}_i(u)$. Let $H \in \mathcal{G}_i(u)$ and $B = N_G(V(H))$. Clearly, $B \subseteq L_{i-1}(u)$. We say that B is the *boundary* of H (in $L_{i-1}(u)$). We also say that $I = L_i(u) \cap V(H)$ is the *interface* of H (in $L_i(u)$). For $i \in \{0, \ldots, s(u) - 1\}$, $\mathcal{B}_i(u)$ is the set of boundaries in $L_i(u)$ of the graphs of $\mathcal{G}_{i+1}(u)$ and $\mathcal{B}(u) = \cup_{i=0}^{s(u)-1} \mathcal{B}_i(u)$. We will use the following result due to Bandelt and Mulder [1], and D'Atri and Moscarini [8].

Lemma 2 ([1,8]). *A connected graph G is distance-hereditary if and only if for every vertex $u \in V(G)$ and every $H \in \mathcal{G}(u)$ with boundary B, the following holds: $N_G(u) \cap V(H) = N_G(v) \cap V(H)$ for $u, v \in B$.*

For the main result of this section, we need the following structural properties of distance-hereditary graphs. All proofs of this section have been removed due to space restrictions.

Lemma 3. *Let G be a connected distance-hereditary graph and $u \in V(G)$. Then for any $B_1, B_2 \in \mathcal{B}(u)$, either $B_1 \cap B_2 = \emptyset$ or $B_1 \subseteq B_2$ or $B_2 \subseteq B_1$.*

Lemma 4. *Let G be a connected distance-hereditary graph, $u \in V(G)$ and let B be an inclusion minimal set of $\mathcal{B}(u)$. If B is an independent set of G, then the vertices of B are false twins.*

Observation 3. *Let G be a graph, and let $X, Y \subseteq V(G)$ be disjoint sets such that every vertex of X is adjacent to every vertex of Y. Then for every vertex cover U of G, $X \subseteq U$ or $Y \subseteq U$.*

Theorem 6. *The maximum number of minimal connected vertex covers of a distance-hereditary graph is at most $2 \cdot 3^{n/3}$, and these can be enumerated in time $O^*(3^{n/3})$.*

Proof outline. Let G be a distance-hereditary graph. If G has no edges, then the claim is trivial. Notice also that the removal of an isolated vertex does not influence connected vertex covers, and if G has two components with at least one edge each, G has no connected vertex cover. Hence, without loss of generality we can assume that G is a connected graph and $n \geq 2$.

Let $u \in V(G)$. We give an algorithm for enumerating all minimal connected vertex covers of G that contain u and upper bound the number of such covers. First we perform breadth-first search of G starting at u and construct $\mathcal{G}(u)$ and $\mathcal{B}(u)$. We construct the set $\mathcal{G}'(u) \subseteq \cup_{i=2}^{s(u)} \mathcal{G}_i(u)$ that contains all $H \in \mathcal{G}(u)$ such that the boundary B of H is an inclusion minimal set of $\mathcal{B}(u)$. Then we give a branching algorithm that we call $\textsc{EnumCVC-D-H}(R, S, F)$, which takes as input an induced subgraph R of G and two disjoint sets $S, F \subseteq V(G)$ such that $u \in S$, and outputs minimal connected vertex covers U of G such that $S \subseteq U \subseteq S \cup F$. To enumerate all minimal connected vertex covers U of G such that $u \in U$, we call $\textsc{EnumCVC-D-H}(G, \{u\}, V(G) \backslash \{u\})$. The algorithm, its correctness proof and the upper bound arguments for the running time and number of minimal connected vertex covers will be given in a full version. These arguments show that the search tree of the algorithm has at most $3^{n/3}$ leaves. This gives an upper bound on the number minimal connected vertex covers of G that contain u. It remains to observe that if $u_1 u_2 \in E(G)$, then every vertex cover of G contains u_1 or u_2. Hence, the number of minimal connected vertex covers of G is upper bounded by the sum of the numbers of minimal connected vertex covers that contain u_1 and u_2 respectively. Observing that every step of the algorithm takes polynomial time, this immediately implies that the algorithm runs in time $O^*(3^{n/3})$.   □

Proposition 1 shows that the upper bound is tight up to a constant factor.

6   Conclusions

The bounds that we have given for chordal graphs and distance-hereditary graphs are tight. While we can hope to improve the other upper bounds of this paper, we conjecture that they exceed $3^{n/3}$. It can be observed that for some classes of graphs of bounded chordality, the number of minimal connected vertex covers becomes polynomial.

Proposition 2. *The number of minimal connected vertex covers of a split graph G is at most n, and these can be enumerated in time $O(n + m)$. The number of minimal connected vertex covers of a cobipartite graph G is at most $n^2/4 + n$, and these can be enumerated in time $O(n^2)$.*

Proof. Notice that if X is a clique of a graph G, then for every vertex cover U of G, either $X \subseteq U$ or $|X \backslash U| = 1$. Let G be a split graph. Without loss of generality we can assume that G is a connected graph with at least two vertices. Let K, I be a partition of $V(G)$ in a clique K and an independent set I and assume that K is an inclusion maximal clique of G. If $V(G) = K$, then G has n minimal connected vertex covers $K \backslash \{v\}$ for $v \in V(G)$. Assume that $I \neq \emptyset$. Then K is a connected vertex cover of G. For $v \in K$, if U is a minimal connected vertex cover of G with $v \notin U$, $U = (K \backslash \{v\}) \cup N_G(v)$. It immediately implies that G has at most n connected vertex covers. Taking into account that a partition K, I can be found in time $O(n + m)$, it follows that the minimal connected vertex covers can be enumerated in time $O(n + m)$.

Let now G be a cobipartite graph. Again, we can assume without loss of generality that G is a connected graph with at least two vertices. If G is a complete graph, then G has n connected vertex covers. Assume that G is not a complete graph, and let K_1, K_2 be a partition of $V(G)$ into two cliques. Let U be a minimal connected vertex cover of G. If $K_1 \subseteq U$, then $U = V(G) \backslash \{v\}$ for $v \in K_2$, and there are at most $|K_2|$ sets of this type. Symmetrically, there are at most $|K_1|$ minimal connected vertex covers U with $K_2 \subseteq U$. If $K_1 \backslash U \neq \emptyset$ and $K_2 \backslash U \neq \emptyset$, then $U = V(G) \backslash \{u, v\}$ for $u \in K_1$ and $v \in K_2$, and G has at most $|K_1||K_2|$ such minimal connected vertex covers. We conclude that G has at most $|K_1| + |K_2| + |K_1||K_2| \leq n^2/4 + n$ minimal connected vertex covers. Clearly, these arguments can be applied to obtain an enumeration algorithm that runs in time $O(n^2)$. □

References

1. Bandelt, H., Mulder, H.M.: Distance-hereditary graphs. J. Comb. Theory, Ser. B **41**(2), 182–208 (1986)
2. Basavaraju, M., Heggernes, P., van't Hof, P., Saei, R., Villanger, Y.: Maximal induced matchings in triangle-free graphs. In: Kratsch, D., Todinca, I. (eds.) WG 2014. LNCS, vol. 8747, pp. 93–104. Springer, Heidelberg (2014)
3. Brandstädt, A., Le, V.B., Spinrad, J.P.: Graph classes: a survey. SIAM Monographs on Discrete Mathematics and Applications, Society for Industrial and Applied Mathematics (SIAM), Philadelphia, PA (1999)
4. Byskov, J.M.: Enumerating maximal independent sets with applications to graph colouring. Oper. Res. Lett. **32**(6), 547–556 (2004)
5. Couturier, J., Heggernes, P., van't Hof, P., Kratsch, D.: Minimal dominating sets in graph classes: combinatorial bounds and enumeration. Theor. Comput. Sci. **487**, 82–94 (2013)
6. Couturier, J.-F., Heggernes, P., van't Hof, P., Villanger, Y.: Maximum number of minimal feedback vertex sets in chordal graphs and cographs. In: Gudmundsson, J., Mestre, J., Viglas, T. (eds.) COCOON 2012. LNCS, vol. 7434, pp. 133–144. Springer, Heidelberg (2012)

7. Cygan, M.: Deterministic parameterized connected vertex cover. In: Fomin, F.V., Kaski, P. (eds.) SWAT 2012. LNCS, vol. 7357, pp. 95–106. Springer, Heidelberg (2012)
8. D'Atri, A., Moscarini, M.: Distance-hereditary graphs, steiner trees, and connected domination. SIAM J. Comput. 17(3), 521–538 (1988)
9. Escoffier, B., Gourvès, L., Monnot, J.: Complexity and approximation results for the connected vertex cover problem in graphs and hypergraphs. J. Discrete Algorithms 8(1), 36–49 (2010)
10. Fomin, F.V., Gaspers, S., Pyatkin, A.V., Razgon, I.: On the minimum feedback vertex set problem: exact and enumeration algorithms. Algorithmica 52(2), 293–307 (2008)
11. Fomin, F.V., Heggernes, P., Kratsch, D.: Exact algorithms for graph homomorphisms. Theor. Comp. Syst. 41(2), 381–393 (2007)
12. Fomin, F.V., Heggernes, P., Kratsch, D., Papadopoulos, C., Villanger, Y.: Enumerating minimal subset feedback vertex sets. Algorithmica 69(1), 216–231 (2014)
13. Fomin, F.V., Kratsch, D.: Exact Exponential Algorithms. Texts in Theoretical Computer Science. An EATCS Series. Springer, Heidelberg (2010)
14. Fomin, F.V., Villanger, Y.: Finding induced subgraphs via minimal triangulations. In: 27th International Symposium on Theoretical Aspects of Computer Science, STACS 2010, pp. 383–394 (2010)
15. Garey, M.R., Johnson, D.S.: Computers and Intractability: A Guide to the Theory of NP-Completeness. W. H. Freeman & Co., New York (1979)
16. Gaspers, S., Mnich, M.: Feedback vertex sets in tournaments. J. Graph Theory 72(1), 72–89 (2013)
17. Golovach, P.A., Heggernes, P., Kratsch, D., Saei, R.: Subset feedback vertex sets in chordal graphs. J. Discrete Algorithms 26, 7–15 (2014)
18. Golumbic, M.C.: Algorithmic Graph Theory and Perfect Graphs. Annals of Discrete Mathematics, vol. 57, 2nd edn. Elsevier Science B.V., Amsterdam (2004)
19. Hujtera, M., Tuza, Z.: The number of maximal independent sets in triangle-free graphs. SIAM J. Discrete Math. 6(2), 284–288 (1993)
20. Johnson, D.S., Papadimitriou, C.H., Yannakakis, M.: On generating all maximal independent sets. Inf. Process. Lett. 27(3), 119–123 (1988)
21. Lawler, E.: A note on the complexity of the chromatic number problem. Inf. Process. Lett. 5(3), 66–67 (1976)
22. Miller, R.E., Muller, D.: A problem of maximum consistent subsets. IBM Research Rep. RC-240. J. T. Watson Research Center, Yorktown Heights, New York, USA (1960)
23. Moon, J.W., Moser, L.: On cliques in graphs. Israel J. Math. 3, 23–28 (1965)
24. Pelsmajer, M.J., Tokazy, J., West, D.B.: New proofs for strongly chordal graphs and chordal bipartite graphs (2004). Unpublished manuscript
25. Tsukiyama, S., Ide, M., Ariyoshi, H., Shirakawa, I.: A new algorithm for generating all the maximal independent sets. SIAM J. Comput. 6(3), 505–517 (1977)

Fast Multiple Order-Preserving Matching Algorithms

Myoungji Han[1], Munseong Kang[2], Sukhyeun Cho[2], Geonmo Gu[1],
Jeong Seop Sim[2], and Kunsoo Park[1][(✉)]

[1] Department of Computer Science and Engineering,
Seoul National University, Seoul, Korea
{mjhan,gmgu,kpark}@theory.snu.ac.kr
[2] Department of Computer Science and Information Engineering,
Inha University, Incheon, Korea
{kmsung1125,csukhyeun}@inha.edu, jssim@inha.ac.kr

Abstract. Given a text T and a pattern P, the order-preserving matching problem is to find all substrings in T which have the same relative orders as P. Order-preserving matching has been an active research area since it was introduced by Kubica et al. [13] and Kim et al. [11]. In this paper we present two algorithms for the multiple order-preserving matching problem, one of which runs in sublinear time on average and the other in linear time on average. Both algorithms run much faster than the previous algorithms.

1 Introduction

Given a text T and a pattern P, the order-preserving matching problem is to find all substrings in T which have the same relative orders as P. For example, given $T = (10, 15, 20, 25, 15, 30, 20, 25, 30, 35)$ and $P = (35, 40, 30, 45, 35)$, P has the same relative orders as the substring $T' = (20, 25, 15, 30, 20)$ of T. In T' (resp. P), the first character 20 (resp. 35) is the second smallest number, the second character 25 (resp. 40) is the third smallest number, the third character 15 (resp. 30) is the smallest number, and so on. This problem has many practical applications such as stock price analysis and musical melody matching. It is naturally generalized to the problem of finding multiple patterns. The order-preserving matching for a single pattern will be called the *single order-preserving matching*, and one for multiple patterns the *multiple order-preserving matching*. In this paper we are concerned with the multiple order-preserving matching problem.

J.S. Sim—This work was supported by the National Research Foundation of Korea (NRF) grant funded by the Korea government (MSIP) (No. 2014R1A2A1A11050337).

K. Park—This research was supported by the Bio &Medical Technology Development Program of the NRF funded by the Korean government, MSIP (NRF-2014M3C9A3063541).

© Springer International Publishing Switzerland 2016
Z. Lipták and W.F. Smyth (Eds.): IWOCA 2015, LNCS 9538, pp. 248–259, 2016.
DOI: 10.1007/978-3-319-29516-9_21

Order-preserving matching was introduced by Kubica et al. [13] and Kim et al. [11], where Kubica et al. [13] defined order relations by order isomorphism of two strings, while Kim et al. [11] defined them explicitly by the sequence of rank values, which they called the *natural representation*. They both proposed $O(n + m \log m)$ time solutions for the single order-preserving matching based on the Knuth-Morris-Pratt algorithm, where n is the length of the text and m is the length of the pattern. Kim et al. [11] also proposed an $O(n \log M)$ time algorithm for the multiple order-preserving matching based on the Aho-Corasick algorithm, where M is the sum of lengths of all the patterns. Henceforth, there has been considerable research on the single and multiple order-preserving matching problems. For the single order-preserving matching, Cho et al. [4] proposed a method to apply the Boyer-Moore bad character rule to order-preserving matching by using the notion of q-grams. Chhabra and Tarhio [3] presented a more practical solution based on filtering. They first encoded input sequences into binary sequences and then applied standard string matching algorithms as a filtering method. Faro and Külekci [7] improved Chhabra and Tarhio's solution by using new encoding techniques which reduced the false positive rate of the filtering step. For the multiple order-preserving matching, Belazzougui et al. [2] theoretically improved the solution of Kim et al. [11] by replacing the underlying data structure by the van-Emde-Boas tree. They achieved randomized $O(n \cdot \min(\log \log n, \sqrt{\frac{\log r}{\log \log r}}, k))$ time for the search, where r is the length of the longest pattern and k is the number of patterns.

Order-preserving matching has been an active research area and many related problems have been studied such as order-preserving suffix trees [6] and order-preserving matching with k mismatches [8]. Kim et al. [10] extended the representations of order relations from binary relations to ternary relations. With their representations, one can modify earlier order-preserving matching algorithms to accommodate strings with duplicate characters, i.e., a number can appear more than once in a string.

In this paper, we present two new algorithms for the multiple order-preserving matching problem which are more efficient on average than the previously proposed algorithms. The algorithms are based on modifications of some conventional pattern matching algorithms such as Wu-Manber [14] and Karp-Rabin [9]. The first algorithm, called Algorithm I, uses the ideas of the Wu-Manber algorithm, and the second algorithm, called Algorithm II, uses the ideas of the Karp-Rabin algorithm and the encoding techniques of Chhabra and Tarhio [3] and Faro and Külekci [7] for fingerprinting. Algorithm I runs in $O(\frac{n}{m} \log M)$ time on average, where n is the length of the text, m is the length of the shortest pattern, and M is the sum of lengths of all the patterns. It is sublinear on average when $m > \log M$. Algorithm II runs in $O(n)$ time on average, assuming that M is polynomial with respect to m. In order to verify practical behaviors of our algorithms, we conducted experiments where the two algorithms were compared with the algorithms of Kim et al. [11] and Belazzougui et al. [2]. Experiments show that our algorithms run much faster in practice.

2   Problem Formulation

Let Σ denote a set of numbers such that a comparison of two numbers can be done in constant time, and let Σ^* denote the set of strings over the alphabet Σ. For a string $x \in \Sigma^*$, let $|x|$ denote the length of x. A string x is described by a sequence of characters $(x[1], x[2], ..., x[|x|])$. Let a substring $x[i..j]$ be $(x[i], x[i + 1], ..., x[j])$ and a prefix x_i be $x[1..i]$. For a character $c \in \Sigma$, let $rank_x(c) = 1 + |\{i : x[i] < c \text{ for } 1 \leq i \leq |x|\}|$.

We use the *natural representation* defined by Kim et al. [11] to compare order relations of two strings. The natural representation is equivalent to *order-isomorphism* defined by Kubica et al. [13], because the natural representation of two strings are identical if and only if they are order-isomorphic.

Definition 1 (Natural Representation [11]). *For a string x of length n, the natural representation is defined as*

$$Nat(x) = (rank_x(x[1]), rank_x(x[2]), ..., rank_x(x[n])).$$

For example, for $x = (30, 40, 30, 45, 35)$, the natural representation is $Nat(x) = (1, 4, 1, 5, 3)$. We will simply say that x matches y if $|x| = |y|$ and $Nat(x) = Nat(y)$.

Order-preserving matching can be defined in terms of the natural representation.

Definition 2 (Single Order-Preserving Matching [11]). *Given a text $T[1..n] \in \Sigma^*$ and a pattern $P[1..m] \in \Sigma^*$, P matches T at position i if $Nat(P) = Nat(T[i - m + 1..i])$. The single order-preserving matching is the problem of finding all the positions of T matched with P.*

Definition 2 is naturally generalized to the multiple order-preserving matching, formally defined in Definition 3.

Definition 3 (Multiple Order-Preserving Matching [11]). *Given a text $T[1..n] \in \Sigma^*$ and a set of patterns $\mathcal{P} = \{P_1, P_2, ..., P_k\}$ where $P_i \in \Sigma^*$ for $1 \leq i \leq k$, the multiple order-preserving matching is the problem of finding all the positions of T matched with any pattern in $\mathcal{P}$.*

There are two other representations in addition to the natural representation for comparing order relations of two strings: *prefix representation* and *nearest neighbor representation*. The prefix representation can be defined as a sequence of rank values of characters in prefixes.

Definition 4 (Prefix Representation [11]). *For a string x, the prefix representation is defined as*

$$Pre(x) = (rank_{x_1}(x[1]), rank_{x_2}(x[2]), ..., rank_{x_{|x|}}(x[|x|])).$$

For example, for $x = (30, 40, 30, 45, 35)$, the prefix representation is $Pre(x) = (1, 2, 1, 4, 3)$. We can compute $Pre(x)$ in time $O(|x| \log |x|)$ for general alphabet using the order-statistic tree [11], which is an augmented version of the red-black tree that supports general order-statistic operations on a dynamic set [5]. The time complexity can be reduced to $O(|x|)$ if the characters can be sorted in $O(|x|)$ time.

Lemma 1. *[4] For two strings x and y where $|x| = |y|$, if x matches y, then $Pre(x) = Pre(y)$.*

The prefix representation has an ambiguity between different strings if they include duplicate characters. For example, when $x = (10, 30, 20)$, and $y = (10, 20, 20)$, the prefix representations of both x and y are $(1, 2, 2)$, whereas their natural representations are different. Kim et al. defined a new representation called the *extended prefix representation* [10] for strings with duplicate characters. We omit the details here.

For the *nearest neighbor representation*, we define $LMax_x[i]$ and $LMin_x[i]$ as follows.

$$LMax_x[i] = \begin{cases} j & \text{if } x[j] = \max\{x[k] : x[k] \le x[i] \text{ for } 1 \le k \le i - 1\} \\ -\infty & \text{if no such } j, \end{cases}$$

$$LMin_x[i] = \begin{cases} j & \text{if } x[j] = \min\{x[k] : x[k] \ge x[i] \text{ for } 1 \le k \le i - 1\} \\ \infty & \text{if no such } j. \end{cases}$$

If there are multiple j's for $LMax_x[i]$ or $LMin_x[i]$, we choose the rightmost one.

Definition 5 (Nearest Neighbor Representation [10, 11]). *For a string x, the nearest neighbor representation is defined as*

$$NN(x) = \left(\binom{LMax_x[1]}{LMin_x[1]}, \binom{LMax_x[2]}{LMin_x[2]}, \cdots, \binom{LMax_x[|x|]}{LMin_x[|x|]} \right).$$

For example, for $x = (30, 40, 30, 45, 30)$, the nearest neighbor representation is as follows.

$$NN(x) = \left(\binom{-\infty}{\infty}, \binom{1}{\infty}, \binom{1}{1}, \binom{2}{\infty}, \binom{3}{3} \right).$$

For convenience, let $x[-\infty] = -\infty$, $x[\infty] = \infty$, $Nat(x)[-\infty] = 0$ and $Nat(x)[\infty] = |x| + 1$ for any string x. Then, $Nat(x)[LMax_x[i]] \le Nat(x)[i] \le Nat(x)[LMin_x[i]]$ holds for $1 \le i \le |x|$.

The time complexity for computing $NN(x)$ is $O(|x| \log |x|)$ [11]. Using this representation, we can check if two strings match in time linear to the size of the input, even when the strings have duplicate characters.

Lemma 2. *[4, 10, 11, 13] Given two strings x and y where $|x| = |y|$, assume $NN(x)$ is computed. Then we can determine whether x matches y in $O(|x|)$ time.*

3   Algorithm I

In this section, we present our first algorithm for the multiple order-preserving matching. Algorithm I is based on the Wu-Manber algorithm, which is widely used for multiple pattern matching. Algorithm I is divided into two steps: the preprocessing step and the searching step.

3.1   Preprocessing Step of Algorithm I

Let m be the length of the shortest pattern, and M be the sum of lengths of all the patterns. We consider only the first m characters of each pattern. Let $\mathcal{P}' = \{P_1', P_2', \cdots, P_k'\}$ where $P_i' = P_i[1..m]$ (this notation is provided only for clarity of exposition). In the preprocessing step, we build a SHIFT table and a HASH table based on $\mathcal{P}'$, which are analogous to those of the Wu-Manber algorithm. However, since we are looking for strings matched with patterns in terms of order-preserving matching, we have to consider the order representations of strings rather than strings themselves for comparison. Consider a block of length b on the text, where $b \leq m$. The SHIFT table determines the shift value based on the prefix representation of the given block. Given a block x, we define

$$l_x = \max\{j : Pre(P_i'[j-b+1..j]) = Pre(x) \text{ for } 1 \leq i \leq k, b \leq j \leq m\} \,.$$

That is, l_x means the position of the rightmost block in any $P_i' \in \mathcal{P}'$ which is likely to match x. Here, the term "is likely to" is used because $Pre(x) = Pre(y)$ does not necessarily mean that x matches y. For convenience, let $l_x = -\infty$ if there is no such block. Then, the SHIFT table is defined as

$$\text{SHIFT}[f(x)] = \min(m - l_x, m - b + 1) \,,$$

where $f(x)$ is a fingerprint mapping a block x to an integer used as an index to the SHIFT table. Using the factorial number system [12], we define $f(x)$ as

$$f(x) = \sum_{i=1}^{b} (Pre(x)[i] - 1) \cdot (i-1)! \,.$$

Note that $f(x)$ maps a block x into a unique integer within the range $[0..b! - 1]$ according to its prefix representation.

Figure 1(a) shows the SHIFT table when there are three patterns. Assume that $b = 3$. Consider the block $T[3..5]$. The rightmost block in $\mathcal{P}'$ whose prefix representation equals that of $T[3..5]$ is $P_1[2..4]$. The fingerprint $f(T[3..5])$ is 3. Thus, SHIFT[3] is $m - 4 = 1$. Note that in the figure, we can safely shift the patterns by 1.

The fingerprint is also used to index the HASH table. HASH[i] contains a pointer to the list of the patterns whose last block in $\mathcal{P}'$ is mapped to the fingerprint i. Figure 1(b) shows the HASH table with the same patterns.

To compute the values of the SHIFT table, we consider each pattern P_i' separately. For each pattern P_i', we compute the fingerprint of each block

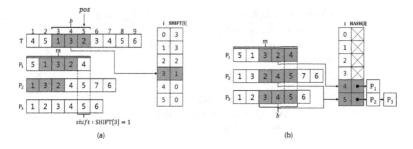

Fig. 1. SHIFT and HASH tables

$P_i'[j - b + 1..j]$ consecutively, and set the corresponding value of the SHIFT table to the minimum between its current value (initially set to $m - b + 1$) and $m - j$. In order to obtain the fingerprint of a block, we have to compute its prefix representation. Once we compute the fingerprint of the first block $Pre(P_i'[1..b])$ using the order-statistic tree, the tree contains the first b characters of P_i'. To compute the prefix representations of the subsequent blocks, we observe that we can compute $Pre(P_i'[j + 1..j + b])$ by taking advantage of the order-statistic tree containing characters of the previous block $P_i'[j..j + b - 1]$. Specifically, we erase $P_i'[j]$ from the tree and insert the new character $P_i'[j + b]$ into the tree. Inserting and deleting an element into the order-statistic tree is accomplished in $O(\log b)$ time since the tree contains $O(b)$ elements. Then we traverse the tree in $O(b)$ time to retrieve the prefix representation of the new block. We repeat this until we reach the last block. When we reach the last block, we map into the HASH table and add P_i into the corresponding list. The whole process is performed for all the patterns. Since there are $O(km)$ blocks, it takes $O(kmb)$ time to construct the SHIFT and HASH tables.

We also precompute the nearest neighbor representations of all the patterns, namely, $NN(P_i)$ for $1 \leq i \leq k$. They are used in the searching step for verifying whether patterns actually match the text. Using the order-statistic tree, they are computed in $O(M \log r)$ time, where r denotes the length of the longest pattern. As a result, the time complexity for the preprocessing step is $O(kmb + M \log r)$.

3.2 Searching Step of Algorithm I

In the searching step, we find all the positions of T matched with any pattern in $\mathcal{P}$. Figure 2 shows the pseudocode of Algorithm I. For the search, we slide a position pos along the text, reading a block of length b, $T[pos - b + 1..pos]$, and computing the corresponding fingerprint i. If $SHIFT(i) > 0$, then we shift the search window to $pos + SHIFT(i)$ and continue the search. Otherwise, $SHIFT(i) = 0$ and there may be a match. Thus we select the list of patterns in $HASH[i]$, and compare each pattern in the list with the text via the nearest neighbor representation. We call this process the verification step. We repeat this until we reach the end of the text.

Algorithm I($P = \{P_1, P_2, \cdots, P_k\}$, $T[1..n]$)

1: $m \leftarrow min_{1 \leq i \leq k}(|P_i|)$
2: Preprocess P and compute SHIFT, HASH, NN
3: $pos \leftarrow m$
4: **while** $pos \leq n$ **do**
5:     $i \leftarrow f(T[pos - b + 1..pos])$
6:     **if** SHIFT$[i] = 0$ **then**
7:         Verify each pattern in HASH$[i]$ via NN
8:             $pos \leftarrow pos + 1$
9:     **else**
10:            $pos \leftarrow pos + $ SHIFT$[i]$
11:     **end if**
12: **end while**

Fig. 2. The pseudocode of Algorithm I

3.3   Average Time for the Search of Algorithm I

We present a simplified analysis of the average running time for the searching step. For the analysis, we assume that there is no duplicate character in any b-length block in strings, i.e., any consecutive b characters in the text and patterns are distinct. Although this assumption restricts the generality of our problem, it is insignificant because: (1) a fairly large alphabet makes the case against the assumption very unlikely to happen; (2) even if it happens, the algorithm still works correctly without a significant impact on the performance in practice. We leave it as an open problem whether the average $O(\frac{n}{m} \log M)$ time can be derived when the strings are totally random, which is more complicated. Now, we assume that each distinct block appears randomly at a given position (i.e., with the same probability). Let us denote $\sigma = |\Sigma|$, then there are $_\sigma P_b$ different possible blocks and the probability of a block to appear is $1/_\sigma P_b$.

Lemma 3. *For two random blocks x and y, where $x, y \in \Sigma^b$ and each has no duplicate character, the probability that $Pre(x) = Pre(y)$ is $\frac{1}{b!}$.*

Recall that Algorithm I determines a shift value according to the prefix representation of a current block on the text.

Lemma 4. *The probability that a random block x leads to a shift value of j, $0 \leq j \leq m - b$, is at most $\frac{k}{b!}$.*

Proof. The necessary condition for the case that x leads to a shift value j is that there exists a pattern P_i' whose block ending at the position $m - j$ belongs to the prefix representation of x. Since there are k patterns, the probability of the necessary condition is $\frac{k}{b!}$. □

Lemma 5. *The expected value of a shift during the search is at least $(m - b + 1)\{1 - \frac{k(m-b+2)}{2b!}\}$.*

Proof. Since all the entries of SHIFT were initialized to $m - b + 1$, the expected value of a shift is $\geq \sum_{j=0}^{m-b} j \cdot \frac{k}{b!} + (m - b + 1)\{1 - (m - b + 1)\frac{k}{b!}\} = (m - b + 1)\{1 - \frac{k(m-b+2)}{2b!}\}$. $\qquad\qquad\square$

We set $b = 1.5 \log M / \log \log M$. Then, by Stirling's approximation [1], we can easily prove that $b! = 2^{b \log b + b \log e + O(\log b)} = \Omega(M)$, and thus the expected value of a shift is at least $\Theta(m)$. Consequently, the average number of iterations of the **while** loop during the search is bounded by $O(\frac{n}{m})$. At each iteration, we compute a fingerprint and the computation takes $O(b \log b) = O(\log M)$ time. Lemma 6 shows that the verification step at each iteration is accomplished in constant time on average.

Lemma 6. *The average cost of the verification step at each iteration is $O(1)$.*

Proof. At each iteration, the probability that a pattern P_i leads to the verification step is $\frac{1}{b!}$ and the cost for the verification for P_i is $O(|P_i|)$ by Lemma 2. Since there are k patterns, the expected cost of the verification step at each iteration is $\sum_{i=1}^{k} \frac{O(|P_i|)}{b!} = \frac{O(M)}{b!} = O(1)$. $\qquad\qquad\square$

Hence, the average time complexity of the searching step is roughly $O(\frac{n}{m} \log M)$.

4   Algorithm II

In this section, we present a simple algorithm that achieves average linear time for search. Algorithm II exploits the ideas of the Karp-Rabin algorithm and the encoding techniques of Chhabra and Tarhio [3] and Faro and Külekci [7] for fingerprinting.

4.1   Fingerprinting in Algorithm II

The Karp-Rabin algorithm is a practical string matching algorithm that makes use of fingerprints to find patterns, and it is important to choose a fingerprint function such that a fingerprint should be efficiently computed and efficiently compared with other fingerprints. Furthermore, the fingerprint function should be suitable for identifying strings in terms of order-preserving matching.

Given an m-length pattern P, Chhabra and Tarhio [3] encode the pattern into a binary sequence $\beta(P)$ of length $m - 1$, where

$$\beta(P)[i] = \begin{cases} 1 & \text{if } P[i] < P[i + 1] \\ 0 & \text{otherwise.} \end{cases}$$

We consider the fingerprint $\beta(P)$ as an $(m - 1)$-bit binary number. We can compute $\beta(P)$ in time $O(m)$.

As m increases, the fingerprint $\beta(P)$ may be too large to work with; we need at least $(m-1)$ bits to represent a fingerprint. To address this issue, we compute the fingerprints as residues modulo a prime number p. According to [9], we choose the prime p pseudorandomly in the range $[1..mn^2]$. With this choice, it is proved that the probability of a single false positive due to the modulo operation while searching is bounded by $2.53/n$, which is negligibly small for sufficiently large n [9].

Faro and Külekci [7] proposed more advanced encoding techniques such as q-NR and q-NO. Instead of comparing between only a pair of neighboring characters, they compared between a set of q characters for computing the relative position of a character. We can compute fingerprints using those techniques similarly to above. We implemented Algorithm II using three encoding techniques, including Chhabra and Tarhio's binary encoding [3], q-NR, and q-NO [7], for fingerprinting. In the following sections, we will describe the algorithm assuming the binary encoding.

4.2   Preprocessing Step of Algorithm II

Again, let $\mathcal{P}' = \{P_1', P_2', \cdots, P_k'\}$ be the set of m-length prefixes of the patterns. In the preprocessing step, we first compute $\beta(P_i')$ for $1 \leq i \leq k$ and build a HASH table. HASH[i] contains a pointer to the list of the patterns whose fingerprints equal i. We also compute NN(P_i) for $1 \leq i \leq k$. In total, the preprocessing step takes $O(M \log r)$ time, where r is the length of the longest pattern. Figure 3(a) shows the HASH table when there are three patterns. We use a prime $p = 7$ in the example.

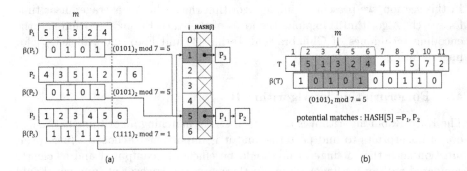

(a)   (b)

Fig. 3. (a) The HASH table (b) An example of the search. For the window $T[2..6]$, the corresponding fingerprint is 5. We check HASH[5], which has P_1, P_2 as elements, and thus verify them via NN.

4.3   Searching Step of Algorithm II

In the searching step, we scan the text T while iteratively computing fingerprints of the successive windows of size m. Figure 4 shows the pseudocode of Algorithm II. We slide a search window $T[i..i+m-1]$ along the text, computing the corresponding fingerprint β. If the list pointed by HASH[β] is not empty, we compare

Algorithm II$(P = \{P_1, P_2, \cdots, P_k\}, T[1..n])$
1: $m \leftarrow min_{1 \leq i \leq k}(|P_i|)$
2: Preprocess P and compute HASH, NN
3: $pos \leftarrow m$
4: **for** $i = 1$ for $n - m + 1$ **do**
5:      $\beta = \beta(T[i..i + m - 1]) \bmod p$
6:      Verify each pattern in HASH$[\beta]$ via NN
7: **end for**

Fig. 4. The pseudocode of Algorithm II

each pattern in the list with the text via its nearest neighbor representation. We call this process the verification step. We repeat this until we reach the end of the text. Figure 3(b) shows an example of the searching step.

4.4   Average Time for the Search of Algorithm II

At each iteration of the **for** loop, we compute the fingerprint β of the search window. Let us denote $\beta_i = \beta(T[i..i + m - 1]) \bmod p$, which is the fingerprint of the i-th search window. We can compute β_1 in time $O(m)$. To compute the fingerprints for the subsequent windows, we observe that we can compute β_{i+1} from β_i using Horner's rule [5], since

$$\beta_{i+1} = (2(\beta_i - H \cdot \beta(T)[i]) + \beta(T)[i + m]) \bmod p \,,$$

where $H = 2^{m-2} \pmod{p}$ is a precomputed value. It is clear that this calculation is done in constant time.

Now, we analyze the time spent to perform the verification step. We assume that the numbers in the text and patterns are statistically independent and uniformly at random. The verification is performed when there is a match between encoded binary strings of the text and patterns. The probability that a 1 appears at a position of an encoded string is $q = (\sigma^2/2 - \sigma/2)/\sigma^2 = (\sigma - 1)/2\sigma$. So the probability of a character match [3] is

$$s = q^2 + (1 - q)^2 = \frac{1}{2} + \frac{1}{2\sigma^2} \,.$$

Since the odd positions of an encoded string are mutually independent, we can (upper) bound the probability of a match between two encoded strings by $s^{(m-1)/2}$. Note that $s \leq 5/8$ for $\sigma \geq 2$.

Lemma 7. *When M is polynomial with respect to m, the average cost of the verification step during the search is $O(1)$.*

Hence, the average time complexity of the searching step is $O(n)$, when M is polynomial with respect to m.

5   Experiments

In order to verify the practical behaviors of our algorithms, we tested them against the previous algorithms based on the Aho-Corasick algorithm: Kim et al.'s [11], and Belazzougui et al.'s [2].[1] Kim et al.'s algorithm is denoted by *KEF*, Belazzougui et al.'s by *BPR*, and Algorithm I by *Alg1*. Algorithm II is denoted by *Alg2*, followed by a notation of the encoding technique adopted for fingerprinting. Specifically, *Alg2_Bin* refers to Algorithm II with Chhabra and Tarhio's binary encoding [3], and *Alg2_NR2* (resp. *Alg2_NO2*) refers to Algorithm II with the q-NR (resp. q-NO) encoding of Faro and Külekci [7] where we set $q = 2$. All algorithms were implemented in C++ and run on a Debian Linux 7(64bit) with Intel Xeon X5672 processor and 32 GB RAM. During the compilation, we used the O3 optimization option.

We tested for a random text T of length $n = 10^6$ searched for $k = 10, 50, 100$ random patterns of length $m = 5, 10, 20, 50, 100$, respectively. All the texts and patterns were selected randomly from an integer alphabet $\Sigma = \{1, 2, \cdots, 1000\}$ (we tested for varying alphabet sizes, but we didn't observe sensible differences in the results). For each combination of k and m, we randomly selected a text and patterns, and then ran each algorithm. We performed this 10 times and measured the average time for the searching step. Table 1 shows the results.

Table 1. Average search times with different values for k and m

k	m	KEF	BPR	$Alg1$	$Alg2\_Bin$	$Alg2\_NR2$	$Alg2\_NO2$
10	5	527.3	1215.1	274.8	**107.6**	164.8	186.5
	10	544.3	1258.2	216.9	**91.5**	148.8	197.6
	20	557.1	1254.8	286.5	**88.4**	155.4	194.8
	50	556.2	1213	**51.1**	65.4	116.7	203
	100	561.7	1244.8	**56.2**	70.4	123.9	206.9
50	5	598	1227.2	647.8	234.1	**215.3**	310
	10	573.8	1238.6	269.6	**100.5**	152.3	194.6
	20	562.9	1244.2	308.6	**114.8**	187.1	216.4
	50	570.7	1239.8	313.5	**113.8**	184.5	226.2
	100	587.6	1271.8	**55.4**	86.1	150.6	227.6
100	5	569	1291.1	674.4	395.6	386.3	**307.3**
	10	629	1304.3	522.4	**81.9**	100.3	150.5
	20	589	1250.2	498.4	**102.6**	164	205.1
	50	605.3	1259.9	103.3	**86**	184.8	225.7
	100	588.9	1247.2	73.2	**53.8**	182	227.5

---

[1] For the implementation of the van-Emde-Boas tree used in [2], we used the source code publicly available at https://code.google.com/p/libveb/.

When m is relatively small, *Alg2* family are the best among the algorithms. As m increases, however, *Alg1* outperforms them. This is due to the increase of the average shift value during the search. The reason that the average shift value increases is that since we set $b = 1.5 \log M / \log \log M$, the block size increases as m increases, and thus the probability that a block appears in the patterns decreases. One thing to note is that as k increases, the point of m where *Alg1* becomes for the first time faster than the *Alg2* family increases. We attribute this to the fact that as k increases, a block appears more often in the patterns, which leads to lower shift values.

References

1. Abramowitz, M., Stegun, I.A.: Handbook of Mathematical Functions. Dover, New York (1972)
2. Belazzougui, D., Pierrot, A., Raffinot, M., Vialette, S.: Single and multiple consecutive permutation motif search. In: Cai, L., Cheng, S.-W., Lam, T.-W. (eds.) Algorithms and Computation. LNCS, vol. 8283, pp. 66–77. Springer, Heidelberg (2013)
3. Chhabra, T., Tarhio, J.: Order-preserving matching with filtration. In: Gudmundsson, J., Katajainen, J. (eds.) SEA 2014. LNCS, vol. 8504, pp. 307–314. Springer, Heidelberg (2014)
4. Cho, S., Na, J.C., Park, K., Sim, J.S.: A fast algorithm for order-preserving pattern matching. Inf. Process. Lett. **115**(2), 397–402 (2015)
5. Cormen, T.H., Leiserson, C.E., Rivest, R.L., Stein, C.: Introduction to Algorithms, 2nd edn. MIT Press, Cambridge (2001)
6. Crochemore, M., et al.: Order-preserving incomplete suffix trees and order-preserving indexes. In: Kurland, O., Lewenstein, M., Porat, E. (eds.) SPIRE 2013. LNCS, vol. 8214, pp. 84–95. Springer, Heidelberg (2013)
7. Faro, S., Külekci, O.: Efficient Algorithms for the Order Preserving Pattern Matching Problem. arXiv preprint arxiv:1501.04001 (2015)
8. Gawrychowski, P., Uznański, P.: Order-preserving pattern matching with k mismatches. In: Kulikov, A.S., Kuznetsov, S.O., Pevzner, P. (eds.) CPM 2014. LNCS, vol. 8486, pp. 130–139. Springer, Heidelberg (2014)
9. Karp, R.M., Rabin, M.O.: Efficient randomized pattern-matching algorithms. IBM J. Res. Dev. **31**(2), 249–260 (1987)
10. Kim, J., Amir, A., Na, J.C., Park, K., Sim, J.S.: On representations of ternary order relations in numeric strings. In: ICABD, pp. 46–52. Springer (2014)
11. Kim, J., Eades, P., Fleischer, R., Hong, S., Iliopoulos, C.S., Park, K., Puglisi, S.J., Tokuyama, T.: Order-preserving matching. Theoret. Comput. Sci. **525**, 68–79 (2014)
12. Knuth, D.E.: The Art of Computer Programming, vol. 2. Seminumerical Algorithms. Addison Wesley, Reading (1998)
13. Kubica, M., Kulczyński, T., Radoszewski, J., Rytter, W., Waleń, T.: A linear time algorithm for consecutive permutation pattern matching. Inf. Process. Lett. **113**(12), 430–433 (2013)
14. Wu, S., Manber, U.: A fast algorithm for multi-pattern searching. Technical report. TR-94-17, Department of Computer Science, University of Arizona (1994)

Minimum Degree Conditions
and Optimal Graphs for Completely
Independent Spanning Trees

Toru Hasunuma[(✉)]

Institute of Socio-Arts and Sciences, Tokushima University,
1-1 Minamijosanjima, Tokushima 770–8502, Japan
hasunuma@tokushima-u.ac.jp

Abstract. Completely independent spanning trees $T_1, T_2, \ldots, T_k$ in a graph G are spanning trees in G such that for any pair of distinct vertices u and v, the k paths in the spanning trees between u and v mutually have no common edge and no common vertex except for u and v. The concept finds applications in fault-tolerant communication problems in a network. Recently, it was shown that Dirac's condition for a graph to be hamiltonian is also a sufficient condition for a graph to have two completely independent spanning trees. In this paper, we generalize this result to three or more completely independent spanning trees. Namely, we show that for any graph G with $n \geq 7$ vertices, if the minimum degree of a vertex in G is at least $n - k$, where $3 \leq k \leq \frac{n}{2}$, then there are $\lfloor \frac{n}{k} \rfloor$ completely independent spanning trees in G. Besides, we improve the lower bound of $\frac{n}{2}$ on the Dirac's condition for completely independent spanning trees to $\frac{n-1}{2}$ except for some specific graph. Our results are theoretical ones, since these minimum degree conditions can be applied only to a very dense graph. We then present constructions of symmetric regular graphs which include optimal graphs with respect to the number of completely independent spanning trees.

1 Introduction

Independent spanning trees rooted at a vertex r of a graph G are spanning trees of G such that for any vertex v different from r, the paths from r to v in the spanning trees are pairwise openly disjoint, i.e., the paths mutually have no common edge and no common vertex except for r and v. There is a famous conjecture on independent spanning trees: There are k independent spanning trees rooted at any vertex of any k-connected graph. This conjecture was proved for $k \leq 4$ (e.g., see [3]) and also for planar graphs [10] but it still remains open for $k \geq 5$. Besides, independent spanning trees find applications in fault-tolerant broadcasting problems in communication networks. Motivated by this application, independent spanning trees have been studied not only from a theoretical point of view but also from a practical point of view. That is, constructions of

This work was supported by JSPS KAKENHI 25330015.

© Springer International Publishing Switzerland 2016
Z. Lipták and W.F. Smyth (Eds.): IWOCA 2015, LNCS 9538, pp. 260–273, 2016.
DOI: 10.1007/978-3-319-29516-9_22

independent spanning trees have been studied for many graph classes related to interconnection networks (e.g., see [14]). There are three variations for independent spanning trees: edge version, directed graph version, and directed edge version (e.g., see [9]). Another variation of independent spanning trees is a stronger version. The notion of completely independent spanning trees strengthens the notion of independent spanning trees [5]; completely independent spanning trees are spanning trees of G such that for any pair of distinct vertices u and v, the paths from u to v in the spanning trees are pairwise openly disjoint. From a practical point of view, completely independent spanning trees have an advantage that we do not need to reconstruct them even if the root vertex which has the data to be broadcasted is changed another vertex, since completely independent spanning trees are independent spanning trees rooted at every vertex of G. From a theoretical point of view, unlike independent spanning trees, it was shown that there is no direct relationship between connectivity and the number of completely independent spanning trees: for any given $k \geq 2$, there exists a k-connected graph which has no two completely independent spanning trees [13]. On the other hand, it is known that there are two completely independent spanning trees in any maximal 4-connected planar graph. It remains unknown whether such a result can be generalized to (non-maximal) 4-connected planar graphs. From a computational point of view, it has been proved that the problem of finding two completely independent spanning trees in a general graph is NP-hard [6]. For interconnection networks such as de Bruijn and Kautz networks, torus networks, chordal rings and WK-recursive networks, the existence of completely independent spanning trees have been investigated [5,7,8,12]. Recently, a sufficient condition for a graph to have two completely independent spanning trees was given in [1]. The statement of the sufficient condition is the same as the famous Dirac's condition, i.e., $\delta(G) \geq \frac{n}{2}$, for a graph with n vertices to be hamiltonian, where $\delta(G)$ denotes the minimum degree of a vertex in G. Afterwards, the result was strengthened; the Ore's condition for a graph to be hamiltonian was also shown to be a sufficient condition for a graph to have two completely independent spanning trees [4].

Motivated by Dirac's condition for a graph to have two completely independent spanning trees, in this paper, we generalize the result for three or more completely independent spanning trees, and also improve the lower bound of $\frac{n}{2}$ on the Dirac's condition. Namely, we show that for any graph G with $n \geq 7$ vertices, if $\delta(G) \geq n - k$, where $3 \leq k \leq \frac{n}{2}$, then there are $\lfloor \frac{n}{k} \rfloor$ completely independent spanning trees in G, and also show that for any graph G with $n \geq 8$ vertices, if $\delta(G) \geq \frac{n-1}{2}$, then there are two completely independent spanning trees in G, except for a specific graph (which will be defined in Sect. 4). Besides, our first result for the case that $k = 3$ improves a recent result [2] on the minimum degree condition for completely independent spanning trees except for the case that $n = 6$. Our results are theoretical ones, since the results can be applied only to a very dense graph which seems to be impractical for communication networks. We then present constructions of symmetric regular graphs which include optimal graphs with respect to the number of completely independent spanning trees. Our proposed graphs have several nice properties such as regularity, symmetry, and fault-tolerance. Thus, it might be useful for a model of a communication network.

This paper is organized as follows. Section 2 presents definitions, notation, and characterizations used in the paper. We generalize and improve the Dirac's condition for completely independent spanning trees in Sects. 3 and 4, respectively. Constructions of symmetric regular graphs which include optimal graphs for completely independent spanning trees are presented in Sect. 5. Section 6 finally concludes the paper.

2   Preliminaries

Let $G = (V, E)$ be a graph. Throughout the paper, a graph means a simple undirected graph unless stated otherwise. The *complement* $\bar{G}$ is the graph with $V(G)$ and in which two distinct vertices u and v are adjacent if and only if they are not adjacent in G. For a bipartite graph $B(V_1, V_2)$ with partite sets V_1, V_2, the *bipartite complement* $\bar{B}(V_1, V_2)$ is the bipartite graph with partite sets V_1, V_2 such that two vertices $u \in V_1$ and $v \in V_2$ are adjacent if and only if they are not adjacent in $B(V_1, V_2)$. The disjoint union of graphs G_1 and G_2 where $V(G_1) \cap V(G_2) = \emptyset$ is denoted by $G_1 \cup G_2$. If two graphs G_1 and G_2 are isomorphic, then we write $G_1 \cong G_2$. A *component* of G is a maximal connected subgraph of G. In particular, a *tree-component* of G is a component which has no cycle. A component of G which is isomorphic to a graph H is called an *H-component* of G. The degree of a vertex in G is denoted by $\deg_G(v)$ and the minimum (respectively, maximum) degree of a vertex in G is denoted by $\delta(G)$ (respectively, $\Delta(G)$). We denote by $N_G(v)$ the neighborhood of a vertex v in G. For $S \subseteq V(G)$, we denote by $\langle S \rangle_G$ (or simply $\langle S \rangle$ if there is no confusion) the subgraph of G induced by S. For $S_1, S_2 \subset V(G)$ such that $S_1 \cap S_2 = \emptyset$, $B_G(S_1, S_2)$ (or simply $B(S_1, S_2)$) is the bipartite subgraph of G with partite sets S_1 and S_2, i.e., $V(B_G(S_1, S_2)) = S_1 \cup S_2$ and $E(B_G(S_1, S_2)) = \{uv \in E(G) \mid u \in S_1, v \in S_2\}$. Two edges in G are *independent* if they have no common end-vertex. An *isolated vertex* of G is a vertex with degree zero in G, while a *leaf* of G is a vertex with degree one in G. An *internal vertex* of a tree T is a vertex which is not a leaf. A *unicyclic graph* is a connected graph which has exactly one cycle. A path, a cycle, and a complete graph with n vertices are denoted by P_n, C_n, and K_n, respectively. Besides, a complete bipartite graph with m vertices in one partite set and n vertices in the other partite set is denoted by $K_{m,n}$. A tree T is called a *caterpillar* if the graph obtained from T by deleting every leaf is a path. For a directed graph D, an arc (u, v) of D is said to be incident from u to v. The *outdegree* of a vertex u in D is the number of arcs incident from u. A *cycle-rooted directed tree* is a weakly connected directed graph whose every vertex has outdegree one. Given a cycle-rooted directed tree, by replacing each arc with a corresponding edge, we have a unicyclic graph. Conversely, given a unicyclic graph, by replacing each edge with an appropriate arc, a cycle-rooted directed tree can be obtained.

In this paper, we show the existence of completely independent spanning trees but we do not directly use the definition since it is a tedious matter to check the openly disjointness for each pair of distinct vertices. Instead, we rely on characterizations for completely independent spanning trees and a graph which has completely independent spanning trees. We briefly explain such characterizations with the essence for their proofs. Let $T_1, T_2, \ldots, T_k$ be spanning trees of a graph G. A characterization of completely independent spanning tree was given in [5]: $T_1, T_2, \ldots, T_k$ are completely independent spanning trees if and only if they are pairwise edge-disjoint and for any vertex v of G, there is at most one spanning tree T_i such that $\deg_{T_i}(v) \geq 2$. The sufficiency part is almost clear, while the necessity part can be shown by contradiction. From this characterization, we can see that the sets $S_1, S_2, \ldots, S_k$ of internal vertices of $T_1, T_2, \ldots, T_k$, respectively, are pairwise disjoint. Besides, $\langle S_i \rangle$ is connected since it is obtained from T_i by deleting all the leaves. Let R be the set of vertices which are leaves in every T_i. If R is not an empty set, then we add every element of R to S_1. Note that $\langle S_1 \rangle$ is still connected. In this way, $V(G)$ is partitioned into $S_1, S_2, \ldots, S_k$ each of which induces a connected subgraph of G. Consider two sets S_i and S_j where $1 \leq i < j \leq k$. Since every vertex of S_i is a leaf of T_j, any vertex of S_i is adjacent to at least one vertex of S_j in $B(S_i, S_j)$. Similarly, since every vertex of S_j is a leaf of T_i, any vertex of S_j is adjacent to at least one vertex of S_i in $B(S_i, S_j)$. By directing each edge $e = uv$ of $B(S_i, S_j)$ where $u \in S_i$ and $v \in S_j$ from u to v (respectively, from v to u) if $e \in E(T_j)$ (respectively, $e \in E(T_i)$), a directed graph $D(S_i, S_j)$ is obtained. Since every vertex of $D(S_i, S_j)$ has outdegree one, every (weakly connected) component of $D(S_i, S_j)$ is a cycle-rooted directed tree. This means that every component of $B(S_i, S_j)$ is not a tree. Conversely, suppose that $V(G)$ is partitioned into subsets $S_1, S_2, \ldots, S_k$ such that every $\langle S_i \rangle$ is connected and for any $1 \leq i < j \leq k$, $B(S_i, S_j)$ contains no tree-component. Then, every component of $B(S_i, S_j)$ has a cycle and thus contains a unicyclic graph. A cycle-rooted directed tree is obtained from a unicyclic graph by appropriately directing each edge. Consequently, we can obtain a directed bipartite subgraph $D'(S_i, S_j)$ in which every vertex has outdegree one. Combining all the edges corresponding arcs incident to a vertex of S_i in $D'(S_i, S_j)$ for $j \neq i$ and a spanning tree of $\langle S_i \rangle$, we have a spanning tree T_i of G. It is not difficult to see that $T_1, T_2, \ldots, T_k$ are completely independent spanning trees. From these observation, a characterization of graphs with k completely independent spanning trees was given in [1]: G has k completely independent spanning trees if and only if $V(G)$ can be partitioned into subsets $S_1, S_2, \ldots, S_k$ such that $\langle S_i \rangle$ is connected for all $1 \leq i \leq k$ and for any $1 \leq i < j \leq k$, $B(S_i, S_j)$ contains no tree-component.

3   Minimum Degree Conditions

Lemma 1. *Let G be a graph with $n \geq 7$ vertices. If $\delta(G) \geq n - 3$, then there are $\lfloor \frac{n}{3} \rfloor$ completely independent spanning trees in G.*

Proof: First consider the case that $n \geq 9$. Since $\Delta(\bar{G}) \leq 2$, $\bar{G}$ consists of a disjoint union of cycles and paths. Let $\bar{G} = (\cup_{1 \leq i \leq p} C_{\ell_i}) \cup (\cup_{p < i \leq p+q} P_{\ell_i})$ such that $\ell_i \geq \ell_j$ for $1 \leq i < j \leq p$. Also, let $C_{\ell_i} = (v_{i,1}, v_{i,2}, \ldots, v_{i,\ell_i}, v_{i,1})$ for $1 \leq i \leq p$ and $P_{\ell_i} = (v_{i,1}, v_{i,2}, \ldots, v_{i,\ell_i})$ for $p < i \leq p+q$. Order lexicographically the vertices of G: $v_{1,1}, v_{1,2}, \ldots, v_{1,\ell_1}, v_{2,1}, v_{2,2}, \ldots, v_{2,\ell_2}, \ldots, v_{p+q,1}, v_{p+q,2}, \ldots, v_{p+q,\ell_{p+q}}$. We call an edge of $\bar{G}$ joining consecutive (respectively, non-consecutive) vertices in this ordering a forward (respectively, backward) edge. Based on the ordering, we define $\sigma : V(G) \mapsto \{0, 1, \ldots, n-1\}$ as $\sigma(v_{i,j}) = \sum_{1 \leq t < i} \ell_t + j - 1$. Partition $V(G)$ into $\lfloor \frac{n}{3} \rfloor$ parts $V_0, V_1, \ldots, V_{\lfloor \frac{n}{3} \rfloor - 1}$ where $V_t = \{v \in V(G) \mid \sigma(v) \bmod \lfloor \frac{n}{3} \rfloor = t\}$ for $0 \leq t < \lfloor \frac{n}{3} \rfloor$. For each t, the induced subgraph $\langle V_t \rangle_{\bar{G}}$ of $\bar{G}$ has no forward edge. Besides, any two backward edges are independent. Thus, each induced subgraph $\langle V_t \rangle_G$ of G is connected. Let $s, t \in \{0, 1, \ldots, \lfloor \frac{n}{3} \rfloor - 1\}$. If $s - t \not\equiv 1$ (mod $\lfloor \frac{n}{3} \rfloor$), then the bipartite complement $\bar{B}(V_s, V_t)$ has no forward edge and it holds that $\delta(B(V_s, V_t)) \geq 2$, i.e., $B(V_s, V_t)$ has no tree-component. Suppose that $s - t \equiv 1$ (mod $\lfloor \frac{n}{3} \rfloor$). If $\Delta(\bar{B}(V_s, V_t)) \leq 1$, $B(V_s, V_t)$ has no tree-component. Consider the case that $\Delta(\bar{B}(V_s, V_t)) = 2$. In this case, $\bar{B}(V_s, V_t)$ has a backward edge xy and two forward edges wx, yz where $x, z \in V_s$ and $w, y \in V_t$. Replace V_s and V_t with $V_s' = (V_s - \{x\}) \cup \{y\}$ and $V_t' = (V_t - \{y\}) \cup \{x\}$, respectively. Then, $\Delta(\bar{B}(V_s', V_t')) = 1$, i.e., $B(V_s', V_t')$ has no tree-component. Although $\langle V_s' \rangle_{\bar{G}}$ has the edge yz, $\langle V_s' \rangle_G$ is still connected since there is no edge adjacent to yz in $\langle V_s' \rangle_{\bar{G}}$. Similarly, $\langle V_t' \rangle_G$ is connected. For any V_r ($r \neq s, t$), this manipulation does not essentially change the structures of $B(V_s, V_r)$ and $B(V_t, V_r)$, i.e., $B(V_s, V_r) \cong B(V_s', V_r)$ and $B(V_t, V_r) \cong B(V_t', V_r)$. Thus, both $B(V_s', V_r)$ and $B(V_t', V_r)$ have no tree-component. There may exist another pair of consecutive parts V_p, V_q with $\Delta(\bar{B}(V_p, V_q)) = 2$. In such a case, by doing the similar manipulation for V_s, V_t, we can obtain a partition of $V(G)$ which induces $\lfloor \frac{n}{3} \rfloor$ completely independent spanning trees.

Let $7 \leq n \leq 8$. Since $\delta(G) \geq n - 3 > \frac{n}{2}$, G has a Hamiltonian cycle $C = (v_1, v_2, \ldots, v_n)$. Let $V_1 = \{v_1, v_2, v_3, v_4\}$ and $V_2 = V(G) - V_1$. Note that there is no isolated vertex in $B(V_1, V_2)$. If $n = 8$, then $\delta(B(V_1, V_2)) \geq 2$ and there is no tree-component in $B(V_1, V_2)$. Suppose that $n = 7$. Then, any vertex in V_2 is not a leaf of $B(V_1, V_2)$. Assume that there exists a tree-component in $B(V_1, V_2)$. Let x and y be leaves of $B(V_1, V_2)$ in V_1. Since $\deg_{\langle V_1 \rangle}(x) = \deg_{\langle V_1 \rangle}(y) = 3$, x and y are not cut-vertices of $\langle V_1 \rangle$. Thus, $\langle V_1 - \{x\} \rangle$ is connected. Clearly, $\langle V_2 \cup \{x\} \rangle$ is also connected. Besides, $B(V_1 - \{x\}, V_2 \cup \{x\})$ contains at most one leaf, since every vertex except for the neighbor of x in V_2 cannot be a leaf of $B(V_1 - \{x\}, V_2 \cup \{x\})$. Therefore, there is no tree-component in $B(V_1 - \{x\}, V_2 \cup \{x\})$. Hence, there are two completely independent spanning trees in G. □

Theorem 1. *Let G be a graph with $n \geq 7$ vertices. Let $3 \leq k \leq \frac{n}{2}$. If $\delta(G) \geq n - k$, then there are $\lfloor \frac{n}{k} \rfloor$ completely independent spanning trees in G.*

Proof: The case that $k = 3$ follows from Lemma 1. Suppose that $k \geq 4$. Let $C = (v_0, v_1, \ldots, v_{n-1})$ be a Hamiltonian cycle of G. Let $r = n - k \lfloor \frac{n}{k} \rfloor$, $p = \lfloor \frac{r}{\lfloor \frac{n}{k} \rfloor} \rfloor$, and $r' = r - p \lfloor \frac{n}{k} \rfloor$. Based on C, we partition $V(G)$ into $\lfloor \frac{n}{k} \rfloor$ parts each of which has at least $k + p$ vertices as follows. Let $V_i = \{v_{(k+p)i}, v_{(k+p)i+1}, \ldots,$

$v_{(k+p)(i+1)-1}\}$ for $i = 0, 1, \ldots, \lfloor \frac{n}{k} \rfloor - r' - 1$ and $V_{\lfloor \frac{n}{k} \rfloor - r' + j} = \{v_{(k+p)(\lfloor \frac{n}{k} \rfloor - r' + j) + j},$
$v_{(k+p)(\lfloor \frac{n}{k} \rfloor - r' + j) + j + 1}, \ldots, v_{(k+p)(\lfloor \frac{n}{k} \rfloor - r' + j) + j + k + p}\}$ for $j = 0, 1, \ldots, r' - 1$. That
is, each V_i consists of vertices with consecutive labels such that $|V_i| = k + p$ for
$0 \le i < \lfloor \frac{n}{k} \rfloor - r'$ and $|V_{i+1}| = k + p + 1$ for $\lfloor \frac{n}{k} \rfloor - r' \le i < \lfloor \frac{n}{k} \rfloor$. Thus, each
$\langle V_i \rangle$ is connected. Note that $\delta(B(V_i, V_j)) \ge p + 1$ for any two parts V_i and V_j since
$\delta(G) \ge n - k$. Thus, if $p \ge 1$, then $B(V_i, V_j)$ has no tree-component. Therefore,
it is sufficient to consider the case that $p = 0$, and in what follows, we assume
that $p = 0$. Besides, notice that if a vertex x in V_i is a leaf of $B(V_i, V_j)$ with the
neighbor x' in V_j, then x is adjacent to every vertex of G except for the vertices
in $V_j - \{x'\}$. Suppose that $B(V_i, V_j)$ contains a tree-component. In what follows,
we present manipulations for exchanging a vertex in V_i and a vertex in V_j so that
for the resultant parts V_i' and V_j', both $\langle V_i' \rangle$ and $\langle V_j' \rangle$ are connected, $B(V_i', V_j')$ has
no tree-component, and $B(V_i', V_t)$ and $B(V_j', V_t)$ have no tree-component for any
$t \in \{0, 1, \ldots, \lfloor \frac{n}{k} \rfloor - 1\} - \{i, j\}$. Applying iteratively the manipulations for each
$B(V_i, V_j)$ which has a tree-component, we finally obtain a partition of G which
induces $\lfloor \frac{n}{k} \rfloor$ completely independent spanning trees.

Let a and b be the numbers of leaves of $B(V_i, V_j)$ in V_i and V_j, respectively.
Without loss of generality, we may assume that $a \ge b$.

Case 1: $a \ge 2$ and $b \ge 2$. Let x_1 and x_2 be leaves of $B(V_i, V_j)$ in V_i. Also, let y_1
and y_2 be leaves of $B(V_i, V_j)$ in V_j. We may assume without loss of generality
that x_1 and y_1 are not adjacent. Let x' and y' be the neighbors of x_1 and y_1
in $B(V_i, V_j)$, respectively. Then, we exchange x_1 and y_1. Namely, we consider
$V_i' = (V_i - \{x_1\}) \cup \{y_1\}$ and $V_j' = (V_j - \{y_1\}) \cup \{x_1\}$. Since there are two
leaves of $B(V_i, V_j)$ in V_i, x_1 is not a cut-vertex of $\langle V_i \rangle$. Similarly, y_1 is not a cut-
vertex of $\langle V_j \rangle$. Thus, both $\langle V_i' \rangle$ and $\langle V_j' \rangle$ are connected. Besides, x_1 is adjacent
to every vertex in $V_i' - \{y_1\}$ and y_1 is adjacent to every vertex in $V_j' - \{x_1\}$.
Every vertex in $V_j' - \{x_1, x'\}$ is adjacent to a vertex in $V_i' - \{y_1\}$ such that
$|V_j' - \{x_1, x'\}| \ge 2$. Therefore, $B(V_i', V_j')$ consists of one component which has a
cycle. Consider $B(V_i', V_t)$ where $t \ne i, j$. The fact that both y_1 and x_2 are leaves
of $B(V_i, V_j)$ implies that y_1 and x_2 are adjacent to every vertex in V_t. This means
that $B(V_i', V_t)$ consists of one component which is not a tree. It can be similarly
checked that $B(V_j', V_t)$ also has no tree-component.

Case 2: $a \ge 2$ and $b \le 1$. Let x_1 and x_2 be leaves of $B(V_i, V_j)$ in V_i. Let x_1'
(respectively, x_2') be the neighbor of x_1 (respectively, x_2) in V_j. Note that x_1'
and x_2' may be identical. Without loss of generality, we may assume that x_1' is
not a leaf of $B(V_i, V_j)$ since $b \le 1$. Let w be a non-cut-vertex of $\langle V_j \rangle$ such that
$w \ne x_1'$. We then exchange x_1 and w. Let $V_i' = (V_i - \{x_1\}) \cup \{w\}$ and $V_j' = (V_j - \{w\}) \cup \{x_1\}$. Both $\langle V_i' \rangle$ and $\langle V_j' \rangle$ are connected. Here, $N_{B(V_i', V_j')}(x_1) = V_i' - \{w\}$,
$\deg_{B(V_i', V_j')}(u) \ge 1$ for any vertex u in V_j', and $\deg_{B(V_i', V_j')}(v) \ge 2$ for any vertex
v in $N_{B(V_i', V_j')}(w) \subset V_j'$. Thus, $B(V_i', V_j')$ consists of one component. Besides,
there exists a vertex z in $V_j' - \{x_1, x_1'\}$ with $\deg_{B(V_i, V_j)}(z) \ge 2$ since $b \le 1$. Such
a vertex z is adjacent to at least two vertices in $V_i' - \{w\}$ in $B(V_i', V_j')$. Therefore,
the component of $B(V_i', V_j')$ has a cycle. Hence, there is no tree-component in
$B(V_i', V_j')$. Consider $B(V_i', V_t)$ where $t \ne i, j$. Since V_i' has the leaf x_2 of $B(V_i, V_j)$,
if there is a vertex $u \ne x_2$ in V_i' with $\deg_{B(V_i', V_t)}(u) \ge 2$, then $B(V_i', V_t)$ consists

of one component with a cycle. If every vertex in $V_i' - \{x_2\}$ is a leaf of $B(V_i', V_t)$. Then, there are at least two vertices in V_i such that they are adjacent to every vertex of V_j in $B(V_i, V_j)$, which contradicts the assumption that $B(V_i, V_j)$ has a tree-component. Thus, $B(V_i', V_t)$ has no tree-component. Similarly, $B(V_j', V_t)$ has no tree-component, since V_j' has another leaf x_1 of $B(V_i, V_j)$.

Case 3: $a = b = 1$. In this case, $B(V_i, V_j)$ contains one tree-component which is a path-component and all the other components in $B(V_i, V_j)$ have no leaf. Let $x \in V_i$ and $y \in V_j$ such that they are leaves of $B(V_i, V_j)$.

Case 3.1: x and y are not adjacent. Let $x' \in V_j$ and $y' \in V_i$ such that x' and y' are the neighbors of x and y, respectively. Now assume that x is a cut-vertex of $\langle V_i \rangle$. Since y' is in the path-component, $\deg_{B(V_i, V_j)}(y') = 2$. Thus, y' is adjacent to every vertex of V_i except for a vertex $z \in V_i$. From the assumption that x is a cut-vertex of $\langle V_i \rangle$, it follows that z is adjacent to x but not adjacent to any other vertex in V_i. This means that $\deg_{B(V_i, V_j)}(z) \geq |V_j| - 1$, i.e., z is adjacent to every vertex of V_j except for y. However, this fact contradicts the assumption that $B(V_i, V_j)$ has a path-component. Thus, x is not a cut-vertex of $\langle V_i \rangle$. Similarly, we can check that y is not a cut-vertex of $\langle V_j \rangle$. Therefore, by exchanging x and y, we have a desirable consequence similar to Case 1 for $B(V_i', V_j')$ and Case 2 for $B(V_i', V_t)$ and $B(V_j', V_t)$ where $t \neq i, j$.

Case 3.2: x and y are adjacent. Suppose that any vertex in $V_i - \{x\}$ is adjacent to every vertex in $V_j - \{y\}$. Namely, $B(V_i - \{x\}, V_j - \{y\})$ is isomorphic to a complete bipartite graph. Let $x' \in V_i - \{x\}$ and $y' \in V_j - \{y\}$. Note that x' and y' are not cut-vertices of $\langle V_i \rangle$ and $\langle V_j \rangle$, respectively. It can be easily checked that $\delta(B((V_i - \{x'\}) \cup \{y'\}, (V_j - \{y'\}) \cup \{x'\})) \geq 2$. Besides, both $B((V_i - \{x'\}) \cup \{y'\}, V_t)$ and $B((V_j - \{y'\}) \cup \{x'\}, V_t)$ have no tree-component for $t \neq i, j$. Suppose that $B(V_i - \{x\}, V_j - \{y\})$ is not isomorphic to a complete bipartite graph. Select nonadjacent vertices $u \in V_i - \{x\}$ and $v \in V_j - \{y\}$. Let $V_i' = (V_i - \{u\}) \cup \{v\}$ and $V_j' = (V_j - \{v\}) \cup \{u\}$. Suppose that $B(V_i', V_j')$ has a tree-component. Except for the case that $B(V_i', V_j')$ has a K_2-component and all the other components have no leaf, we can apply Cases 1, 2, or 3.1 to $B(V_i', V_j')$. Consider the exceptional case. Suppose that the K_2-component of $B(V_i', V_j')$ consists of adjacent vertices $w \in V_i'$ and $z \in V_j'$. Then w (respectively, z) is adjacent to v (respectively, u), and w (respectively, z) is not adjacent to u (respectively, v) such that $\deg_{B(V_i, V_j)}(w) = \deg_{B(V_i, V_j)}(z) = 2$. Let $V_i'' = (V_i - \{w\}) \cup \{v\}$ and $V_j'' = (V_j - \{v\}) \cup \{w\}$. Then, z is a leaf of $B(V_i'', V_j'')$ but the neighbor u is not a leaf of $B(V_i'', V_j'')$. Hence, either $B(V_i'', V_j'')$ has no tree-component or the manipulations of the other cases can be applied to obtain a desirable consequence. When $B(V_i'', V_j'')$ has no tree-component, it also holds $B(V_i'', V_t)$ and $B(V_i'', V_t)$ have no tree-component since V_i'' and V_j'' have x and y, respectively.   □

Note that the lower bound on k in Theorem 1 cannot be improved to 2. For $k = 2$, the same statement holds only if $n = 5$. For any fixed k which is independent of n, the condition $\delta(G) \geq n - k$ itself is preserved while deleting any number of vertices (but not all) from G. Thus, the following corollary holds.

Corollary 1. *Let G be a graph with $n \geq 7$ vertices with $\delta(G) \geq n - k$. Let k be a fixed integer such that $3 \leq k \leq \frac{n}{2}$. For any $S \subset V(G)$ where $|S| \leq \min\{n - 7, n - 2k\}$, $G - S$ has $\lfloor \frac{n - |S|}{k} \rfloor$ completely independent spanning trees.*

By strengthen the minimum degree condition in Theorem 1, we can easily show the following propositions. Propositions 1 and 2 follow from the facts that $\delta(B(V_i, V_j)) \geq 2$ and $\delta(\langle V_i \rangle) \geq \frac{|V_i| - 1}{2}$ for a partition $(V_1, V_2, \ldots, V_t)$ of $V(G)$ satisfying the assumptions, respectively. These results are obviously weaker than Theorem 1, but they have additional properties that might be useful to obtain other theoretical results.

Proposition 1. *Let G be a graph with n vertices. Let $2 \leq k \leq \frac{n}{2}$. If $\delta(G) \geq n - k + 1$, then every consecutive partition of a Hamiltonian cycle in G which has $\lfloor \frac{n}{k} \rfloor$ parts with at least k vertices induces $\lfloor \frac{n}{k} \rfloor$ completely independent spanning trees in G.*

Proposition 2. *Let G be a graph with n vertices. Let $3 \leq k \leq \frac{n}{2}$. If $\delta(G) \geq n - \frac{k+1}{2}$, then every partition of $V(G)$ which has $\lfloor \frac{n}{k} \rfloor$ parts with at least k vertices induces $\lfloor \frac{n}{k} \rfloor$ completely independent spanning trees in G.*

4   An Improved Degree Condition

Let W_n be the graph obtained from two complete graphs $K_{\lfloor \frac{n+1}{2} \rfloor}$ and $K_{\lceil \frac{n+1}{2} \rceil}$ by identifying one vertex of $K_{\lfloor \frac{n+1}{2} \rfloor}$ and one vertex of $K_{\lceil \frac{n+1}{2} \rceil}$.

Theorem 2. *Let G be a graph with $n \geq 8$ vertices such that $G \not\cong W_n$. If $\delta(G) \geq \frac{n-1}{2}$, then there are two completely independent spanning trees in G.*

Proof: When n is even, the proposition follows from Theorem 1. Thus, it is sufficient to consider the case that $n = 2k - 1$ where $k \geq 5$. Suppose that $G \not\cong W_n$ and $\delta(G) \geq \frac{n-1}{2} = k - 1$. From the assumption of the minimum degree of a vertex in G, G has a Hamiltonian path. Let $P = (v_1, v_2, \ldots, v_n)$ be a Hamiltonian path of G. Based on P, we divide $V(G)$ into $V_1 = \{v_1, v_2, \ldots, v_{k-1}\}$ and $V_2 = \{v_k, v_{k+1}, \ldots, v_{2k-1}\}$. Then, $\langle V_1 \rangle$ and $\langle V_2 \rangle$ are both connected. Note that $\deg_{B(V_1, V_2)}(u) \geq 1$ for any vertex $u \in V_1$, but in V_2, there may exist an isolated vertex of $B(V_1, V_2)$. Note that an isolated vertex v of $B(V_1, V_2)$ is adjacent to every vertex in $V_2 - \{v\}$. Thus, if there is an isolated vertex v of $B(V_1, V_2)$, then every vertex in $V_2 - \{v\}$ is not a cut-vertex of $\langle V_2 \rangle$.

We first show that the case that an isolated vertex exists can be reduced to the case that no isolated vertex exists. Suppose that there is an isolated vertex of $B(V_1, V_2)$ in V_2. If there exists a non-isolated vertex w of $B(V_1, V_2)$ in V_2 such that w is not adjacent to a leaf of $B(V_1, V_2)$ in V_1, then by moving w to V_1, we have a desirable bipartition. Namely, both $\langle V_1 \cup \{w\} \rangle$ and $\langle V_2 - \{w\} \rangle$ are connected and $B(V_1 \cup \{w\}, V_2 - \{w\})$ has no isolated vertex. Suppose that every non-isolated vertex of $B(V_1, V_2)$ in V_2 is adjacent to a leaf of $B(V_1, V_2)$ in V_1.

Since $G \not\cong W_n$, there are at least two non-isolated vertices of $B(V_1, V_2)$ in V_2. Let u and v be non-isolated vertices in V_2. Also let x and y be the leaves in V_1 adjacent to u and v, respectively. If u is not a leaf, then by exchanging x and v, a desirable bipartition is obtained. Note that x is not a cut-vertex of $\langle V_1 \rangle$ and $xy \in E(G)$. Thus, $\langle (V_1 - \{x\}) \cup \{v\} \rangle$ and $\langle (V_2 - \{v\}) \cup \{x\} \rangle$ are connected such that the corresponding bipartite graph has no isolated vertex. Similarly, if v is not a leaf, then we can obtain a desirable bipartition. Suppose that every non-isolated vertex of $B(V_1, V_2)$ in V_2 is a leaf of $B(V_1, V_2)$. We can select two independent edges $x_1 u_1, x_2 u_2 \in E(G)$ where $x_1, x_2 \in V_1$ and $u_1, u_2 \in V_2$ such that $u_1 u_2 \in E(G)$. We then exchange x_1 and u_2. Since u_2 is adjacent to u_1, in the resultant bipartite graph, u_1 is not an isolated vertex. Hence, we have a desirable bipartition.

In what follows, we suppose that $|V_1| = k - 1$, $|V_2| = k$, both $\langle V_1 \rangle$ and $\langle V_2 \rangle$ are connected, $B(V_1, V_2)$ has a tree-component, and there is no isolated vertex of $B(V_1, V_2)$. We show that the bipartition (V_1, V_2) can be modified so that the resultant bipartition induces two completely independent spanning trees. Let a and b be the numbers of leaves of $B(V_1, V_2)$ in V_1 and V_2, respectively.

Case 1: $b \geq 2$. Let u_1 and u_2 be leaves of $B(V_1, V_2)$ in V_2. Then, $\deg_{\langle V_2 \rangle}(u_1) \geq k - 2$ and $\deg_{\langle V_2 \rangle}(u_2) \geq k - 2$. Assume that both u_1 and u_2 are cut-vertices of $\langle V_2 \rangle$. Then, there exist two vertices v_1 and v_2 such that v_1 (respectively, v_2) is adjacent only to u_1 (respectively, u_2) in $\langle V_2 \rangle$. If there is another leaf $u_3(\neq u_1, u_2)$, then such vertices v_1 and v_2 cannot exist. Thus, $b = 2$. Besides, it holds that $\deg_{B(V_1, V_2)}(v_1) \geq k - 2$ and $\deg_{B(V_1, V_2)}(v_2) \geq k - 2$. If $N_{B(V_1, V_2)}(v_1) \cup N_{B(V_1, V_2)}(v_2) = V_1$, then there is no tree-component of $B(V_1, V_2)$. Therefore, $N_{B(V_1, V_2)}(v_1) = N_{B(V_1, V_2)}(v_2) = V_1 - \{w\}$ for some vertex w such that either w and one of u_1 and u_2 induce a K_2-component, or w, u_1, and u_2 induce a P_3-component. In each case, by moving a vertex $y \in V_2 - \{u_1, u_2, v_1, v_2\}$ to V_1, the resultant bipartition induces two completely independent spanning trees. Hence, we can suppose that one of the leaves of $B(V_1, V_2)$ in V_2 is not a cut-vertex of $\langle V_2 \rangle$.

Case 1.1: $a \geq 2$. Let x and y be leaves of $B(V_1, V_2)$ in V_1 and V_2 respectively, such that y is not a cut-vertex of $\langle V_2 \rangle$ and $xy \notin E(G)$. Let $V_1' = (V_1 - \{x\}) \cup \{y\}$ and $V_2' = (V_2 - \{y\}) \cup \{x\}$. If there is no isolated vertex of $B(V_1', V_2')$, then this bipartition is a desirable one. Suppose that there exists an isolated vertex z of $B(V_1', V_2')$ in V_2. Note that $xz \in E(G)$ and $yz \notin E(G)$ such that z is a leaf of $B(V_1, V_2)$. If there exists a vertex v in $V_2' - \{x, z\}$ with $\deg_{\langle V_2' \rangle}(v) \geq 2$, then, by moving v to V_1', we have a desirable bipartition. Assume that every vertex in $V_2' - \{x, z\}$ has degree one in $\langle V_2' \rangle$. This means every vertex in $V_2' - \{x, z\}$ has degree at least $k - 3$ in $B(V_1, V_2)$. Let w be the neighbor of y in V_1. If w is a leaf of $B(V_1, V_2)$, then it can be checked that the bipartition $(\{x, z, w, y\}, V(G) - \{x, z, w, y\})$ induces two completely independent spanning trees. If w is not a leaf of $B(V_1, V_2)$, then by selecting a leaf $x^*(\neq x)$ of $B(V_1, V_2)$ in V_1 and exchanging x^* and y, we have a desirable bipartition, i.e., $((V_1 - \{x^*\}) \cup \{y\}, (V_2 - \{y\}) \cup \{x^*\})$ induces completely independent spanning trees.

Case 1.2: $a \leq 1$. Let y be a leaf of $B(V_1, V_2)$ in V_2 such that y is not a cut-vertex of $\langle V_2 \rangle$. Also, let w be the neighbor of y in V_1. Move y to V_1, i.e., let $V_1' = V_1 \cup \{y\}$ and $V_2' = V_2 - \{y\}$. Since $\deg_{B(V_1', V_2')}(y) \geq k - 2$, the resultant bipartition is a desirable one except for the following four cases:

- Case A: w is an isolated vertex of $B(V_1', V_2')$.
- Case B: w and $y' \in V_2' - N_{B(V_1', V_2')}(y)$ induce a K_2-component of $B(V_1', V_2')$.
- Case C: a leaf z of $B(V_1, V_2)$ in V_1 and y' induce a K_2-component of $B(V_1', V_2')$.
- Case D: w, y', and z induce a P_3-component of $B(V_1', V_2')$.

Note that Cases B, C, and D do not occur if $N_{B(V_1', V_2')}(y) = V_2'$ and Cases C and D do not occur if $a = 0$. In Case A, it is sufficient for obtaining a desirable bipartition to move a vertex $v(\neq w, y)$ with degree at least two in $\langle V_1' \rangle$ which is not adjacent to y' if y' exists and y' is a leaf of $B(V_1', V_2')$, to V_2'. In Case B (respectively, Cases C and D), it is sufficient to move a non-cut-vertex v of $\langle V_1' \rangle$ where $v \neq w, y$ (respectively, $v \neq w, y, z$) with degree at least two in $\langle V_1' \rangle$ which is adjacent to w (respectively, z), to V_2'. In each case, there are at least $k - 4 \geq 1$ such vertices if we ignore the restriction on the degree. More precisely, in Cases A and B, w has at least $k - 3$ neighbors and at least $k - 4$ neighbors in V_1, respectively, which satisfy the above conditions except for the degree condition in $\langle V_1' \rangle$, while in both Cases C and D, z has at least $k - 3$ such neighbors in V_1. Now, assume that all such vertices have degree one in $\langle V_1' \rangle$. Then, all such vertices have degree at least $k - 2$ in $B(V_1', V_2')$ and also in $B(V_1, V_2)$. In particular, for Case B, if the number of non-cut-vertices adjacent to w except for y is equal to $k - 4$, then there exists a leaf of $\langle V_1' \rangle$ which is not adjacent to w. Thus, there are at least $k - 3$ leaves of $\langle V_1' \rangle$ for Case B. Therefore, the structure of $B(V_1, V_2)$ in each case can be stated as follows:

- Case A: $B(V_1, V_2)$ has exactly two components such that one contains $K_{k-3, k-2}$ and the other is a K_2-component consisting of w and y.
- Case B: $B(V_1, V_2)$ has exactly two components such that one contains $K_{k-3, k-2}$ and the other is a P_3-component consisting of w, y, and y'.
- Case C: $B(V_1, V_2)$ has exactly two components such that one contains $K_{k-3, k-2}$ and the other is a K_2-component consisting of z and y'.
- Case D: $B(V_1, V_2)$ has exactly two components such that one is $K_{k-3, k-2}$-component and the other is a P_4-component consisting of w, y, y', and z.

The structure of $B(V_1, V_2)$ in Case D contradicts the assumption that $b \geq 2$. Consider Case A in which y' is not a leaf of $\langle V_2 \rangle$. Let y'' be a leaf of $\langle V_2 \rangle$ such that $y'' \neq y$. Since $\deg_{\langle V_2 \rangle}(y) \geq k - 2$ and $\deg_{\langle V_2 \rangle}(y'') \geq k - 2$, there are at least $k - 4 \geq 1$ vertices adjacent to both y and y'' such that they are not cut-vertices of $\langle V_2 \rangle$. By moving one of such $k - 4$ vertices to V_1, we have a desirable bipartition. Next, consider Case A in which y' is a leaf of $\langle V_2 \rangle$. Since $yy' \notin E(G)$, there are $k - 2$ non-cut-vertices which are adjacent to both y and y'. Then, by moving one of such $k - 2$ vertices except for a vertex which is adjacent to the neighbor v of y' in $B(V_1, V_2)$ if $\deg_{B(V_1, V_2)}(v) = 2$, a desirable bipartition is obtained. In Case B, by moving a vertex in $V_2 - \{y, y'\}$ which is not a cut-vertex of $\langle V_2 \rangle$, and is adjacent to both y and y' but not adjacent to a leaf of $B(V_1, V_2)$ in V_1, we have a desirable bipartition. Similarly, in Case C, by moving a vertex in $V_2 - \{y, y'\}$ which is not a cut-vertex of $\langle V_2 \rangle$, and is adjacent to both y and y' but not adjacent to w, a desirable bipartition is obtained. In both Case B and Case C, y' is a leaf of $\langle V_2 \rangle$. Thus, there are at least $k - 2$ non-cut-vertices of $\langle V_2 \rangle$ which is adjacent to both y and y'. Hence, there are least $k - 3$ vertices for producing a desirable bipartition.

Case 2: $a \geq 2$ and $b \leq 1$. We can select a leaf x of $B(V_1, V_2)$ in V_1 whose neighbor in V_2 is not a leaf of $B(V_1, V_2)$. By moving x to V_2, we have a desirable bipartition.

Case 3: $a = b = 1$. Let $x \in V_1$ and $y \in V_2$ be the leaves of $B(V_1, V_2)$.

Case 3.1: $B(V_1, V_2)$ has a P_t-component where $t \geq 6$. Similar to Case 3.1 in the proof of Theorem 1, it is shown that x is not a cut-vertex of $\langle V_1 \rangle$. Thus, we can move x to V_2 and obtain a desirable bipartition.

Case 3.2: $B(V_1, V_2)$ has a P_4-component. Let $x' \in V_1$ and $y' \in V_2$ be internal vertices of the path-component. Namely, $xy', y'x', x'y \in E(G)$. If x is not a cut-vertex of $\langle V_1 \rangle$, then we can move it to V_2. Suppose that x is a cut-vertex of $\langle V_1 \rangle$. Then, there exists a vertex $z \in V_1 - \{x, x'\}$ such that $\deg_{\langle V_1 \rangle}(z) = 1$. This means that $N_{B(V_1, V_2)}(z) = V_2 - \{y, y'\}$. Let $w \in V_1 - \{x, x', z\}$. Then, w is adjacent to x and x'. If $k \geq 6$, then by moving w to V_2, we have a desirable bipartition. Consider the case that $k = 5$. Since any vertex in V_2 except for y has degree at least two in $B(V_1, V_2)$, we can see that $B(V_1, V_2)$ has two components such that one is P_4-component and the other is $K_{2,3}$-component. Then, by selecting a non-cut-vertex of $\langle V_2 \rangle$ in $V_2 - \{y, y'\}$ which is adjacent to y and moving it to V_1, we have a desirable bipartition. Note that any vertex in $V_2 - \{y, y'\}$ has degree at least two in $\langle V_2 \rangle$.

Case 3.3: $B(V_1, V_2)$ has a K_2-component. If there exists a vertex $u(\neq x)$ in V_1 and a non-cut-vertex $v(\neq y)$ of $\langle V_2 \rangle$ such that $uv \notin E(G)$, then similar to Case 3.2 in the proof of Theorem 1, by exchanging u and v, either we have a desirable bipartition or we can apply the manipulations in other cases. Suppose that every non-cut-vertex of $\langle V_2 \rangle$ is adjacent to every vertex in $V_1 - \{x\}$. Since $\deg_{\langle V_2 \rangle}(y) \geq k - 2$, there are at least $k - 3$ non-cut-vertices of $\langle V_2 \rangle$ which are adjacent to y. Let $u' \in V_1 - \{x\}$ and $v' \in V_2 - \{y\}$ such that v' is a non-cut-vertex adjacent to y. By exchanging u' and v', a desirable bipartition is obtained. $\square$

5   Optimal Graphs for Completely Independent Spanning Trees

Definition 1. *For $n \geq k \geq 2$, $H(n, k)$ is defined to be the graph with the vertex set $\{0, 1, 2, \ldots, n - 1\}$ and in which two distinct vertices x and y are adjacent if and only if $y \in \{(x + k\ell) \bmod n, (x + k\ell + 1) \bmod n\}$ for $0 \leq \ell \leq \frac{1}{2} \lfloor \frac{n}{k} \rfloor$, or $y = (x + \frac{k}{2} \lceil \frac{n}{k} \rceil) \bmod n$ when $\lfloor \frac{n}{k} \rfloor$ is odd and k does not divide n.*

Theorem 3. *There are $\lfloor \frac{n}{k} \rfloor$ completely independent spanning trees in $H(n, k)$. In particular, if k divides n, then $H(n, k)$ has $\frac{n}{k}$ isomorphic completely independent spanning trees.*

Proof: Let I be an interval on $V(H(n,k))$, i.e., I consists of consecutive numbers modulo n. Let v be a vertex in $H(n,k)$. By definition, it can be checked that if $v \notin I$ and $|I| \geq k$, then v is adjacent to at least two vertices in I. This means that for any two disjoint intervals S and T on $V(H(n,k))$ with $|S| \geq k$ and $|T| \geq k$, $\delta(B_{H(n,k)}(S,T)) \geq 2$. Therefore, $B_{H(n,k)}(S,T)$ has no tree-component. Besides, both $\langle S \rangle_{H(n,k)}$ and $\langle T \rangle_{H(n,k)}$ are clearly connected from the definition. Hence, $H(n,k)$ has $\lfloor \frac{n}{k} \rfloor$ completely independent spanning trees. To prove the second statement, we show that for any two intervals $S = \{p, p+1, \ldots, p+k-1\}$ and $T = \{q, q+1, \ldots, q+k-1\}$ such that $S \cap T = \emptyset$ and $\min\{q - p \bmod n, p - q \bmod n\} \leq k\lfloor \frac{1}{2}\lfloor \frac{n}{k} \rfloor \rfloor$, $B_{H(n,k)}(S,T)$ is hamiltonian. Without loss of generality, we may assume that $S = \{0, 1, \ldots, k-1\}$ and $T = \{i, i+1, \ldots, i+k-1\}$, where $k \leq i \leq k\lfloor \frac{1}{2}\lfloor \frac{n}{k} \rfloor \rfloor$. Let $i+t \equiv 0 \bmod k$ such that $0 \leq t < k$. By definition, we can see that $0(i+t), 0(i+t+1), 1(i+t+1), 1(i+t+2), \ldots, (k-1)(i+t+k-1), (k-1)(i+t+k) \in E(H(n,k))$. Since $k-1$ is also adjacent to $i+t$, there exists a Hamiltonian cycle $(i+t, 0, i+t+1, 1, \ldots, i+k-1, k-t-1, i+k, k-t, i+k+1, \ldots, k-2, i+t+k-1, k-1, i+t)$ in $H(n,k)$. Here, for each $0 \leq j < t$, both $k-t-1+j$ and $k-t+j$ are adjacent to $i+j$, since they are adjacent to $i+k+j$. This means that there exists a Hamilton cycle in $B_{H(n,k)}(S,T)$. Suppose that k divides n. Then, for $0 \leq i < \frac{n}{k}$, by setting $V_i = \{ki, ki+1, \ldots, k(i+1)-1\}$, $B_{H(n,k)}(V_i, V_j)$ has a Hamiltonian cycle for any $0 \leq i < j < \frac{n}{k}$. Based on each Hamiltonian cycle $C(i,j)$ in $B_{H(n,k)}(V_i, V_j)$, we can construct $\frac{n}{k}$ completely independent spanning trees each of which is isomorphic to the caterpillar obtained from a path P_k by adding $\frac{n}{k} - 1$ leaves to each vertex of P_k. □

Lemma 2. *Suppose that $k \geq 3$. If $\lfloor \frac{n}{k} \rfloor$ is even, then $H(n,k)$ is $2\lfloor \frac{n}{k} \rfloor$-regular (respectively, $(2\lfloor \frac{n}{k} \rfloor + 1)$-regular, $(2\lfloor \frac{n}{k} \rfloor + 2)$-regular) if $n \bmod k = 1$ (respectively, $n \bmod k \in \{0, 2\}$, $n \bmod k \notin \{0, 1, 2\}$). If $\lfloor \frac{n}{k} \rfloor$ is odd, then $H(n,k)$ is $2\lfloor \frac{n}{k} \rfloor$-regular (respectively, $(2\lfloor \frac{n}{k} \rfloor + 2)$-regular) if $n \bmod k \in \{0, 1\}$ (respectively, $n \bmod k \notin \{0, 1\}$).*

By definition, we can show the above lemma. If $H(n,k)$ is $2\lfloor \frac{n}{k} \rfloor$-regular or $(2\lfloor \frac{n}{k} \rfloor + 1)$-regular, then the number of completely independent spanning trees in $H(n,k)$ is optimal in the sense that any r-regular graph cannot have $\lfloor \frac{r}{2} \rfloor + 1$ completely independent spanning trees except for the complete graph with an even number of vertices. Therefore, from Lemma 2 and Theorem 3, we have the following corollaries.

Corollary 2. *Suppose that $k \geq 3$. If $\lfloor \frac{n}{k} \rfloor$ is even and $n \bmod k \in \{0, 1, 2\}$, or $\lfloor \frac{n}{k} \rfloor$ is odd and $n \bmod k \in \{0, 1\}$, then $H(n,k)$ is optimal with respect to the number of completely independent spanning trees.*

Corollary 3. *For any $n \geq 7$, $H(n, \lfloor \frac{n}{2} \rfloor)$ is optimal for two completely independent spanning trees. For any $n \geq 9$ where $n \not\equiv 2 \pmod 3$, $H(n, \lfloor \frac{n}{3} \rfloor)$ is optimal for three completely independent spanning trees. For any $n \geq 12$ where $n \not\equiv 3 \pmod 4$, $H(n, \lfloor \frac{n}{4} \rfloor)$ is optimal for four completely independent spanning trees.*

In some special cases, we can modify $H(n,k)$ while keeping the same number of completely independent spanning trees. For the case that $\lfloor \frac{n}{k} \rfloor$ is odd and k

does not divide n, let $H'(n,k)$ be the graph obtained from $H(n,k)$ by deleting the edges $x(x+\frac{k}{2}\lceil\frac{n}{k}\rceil \bmod n)$ for $0 \le x < n$ and adding the edges $x(x+\lceil\frac{n}{2}\rceil-1 \bmod n)$ for $0 \le x < n$. For $n = 2tk$ where $k \ge 3$ and $t \ge 1$, let $H^*(n,k)$ be the graph obtained from $H(n,k)$ by deleting the edges $x(x+\frac{n}{2} \bmod n)$ for $0 \le x < n$. For $n = (2t+1)k+2$ where k is even, $k \ge 3$, and $t \ge 1$, let $H^\star(n,k)$ be the graph obtained from $H(n,k)$ by deleting the edges $x(x+\frac{k}{2}\lceil\frac{n}{k}\rceil \bmod n)$ for $0 \le x < n$ and adding the edges $x(x+\lfloor\frac{n}{2}\rfloor \bmod n)$ for $0 \le x < n$.

Proposition 3. *The graphs $H'(n,k)$, $H^*(n,k)$, and $H^\star(n,k)$ have $\lfloor\frac{n}{k}\rfloor$ completely independent spanning trees.*

Note that $H^*(n,k)$ and $H^\star(n,k)$ are $2\lfloor\frac{n}{k}\rfloor$-regular and $(2\lfloor\frac{n}{k}\rfloor+1)$-regular, respectively. Thus, $H^*(n,k)$ and $H^\star(n,k)$ are optimal graphs with respect to the number of completely independent spanning trees.

6   Concluding Remarks

In this paper, we have generalized and improved the Dirac's condition for completely independent spanning trees. We have also presented constructions of optimal graphs for completely independent spanning trees. In order to decrease our lower bound on the minimum degree, it would be helpful to assume some additional properties. For example, by assuming regularity and 2-connectedness, the lower bound on the minimum degree for the existence of a Hamiltonian cycle can be improved from $\frac{n}{2}$ to $\frac{n}{3}$ [11].

References

1. Araki, T.: Dirac's condition for completely independent spanning trees. J. Graph Theory **77**, 171–179 (2014)
2. Chang, H.-Y., Wang, H.-L., Yang, J.-S., Chang, J.-M.: A note on the degree condition of completely independent spanning trees. IEICE Trans. **98–A**, 2191–2193 (2015)
3. Curran, S., Lee, O., Yu, X.: Finding four independent trees. SIAM J. Comput. **35**, 1023–1058 (2006)
4. Fan, G., Hong, Y., Liu, Q.: Ore's condition for completely independent spanning trees. Discrete Appl. Math. **177**, 95–100 (2014)
5. Hasunuma, T.: Completely independent spanning trees in the underlying graph of a line digraph. Discrete Math. **234**, 149–157 (2001)
6. Hasunuma, T.: Completely independent spanning trees in maximal planar graphs. In: Kučera, L. (ed.) WG 2002. LNCS, vol. 2573, pp. 235–245. Springer, Heidelberg (2002)
7. Hasunuma, T.: Structural properties of subdivided-line graphs. J. Discrete Algorithms **31**, 69–86 (2015)
8. Hasunuma, T., Morisaka, C.: Completely independent spanning trees in torus networks. Networks **60**, 59–69 (2012)
9. Hasunuma, T., Nagamochi, H.: Independent spanning trees with small depths in iterated line digraphs. Discrete Appl. Math. **110**, 189–211 (2001)

10. Huck, A.: Independent trees in graphs. Graphs Combin. **10**, 29–45 (1994)
11. Jackson, B.: Hamilton cycles in regular 2-connected graphs. J. Combin. Theory B **29**, 27–46 (1980)
12. Pai, K.-J., Yang, J.-S., Yao, S.-C., Tang, S.-M., Chang, J.-M.: Completely independent spanning trees on some interconnection networks. IEICE Trans. **97–D**, 2514–2517 (2014)
13. Péterfalvi, F.: Two counterexamples on completely independent spanning trees. Discrete Math. **312**, 808–810 (2012)
14. Tang, S.-M., Yang, J.-S., Wang, Y.-L., Chang, J.-M.: Independent spanning trees on multidimensional torus networks. IEEE Trans. Comput. **59**, 93–102 (2010)

A Faster FPTAS for the Unbounded Knapsack Problem

Klaus Jansen[1] and Stefan E.J. Kraft[1(✉)]

Department of Computer Science, Kiel University, 24098 Kiel, Germany
{kj,stkr}@informatik.uni-kiel.de

Abstract. The Unbounded Knapsack Problem (UKP) is a well-known variant of the famous 0-1 Knapsack Problem (0-1 KP). In contrast to 0-1 KP, an arbitrary number of copies of every item can be taken in UKP. Since UKP is NP-hard, fully polynomial time approximation schemes (FPTAS) are of great interest. Such algorithms find a solution arbitrarily close to the optimum $\text{OPT}(I)$, i.e. of value at least $(1 - \varepsilon)\text{OPT}(I)$ for $\varepsilon > 0$, and have a running time polynomial in the input length and $\frac{1}{\varepsilon}$. For over thirty years, the best FPTAS was due to Lawler with a running time in $O(n + \frac{1}{\varepsilon^3})$ and a space complexity in $O(n + \frac{1}{\varepsilon^2})$, where n is the number of knapsack items. We present an improved FPTAS with a running time in $O(n + \frac{1}{\varepsilon^2} \log^3 \frac{1}{\varepsilon})$ and a space bound in $O(n + \frac{1}{\varepsilon} \log^2 \frac{1}{\varepsilon})$. This directly improves the running time of the fastest known approximation schemes for Bin Packing and Strip Packing, which have to approximately solve UKP instances as subproblems.

1 Introduction

An instance I of the Knapsack Problem (KP) consists of a list of n items $a_1, \ldots, a_n$, $n \in \mathbb{N}$, where every item has a profit $p_j \in (0, 1]$ and a size $s_j \in (0, 1]$. Moreover, we have the knapsack size $c = 1$. In the 0-1 Knapsack Problem (0-1 KP), a subset $V \subset \{a_1, \ldots, a_n\}$ has to be chosen such that the total profit of V is maximized and the total size of the items in V is at most c. Mathematically, the problem is defined by $\max\{\sum_{j=1}^{n} p_j x_j \mid \sum_{j=1}^{n} s_j x_j \leq c; x_j \in \{0, 1\} \, \forall j\}$. In this paper, we focus on the unbounded variant (UKP) where an arbitrary number of copies of every item is allowed, i.e. $\max\{\sum_{j=1}^{n} p_j x_j \mid \sum_{j=1}^{n} s_j x_j \leq c; x_j \in \mathbb{N} \, \forall j\}$.

1.1 Known Results

The 0-1 Knapsack Problem and other variants of KP are well-known NP-hard problems [5]. They can be optimally solved in pseudo-polynomial time by dynamic programming [1,17]. Furthermore, fully polynomial time approximation schemes (FPTAS) are known for different variants of KP. An FPTAS is

Research supported by DFG project JA612/14-2, "Entwicklung und Analyse von effizienten polynomiellen Approximationsschemata für Scheduling- und verwandte Optimierungsprobleme".

© Springer International Publishing Switzerland 2016
Z. Lipták and W.F. Smyth (Eds.): IWOCA 2015, LNCS 9538, pp. 274–286, 2016.
DOI: 10.1007/978-3-319-29516-9_23

a family of algorithms $(A_\varepsilon)_{\varepsilon > 0}$, where for every $\varepsilon > 0$ the algorithm A_ε finds for a given instance I a solution of profit $A_\varepsilon(I) \geq (1 - \varepsilon)\text{OPT}(I)$. The value $\text{OPT}(I)$ denotes the optimal value for I. FPTAS have a running time polynomial in $\frac{1}{\varepsilon}$ and the input length.

The first FPTAS for 0-1 KP was presented by Ibarra and Kim [7]. Lawler improved the running time in his seminal paper [20]. In 1981, Magazine and Oguz [22] presented a method to decrease the space complexity of the dynamic program. The currently fastest known algorithm is due to Kellerer and Pferschy [15–17, pp. 166–183] with a space complexity in $O(n + \frac{1}{\varepsilon^2})$ and a running time in $O(n \min\{\log n, \log \frac{1}{\varepsilon}\} + \frac{1}{\varepsilon^2} \log(\frac{1}{\varepsilon}) \cdot \min\{n, \frac{1}{\varepsilon} \log(\frac{1}{\varepsilon})\})$. For $n \in \Omega(\frac{1}{\varepsilon} \log \frac{1}{\varepsilon})$, this is in $O(n \log(\frac{1}{\varepsilon}) + \frac{1}{\varepsilon^3} \log^2(\frac{1}{\varepsilon}))$.

For UKP, Ibarra and Kim [7] presented the first FPTAS by extending their 0-1 KP algorithm. This UKP algorithm has a running time in $O(n + \frac{1}{\varepsilon^4} \log \frac{1}{\varepsilon})$ and a space complexity in $O(n + \frac{1}{\varepsilon^3})$. Kellerer et al. [17, pp. 232–234] have moreover described an FPTAS with a time complexity in $O(n \log(n) + \frac{1}{\varepsilon^2}(n + \log \frac{1}{\varepsilon}))$ and a space bound in $O(n + \frac{1}{\varepsilon^2})$. In 1979, Lawler [20] presented his FPTAS with a running time in $O(n + \frac{1}{\varepsilon^3})$ and a space complexity in $O(n + \frac{1}{\varepsilon^2})$. For $n \in \Omega(\frac{1}{\varepsilon})$, this is still the best known FPTAS.

The study of KP is not only interesting in itself, but also motivated by column generation for optimization problems like the famous Bin Packing Problem (BP) and Strip Packing Problem (SP). In the former problem, a set J of n items of size in $(0, 1]$ has to be packed in as few unit-sized bins as possible. In the latter problem, a set J of n rectangles of width $(0, 1]$ and height $(0, 1]$ has to be packed in a strip of unit width such that the height of the packing is minimized. Many algorithms for optimization problems like Bin Packing have to solve linear programs (LPs), but enumerating all columns of the linear programs would take too much time. One way to avoid this is the consideration of the dual of the LP and to (approximately or exactly) solve a separation problem, which can e.g. be KP, to find violated inequalities of the dual. Examples can be found in [6,13].

Since Bin Packing and Strip Packing are NP-complete [5], several approximation algorithms have been found for both problems. However, no efficient (i.e. polynomal-time) algorithm A for BP or SP can achieve $A(J) \leq c \cdot \text{OPT}(J)$ for $c < \frac{3}{2}$ and all problem instances J unless P = NP [5]. So-called asymptotic fully polynomial-time approximation schemes (AFPTAS) $(A_\varepsilon)_{\varepsilon > 0}$ are therefore especially interesting. They find for every $\varepsilon > 0$ and instance J a solution of value at most $(1 + \varepsilon)\text{OPT}(J) + f(\frac{1}{\varepsilon})$, and have a running time polynomial in the input length and $\frac{1}{\varepsilon}$. Roughly speaking, AFPTAS achieve an approximation ratio of $c = (1 + \varepsilon)$ for large problem instances.

For Bin Packing, the first AFPTAS was presented by Karmarkar and Karp [13] with $f(\frac{1}{\varepsilon}) = O(\frac{1}{\varepsilon^2})$. In 1991, Plotkin et al. [23] described an improved algorithm with a smaller additive term $f(\frac{1}{\varepsilon}) = O(\frac{1}{\varepsilon} \log(\frac{1}{\varepsilon}))$ and a time complexity in $O(\frac{1}{\varepsilon^6} \log^6(\frac{1}{\varepsilon}) + \log(\frac{1}{\varepsilon})n)$. The AFPTAS by Shachnai and Yehezkely [24] has the same additive term and a running time in $O(\frac{1}{\varepsilon^6} \log^3(\frac{1}{\varepsilon}) + \log(\frac{1}{\varepsilon})n)$ for general instances. Currently, the AFPTAS in [9] has the smallest additive term $f(\frac{1}{\varepsilon}) = O(\log^2 \frac{1}{\varepsilon})$ and the fastest running time in $O(\frac{1}{\varepsilon^6} \log \frac{1}{\varepsilon} + \log(\frac{1}{\varepsilon})n)$.

The first AFPTAS for Strip Packing was presented by Kenyon and Rémila [18] with $f(\frac{1}{\varepsilon}) = O(\frac{1}{\varepsilon^2})$. Bougeret et al. [2] and Sviridenko [25] independently improved the additive term to $f(\frac{1}{\varepsilon}) = O(\frac{1}{\varepsilon} \log \frac{1}{\varepsilon})$. The algorithm in [2] needs time in $O(\frac{1}{\varepsilon^6} \log(\frac{1}{\varepsilon}) + n \log n)$, which is the currently fastest known AFPTAS.

Both algorithms in [2,9] solve UKP instances for column generation. A faster FPTAS for UKP therefore directly yields faster AFPTAS for Bin Packing and Strip Packing.

1.2 Our Result

We have derived an improved FPTAS for UKP that is faster and needs less space than previously known algorithms.

Theorem 1. *There is an FPTAS for UKP with a running time in $O(n + \frac{1}{\varepsilon^2} \log^3(\frac{1}{\varepsilon}))$ and a space complexity in $O(n + \frac{1}{\varepsilon} \log^2(\frac{1}{\varepsilon}))$.*

Not only the improved running time, but also the improved space complexity is interesting because "for higher values of $\frac{1}{\varepsilon}$ the space requirement is usually considered to be a more serious bottleneck for practical applications than the running time" [17, p. 168]. Nevertheless, the improved time complexity has direct practical consequences. Let $KP(d, \varepsilon)$ be the running time to find a $(1-\varepsilon)$ approximate solution to a UKP instance with d items. The Bin Packing algorithm in [9] has the running time $O(KP(d, \frac{\bar{\varepsilon}}{6}) \cdot \frac{1}{\varepsilon^3} \log \frac{1}{\varepsilon} + \log(\frac{1}{\varepsilon})n)$ if we assume that $KP(d, \frac{\bar{\varepsilon}}{6}) \in \Omega(\frac{1}{\varepsilon^2})$ (where $\bar{\varepsilon} \in \theta(\varepsilon)$ and $d \in O(\frac{1}{\varepsilon} \log \frac{1}{\varepsilon})$). By using the new FPTAS for UKP, we get a faster AFPTAS:

Corollary 2. *There is an AFPTAS $(A_\varepsilon)_{\varepsilon>0}$ for Bin Packing that finds for $\varepsilon \in (0, \frac{1}{2}]$ a packing of J in $A_\varepsilon(J) \leq (1 + \varepsilon)\mathrm{OPT}(J) + O(\log^2(\frac{1}{\varepsilon}))$ bins. Its running time is in $O(\frac{1}{\varepsilon^5} \log^4 \frac{1}{\varepsilon} + \log(\frac{1}{\varepsilon})n)$.*

Similarly, the Strip Packing algorithm in [2] (see also [8]) has a running time in $O(d(\frac{1}{\varepsilon^2} + \ln d) \max\{KP(d, \frac{\bar{\varepsilon}}{6}), d \ln \ln(\frac{d}{\varepsilon})\} + n \log n)$ where again $d \in O(\frac{1}{\varepsilon} \log \frac{1}{\varepsilon})$ and $\bar{\varepsilon} \in \theta(\varepsilon)$. The new FPTAS yields the following improved AFPTAS:

Corollary 3. *There is an AFPTAS $(A_\varepsilon)_{\varepsilon>0}$ for Strip Packing that finds a packing for J of total height $A_\varepsilon(J) \leq (1 + \varepsilon)\mathrm{OPT}(J) + O(\frac{1}{\varepsilon} \log(\frac{1}{\varepsilon}))$. Its running time is in $O(\frac{1}{\varepsilon^5} \log^4 \frac{1}{\varepsilon} + n \log n)$.*

Note that the full version of this paper is available on arXiv [11].

1.3 Techniques

Most algorithms for UKP [7,17,20] rely on 0-1 KP algorithms. The 0-1 KP algorithms determine a first lower bound P_0 for $\mathrm{OPT}(I)$. Based on a threshold T depending on P_0, the items are partitioned into large(-profit) items with $p_j \geq T$ and small(-profit) items with $p_j < T$. A subset of the large items is taken, which is sufficient for an approximate solution. Its profits are then scaled and the well-known dynamic programming by profits applied to the subset. All combinations of large items (packed by the dynamic program) and small items (which are greedily added) are checked and the best one returned. For UKP, copies of the items in the reduced large item set are taken to transform the UKP instance into a 0-1 KP instance.

Our algorithm also first reduces the number of large items. However, we further preprocess the remaining large items by taking advantage of the unboundedness: large items of similar profit $[2^k T, 2^{k+1} T)$ are iteratively combined ("glued") together to larger items. Apart from two special cases that can be easily solved, we prove for this new set G a structure property: there are approximate solutions where at most one large item from every interval $[2^k T, 2^{k+1} T)$ is used, i.e. only $O(\log \frac{1}{\varepsilon})$ items in total. As a next step, a large item $a_{\text{eff}-c}$ that consists of several copies of the most efficient small item a_{eff} is introduced. We prove that there are now approximate solutions to the large items $G \cup \{a_{\text{eff}-c}\}$ of cardinality $O(\log \frac{1}{\varepsilon})$ and that additionally use at least one item of profit at least $\frac{1}{4} P_0$. Instead of exact dynamic programming, we use approximate dynamic programming: the profits in $[\frac{1}{4} P_0, 2P_0]$ are divided into intervals of equal length. During the execution of the dynamic program, we eliminate dominated solutions and store for each interval at most one solution of smallest size. The combination of approximate dynamic programming with the structure properties yields the considerable improvement in the running time and the space complexity. The algorithm then returns the best combination of large items (packed by the dynamic program) and copies of the small item a_{eff} (added greedily).

2 Preliminaries

We introduce some useful notation. The profit of an item a is denoted by $p(a)$ and its size by $s(a)$. If $a = a_j$, we also write $p(a_j) = p_j$ and $s(a_j) = s_j$. Let $V = \{x_a : a \mid a \in I, x_a \in \mathbb{N}\}$ be a multiset of items, i.e. a subset of items in I with their multiplicities. We naturally define the total profit $p(V) := \sum_{x_a > 0} p(a) x_a$ and the total size $s(V) := \sum_{x_a > 0} s(a) x_a$.

Let $v \leq c = 1$ be a part of the knapsack. The corresponding optimum profit is denoted by $\text{OPT}(I, v) = \max\{\sum_a p(a) x_a \mid \sum_a s(a) x_a \leq v; \ a \in I; \ x_a \in \mathbb{N}\}$. Obviously, $\text{OPT}(I) = \text{OPT}(I, c)$ holds.

Finally, we assume throughout the paper that basic arithmetic operations as well as computing the logarithm can be performed in $O(1)$.

2.1 A First Approximation

We present a simple approximation algorithm for $\text{OPT}(I)$. Take the most efficient item $a_{\text{meff}} := \arg\max_{a \in I} \frac{p(a)}{s(a)}$. Fill the knapsack with as many copies of a_{meff} as possible, i.e. take $\lfloor \frac{c}{s(a_{\text{meff}})} \rfloor \overset{c=1}{=} \lfloor \frac{1}{s(a_{\text{meff}})} \rfloor$ copies of a_{meff}. [17, p. 232, 20]:

Theorem 4. *We have* $P_0 := p(a_{\text{meff}}) \cdot \lfloor \frac{c}{s(a_{\text{meff}})} \rfloor \geq \frac{1}{2}\text{OPT}(I)$. *The value* P_0 *can be found in time* $O(n)$ *and space* $O(1)$.

From now on, we assume without loss of generality that $\varepsilon \leq \frac{1}{4}$ and $\varepsilon = \frac{1}{2^{\kappa}-1}$ for $\kappa \in \mathbb{N}$. Otherwise, we replace ε by the corresponding $\frac{1}{2^{\kappa}-1}$ such that $\frac{1}{2^{\kappa}-1} \leq \varepsilon < \frac{1}{2^{\kappa}-2}$. Note that $\log_2(\frac{2}{\varepsilon}) = \kappa$ holds. Similar to Lawler [20], we introduce the threshold T and a constant K:

$$T := \frac{1}{2}\varepsilon P_0 = \frac{1}{2}\frac{1}{2^{\kappa}-1}P_0 \quad \text{and} \quad K := \frac{\varepsilon}{4}\frac{1}{\log_2(\frac{2}{\varepsilon})+1}T = \frac{1}{4}\frac{1}{\kappa+1}\frac{1}{2^{\kappa}-1}T . \quad (1)$$

3   Reducing the Items

We first partition the items into large(-profit) and small(-profit) items, and only keep the most efficient small item: $I_L := \{a \in I \mid p(a) \geq T\}$, $I_S := I \setminus I_L$, and $a_{\text{eff}} := \arg\max\{\frac{p(a)}{s(a)} \mid p(a) < T\}$. The construction of these sets is trivial.

Theorem 5. *The sets* I_L, I_S *and the item* a_{eff} *can be found in time* $O(n)$ *and space* $O(n)$. *This is also the space needed to save* I_L.

Similar to Lawler, we now reduce the item set I_L. First, we partition the interval of large item profits $[T, 2P_0)$ into $L^{(k)} := [2^k T, 2^{k+1}T)$ for $k \in \{0, \ldots, \kappa+1\}$. Note that $L^{(\kappa)} = [2^{\kappa}T, 2^{\kappa+1}T) = [2^{\kappa}\frac{1}{2}\frac{1}{2^{\kappa}-1}P_0, 2^{\kappa+1}\frac{1}{2}\frac{1}{2^{\kappa}-1}P_0) = [P_0, 2P_0)$. For convenience, we set $L^{(\kappa+1)} := \{2P_0\}$. We further split the $L^{(k)}$ into disjoint sub-intervals, each of length $2^k K$: $L_{\gamma}^{(k)} := [2^k T + \gamma \cdot 2^k K, 2^k T + (\gamma+1)2^k K)$ for $\gamma \in \{0, \ldots, 2^{\kappa+1}(\kappa+1) - 1\}$. Note that indeed $L^{(k)} = \bigcup_{\gamma} L_{\gamma}^{(k)}$ holds. Similar to above, we set $L_0^{(\kappa+1)} := \{2P_0\}$.

We only keep the smallest item a for every profit interval $L_{\gamma}^{(k)}$, again similar to Lawler [20]. We will see that these items are sufficient for an approximation.

Definition 6. *For an item* a *with* $p(a) \geq T$, *let* $k(a) \in \mathbb{N}$ *be the interval such that* $p(a) \in L^{(k(a))}$ *and* $\gamma(a) \in \mathbb{N}$ *be the sub-interval such that* $p(a) \in L_{\gamma(a)}^{(k(a))}$. *Let* $a_{\gamma}^{(k)}$ *be the smallest item for the profit interval* $L_{\gamma}^{(k)}$, *i.e. we set* $a_{\gamma}^{(k)} := \arg\min\{s(a) \mid a \in I_L \text{ and } p(a) \in L_{\gamma}^{(k)}\}$ *for all* k *and* γ.

Let $I_{L,\text{red}} := \bigcup_k \bigcup_{\gamma} \{a_{\gamma}^{(k)}\}$ be the reduced set of large items. As in [20], we can prove that $I_{L,\text{red}}$ is sufficient for an approximation.

Lemma 7. *Let* $0 \leq v \leq c = 1$. *Then* $\text{OPT}(\{a_{\text{eff}}\}, c - v) \geq \text{OPT}(I_S, c - v) - T$ *and* $\text{OPT}(I_{L,\text{red}}, v) \geq (1 - \frac{\varepsilon}{4}\frac{1}{\log_2(\frac{2}{\varepsilon})+1})\text{OPT}(I_L, v)$.

Theorem 8. *The set* $I_{L,\text{red}}$ *has* $O(\frac{1}{\varepsilon}\log^2\frac{1}{\varepsilon})$ *items. It can be constructed in time* $O(n + \frac{1}{\varepsilon}\log^2\frac{1}{\varepsilon})$ *and space* $O(\frac{1}{\varepsilon}\log^2\frac{1}{\varepsilon})$, *including the space to save* $I_{L,\text{red}}$.

Remark 9. If there is one item a with the profit $p(a) = 2P_0$, i.e. whose profit attains the upper bound, one optimum solution obviously consists of this single item. During the partition of I into I_L and I_S, it can easily be checked whether such an item is contained in I. Since the algorithm can directly stop if this is the case, we will from now on assume without loss of generality that such an item does not exist and that $a_0^{(\kappa+1)} = \emptyset$.

4 A Simplified Solution Structure

In this section, we will transform $I_{L,\text{red}}$ into a new instance G whose optimum $\text{OPT}(G, v)$ is only slightly smaller than $\text{OPT}(I_{L,\text{red}}, v)$ and where the corresponding solution has a special structure. This new transformation will allow us later to faster construct the approximate solution. First, we define $I^{(k)} := \{a \in I_{L,\text{red}} \mid p(a) \in L^{(k)}\} = \{a \in I_{L,\text{red}} \mid p(a) \in [2^k T, 2^{k+1} T)\}$. Note that the items are already partitioned into the sets $I^{(k)}$ because of the way $I_{L,\text{red}}$ has been constructed.

Definition 10. *Let a_1, a_2 be two knapsack items with $s(a_1) + s(a_2) \le c$. The gluing operation $\oplus$ combines them into a new item $a_1 \oplus a_2$ with $p(a_1 \oplus a_2) = p(a_1) + p(a_2)$ and $s(a_1 \oplus a_2) = s(a_1) + s(a_2)$.*

Thus, the gluing operation is only defined on pairs of items whose combined size does not exceed c.

The basic idea for the new instance G is as follows: we first set $G^{(0)} := I^{(0)}$. Then, we construct $a_1 \oplus a_2$ for all $a_1, a_2 \in G^{(0)}$ (including the case $a_1 = a_2$), which yields the item set $H^{(1)} := \{a_1 \oplus a_2 \mid a_1, a_2 \in G^{(0)}\}$. Note that we have $p(a_1 \oplus a_2) \in [2T, 4T) = L^{(1)}$. For every profit interval $L_\gamma^{(1)}$, we keep only the item of smallest size in $I^{(1)} \cup H^{(1)}$, which yields the item set $G^{(1)}$. This procedure is iterated for $k = 1, \ldots, \kappa - 1$: the set $G^{(k)}$ contains the items with a profit in $[2^k T, 2^{k+1} T) = L^{(k)}$. Gluing like above yields the item set $H^{(k+1)}$ with profits in $[2^{k+1} T, 2^{k+2} T) = L^{(k+1)}$. By taking again the smallest item in $H^{(k+1)} \cup I^{(k+1)}$ for every $L_\gamma^{(k+1)}$, the set $G^{(k+1)}$ is derived. The item in $G^{(k)}$ with a profit in $L_\gamma^{(k)}$ is denoted by $\tilde{a}_\gamma^{(k)}$ for every k and γ.

We finish when $G^{(\kappa)}$ has been constructed. We are in the case where $I^{(\kappa+1)} = \emptyset$, i.e. $a_0^{(\kappa+1)} = \emptyset$, and we will see at the beginning of Sect. 5 that it is not necessary to construct $G^{(\kappa+1)}$ from the items in $G^{(\kappa)}$. Hence, we also have $\tilde{a}_0^{(\kappa+1)} = \emptyset$.

Note that we may glue items together that already consist of glued items. For backtracking, we save for every $\tilde{a}_\gamma^{(k)}$ which two items in $G^{(k-1)}$ have formed it or whether $\tilde{a}_\gamma^{(k)}$ has already been an item in $I^{(k)}$.

Remark 11. One item $\tilde{a}_\gamma^{(k)}$ is in fact the combination of several items in $I_{L,\text{red}}$. The profit and size of $\tilde{a}_\gamma^{(k)}$ is equal to the total profit and size of these items. The $\tilde{a}_\gamma^{(k)}$ represent feasible item combinations because an arbitrary number of item copies can be taken in UKP.

We define the item set $G := \bigcup_{k=0}^{\kappa} G^{(k)}$.

Definition 12. *Let I' be a set of knapsack items with $p(a) \geq T$ for every $a \in I'$. For a knapsack volume $v \leq c$ and $k_0 \in \{0, \ldots, \kappa\}$, a solution is structured for $k = k_0$ if it fits into v and uses for every $k \in \{0, \ldots, k_0\}$ at most one item with a profit in $L^{(k)}$. We denote by $\mathrm{OPT}_{\leq k_0}(I', v)$ the corresponding optimum profit.*

The solution for $\mathrm{OPT}_{\leq k_0}\left(G^{(0)} \cup \ldots \cup G^{(k_0)} \cup G^{(k_0+1)} \cup I^{(k_0+2)} \cup \ldots \cup I^{(\kappa)}, v\right)$ fits into the volume v, and it uses only one item from every $G^{(k)}$ for $k \in \{0, \ldots, k_0\}$.

Theorem 13. *For $v \leq c$ and $k_0 \in \{0, \ldots, \kappa - 1\}$, we have the following bound:*
$$\mathrm{OPT}_{\leq k_0}\left(\bigcup_{k=0}^{k_0+1} G^{(k)} \cup \bigcup_{k=k_0+2}^{\kappa} I^{(k)}, v\right) \geq \left(1 - \frac{\varepsilon}{4} \frac{1}{\log_2(\frac{2}{\varepsilon})+1}\right)^{k_0+1} \mathrm{OPT}(I_{L,\mathrm{red}}, v).$$

Lemma 14. *We have $\mathrm{OPT}(G \cup \{a_{\mathrm{eff}}\}) \leq \mathrm{OPT}(I_{L,\mathrm{red}} \cup I_S) \leq \mathrm{OPT}(I_L \cup I_S) = \mathrm{OPT}(I) \leq 2P_0$.*

Up to now, we have (only) reduced the original item set I to $G \cup \{a_{\mathrm{eff}}\}$.

Lemma 15. *Assume as mentioned in Remark 9 that $a_0^{(\kappa+1)} = \emptyset$. Consider the optimum structured solutions to $G \cup \{a_{\mathrm{eff}}\}$ for $k_0 = \kappa - 1$ (see Definition 12). This means that at most one item is used from every $G^{(k)}$ for $k \in \{0, \ldots, \kappa - 1\}$. (The item a_{eff} has a profit $p(a_{\mathrm{eff}}) < T$ such that it does not have to satisfy any structural conditions.) Then there are two possible cases:*

- *One solution uses (at least) two items in $G^{(\kappa)}$. This is the case if and only if the optimum for $G \cup \{a_{\mathrm{eff}}\}$ is $2P_0$, and the solution consists of two item copies of the item $\tilde{a}_0^{(\kappa)}$ with $p(\tilde{a}_0^{(\kappa)}) = P_0$.*
- *Every solution uses at most one item in $G^{(\kappa)}$. Then, $\mathrm{OPT}_{\leq \kappa-1}(G, v') = \mathrm{OPT}_{\leq \kappa}(G, v')$ holds for all values $0 \leq v' \leq c$, and there is a value $0 \leq v \leq c$ such that $\mathrm{OPT}_{\leq \kappa}(G, v) + \mathrm{OPT}(\{a_{\mathrm{eff}}\}, c - v) = \mathrm{OPT}_{\leq \kappa-1}(G, v) + \mathrm{OPT}(\{a_{\mathrm{eff}}\}, c - v) \geq \left(1 - \frac{\varepsilon}{4} \frac{1}{\log_2(\frac{2}{\varepsilon})+1}\right)^{\kappa+1} \mathrm{OPT}(I) - T$.*
  Moreover, $\mathrm{OPT}_{\leq \kappa}(G, v)$ uses at least one item in $G^{(\kappa-2)} \cup G^{(\kappa-1)} \cup G^{(\kappa)}$, and/or we have $\mathrm{OPT}(\{a_{\mathrm{eff}}\}, c - v) \geq \frac{1}{4}P_0$.

Definition 16. *Take $\lceil \frac{P_0/4}{p(a_{\mathrm{eff}})} \rceil$ items a_{eff}. If their total size is at most c, they are glued together to $a_{\mathrm{eff}-c}$.*

Obviously, $a_{\mathrm{eff}-c}$ consists of the smallest number of items a_{eff} whose total profit is at least $\frac{P_0}{4}$. Moreover, $a_{\mathrm{eff}-c}$ is a large item.

Definition 17. *Take a knapsack volume $v \leq c$. Consider the following solutions to $G \cup \{a_{\mathrm{eff}-c}\}$ of size at most v: they are structured for $k = \kappa$, i.e. they use for every $k \in \{0, \ldots, \kappa\}$ at most one item in $G^{(k)}$. Moreover, they use the item $a_{\mathrm{eff}-c}$ at most once and at least one item $\tilde{a} \in G^{(\kappa-2)} \cup G^{(\kappa-1)} \cup G^{(\kappa)} \cup \{a_{\mathrm{eff}-c}\}$. Hence, these solutions have a profit of at least $p(\tilde{a}) \geq 2^{\kappa-2}T = \frac{1}{4}P_0$.*

These special solutions are called structured solutions with a lower bound *(on the profit). The optimal profit for such solutions of total size at most v is denoted by $\mathrm{OPT}_{\mathrm{St}}(G \cup \{a_{\mathrm{eff}-c}\}, v)$. If v is too small such that such a solution does not exist, we set $\mathrm{OPT}_{\mathrm{St}}(G \cup \{a_{\mathrm{eff}-c}\}, v) = 0$.*

Theorem 18. *In the second case of Lemma 15, there is one $0 \leq v \leq c$ such that* $\mathrm{OPT}_{\mathrm{St}}(G \cup \{a_{\mathrm{eff}-c}\}, v) + \mathrm{OPT}(\{a_{\mathrm{eff}}\}, c - v) \geq (1 - \frac{\varepsilon}{4} \frac{1}{\log_2(\frac{2}{\varepsilon})+1})^{\kappa+1}\mathrm{OPT}(I) - T$.

So far, we have not constructed an actual solution. We have only shown in Theorem 18 that there is a solution to $G \cup \{a_{\mathrm{eff}-c}\} \cup \{a_{\mathrm{eff}}\}$ that is close to $\mathrm{OPT}(I)$ and that is a structured solution with a lower bound.

Theorem 19. *The cardinality of $G^{(k)}$ is in $O(\frac{1}{\varepsilon} \log \frac{1}{\varepsilon})$, i.e. G has $O(\frac{1}{\varepsilon} \log^2 \frac{1}{\varepsilon})$ items. The set G can be constructed in time $O(\frac{1}{\varepsilon^2} \log^3(\frac{1}{\varepsilon}))$ and space $O(\frac{1}{\varepsilon} \log^2 \frac{1}{\varepsilon})$, which also includes the space to store G and the backtracking information. The item $a_{\mathrm{eff}-c}$ can be constructed in time $O(1)$.*

5   Finding an Approximate Structured Solution by Dynamic Programming

The previous section has shown that there are three cases: the instance I has one item of profit $2P_0$ (Remark 9), G has one item of profit P_0 and size at most $\frac{c}{2}$, or there is an approximate structured solution to $G \cup \{a_{\mathrm{eff}-c}\} \cup \{a_{\mathrm{eff}}\}$ with a lower bound (see Theorem 18). As the first two cases can be easily checked, we will from now on assume that we are in the third case: a solution uses at most one item from every $G^{(k)}$ for $k \in \{0, \ldots, \kappa\}$ as well as $a_{\mathrm{eff}-c}$ at most once. At the same time, at least one item $\tilde{a} \in G^{(\kappa-2)} \cup G^{(\kappa-1)} \cup G^{(\kappa)} \cup \{a_{\mathrm{eff}-c}\}$ is chosen.

We use dynamic programming to find for all $0 \leq v \leq c$ the corresponding set of large items $V \subseteq G \cup \{a_{\mathrm{eff}-c}\}$ with $s(V) \leq v$. For convenience, let $G^{(\kappa+1)} := \{a_{\mathrm{eff}-c}\}$. We introduce tuples (p, s, k) similar to Lawler [20]. For profit p with $0 \leq p \leq 2P_0$ and size $0 \leq s \leq c$, the tuple (p, s, k) states that there is an item set of size s whose total profit is p. Moreover, the set has only items in $G^{(k)} \cup \cdots \cup G^{(\kappa+1)}$ and respects the structure above.

The dynamic program is quite simple: start with the tuple set $F^{(\kappa+2)} := \{(0, 0, \kappa + 2)\}$. For $k = \kappa + 1, \ldots, \kappa - 2$, the tuples in $F^{(k)}$ are recursively constructed by $F^{(k)} := \{(p, s, k) \mid (p, s, k + 1) \in F^{(k+1)}\} \cup \{(p + p(\tilde{a}), s + s(\tilde{a}), k) \mid (p, s, k + 1) \in F^{(k+1)}, \tilde{a} \in G^{(k)}, s + s(\tilde{a}) \leq c\}$. As $(0, 0, k + 1) \in F^{(k+1)}$, the set $F^{(k)}$ also contains the entries $(p(\tilde{a}), s(\tilde{a}), k)$ for $\tilde{a} \in G^{(k)}$ if $k \in \{\kappa + 1, \ldots, \kappa - 2\}$. For $k \in \{\kappa - 3, \ldots, 0\}$, this tuple $(0, 0, k + 1)$ is no longer considered to form the new tuples, which guarantees that tuples of the form $(p + p(\tilde{a}), s + s(\tilde{a}), k)$ for $\tilde{a} \in G^{(k)}$ have $p, s \neq 0$. The recursion becomes $F^{(k)} := \{(p, s, k) \mid (p, s, k + 1) \in F^{(k+1)}\} \cup \{(p + p(\tilde{a}), s + s(\tilde{a}), k) \mid (p, s, k + 1) \in F^{(k+1)} \setminus \{(0, 0, k + 1)\}, \tilde{a} \in G^{(k)}, s + s(\tilde{a}) \leq c\}$. The actual item set corresponding to (p, s, k) can be reconstructed by saving backtracking information.

Definition 20. *A tuple (p_2, s_2, k) is dominated by (p_1, s_1, k) if $p_2 \leq p_1$ and $s_2 \geq s_1$.*

As in [20], dominated tuples $(p, s, k + 1)$ are now removed from $F^{(k+1)}$ before $F^{(k)}$ is constructed. It is not difficult to see that a non-dominated tuple (p, s, k) is optimal, i.e. the profit p can only be obtained with items of size at least s if items in $G^{(k)}, \ldots, G^{(\kappa+1)}$ are considered.

Lemma 21. *A tuple $(p, s, k) \in F^{(k)}$ stands for a structured solution with a lower bound (see Definition 17). Therefore, we have $p \geq 2^{\kappa-2}T$ if $p > 0$. For every $v \leq c$, there is a tuple $(p, s, 0) \in F^{(0)}$ with $p = \mathrm{OPT}_{\mathrm{St}}(G \cup \{a_{\mathrm{eff}-c}\}, v)$ and $s \leq v$.*

While the dynamic program above constructs the desired tuples, their number may increase dramatically until $F^{(0)}$ is obtained. We therefore use approximate dynamic programming for the tuples with profits in $[\frac{1}{4}P_0, 2P_0]$. This method is inspired by the dynamic programming used in [14] (see also [17, pp. 97–112]).

Definition 17 and Lemma 21 state that a tuple (p, s, k) with $p > 0$ satisfies $p \geq 2^{\kappa-2}T$. Apart from $(0, 0, k)$, all tuples have therefore profits in the interval $[2^{\kappa-2}T, 2P_0] \overset{(1)}{=} [\frac{1}{4}P_0, 2P_0] = [2^{\kappa-2}T, \dots, 2^{\kappa+1}T]$. We partition this interval into sub-intervals of length $2^{\kappa-2}K$. We get the partitioning $[2^{\kappa-2}T, 2P_0] = \bigcup_{\xi=0}^{\xi_0} [2^{\kappa-2}T + \xi \cdot 2^{\kappa-2}K, \ 2^{\kappa-2}T + (\xi+1)2^{\kappa-2}K) \cup \{2P_0\} =:$ $\bigcup_{\xi=0}^{\xi_0} \tilde{L}_\xi^{(\kappa-2)} \cup \tilde{L}_{\xi_0+1}^{(\kappa-2)}$ for $\xi_0 := 7(\kappa+1)2^{\kappa+1} - 1$. (A short calculation shows that $2^{\kappa-2}T + (\xi_0 + 1)2^{\kappa-2}K = 2P_0$.) The approximate dynamic program keeps for every $\xi \in \{0, \dots, \xi_0 + 1\}$ only the tuple (p, s, k) with $p \in \tilde{L}_\xi^{(\kappa-2)}$ that has the smallest size s. The dominated tuples are removed when all tuples for k have been constructed. The sets of these non-dominated tuples are denoted by $D^{(k)}$. For convenience, $(p(\xi), s(\xi), k) \in D^{(k)}$ denotes the smallest tuple with a profit in $\tilde{L}_\xi^{(\kappa-2)}$. We again save the backtracking information during the execution of the dynamic program.

Lemma 22. *Let $\tilde{D}^{(k)}$ be the set $D^{(k)}$ before the dominated entries are removed. A tuple $(p, s, k) \in \tilde{D}^{(k)}$ for $k = \kappa+1, \dots, 0$ stands for a structured solution with a lower bound. Therefore, we have $p \geq 2^{\kappa-2}T$ if $p > 0$. This is also true for $(p, s, k) \in D^{(k)}$.*

Theorem 23. *Let $k \in \{0, \dots, \kappa + 1\}$. For every (non-dominated) entry $(\bar{p}, \bar{s}, k) \in F^{(k)}$, there is a tuple $(p, s, k) \in D^{(k)}$ such that $p \geq (1 - \frac{\varepsilon}{4} \frac{1}{\log_2(\frac{2}{\varepsilon})+1})^{\kappa-k+1} \bar{p}$ and $s \leq \bar{s}$.*

Proof (Sketch). The theorem is proved by induction for $k = \kappa + 1, \dots, 0$. For $(0, 0, k) \neq (\bar{p}, \bar{s}, k) \in F^{(k)}$, there are the two cases $(\bar{p}, \bar{s}, k+1) \in F^{(k+1)}$ and $(\bar{p}, \bar{s}, k+1) \notin F^{(k+1)}$. We only show the first case. There is by the induction hypothesis a tuple $(p_1, s_1, k+1) \in D^{(k+1)}$ such that $p_1 \geq \bar{p} \cdot (1 - \frac{\varepsilon}{4} \frac{1}{\log_2(\frac{2}{\varepsilon})+1})^{\kappa-(k+1)+1}$ and $s_1 \leq \bar{s}$. Note that this implies $(p_1, s_1, k+1) \neq (0, 0, k+1)$ and therefore $p_1 \geq 2^{\kappa-2}T$ by Lemma 22. Let ξ_1 be the index such that $p_1 \in \tilde{L}_{\xi_1}^{(\kappa-2)}$. In the dynamic program, $(p_1, s_1, k+1)$ yields the tuple (p_1, s_1, k), which may only be replaced in $\tilde{D}^{(k)}$ by a tuple of smaller size, but with a profit still in $\tilde{L}_{\xi_1}^{(\kappa-2)}$. Thus, there must be a tuple $(p_2, s_2, k) \in \tilde{D}^{(k)}$ with $s_2 \leq s_1$ and $p_2 \in \tilde{L}_{\xi_1}^{(\kappa-2)}$. Let now $(p, s, k) \in D^{(k)}$ be the tuple that dominates (p_2, s_2, k) (which can of course be (p_2, s_2, k) itself), i.e. $p \geq p_2$ and $s \leq s_2$. For the profit, we have $p \geq p_2 \geq p_1 - 2^{\kappa-2}K \overset{p_1 \neq 0}{=} p_1 \cdot (1 - \frac{2^{\kappa-2}K}{p_1}) \overset{\text{Lem. 22}}{\geq} p_1 \cdot (1 - \frac{2^{\kappa-2}K}{2^{\kappa-2}T}) \overset{(1)}{=}$

$p_1 \cdot (1 - \frac{\varepsilon}{4} \frac{1}{\log_2(\frac{2}{\varepsilon})+1}) \geq \bar{p} \cdot (1 - \frac{\varepsilon}{4} \frac{1}{\log_2(\frac{2}{\varepsilon})+1})^{\kappa-k+1}$. The lower bound on the profit is therefore true for (p, s, k). We have $s \leq s_2 \leq s_1 \leq \bar{s}$ for the size. Note that the glued item set G with its structured solution (Theorem 13) allows for the proof of Lemma 15, the introduction of $a_{\text{eff}-c}$, and the proof of Theorem 18. This guarantees that $p \geq 2^{\kappa-2}T$ (see Lemmas 21 and 22), which is essential to the proof of this theorem. Otherwise, the approximate dynamic programming would need for the same approximation ratio profit sub-intervals like $\tilde{L}_\xi^{(\kappa-2)}$ of length smaller than $2^{\kappa-2}K$, and it would have to save more tuples. Both would increase the time and space complexity. This is also true for the second case, where Lemma 22 is also essential.

Corollary 24. *For every $v \leq c$, there is a tuple $(p, s, 0) \in D^{(0)}$ such that $s \leq v$ and $p \geq (1 - \frac{\varepsilon}{4} \frac{1}{\log_2(\frac{2}{\varepsilon})+1})^{\kappa+1} \text{OPT}_{\text{St}}(G \cup \{a_{\text{eff}-c}\}, v)$.*

Theorem 25. *All tuple sets $D^{(k)}$ for $k = \kappa+1, \ldots, 0$ can be constructed in time $O(\frac{1}{\varepsilon^2} \log^3 \frac{1}{\varepsilon})$. The space needed for the dynamic program and for saving the $D^{(k)}$ as well as the backtracking information is in $O(\frac{1}{\varepsilon} \log^2 \frac{1}{\varepsilon})$.*

6   The Algorithm

We can now put together the entire approximation algorithm.

Input: Item set I
Output: Profit P, solution set J
Determine P_0 and define T, K;
Partition the items into I_L and I_S and find a_{eff};
if *Item a with $p(a) = 2P_0$ found during the partitioning* **then**
| **return** $2P_0, \{a\}$;

Reduce I_L to $I_{L,\text{red}}$ as shown in Sect. 3;
Construct G and the item $a_{\text{eff}-c}$ like in Sect. 4;
if $p(\tilde{a}_0^{(\kappa)}) = P_0$ *and* $s(\tilde{a}_0^{(\kappa)}) \leq \frac{c}{2}$ **then**
| Recursively undo the gluing of $\tilde{a}_0^{(\kappa)}$ to get the item set J'. Let J be the set
| consisting of two copies of every item in J';
| **return** $2P_0, J$;

Construct the tuple sets $D^{(\kappa+1)}, \ldots, D^{(0)}$ as described in Sect. 5;
Find $(p, s, 0) \in D^{(0)}$ such that
$P := p + \text{OPT}(\{a_{\text{eff}}\}, c - s) = \max_{(p', s', 0) \in D^{(0)}} p' + \text{OPT}(\{a_{\text{eff}}\}, c - s')$;
Backtrack the tuple $(p, s, 0)$ to find the corresponding structured solution with a
lower bound $J' \subset G \cup \{a_{\text{eff}-c}\}$;
Recursively undo the gluing of all $\tilde{a} \in J'$ and add these items to the solution set
J;
Add the items of $\text{OPT}(\{a_{\text{eff}}\}, c - s)$ to J;
return P, J ;

Algorithm 1. The complete algorithm

Theorem 26. *Algorithm 1 finds a solution of value at least* $(1 - \varepsilon)\mathrm{OPT}(I)$.

Theorem 27. *The algorithm has a running time in* $O(n + \frac{1}{\varepsilon^2} \log^3 \frac{1}{\varepsilon})$ *and needs space in* $O(n + \frac{1}{\varepsilon} \log^2 \frac{1}{\varepsilon})$.

7 Concluding Remarks

The most important steps in this algorithm are the creation of the item set G by gluing and the introduction of $a_{\mathrm{eff}-c}$. This guarantees the existence of an approximate structured solution with a lower bound (see Definition 17). Therefore, the approximate dynamic program has to store less tuples (p, s, k) than in the case without the structure.

We [19] have extended our algorithm to the Unbounded Knapsack Problem with Inversely Proportional Profits (UKPIP) introduced in [10]. Here, several knapsack sizes $0 < c_1 < \ldots < c_M = 1$ are given, and the profit of an item counts as p_j/c_l if packed in c_l. The goal is to find the best knapsack size and the corresponding solution of maximum profit. UKPIP is used for column generation in our AFPTAS for Variable-Sized Bin Packing [9] where several bin sizes are given and the goal is to minimize the total volume of the bins used. The faster FPTAS for UKPIP yields a faster AFPTAS for Variable-Sized Bin Packing.

There are interesting open questions. As stated in Subsect. 1.2, the space complexity is a more serious bottleneck than the running time. Recently, Lokshtanov and Nederlof [21] showed that the 0-1 KP and the Subset Sum Problem have a pseudo-polynomial time and only polynomial space algorithm. Subset Sum is a special case of Knapsack where the profit of an item is equal to its size, i.e. $p_j = s_j$. Moreover, it was shown that Unary Subset Sum is in Logspace [3,12]. Gál et al. [4] described an FPTAS for Subset Sum whose space complexity is in $O(\frac{1}{\varepsilon})$, i.e. which does not depend on the actual input size, and whose running time is in $O(\frac{1}{\varepsilon} n(n + \log n + \log \frac{1}{\varepsilon}))$. Can any of these results be further extended to improve the space complexity of an UKP FPTAS? Finally, it is open whether the ideas presented in this paper can be extended to the normal 0-1 KP or other KP variants as well as used for column generation of other optimization problems. The currently fastest known algorithm for 0-1 KP is due to Kellerer and Pferschy [15–17]. We mention in closing that by using the same approach similar improved approximation algorithms can be expected for various Packing and Scheduling Problems, e.g. for Bin Covering, Bin Packing with Cardinality Constraints, Scheduling Multiprocessor Tasks and Resource-constrained Scheduling.

References

1. Bellman, R.E.: Dynamic Programming. Princeton University Press, Princeton (1957)
2. Bougeret, M., Dutot, P.-F., Jansen, K., Robenek, C., Trystram, D.: Approximation algorithms for multiple strip packing and scheduling parallel jobs in platforms. Discrete Math. Alg. Appl. **3**(4), 553–586 (2011)

3. Elberfeld, M., Jakoby, A., Tantau, T.: Logspace versions of the theorems of Bodlaender and Courcelle. Technical report 62. Electronic Colloquium on Computational Complexity (ECCC), 2010. First published in 51th Annual Symposium on Foundations of Computer Science (FOCS 2010), pp. 143–152. IEEE Computer Society (2010)
4. Gál, A., Jang, J., Limaye, N., Mahajan, M., Sreenivasaiah, K.: Space-Efficient Approximations for Subset Sum. Technical report 180. Electronic Colloquiumon Computational Complexity (ECCC) (2014)
5. Garey, M.R., Johnson, D.S.: Computers and Intractability: A Guide to the Theory of NP-Completeness. W.H. Freeman and Company, New York (1979)
6. Gilmore, P., Gomory, R.: A linear programming approach to the cutting stock problem. Oper. Res. **9**(6), 849–859 (1961)
7. Ibarra, O.H., Kim, C.E.: Fast approximation algorithms for the knapsack and sum of subset problem. J. ACM **22**, 463–468 (1975)
8. Jansen, K.: Approximation algorithms for min-max and max-min resource sharing problems, and applications. In: Bampis, E., Jansen, K., Kenyon, C. (eds.) Efficient Approximation and Online Algorithms. LNCS, vol. 3484, pp. 156–202. Springer, Heidelberg (2006)
9. Jansen, K., Kraft, S.: An improved approximation scheme for variable-sized bin packing. In: Sassone, V., Widmayer, P., Rovan, B. (eds.) MFCS 2012. LNCS, vol. 7464, pp. 529–541. Springer, Heidelberg (2012)
10. Jansen, K., Kraft, S.: An improved knapsack solver for column generation. In: Bulatov, A.A., Shur, A.M. (eds.) CSR 2013. LNCS, vol. 7913, pp. 12–23. Springer, Heidelberg (2013)
11. Jansen, K., Kraft, S.E.J.: A faster FPTAS for the unbounded knapsack problem (2014). arXiv:1504.04650
12. Kane, D.M.: Unary subset-sum is in logspace (2010). arXiv:1012.1336
13. Karmarkar, N., Karp, R.M.: An efficient approximation scheme for the one-dimensional bin-packing problem. In: 23rd Annual Symposium on Foundations of Computer Science (FOCS 1982), pp. 312–320. IEEE Computer Society (1982)
14. Kellerer, H., Mansini, R., Pferschy, U., Speranza, M.G.: An efficient fully polynomial approximation scheme for the Subset-Sum Problem. J. Comput. Syst. Sci. **66**(2), 349–370 (2003)
15. Kellerer, H., Pferschy, U.: A new fully polynomial time approximation scheme for the knapsack problem. J. Comb. Optim. **3**(1), 59–71 (1999)
16. Kellerer, H., Pferschy, U.: Improved dynamic programming in connection with an FTPAS for the knapsack problem. J. Comb. Optim. **8**(1), 5–11 (2004)
17. Kellerer, H., Pferschy, U., Pisinger, D.: Knapsack Problems. Springer, Heidelberg (2004)
18. C. Kenyon, E. Rémila.: A near-optimal solution to a two-dimensional cutting stock problem. Math. Oper. Res. 25(4), 645–656 (2000). First published as "Approximate Strip Packing". In: 37th Annual Symposium on Foundations of Computer Science (FOCS 1996), pp. 31–36. IEEE Computer Society (1996)
19. Kraft, S.E.J.: Improved approximation algorithms for packing and scheduling problems. Ph.D thesis. Christian-Albrechts-Universität zu Kiel (2015)
20. Lawler, E.L.: Fast approximation algorithms for knapsack problems. Math. Oper. Res. **4**(4), 339–356 (1979)
21. Lokshtanov, D., Nederlof, J.: Saving space by algebraization. In: Proceedings of the 42nd ACM Symposium on Theory of Computing, STOC 2010, pp. 321–330. ACM (2010)

22. Magazine, M.J., Oguz, O.: A fully polynomial approximation algorithm for the 0–1 knapsack problem. Eur. J. Oper. Res. **8**(3), 270–273 (1981)
23. Plotkin, S.A., Shmoys, D.B., Tardos, É.: Fast approximation algorithms for fractional packing and covering problems. Math. Oper. Res. 20, 257–301 (1995). First published in 32nd Annual Symposium on Foundations of Computer Science (FOCS 1991), pp. 495–504. IEEE Computer Society (1991)
24. Shachnai, H., Yehezkely, O.: Fast asymptotic FPTAS for packing fragmentable items with costs. In: Csuhaj-Varjú, E., Ésik, Z. (eds.) FCT 2007. LNCS, vol. 4639, pp. 482–493. Springer, Heidelberg (2007)
25. Sviridenko, M.: A note on the Kenyon-Remila strip-packing algorithm. Inf. Process. Lett. **112**(1–2), 10–12 (2012)

Computational Complexity
of Distance Edge Labeling

Dušan Knop and Tomáš Masařík [✉]

Department of Applied Mathematics, Faculty of Mathematics and Physics,
Charles University, Prague, Czech Republic
{knop,masarik}@kam.mff.cuni.cz

Abstract. The problem of DISTANCE EDGE LABELING is a variant of
DISTANCE VERTEX LABELING (also known as $L_{2,1}$ labeling) that has been
studied for more than twenty years and has many applications, such as
frequency assignment.

The DISTANCE EDGE LABELING problem asks whether the edges of
a given graph can be labeled such that the labels of adjacent edges differ
by at least two and the labels of edges at distance two differ by at least
one. Labels are chosen from the set $\{0, 1, \ldots, \lambda\}$ for λ fixed.

We present a full classification of its computational complexity—a
dichotomy between the polynomially solvable cases and the remaining
cases which are NP-complete. We characterise graphs with $\lambda \leq 4$ which
leads to a polynomial-time algorithm recognizing the class and we show
NP-completeness for $\lambda \geq 5$ by several reductions from MONOTONE NOT
ALL EQUAL 3-SAT.

Keywords: Computational complexity · Distance labeling · Line-
graphs

1 Introduction

We study the computational complexity of the distance edge-labeling problem.
This problem belongs to a wider class of problems that generalize the graph
coloring problem. The task is to assign a set of colors to each vertex, such that
whenever two vertices are adjacent, their colors differ from each other. For a
survey about this famous graph problem and related algorithms, see [1].

We are interested in the so-called *distance labeling*. In this generalization of
the former problem the condition enforcing different colors is extended and takes
into account also the second neighborhood of a vertex (or an edge). The second
neighborhood is the set of vertices (or edges) at distance at most 2. For a survey

Research was supported by the project Kontakt LH12095, by the project SVV-2015-
260223 and by the project CE-ITI P202/12/G061 of GAČR.
Full version of the paper can be found on arXive: http://arxiv.org/abs/1508.01014
D. Knop— Author was supported by the project GAUK 1784214.

© Springer International Publishing Switzerland 2016
Z. Lipták and W.F. Smyth (Eds.): IWOCA 2015, LNCS 9538, pp. 287–298, 2016.
DOI: 10.1007/978-3-319-29516-9_24

about distance labelings, we refer to the article by Tiziana Calamoneri [2], as well as her online survey [3].

Graph distance labeling has been first studied by Griggs and Yeh [4,5] in 1992. The problem has many applications, the most important one being frequency assignment [6]. The complexity of $L_{2,1}$ labeling for a fixed parameter λ has been established in [7]. They show a dichotomy between polynomial cases for $\lambda \leq 3$ and NP-complete cases for $\lambda \geq 4$.

Moreover, for the usual graph coloring problem there is a theorem of Vizing [8], which states that for the edge-coloring number $\chi'(G)$ it holds that $\Delta \leq \chi' \leq \Delta + 1$, where Δ is the maximum degree of the graph. For $L_{2,1}$ labeling there is a general bound due to Havet et al. [9], namely $\lambda \leq \Delta^2$, for $\Delta \geq 79$.

Before we proceed to the formal definition of the corresponding decision problem, we give several definitions of a labeling mapping of a graph and of the minimal distance edge-labeling number. Note that the distance edge-labeling is equivalent to the distance vertex-labeling of the associated line-graphs. A *line-graph* $L(G)$ is a graph derived from another graph G such that vertices of $L(G)$ are edges of G and two vertices a, b of $L(G)$ are connected by an edge whenever a, b (as edges of G) are adjacent. We define the *distance* between edges of a graph as their distance in the corresponding line-graph.

Definition 1 (Edge-labeling mapping). *Let $G(V, E)$ be a graph. A mapping $f'_{2,1}\colon E \to \mathbb{N}$ is an edge-labeling, if it satisfies:*

- $|f'_{2,1}(e) - f'_{2,1}(e')| \geq 2$ *for neighboring edges (i.e. those in the distance one),*
- $|f'_{2,1}(e) - f'_{2,1}(e')| \geq 1$ *for edges at distance two.*

As usual, we are interested in a labeling that minimizes the number of labels used by a feasible labeling.

Definition 2 (Minimum distance edge-labeling). *Let G be a graph and $f'_{2,1}$ an edge-labeling mapping, we define the graph parameter $\lambda'_{2,1}$ as:*

$$\lambda'_{2,1}(G) := \min_{f'_{2,1}} \max_{e \in E} f'_{2,1}(e).$$

The size of the range of a (not necessarily optimal) edge-labeling mapping $f'_{2,1}$ is called the *span*.

Definition 3 (DISTANCE EDGE LABELING problem (also known as $L'_{2,1}$)).

Input:	*A graph G.*
Parameter:	$\lambda \in \mathbb{N}$.
Question:	*Is $\lambda'_{2,1}(G) \leq \lambda$?*

1.1 Our Results

Our main result is the following theorem about the dichotomy of the DISTANCE EDGE LABELING problem.

Theorem 1 (Dichotomy of distance edge-labeling). *The problem* $L'_{2,1}$ *is polynomial-time solvable if and only if* $\lambda \leq 4$. *Otherwise it is* NP-*complete.*

We derive Theorem 1 as a combination of Theorem 4 that describes all graphs with $\lambda'_{2,1} \leq 4$ and Theorem 6 presenting the NP-completeness result. Note that our Theorem 6 also extends to the following inapproximability result:

Corollary 1. *The* DISTANCE EDGE LABELING *problem cannot be approximated within a factor of* $6/5 - \varepsilon$, *unless* P = NP.

Moreover, according to [10], the proof implies that the DISTANCE EDGE LABELING is paraNP-hard while parameterized by its span.

1.2 Preliminaries

We state several basic and well-known observations with the connection to Definition 3, as well as some notation used in this paper.

For further standard notation in graph theory, we refer to the monograph [11]. The first observation gives a trivial lower-bound on $\lambda'_{2,1}(G)$.

Observation 2 (Max-degree lower-bound). *Let* G *be a graph and let* Δ *be its maximum degree. Then* $\lambda'_{2,1}(G) \geq 2(\Delta - 1)$.

Note that this observation gives also an upper bound on the max-degree of a graph G with $\lambda'_{2,1}(G) \leq \lambda$ for a given $\lambda \in \mathbb{N}$.

Observation 3 (The symmetry of distance labeling). *Let* G *be a graph, a mapping* $f \colon E \to \mathbb{N}$ *be a (not necessarily optimal) labeling with span* λ. *Then also the mapping* $f'(e) := \lambda - f(e)$ *is a valid labeling with the same span.*

We call such a derived labeling of the edges of a graph a λ-*inversion*.

2 Polynomial Cases

In this section we give a full description of graphs admitting a labeling with small number of labels, in particular graphs G with $\lambda'_{2,1}(G) \leq 4$. Moreover, these graphs can be recognized in polynomial time. This leads to Theorem 4, which is the main result of this section.

For the ease of presentation we split the proof and statement of the Theorem 4 into several lemmas, each for a particular value of $\lambda'_{2,1}(G)$.

Theorem 4 (Polynomial cases of distance edge-labeling). *For any graph* G *and for* $\lambda = 0, 1, 2, 3, 4$ *the* DISTANCE EDGE LABELING *problem* $\lambda'_{2,1}(G) = \lambda$ *(or* $\lambda'_{2,1}(G) \leq \lambda$*) can be solved in polynomial time. Moreover, it is possible to compute such a labeling in polynomial time.*

Without loss of generality we deal with connected simple undirected graphs.

First observe, that for $\lambda < 4$ the graph cannot contain a vertex of degree 3. We use P_i as a symbol for the *path* on i vertices.

Lemma 1 (Graphs with $\lambda'_{2,1}(G) \leq 3$).

- *The only graphs with $\lambda'_{2,1}(G) = 0$ are P_1 or P_2.*
- *There is no graph with $\lambda'_{2,1}(G) = 1$.*
- *The only graph with $\lambda'_{2,1}(G) = 2$ is P_3.*
- *Finally, graphs with $\lambda'_{2,1}(G) = 3$ are P_4 and P_5.*

When $\lambda = 4$, the graph may contain vertices with degree 3. We call a vertex *hairy* if it is of degree 3 and at least one of its neighbors is of degree 1. We call this degree one vertex, together with the connecting edge, *pendant*. Note that any vertex of degree 3 in a graph G satisfying $\lambda(G) = 4$ cannot have all its neighbors of degree 2 or greater. It is easy to see that there is no labeling of span 4 of such a graph. We say that two hairy vertices are *consecutive*, if there is no other hairy vertex on a path between them or if there is the only hairy vertex on a cycle. In this particular case the vertex is consecutive to itself.

For the purpose of the following lemmas, we say that a graph is a *generalized cycle* if it is a cycle with several (possibly 0) pendant edges. We say that a graph is a *generalized path* if it is a path with several (possibly 0) pendant edges. All observations made in the last paragraphs imply the following lemma:

Lemma 2. *Let G be a graph satisfying $\lambda'_{2,1}(G) \leq 4$, then G is either a generalized path or a generalized cycle.*

On the contrary not every generalized cycle or path has $\lambda'_{2,1} \leq 4$. The following lemmas state all the conditions for a generalized cycle or path to satisfy $\lambda'_{2,1} \leq 4$.

Notation in the proofs. Both proofs are done by a case analysis. For generalized cycles and paths the idea is to label path or cycle while there is the possibility to label all the pendant edges. To do so, we use sequences of numbers representing labels on edges. Note that it follows from Observation 2 that only numbers $0, 2, 4$ can occur around a hairy vertex and any pendant vertex must get label 2. For labelings we use sequences of numbers describing labels of consecutive edges and a symbol "|" for a hairy vertex—so there is a pendant edge on a vertex with label 2. This gives us immediately the following observation.

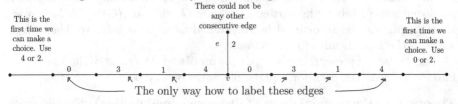

Observation 5 (The labeling of a hairy vertex and its neighborhood). *The neighborhood of a hairy vertex can be labeled only by a sequence $0314|0314$ or its λ-inversion $4130|4130$.*

Lemma 3. *Let $G = (V, E)$ be a generalized path. Let W be the set of all hairy vertices. Then $\lambda'_{2,1}(G) \leq 4$ if and only if for every consecutive pair $u, v \in W$ their distance $d = d_G(u, v)$ is either 4, or at least 8.*

Proof. We need to show that each sequence can be correctly labeled or that it is impossible to label it at all.

The easier fact is the existence of correct labelings. Sequences $|0314|(d = 4)$, $|031420314|(d = 9), |0314204130|(d = 10), |03140240314|(d = 11)$ can be extended by a sequence 0314 at the beginning to get sequences of length at least 8.

Now we have to show that there are no valid sequences of length $1, 2, 3, 5, 6, 7$. Observation 5 banns immediately sequences of length $1, 2, 3$. Furthermore, the same observation also implies that there is no chance to overlap two sequences which is necessary to get lengths $5, 6$ or 7.   □

Lemma 4. *Let $G = (V, E)$ be a generalized cycle. Let W be the set of all hairy vertices. Then $\lambda'_{2,1}(G) \leq 4$ if and only if for every consecutive pair $u, v \in W$ their distance $d = d_G(u, v)$ fulfills one of the following:*

- *$d = 4, 8, 9$ or $d \geq 11$,*
- *if there exists a consecutive pair with $d = 10$, then there is even number of such consecutive pairs, or there exists a consecutive pair with $d = 13, 14, 16$ or greater.*

Firstly it is easy to observe that cycles of any length without a hairy vertex can be labeled correctly.

The proof of the first part is similar to Lemma 3, except for the sequences of length 10. Because such a sequence cannot be connected via hairy vertex to any sequence presented in the proof of Lemma 3, unless we use a λ-inversion of some of them. So in the proof of the second part we need to show two things:

- The only labeling of a sequence of length 10 is the one already presented.
- The sequences of length less than or equal to 12 and 15 do not have a labeling that starts and ends by the label 0, while sequences of all other possible lengths admit such a labeling.

These arguments can be proved by a case analysis which we ommit due to space limitations.

3   NP-complete Cases

Theorem 6. *The problem* DISTANCE EDGE LABELING *is* NP*-complete for every fixed $\lambda \geq 5$.*

The proof of the hardness result is done for every $\lambda \geq 5$. However as there is a natural difference between odd and even λ, the proof is divided according to the parity of λ to two basic general cases. The proof of the even (odd) part is contained in Subsects. 3.2 and 3.3 respectively.

Furthermore, as the gadgets developed to carry the labeling does not work for small cases, we have to exclude the borderline values $\lambda = 5, 6, 7$ from the general proof. Due to space limitations, we ommit these proofs here. Furthermore, we also omit the full correctness of gadgets depicted in figures.

Our basic reduction tool is the MONOTONE NOT ALL EQUAL 3-SAT problem which all cases are reduced from. We say a formula φ is a *3-MCNF (monotone conjunctive normal form)* if it is a conjunction of clauses with exactly 3 logical variables without negations.

Definition 4 (MONOTONE NOT ALL EQUAL 3-SAT **problem (also known as** MNAE-3-SAT)).

Input:	*A 3-MCNF formula φ.*
Question:	*Is it possible to find an assignment such that each clause has at least one literal set to true and at least one literal set to false?*

This problem is a specialized version of NAE-3-SAT, which was shown to be NP-complete by Schaefer [12] by a more general argument about CSP's. We can find MNAE-3-SAT in the list of NP-complete problems in the monograph of Garey and Johnson [13].

The reduction procedure. For a 3-MCNF formula φ and positive integer $\lambda \geq 5$ we show how to build a graph G_φ^λ. We will ensure that $\lambda'_{2,1}(G_\varphi^\lambda) \leq \lambda$ if and only if the answer to the question of MNAE-3-SAT problem is "YES". In our proofs the main focus is to prove the correspondence between a satisfying assignment to the variables of φ and the λ-labeling of the graph G_φ^λ. We call this the *correctness of a gadget.*

Definition 5 (**Odd and Even sets**). *For any $\lambda \in \mathbb{N}$ we define two subsets of the set $\{0, \ldots, \lambda\}$. The* odd subset $\mathbb{O} = \{l \in \mathbb{N} : l \leq \lambda, l \text{ odd}\}$ *and the* even subset $\mathbb{E} = \{l \in \mathbb{N} : l \leq \lambda, l \text{ even}\}$.

Example 1 Take $\lambda = 10$ (even). Now according to Observation 2, the maximum possible degree of a vertex in a graph admitting a distance labeling with λ labels is 6. Moreover, only labels from the set $\mathbb{E}$ can appear on edges incident with such a vertex.

3.1   Basic Lemmas

We state here some auxiliary lemmas that are used in our reductions.

Lemma 5 (**Labeling of edges incident to a maximum degree vertex**). *Let $\lambda \in \mathbb{N}$, let G be a graph with $\lambda'_{2,1}(G) \leq \lambda$ and its maximum degree vertex v. Then:*

even λ: *If $deg(v) = \frac{\lambda}{2} + 1$ then vertex v has its incident edges labeled by labels from the set $\mathbb{E}$.*

odd λ: *If $deg(v) = \frac{\lambda+1}{2}$ then a vertex v has its incident edges labeled by labels from the one of the sets: $\mathbb{O}$, $\mathbb{O}\backslash\{1\} \cup \{0\}$, $\mathbb{E}$ or $\mathbb{E}\backslash\{\lambda - 1\} \cup \{\lambda\}$.*

Lemma 6 (Adjacent vertices with maximum degree, even span version). *Let $\lambda \in \mathbb{N}$, λ even and let $G = (V, E)$ be a graph with $\lambda'_{2,1}(G) \leq \lambda$. Take two neighboring vertices $u, v \in V$ such that $deg(u) = \frac{\lambda}{2} + 1$, $deg(v) = \frac{\lambda}{2}$ and $\{u, v\} \in E$.*

Then there are only two possibilities:

- *The edge $\{u, v\}$ is labeled by 0, all the edges incident to u are labeled by the elements from the set $\mathbb{E}\backslash\{0\}$ and finally all the edges incident to v are labeled by the elements from the set $\mathbb{O}\backslash\{1\}$.*
- *The edge $\{u, v\}$ is labeled by λ, all the edges incident to u are labeled by the elements from the set $\mathbb{E}\backslash\{\lambda\}$ and finally all the edges incident to v are labeled by the elements from the set $\mathbb{O}\backslash\{\lambda - 1\}$.*

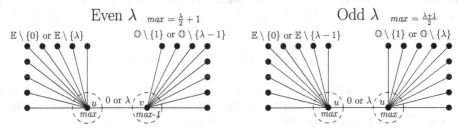

Lemma 7 (Adjacent vertices with maximum degree, odd span version). *Let $\lambda \in \mathbb{N}$, λ odd and let $G = (V, E)$ be a graph with $\lambda'_{2,1}(G) \leq \lambda$. Take two neighboring vertices $u, v \in V$ such that $deg(u) = deg(v) = \frac{\lambda+1}{2}$.*

Then there are only two possibilities:

- *The edge $\{u, v\}$ is labeled by 0, all the edges incident to u are labeled by the elements from the set $\mathbb{E}\backslash\{0\}$ and finally all the edges incident to v are labeled by the elements from the set $\mathbb{O}\backslash\{1\}$.*
- *The edge $\{u, v\}$ is labeled by λ, all the edges incident to u are labeled by the elements from the set $\mathbb{E}\backslash\{\lambda - 1\}$ and finally all the edges incident to v are labeled by the elements from the set $\mathbb{O}\backslash\{\lambda\}$.*

Proof of both lemmas above is an easy application of Lemma 5.

Notation in gadgets. We further use max as the number for the maximum degree in graph G with $\lambda'_{2,1}(G) \leq \lambda$. We also use directed edges in gadget graphs. An outgoing edge represents an *output*, while an ingoing edge represents an *input* to the gadget. We build all the gadgets so that the labels on output edges can take only several values.

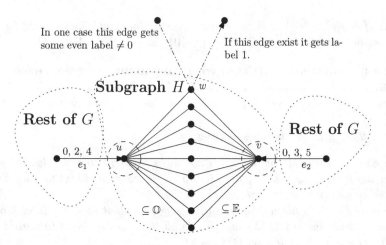

In one case this edge gets some even label $\neq 0$

If this edge exist it gets label 1.

Subgraph H w

Rest of G

Rest of G

$0, 2, 4$ u v $0, 3, 5$

e_1 e_2

$\subseteq \mathbb{O}$ $\subseteq \mathbb{E}$

Lemma 8 (A correct labeling of joint even and odd part). *Let* $\lambda \in \mathbb{N}$, *let* G *be a graph with* $\lambda'_{2,1}(G) \leq \lambda$ *and* H *be its subgraph represented by complete bipartite graph* $K_{2,\max -1}$ *such that:*

- *The only two edges connecting* $G \backslash H$ *to* H *are* e_1 *and* e_2, *where* $u \in e_1$ *and* $v \in e_2$.
- *The graph* H *contains vertices* $u \neq v$, $deg_G(u) = deg_G(v) \geq 4$ *and their common neighbors, call them* N. *Vertices from* N *are not adjacent, but exactly one of them* w *may have zero, one or two other neighbors outside* H.
- *Moreover, each edge* $\{u, z\}, z \in N$ *can be labeled only by odd labels* $(\mathbb{O})$ *and each edge* $\{v, z\}, z \in N$ *can be labeled only by even labels* $(\mathbb{E})$ *and has no other condition on them from the rest of* G. *(It's essential that they can be labeled by arbitrary label of appropriate set except the labels of edges* e_1 *and* e_2.)*

We have four cases which depends on labels of e_1 *and* e_2, *on the degree of* u *and* v *and on the number of neighbors of* w. *If one of the following cases happen:*

    I. *Both* e_1, e_2 *have label 0,* $deg_G(u) = deg_G(v) = \max$ *and the vertex* w *has one output edge. (for* λ *odd)*

    II. *Both* e_1, e_2 *have label 0,* $deg_G(u) = deg_G(v) = \max -1$ *and vertex* w *has two output edges. (for* λ *even)*

    III. *The edge* e_1 *has label 2 and edge* e_2 *has label 3 and* $deg_G(v) = deg_G(u) = \max -1$. *(for* λ *odd)*

    IV. *The edge* e_1 *has label 4 and edge* e_2 *has label 5 and* $deg_G(v) = deg_G(u) = \max -1$. *(for* λ *odd)*

Then all edges incident to vertices of N *can be labeled correctly.*

    I. *The output edge incident to* w *has to have a label 1.*

    II. *The output edges incident to* w *has to have 1 and some* $s \neq 0$ *even.*

Due to space limitations we have to omit the full proof of this technical lemma here. The idea of the proof is to construct an auxiliary bipartite graph.

Each edge of H is labeled by some label from the correct set and it is represented by a vertex. Two vertices are connected whenever they be incident in graph H without breaking condition of a correct labeling. It can be shown that such graph is almost k-regular for some k. Moreover we can delete some edges from that graph and then it becomes k-regular. Then we can found perfect matching using Hall marriage theorem [14].

The Labeling of the output edge is then easy to show because label 1 is the only unused label and it cannot be placed anywhere else. The other edge incident to the vertex w has an arbitrary nonzero even label and we have exactly one left.

The main reductions proof idea. We would like to give a reader the general idea used in proofs of all cases. We will develop some gadgets to model the two parts of the input of MNAE-3-SAT. Namely the logical variables and the formula itself, which we model clause by clause. Moreover, in general-case reductions we need some middle-pieces to glue them together.

To prove that the gadget for a variable works correctly we need to check that there is no any other labeling of output edges in the *variable gadget* than the one described in the image, or its λ-inversion. Note that the only possible labels on an output edge are 0 (or 1) and λ (or $\lambda - 1$)—these will represent the logical value of the variable. For now on, we omit the λ-inversion case in the proof. Every variable gadget contains a part with an output edge such that it is possible to repeat it arbitrarily—we call this part *repeatable*.

For a clause, we use a gadget for a given span with exactly 3 input edges. This *clause gadget* has to admit a labeling whenever at most two input edges represents the same logical value. On the other hand it does not admit a labeling when all input edges represents the same logical value.

3.2 Even $\lambda \geq 8$

We divide the *variable gadget* into three parts. The *initial part* and the *ending part* are only technical support for starting and ending process correctly. The main work is done in the *repeatable part*.

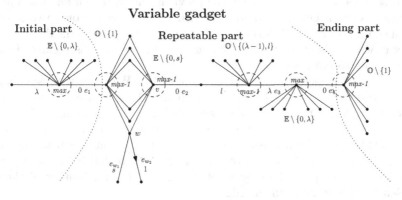

Variable gadget

By Lemma 6 the label of e_1 is 0. Now we have two possibilities (sets) how to label all the edges incident to the vertex v: $\mathbb{E} \cup \{1\} \backslash \{0, 2\}$ and $\mathbb{E} \backslash \{s \in \mathbb{E}\}$. If we label edges incident to v from the set $\mathbb{E} \cup \{1\} \backslash \{0, 2\}$ it is impossible to label both edges e_{w_1}, e_{w_2} incident to the vertex w, because we need to use both 0, 2 labels on them. But the label 0 is already used for the edge e_1 which is at distance two. While if we label these edges from the set $\mathbb{E} \backslash \{s \in \mathbb{E}\}$, in this case it is possible to label the output edge by s or by 1.

Later the middle-piece gadget further restricts the output, so that the only possible label is 1.

To prove that it is correct we use Lemma 8 part II.

Edges e_3 and e_4 need to have labels 0 or λ by Lemma 6. As e_3 is in distance two to e_2 and e_2 is labeled by 0 implies that e_3 cannot have label 0.

The *middle-piece gadget* gives us only two possible outputs: 2 or 0. This is because Lemma 5. Moreover, this implies that the only possible labeling of input edges is by the label 1.

The output of the middle-piece gadget is plugged into the *clause gadget*.

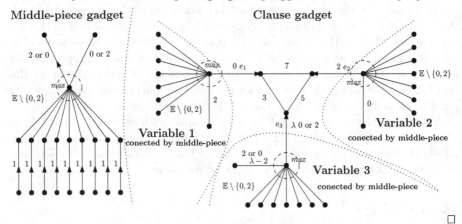

    □

3.3   Odd $\lambda \geq 9$

This case is more complicated than the previous one. A reason for this is in the difference between Lemmas 7 and 6. In either case there are only two possible labelings, but in Lemma 7 the degree of the vertex u equals to the degree of the vertex v, while this is not true in Lemma 6 and so we can distinguish them in the even case shown before.

We start with correctness of the *variable gadget*. We prove that neighboring edges of vertex v are labeled by labels from $\mathbb{O} \backslash \{1\} \cup \{0\}$. We proceed by contradiction. Suppose that these edges are labeled by $\mathbb{E}$ (according to Lemma 5 this is the only other option) then edges incident to the vertex u has labels from $\mathbb{O} \backslash \{1\} \cup \{0\}$. Then exists the edge $e = \{v, z\}$ that is labeled by some odd $l \neq \lambda$. So the neighborhood of the vertex z can be labeled either by a set $\mathbb{E} \backslash \{0, 2, l-1, l+1\} \cup \{1\}$ or by a set $\mathbb{E} \backslash \{0, l-1, l+1\}$. Neither of them is sufficiently large to label all the edges.

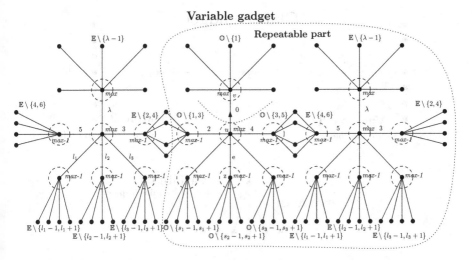

The correctness of the other labeling is shown in the image.

Lemma 8 parts III. and IV. ensures that it is possible to repeat the repeatable part of the gadget. Note that the repeatable part consists of two identical parts, but it is possible to use only one of them as an output, because these parts are labeled λ-symmetrically.

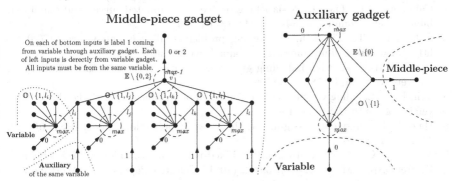

The correctness of the *auxiliary gadget* is described in Lemma 8 part I. The purpose of this gadget is to create an edge with label 1.

The *middle-piece gadget* has two kinds of inputs. Both kinds of inputs correspond to the *variable gadget*, but one of them is connected to the middle-piece through the *auxiliary gadget*.

The edges incident to the vertex v can by labeled only by labels from the set $\{\mathbb{E}\}$. This is ensured by the variable inputs, because they contains each label from the set $\mathbb{O}\backslash\{1\}$ and also by auxiliary inputs containing label 1. Note, that we can create as many such inputs as it is needed. Moreover, the label 1 forbids labels 0 and 2 anywhere besides the output edge.

Each output from the *middle-piece gadget* is plugged into the *clause gadget* in the following way, which completes the proof.

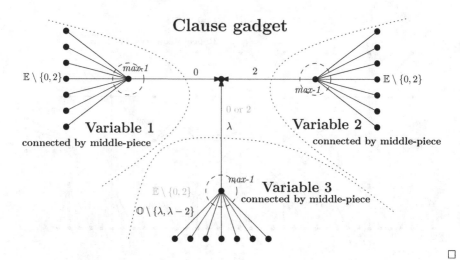

References

1. Formanowicz, P., Tanaś, K.: A survey of graph coloring - its types, methods and applications. Found. Comput. Decis. Sci **37**(3), 223–238 (2012)
2. Calamoneri, T.: The L(h, k)-labelling problem: an updated survey and annotated bibliography. Comput. J. **54**(8), 1344–1371 (2011)
3. Calamoneri, T.: The L(h, k)-labelling problem (online updated survey) (2013). http://wwwusers.di.uniroma1.it/~calamo/survey.html
4. Griggs, J.R., Yeh, R.K.: Labelling graphs with a condition at distance 2. SIAM J. Discrete Math. **5**(4), 586–595 (1992)
5. Yeh, R.K.: Labeling Graphs with a Condition of Distance Two. Ph.D. thesis, University of South Carolina (1990)
6. Hale, W.K.: Frequency assignment: theory and applications. Proc. IEEE **68**(12), 1497–1514 (1980)
7. Fiala, J., Kloks, T., Kratochvíl, J.: Fixed-parameter complexity of λ-labelings. Discrete Appl. Math. **113**(1), 59–72 (2001). Selected Papers: 12th Workshop on Graph-Theoretic Concepts in Computer Science.
8. Vizing, V.G.: On an estimate of the chromatic class of a p-graph. Diskret. Analiz **3**(7), 25–30 (1964)
9. Havet, F., Reed, B., Sereni, J.S.: Griggs and yeh's conjecture and $l(p, 1)$-labellings. SIAM J. Discrete Math. **26**(1), 145–168 (2012)
10. Flum, J., Grohe, M.: Parameterized Complexity Theory. Texts in Theoretical Computer Science. An EATCS Series. Springer, Heidelberg (2006)
11. Diestel, R.: Graph Theory. Electronic library of mathematics. Springer, New York (2006)
12. Schaefer, T.J.: The complexity of satisfiability problems. In: Proceedings of the Tenth Annual ACM Symposium on Theory of Computing, STOC 1978, pp. 216–226. ACM, New York (1978)
13. Garey, M.R., Johnson, D.S.: Computers and Intractability: A Guide to the Theory of NP-Completeness. W. H. Freeman & Co., New York (1979)
14. Hall, P.: On representatives of subsets. J. Lonon Math. Society **101**, 26–30 (1935)

1.5-Approximation Algorithm for the 2-Convex Recoloring Problem

Reuven Bar-Yehuda[1], Gilad Kutiel[1]($\boxtimes$), and Dror Rawitz[2]

[1] Department of Computer Science, Technion, Haifa, Israel
{reuven,gkutiel}@cs.technion.ac.il
[2] Faculty of Engineering, Bar Ilan University, Ramt-Gan, Israel
dror.rawitz@biu.ac.il

Abstract. Given a graph $G = (V, E)$, a coloring function $\chi : V \to C$, assigning each vertex a color, is called *convex* if, for every color $c \in C$, the set of vertices with color c induces a connected subgraph of G. In the CONVEX RECOLORING problem a colored graph G_χ is given, and the goal is to find a convex coloring χ' of G that *recolors* a minimum number of vertices. The 2-CONVEX RECOLORING problem (2-CR) is the special case, where the given coloring χ assigns the same color to at most two vertices. 2-CR is known to be NP-hard even if G is a path.

We show that weighted 2-CR problem cannot be approximated within any ratio, unless $P = NP$. On the other hand, we provide an alternative definition of (unweighted) 2-CR in terms of maximum independent set of paths, which leads to a natural greedy algorithm. We prove that its approximation ratio is $\frac{3}{2}$ and show that this analysis is tight. This is the first constant factor approximation algorithm for a variant of CR in general graphs. For the special case, where G is a path, the algorithm obtains a ratio of $\frac{5}{4}$, an improvement over the previous best known approximation. We also consider the problem of determining whether a given graph has a convex recoloring of size k. We use the above mentioned characterization of 2-CR to show that a problem kernel of size $4k$ can be obtained in linear time and to design a $O(|E|) + 2^{O(k \log k)}$ time algorithm for parametrized 2-CR.

1 Introduction

Let $G = (V, E)$ be a graph and let $\chi : V \to C$ be a coloring function, assigning each vertex in V a color in C. We say that χ is a *convex coloring* of G, if for every color $c \in C$, the vertices with color c induce a connected sub-graph of G. In the CONVEX RECOLORING problem (abbreviated CR), we are given a colored graph G_χ, and we wish to find a recoloring of a minimum number of vertices of G, such that the resulting coloring is convex. That is, the goal is to find a convex coloring χ', that minimizes the size of the set $\{v : \chi(v) \neq \chi'(v)\}$. The t-CONVEX RECOLORING problem (t-CR) is the special case, in which the given coloring assigns the same color to at most t vertices in G.

D. Rawitz—Supported in part by the Israel Science Foundation (grant no. 497/14).

© Springer International Publishing Switzerland 2016
Z. Lipták and W.F. Smyth (Eds.): IWOCA 2015, LNCS 9538, pp. 299–311, 2016.
DOI: 10.1007/978-3-319-29516-9_25

The CONVEX RECOLORING problem (CR) in trees was introduced by Moran and Snir [10] and was motivated by its relation with the concept of *perfect phylogeny*. They proved that the problem is NP-hard [6], even when the given graph is a simple path. Later, Kanj et al. [6] showed that 2-CR is also NP-hard on paths. Applications of CR in general graphs, such as multicast communication, were described by Kammer and Tholey [5]. Many variants of the problem have been intensively investigated. The differences between one variant to another can be related to

- The structure of the given graph G. The given graph can be a simple path, a tree, a bounded treewidth graph, a general graph, and others.
- Constraints on the coloring function χ. In a t-coloring at most t colors are used to color a graph, while in a t-CR instance at most t vertices are colored using the same color.
- Type of weight function. In the weighted case, each vertex is associated with a weight, and the weight of a recoloring is the total weight of recolored vertices. In the unweighted case the weight of the solution is the number of recolored vertices. In a third variant, referred to as *block recoloring* [5], a cost is incurred for a color c if at least one vertex of color c was recolored.

Since CR was shown to be NP-hard it was natural to try to design both approximation algorithms and parameterized algorithms.

Moran and Snir [9] presented a 2-approximation algorithm for CR in paths and a 3-approximation algorithm for CR in trees. Both algorithms work for the problem with costs. Bar-Yehuda et al. [2] improved the latter by providing a $(2 + \varepsilon)$-approximation algorithm for CR in trees. This result was later extended to bounded treewidth graphs by Kammer and Tholey [5]. Recently, Lima and Wakabayashi [8] gave a $\frac{3}{2}$-approximation algorithm for unweighted 2-CR in simple paths.

On the negative side, Kammer and Tholey [5] proved that if vertex weights are either 0 or 1, then 2-CR has no polynomial time approximation algorithm with a ratio of size $(1 - o(1)) \ln \ln n$ unless NP $\subseteq$ DTIME($n^{O(\log \log n)}$). In Sect. 2 we show that this variant of the problem can not be approximated at all. Campêlo et al. [4] showed that, for $t \geq 2$, CR is NP-hard on t-colored grids. They also proved that there is no polynomial time approximation algorithm within a factor of $c \ln n$ for some constant $c > 0$, unless $P = NP$, for unweighted CR in bipartite graphs with 2-colorings.

Moran and Snir [10] presented an algorithm for CR whose running time is $O(n^4 \cdot k(\frac{k}{\log k})^k)$, where k is the number of recoloring in an optimal solution. Razgon [12] gave a $2^{O(k)}$poly(n) time algorithm for CR in trees. Ponta et al. [11] designed several algorithms for different variants of CR in trees. For the unweighted case, they gave a $O(3^b \cdot b \cdot n)$ time algorithm, where b is the number of colors that do not induce a connected subtree, and it is bounded from above by $2k$. Bar-Yehuda et al. [2] provided an algorithm with an upper bound of $O(n^2 + n \cdot k2^k)$ on the running time, but it is not hard to verify that this bound can be improved to $O(n \cdot k2^k)$. Bodlaender et al. [3] showed that CR admits a kernel of size $O(k^2)$ in trees. Bachoore and Bodlaender [1] presented an algorithm for leaf-colored trees with a running time of $O(4^k \cdot n)$. Campêlo et al. [4] proved that, for $t \geq 2$, CR is $W[2]$-hard in t-colored graphs.

Our Results. Sect. 2 contains our hardness result for the weighted version of 2-CR. In Sect. 3 we provide an alternative definition of 2-CR in terms of maximal independent set of paths. We first show that we can focus on a specific type of recoloring, called a *path recoloring*, in which each color induces a path. Then we show that finding a path recoloring can be translated into finding an independent set of paths. In Sect. 4 we present a greedy algorithm for (unweighted) 2-CR in general graphs that is based on iteratively adding a shortest path to the current independent set of paths. We provide a tight analysis for the algorithm and show that its approximation ratio is $\frac{3}{2}$. This is the first time a constant-ratio approximation algorithm is given for a variant of CR in general graphs. We also show that when G is a simple path, the same algorithm yields a $\frac{5}{4}$-approximation, improving the previous best known approximation ratio by Lima and Wakabayashi [8]. In Sect. 5 we use the above mentioned characterization of 2-CR to show that a problem kernel of size $4k$ can be obtained in linear time. This leads to a $O(|E|) + 2^{O(k \log k)}$ time algorithm for 2-CR parameterized by the number of color changes k.

2 Hardness Result

In this section we prove that the weighted version of 2-CR cannot be approximated within any multiplicative ratio, unless $P = NP$. We do so using a simple reduction from the DISJOINT CONNECTING PATHS problem.

In the DISJOINT CONNECTING PATHS problem the input consists of an undirected graph $G = (V, E)$ and a set of pairs $\{(s_1, t_1), \ldots, (s_k, t_k)\}$, and the question is whether there are k vertex-disjoint paths that connect s_i to t_i. This problem is known to be NP-hard [7].

Theorem 1. *The weighted version of* 2-CR *cannot be approximated within any multiplicative ratio, unless* $P = NP$.

Proof. Let G and $\{(s_1, t_1), \ldots, (s_k, t_k)\}$ be an instance of DISJOINT CONNECTING PATHS. We constructed a weighted 2-CR instance as follows. First, we use the same graph G. Let $\chi(v) = c_0$, if $v \notin \{s_1, \ldots, s_k\} \cup \{t_1, \ldots, t_k\}$, and let $\chi(s_i) = \chi(t_i) = c_i$, for every i, where $c_0, \ldots, c_k$ are distinct colors. Define a weight function w as follows: $w(v) = 1$, if $u \in \{s_1, \ldots, s_k\} \cup \{t_1, \ldots, t_k\}$, and $w(v) = 0$, otherwise. It is not hard to verify that k disjoint paths exist in G if and only if the minimum convex recoloring costs zero. Hence, it is NP-hard to decide whether a zero cost solution exists.   □

3   Properties of Optimal Recolorings

In this section we introduce a special type of convex recoloring, called *path-recoloring*. Path-recolorings are more constrained than general recolorings and thus are simpler to understand and analyze. Nevertheless, we show that there is always an optimal convex recoloring that is a path-recoloring. Based on the above, we give an alternative definition to 2-CR in terms of independent set of paths. From now on, we only consider the 2-CR problem, in particular, whenever we mention a colored graph, we refer to a 2-CR instance.

Given a colored graph G_χ, if two vertices in the colored graph have the same color we call them a *pair*, if they are connected with an edge then they are a *connected pair*, otherwise they are a *disconnected pair*. Any vertex with a unique color c is a *singleton*, we call c a *singleton color*. We denote by $G_\chi[c]$ the subgraph induced by the set of vertices $\{v : \chi(v) = c\}$. Figure 1 depicts these concepts.

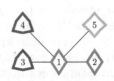

Fig. 1. In this colored graph, vertex 5 is a singleton, vertices 1 and 2 are a *connected pair*, and vertices 3 and 4 are a disconnected pair (Color figure online).

3.1   Path-Recoloring

As a first step we show that it may be assumed that all colors retain at least one representative. In other words, we show that there is always an optimal convex recoloring χ' that does not recolor singletons and recolors at most one vertex of every pair.

Given a colored graph G_χ and a recoloring χ', a vertex v *retains* its color if $\chi(v) = \chi'(v)$. We say that χ' *retains* a pair p, if both vertices of p retains their color. The recoloring χ' *retains* a color $c \in C$, if there exists a vertex $v \in G$ such that $\chi'(v) = \chi(v) = c^1$. If a recoloring retains all the colors of a graph, we refer to it as a *retains-all* recoloring. Observe that a retains-all recoloring does not recolor singletons. We show that there exists a retains-all optimal recoloring.

Lemma 1. *For every colored graph G_χ, there exists a retains-all optimal convex recoloring.*

Proof. Consider an optimal convex recoloring χ' that retains a maximum number of colors over all optimal, convex recolorings of G. Assume for contradiction that χ' does not retain a color c, and let v be a vertex such that $\chi(v) = c$ and $\chi'(v) = c' \neq c$. Without loss of generality we assume that c is not used by χ' (otherwise, we can recolor each vertex in $G_{\chi'}[c]$ using a new unique color^2). We define a recoloring χ'' by considering $G_{\chi'}[c']$. First, define $\chi''(v) = c$. Next, if $G_{\chi'}[c']$

---

[1] Note that this definition is a different than the one given in [6].

[2] Unique colors are used for simplicity. The new colors can be replaced by original colors, by iteratively recoloring a vertex with unique color using the color of an adjacent vertex which is colored by an original color.

contains a vertex u such that $\chi(u) = c'$, then define $\chi''(u) = c'$. Each vertex of the remaining vertices in $G_{\chi'}[c']$ is colored by χ'' using a unique new color. Observe that if there exists a second vertex $u' \neq u$ such that $\chi'(u') = \chi(u') = c'$, then χ'' recolors it. Finally, $\chi''(x) = \chi'(x)$ for any vertex x not in $G_{\chi'}[c']$.

χ'' is convex, since all vertices in $G_{\chi'}[c']$ are colored by different colors. χ'' recolors at most as many vertices as χ', since it may recolor u's mate (if it exists), but it avoids the recoloring of v. Finally, χ'' retains more colors than χ', since it retains c. Thus, we obtained an optimal recoloring that retains more colors than χ'. A contradiction. □

Next we show that we need not recolor connected pairs.

Lemma 2. *For every colored graph G_χ there exists a retains-all, optimal, convex recoloring that does not recolor any connected pair.*

Proof. Consider an optimal retains-all convex recoloring χ' that retains the maximum number of connected pairs, over all optimal, retains-all recolorings of G. Assume for contradiction that χ' recolors one of the vertices of a connected pair $\{u, v\}$, that is, w.l.o.g., $\chi(u) = c$ and $\chi'(v) = c' \neq c$. We can use a similar argument to the one used in the proof of Lemma 1. The difference is in the definition of χ''. If $G_{\chi'}[c']$ contains a connected pair u_1, u_2 such that $\chi(u_1) = \chi(u_2) = c'$, then we define $\chi''(u_1) = \chi''(u_2) = c'$. Observe that χ'' is convex since all vertices in $G_{\chi'}[c']$ are colored using different colors, with the exception of u and v which are colored by c, and u_1 and u_2 which are colored by c' if they exist. Observe also that χ'' recolors at most as many vertices as χ', and retains more connected pairs than χ'. A contradiction. □

We are now ready to define *path-recolorings*. Given a colored graph G_χ and a convex recoloring χ', we say that χ' *path-recolors* G with respect to $c \in C$ if the vertices of $G_{\chi'}[c]$ form a simple path: $u, \ldots, v$ such that $\chi(u) = \chi(v) = c$. A special case of this definition is when $G_{\chi'}[c]$ is a single vertex v and $\chi(v) = c$. We say that χ' is a *path-recoloring* if:

1. χ' does not recolor any connected pair, and
2. χ' path-recolors G with respect to every $c \in C$.

Clearly, every path recoloring, also retains all colors.

Lemma 3. *For every colored graph G_χ there exists an optimal recoloring that is a path recoloring.*

Proof. Let χ' be an optimal, retains-all recoloring that does not recolor any connected pair (whose existence was shown in Lemma 2) that path-recolors G with the maximum possible number of colors in C. Assume for contradiction that χ' is not a path-recoloring. Hence there is a color c such that χ' does not path-recolor G with c. Consider $G_{\chi'}[c]$, and assume for now that there are two vertices u and v in $G_{\chi'}[c]$ such that $\chi(u) = \chi(v) = c$. Fix a simple path from u to v in $G_{\chi'}[c]$, and let χ'' be identical to χ' with the following modification: χ''

assigns a unique color for every vertex in $G_{\chi'}[c]$ that is not on the simple path from u to v. Clearly, χ'' is an optimal recoloring that path-recolors G with more colors than χ'. A contradiction.

Now, if there is at most one vertex v in $G_{\chi'}[c]$ such that $\chi(v) = c$ then consider a recoloring χ'' that is identical to χ' except it assigns a unique color to every vertex in $G_{\chi'}[c]$ that is not v. This time, again, we found an optimal recoloring that path-recolors G with more colors than χ'. A contradiction.   □

Henceforth, whenever we refer to a recoloring, we assume that it is a path-recoloring, in particular, we assume that for every disconnected pair in G with color c, a recoloring either: (i) colors a path between the disconnected pair with a color c, (ii) colors exactly one of its vertices.

3.2   Path Independence

When a recoloring colors a path between disconnected pair, we refer to the path as a *colored path*. Let D be the set of all disconnected pairs in G, and denote by I the set of colored paths in $G_{\chi'}$, then the following lemma holds:

Lemma 4. *Given a colored graph G_χ, a path-recoloring χ' recolors exactly $|D| - |I|$ vertices.*

Proof. By definition, χ' does not recolor any of the pairs that form the endpoints of paths in I. χ' must recolor exactly one vertex of every other disconnected pair: if it recolors both vertices then the color of this pair is not retained, and if it recolors none of them then convexity does not hold.   □

Given a colored graph G_χ and a path p, let $V(p)$ be the set of vertices on the path and let $\chi(p)$ be the set of colors assigned to vertices on this path, i.e. $\chi(p) = \{\chi(v) : v \in V(p)\}$. Given two paths p_1 and p_2 in G_χ:

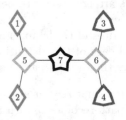

- p_1 and p_2 are in *direct conflict* if $V(p_1) \cap V(p_2) \neq \emptyset$,
- p_1 and p_2 are in *indirect conflict* if $\chi(p_1) \cap \chi(p_2) \neq \emptyset$.

Fig. 2. In this colored graph, the paths (4, 6, 3) and (1, 5, 2) are colorable, while the path (6, 7, 5) is not. Path (4, 6, 3) is in indirect conflict with path (1, 5, 2) and in direct conflict with path (6, 7, 5) (Color figure online)

p_1 and p_2 are in *conflict* if they are either in a direct or indirect conflict (observe that two paths can be both in direct and indirect conflicts). If two paths are not in conflict, then they are *independent*. Given a set of paths I, we say that this set is *independent* if it is pairwise independent, that is, if every two paths $p_1, p_2 \in I$ are independent. A path $u, \ldots, v$ in G is called *colorable* if u and v form a disconnected pair and the path does not contain singletons nor vertices of connected pairs. Figure 2 depicts these concepts.

In the next lemma we show that the set of colored paths that is induced by a path-recoloring is an independent set of colorable paths.

Lemma 5. *Let G_χ be a colored graph. Also, let χ' be a path-recoloring, and let I denote the set of colored paths (w.r.t. χ'). Then I is an independent set of colorable paths with respect to χ.*

Proof. Obviously, two colored paths in I cannot be in direct conflict. Assume for contradiction that there are two paths $p_1, p_2 \in I$ that are in indirect conflict, that is, there is a color $c \in \chi(p_1) \cap \chi(p_2)$. It follows that χ' recolors two vertices of the same color, and we get a contradiction since χ' does not retain all colors. Finally, it follows that all colored paths are colorable with respect to χ, or otherwise χ' must recolor a singleton or a connected pair. $\square$

We say that a set of paths, I, *covers* a pair if at least one of the vertices of the pair belongs to one of the paths in I.

Lemma 6. *For any independent set of colorable paths I in G_χ, there exists a convex path-recoloring χ' of G, where I is the set of colored path.*

Proof. Consider a recoloring χ' that colors every colorable path in I using the color of its endpoints and assigns a unique new color for every disconnected pair that is not covered by I. We first show that χ' is a path recoloring. Since I is independent, no two paths in I contain the same color, moreover, every path in I is colorable, thus it does not contain singletons, and it follows that χ' retains all colors. Also, every path in I does not contain any vertices of connected pairs, thus I does not recolor vertices of connected pairs. Finally, by the construction of χ', it is not hard to verify that for every color $c \in C$, the subgraph $G_{\chi'}[c]$ is either a simple path or a single vertex. Further observe that for every color $c \in \text{Image}(\chi')$, $G_{\chi'}[c]$ is either a simple path or a single vertex, and it follows that χ' is a convex recoloring. Finally, by the construction of χ', it is straightforward to verify that I is the set of colored paths induced by χ'. $\square$

The following is obtained due to Lemmas 4, 5 and 6.

Theorem 2. *Given a colored graph G_χ the cost of an optimal (path-)recoloring is $|D| - s$ if and only if the size of the maximum independent set of colorable paths is s.*

Theorem 2 suggests an alternative definition to 2-CR: given a colored graph G_χ, find a maximum independent set of colorable paths in G_χ.

4 Greedy Algorithm

In this section we describe a natural greedy algorithm to construct a maximal independent set of colorable paths, we discuss how this algorithm can be implemented, and define the corresponding path-recoloring it computes.

While we are not attempting to achieve an approximation to the size of the maximum independent set of colorable paths, the alternative definition given at the end of the previous section leads us to a natural greedy algorithm: choose the shortest colorable path that is not in conflict with colorable paths already been chosen and add it to an independent set of colorable paths. A formal description of this algorithm is given in Algorithm 1.

Algorithm 1. Greedy algorithm for 2-CR

$I \leftarrow \emptyset$
while there is a colorable path, not in conflict with I **do**
    add to I a shortest colorable path, not in conflict with I
end while
return the path-recoloring corresponding to I

We now describe how a shortest colorable path can be found. To do that, at the initialization of the algorithm, all singletons and connected pairs should be removed from the graph. Every colorable path that is added to I should be also removed from the graph. In addition, after each path removal, one should also remove all vertices with a unique color in the remaining graph. On the remaining graph, a shortest path, between two vertices of the same color, is guaranteed to be colorable and independent of I. In particular, it cannot contain two vertices that are colored by the same color, since such a path is not a shortest path.

Analysis. We now analyze the greedy algorithm and prove that it recolors at most $\frac{3}{2} \cdot k$ vertices, where k is the minimum number of vertices that must be recolored to achieve a convex coloring. We also show that the analysis is tight even in trees. For the special case when G is a simple path, we show that the greedy algorithm achieves a $\frac{5}{4}$ approximation ratio. We show that the analysis is tight for this special case as well.

Recall that D is the set of disconnected pairs in G, $I \subseteq D$ is the set of independent paths colored by the greedy algorithm, and let $I^* \subseteq D$ be the set of independent paths colored by an arbitrary, but fixed, optimal path-recoloring. Define $\alpha := \frac{|I^*|}{|I|}$, then by Lemma 4 the ratio between the number of vertices recolored by the greedy algorithm and the number of vertices recolored by an optimal path-recoloring is: $r := \frac{|D|-|I|}{|D|-|I^*|} = \frac{|D|-|I|}{|D|-\alpha|I|}$. We now analyze the relationship between $|I|$, $|D|$ and α.

Let ℓ_p be the number of vertices on a path p, and observe that:

Lemma 7. *If $p' \in I^* \backslash I$ then there is a path $p \in I$ that is in conflict with p', and $\ell_p \leq \ell_{p'}$.*

Proof. Consider the running of the greedy algorithm at the most recent point, where I contains no path with a distance greater than $\ell_{p'}$. If, at this point, there is no path in I that is in conflict with p' then nothing prevents the greedy algorithm from adding p' to I.   □

Next, we assign each optimal path to a greedy path. Given a path $p' \in I^*$, the *conflict source* of p' is p' itself if $p' \in I$, otherwise it is an arbitrary shortest path in I that is in conflict with p'. Notice that this is a many-to-one assignment, that is, several optimal paths can be assigned to a single greedy path. For a path $p \in I$, the set of paths in I^* such that p is their conflict source is denoted as $N(p)$. The members of $N(p)$ are called the *neighbours* of p.

Due to Lemma 7 we have that:

Observation 1. *For every path $p \in I$, if $p' \in N(p)$, then $\ell_p \le \ell_{p'}$.*

Denote $d_p := |N(p)|$ and refer to d_p as the *degree* of p. Our next goal is to find an upper bound to d_p, for $p \in I$.

Lemma 8. $d_p \le \ell_p - 1$.

Proof. Given a path $p \in I$, we show that the number of paths in I^* that are in conflict with p is at most $\ell_p - 1$. Associate a vertex in p to every path that is in conflict with p. For a path that is in direct conflict with p associate a common vertex of these paths, for a path that is in indirect conflict due to color c, associate the vertex that has color c in p. Recall that no two paths in I^* are in conflict, and observe that associating the same vertex to more than one path implies the existence of such conflict. Finally, only one end point of p can be associated with a path in I^*, or otherwise, I^* is not independent. $\square$

In what follows we obtain two lower bounds on $|D|$, which translate into two upper bounds on the approximation ratio r.

Lemma 9. $|D| \ge 2|I^*|$.

Proof. Every path in I^* has length of at least 3, or otherwise it is a connected pair and the considered recoloring is not a path-recoloring. Thus, we can observe the existence of at least 2 disconnected pairs for every path in I^*. We do not count the same disconnected pair twice, or otherwise, I^* is not independent. $\square$

Observation 2. $\sum_{p \in I} d_p = |I^*| = \alpha \cdot |I|$

Proof. By definition every path $p' \in I^*$ has one, and only one, conflict source in I, thus, there exists exactly one path $p \in I$ such that $p' \in N(p)$. $\square$

Lemma 10. $\frac{\sum_{p \in I} d_p^2}{|I|} \ge \alpha^2$.

Proof. Due to Observation 2 we have that the average degree of paths in I is α, i.e. $\frac{\sum_{p \in I} d_p}{|I|} = \alpha$. The lemma follows from the generalized mean inequality. $\square$

Lemma 11. $|D| \ge \alpha^2 |I|$.

Proof. Let v be a vertex in a path $p' \in I^*$, then v must be a part of a disconnected pair, otherwise v is a singleton or part of a connected pair and thus the considered recoloring is not a path-recoloring. Observe, also, that aside from the two endpoints of each path, no two vertices of the same disconnected pair can be present in I^* (or otherwise I^* is not independent). Thus, we can count $\ell_{p'} - 1$ disconnected pairs for every path p' in I^*. Hence

$$|D| \geq \sum_{p' \in I^*} (\ell_{p'} - 1) = \sum_{p' \in I^*} \ell_{p'} - |I^*| .$$

Consider a path $p \in I$, recall from Observation 1 that if $p' \in N(p)$ then $\ell_{p'} \geq \ell_p$. Hence, $\sum_{p' \in N(p)} \ell_{p'} \geq d_p \cdot \ell_p$, and it follows that

$$\sum_{p' \in I^*} \ell_{p'} \geq \sum_{p \in I} d_p \cdot \ell_p \geq \sum_{p \in I} d_p(d_p + 1) = \sum_{p \in I} d_p^2 + |I^*| ,$$

where the second inequality is due to Lemma 8 and the equality is due to Observation 2. Hence, by Lemma 10 we have that $|D| \geq \sum_{p \in I} d_p^2 \geq \alpha^2 |I|$. □

We can now obtain two upper bounds on the approximation ratio r as a function of α.

Theorem 3. *The greedy algorithm is a $\frac{3}{2}$-approximation algorithm for 2-CR.*

Proof. Using Lemma 11 and the fact that $\alpha \geq 1$ we get that

$$r = \frac{|D| - |I|}{|D| - \alpha \cdot |I|} \leq \frac{\alpha^2 \cdot |I| - |I|}{\alpha^2 \cdot |I| - \alpha \cdot |I|} = \frac{\alpha^2 - 1}{\alpha^2 - \alpha} = \frac{\alpha + 1}{\alpha} ,$$

and from Lemma 9 we get that

$$r = \frac{|D| - |I|}{|D| - \alpha \cdot |I|} \leq \frac{2\alpha \cdot |I| - |I|}{2\alpha \cdot |I| - \alpha \cdot |I|} = \frac{2\alpha - 1}{\alpha} .$$

Putting the two bound together, it follows that $r \leq \frac{1}{\alpha} \cdot \min\{\alpha + 1, 2\alpha - 1\} \leq \frac{3}{2}$ □

We show that our analysis is tight even for colored trees, using the instance depicted in Fig. 3, which consists of a colored graph where the greedy algorithm might recolor $\frac{3}{2}$ times more vertices than the optimal convex recoloring. We note that one can simply duplicate the instance using new colors for each copy in order to construct an arbitrary large graph with the same approximation ratio.

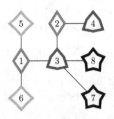

Fig. 3. Greedy might choose to color the path $(1, 3, 2)$, then it must recolor one of the vertices $\{5, 6\}$ and one of the vertices $\{7, 8\}$, a total of three recolored vertices, while an optimal recoloring can color two paths: $(5, 1, 6)$ and $(7, 3, 8)$, a total of two recolored vertices (Color figure online).

Simple Paths. We show that for the special case where G is a path, the greedy algorithm achieves an even better approximation ratio. To do that we start by observing that, in the case of a path, we can replace Lemma 8 with the following lemma:

Lemma 12. $d_p \leq \ell_p - 2$.

Proof. Given a path $p \in I$, we show that the number of paths in I^* that are in conflict with p is at most $\ell_p - 2$. This is true, because if G is a path, then for every path that is in conflict with p, we can now associate a vertex that is not one of the endpoints of p. If the conflict is direct, then the two paths must overlap, thus associate the other path's endpoint. If the conflict is indirect, then this must be due to some color other than the one on the endpoints of p, or else this is a direct conflict. $\qquad\square$

Using the above lemma we can strengthen Lemma 11.

Lemma 13. $|D| \geq (\alpha^2 + \alpha)|I|$.

Proof. The proof is similar to the proof of Lemma 11, where the main difference is that

$$\sum_{p' \in I^*} \ell_{p'} \geq \sum_{p \in I} d_p \cdot \ell_p \geq \sum_{p \in I}(d_p^2 + 2d_p) = \sum_{p \in I} d_p^2 + 2|I^*| \ ,$$

which means that

$$|D| \geq \sum_{p' \in I^*} \ell_{p'} - |I^*| \geq \sum_{p \in I} d_p^2 + |I^*| = \sum_{p \in I} d_p^2 + \alpha|I| \geq (\alpha^2 + \alpha)|I| \ ,$$

and the lemma follows. $\qquad\square$

In this case only one upper bound suffices.

Theorem 4. *Greedy is a $\frac{5}{4}$-approximation algorithm for 2-CR on paths.*

Proof. Using Lemma 13 we get that

$$r = \frac{|D| - |I|}{|D| - \alpha \cdot |I|} \leq \frac{(\alpha^2 + \alpha)|I| - |I|}{(\alpha^2 + \alpha)|I| - \alpha|I|} = \frac{\alpha^2 + \alpha - 1}{\alpha^2} \leq \frac{5}{4},$$

as required. $\qquad\square$

To see that our analysis is tight, consider the instance depicted in Fig. 4. On this colored path, the greedy algorithm might recolor $\frac{5}{4}$ times more vertices than the optimal recoloring.

Fig. 4. The greedy algorithm might choose to color the path $(2, 3, 4, 5)$, then it must recolor one of the vertices $\{1, 10\}$, one of $\{6, 12\}$, and one of $\{8, 11\}$, a total of five recolored vertices, while an optimal recoloring can recolor the paths $(4, 5, 6, 7)$ and $(8, 9, 10, 11)$, a total of four recolored vertices (Color figure online).

5   Parameterized Complexity

Let k be the minimum number of vertices that has to be recolored. In this section, we describe an exact algorithm for 2-CR with a running time of $O(|E|)+2^{O(k \log k)}$.

Theorem 5. *There is a $4k$ kernel for 2-CR that can be computed in linear time, where the parameter k is the size of an optimal solution.*

Proof. Recall from Lemma 4 that $k = |D| - |I^*|$. From Lemma 9 we know that $|D| \geq 2|I^*|$ and we can conclude that $|D| \leq 2k$. That is, the number of disconnected pairs in any colored graph is at most twice the minimum number of vertices needed to be recolored by any convex recoloring. Finally, Lemma 3 states that an optimal solution that recolors only disconnected pairs exists. Thus, given a colored graph, we can reduce, in linear time, the number of vertices in the graph to no more than $4k$ by removing all singletons and connected pairs. $\square$

We use kernelization to obtain a parameterized algorithm for 2-CR.

Theorem 6. *There is an $O(|E|) + 2^{O(k \log k)}$ time algorithm for 2-CR, where the parameter k is the size of an optimal solution.*

Proof. First reduce the size of the graph to no more than $4k$ vertices, as described in the proof of Theorem 5. A brute force search algorithm that test every possible recoloring can now be used to find an optimal one. Such algorithm need to consider $2k$ colors for every vertex of the $4k$ vertices, thus at most $2k^{4k}$ coloring exists. Each proposed recoloring can be tested in linear time, so the overall running time of the algorithm is $O(k^2 \cdot 2k^{4k}) = 2^{O(k \log k)}$ $\square$

References

1. Bachoore, E.H., Bodlaender, H.L.: Convex recoloring of leaf-colored trees. Technical report UU-CS–010, Department of Information and Computing Sciences, Utrecht University (2006)
2. Bar-Yehuda, R., Feldman, I., Rawitz, D.: Improved approximation algorithm for convex recoloring of trees. Theory Comput. Syst. **43**(1), 3–18 (2008)
3. Bodlaender, H.L., Fellows, M.R., Langston, M.A., Ragan, M.A., Rosamond, F.A., Weyer, M.: Quadratic kernelization for convex recoloring of trees. Algorithmica **61**(2), 362–388 (2011)
4. Campêlo, M.B., Huiban, C.G., Sampaio, R.M., Wakabayashi, Y.: On the complexity of solving or approximating convex recoloring problems. In: Du, D.-Z., Zhang, G. (eds.) COCOON 2013. LNCS, vol. 7936, pp. 614–625. Springer, Heidelberg (2013)
5. Kammer, F., Tholey, T.: The complexity of minimum convex coloring. Discrete Appl. Math. **160**(6), 810–833 (2012)
6. Kanj, I.A., Kratsch, D.: Convex recoloring revisited: complexity and exact algorithms. In: Ngo, H.Q. (ed.) COCOON 2009. LNCS, vol. 5609, pp. 388–397. Springer, Heidelberg (2009)

7. Karp, R.M.: On the computational complexity of combinatorial problems. Networks **5**, 45–68 (1975)
8. Lima, K.R., Wakabayashi, Y.: Convex recoloring of paths. Discrete Appl. Math. **164**, 450–459 (2014)
9. Moran, S., Snir, S.: Efficient approximation of convex recolorings. J. Comput. Syst. Sci. **73**(7), 1078–1089 (2007)
10. Moran, S., Snir, S.: Convex recolorings of strings and trees: Definitions, hardness results and algorithms. J. Comput. Syst. Sci. **74**(5), 850–869 (2008)
11. Ponta, O., Hüffner, F., Niedermeier, R.: Speeding up dynamic programming for some NP-hard graph recoloring problems. In: Agrawal, M., Du, D.-Z., Duan, Z., Li, A. (eds.) TAMC 2008. LNCS, vol. 4978, pp. 490–501. Springer, Heidelberg (2008)
12. Razgon, I.: A $2^{O(k)}$poly(n) algorithm for the parameterized convex recoloring problem. Inf. Process. Lett. **104**(2), 53–58 (2007)

Computing the BWT and the LCP Array in Constant Space

Felipe A. Louza$^{(\boxtimes)}$ and Guilherme P. Telles

Institute of Computing, University of Campinas, Campinas, São Paulo, Brazil
{louza,gpt}@ic.unicamp.br

Abstract. We show how to modify the in-place Burrows-Wheeler transform (BWT) algorithm proposed by Crochemore *et al.* [4, 5] to also compute the longest common prefix (LCP) array. Our algorithm runs in quadratic time, as its predecessor, constructing both the BWT and the LCP array using just $O(1)$ additional space. It is supported by interesting properties of the BWT and of the LCP array and inherits its predecessor simplicity.

Keywords: Burrows-Wheeler transform · Longest common prefix array · Constant space algorithms

1 Introduction

The suffix array [9,16], the longest common prefix (LCP) array and the Burrows-Wheeler transform (BWT) [3] are in the core of string matching problems as building blocks of both practical solutions and theoretical improvements in algorithms [1,10,20,21].

There are many suffix array construction algorithms (SACAs) (see [6,23] for reviews) including linear time [12,14,15,19] and some that compute the LCP array during the suffix sorting [7]. There are linear time LCP array construction algorithms that take both the text and the suffix array as input [11,13,17] and others that take only the BWT as input [2]. The BWT can either be obtained from the suffix array and the text or can be computed directly from the text [22], both in linear time [20].

In 2007, Franceschini and Muthukrishnan [8] introduced the first constant space SACA, that is, it uses $O(1)$ additional memory apart from the input text and by the output suffix array. Its running time is $O(n \log n)$ for an input text of length n from an unbounded alphabet. Recently, Nong [18] presented a linear time and constant space SACA for constant size alphabets.

In 2013, Crochemore *et al.* [4,5] introduced the first in-place BWT algorithm, that is, it uses $O(1)$ additional memory apart from the memory used by the input text, which is overwritten by the output BWT. In other words, the input text is not read-only, it is directly permuted into the output BWT. Its running time is $O(n^2)$ for unbounded alphabets, but its elegance and simplicity make it very attractive from a theoretical point of view.

© Springer International Publishing Switzerland 2016
Z. Lipták and W.F. Smyth (Eds.): IWOCA 2015, LNCS 9538, pp. 312–320, 2016.
DOI: 10.1007/978-3-319-29516-9_26

In this article we introduce an extension to the in-place BWT algorithm [4,5] that computes both the BWT and the LCP array at the same time in constant space. We consider the model in which the input text is stored in an array of n symbols, which are overwritten by the n symbols of the output BWT and in which the output LCP array is stored in another array of n integers. Our algorithm runs in quadratic time using $O(1)$ additional memory (it uses eight integer variables, as shown in appendix) and, as its predecessor, is quite simple.

We are aware that given the input text of length n from an unbounded alphabet, one could construct the suffix array in $O(n \log n)$ time and constant space [8] and then easily compute the LCP array in $O(n^2)$ time and constant space, overwriting the suffix array. Thus, one could use this procedure and the algorithm by Crochemore et al. to compute both the BWT and the LCP array in constant space. However, our algorithm is equivalent in time complexity and much simpler than such procedure. The C implementation of the algorithm fits in a single page.

The rest of the article is organized as follows. In Sect. 2 we introduce concepts and notation. In Sect. 3 we present the in-place BWT algorithm. In Sect. 4 we present our algorithm and in Sect. 5 we conclude the article.

2   Definitions and Notation

Let Σ be an ordered alphabet of σ symbols. The alphabet Σ can be unbounded. We denote the set of every nonempty string of symbols in Σ by Σ^*. Let $ be a symbol not in Σ that precedes every symbol in Σ. We define $\Sigma^\$ = \{T\$ | T \in \Sigma^*\}$. We use the symbol $<$ for the lexicographic order relation between strings.

The i-th symbol in a string T will be denoted by $T[i]$. Let $T = T[0]T[1] \ldots T[n-1]$ be a string of length n. A substring of T will be denoted by $T[i,j] = T[i] \ldots T[j]$, $0 \le i \le j < n$. A prefix of T is a substring of the form $T[0,k]$ and a suffix is a substring of the form $T[k, n-1]$, $0 \le k < n$. The suffix $T[k, n-1]$ will be denoted by T_k.

A suffix array for a string provides the lexicographic order for all its suffixes. Formally, a suffix array SA for a string $T \in \Sigma^\$$ of size n is an array of integers $\mathsf{SA} = [i_0, i_1, \ldots, i_{n-1}]$ such that $T_{i_0} < T_{i_1} < \ldots < T_{i_{n-1}}$ [9,16].

Let $lcp(S, T)$ be the length of the longest common prefix of two strings S and T in $\Sigma^\$$. The LCP array for T stores the value of lcp for suffixes pointed by consecutive positions of a suffix array. We define $\mathsf{LCP}[0] = 0$ and $\mathsf{LCP}[i] = lcp(T_{\mathsf{SA}[i]}, T_{\mathsf{SA}[i-1]})$ for $1 \le i < n$.

The BWT of a string T can be constructed listing all the n circular shifts of T, lexicographically sorting them, aligning the shifts columnwise and taking the last column [3]. The BWT is reversible and tends to group similar symbols in runs. It may also be defined in terms of the suffix array, to which it is closely related. Let the BWT of a string T be denoted simply by BWT. We define $\mathsf{BWT}[i] = T[\mathsf{SA}[i] - 1]$ if $\mathsf{SA}[i] \ne 0$ or $\mathsf{BWT}[i] = \$$ otherwise.

The first column of the conceptual matrix of the BWT will be referred to as F, and the last column will be referred to as L. The LF-mapping property of

	circular shifts	sorted circular shifts				sorted suffixes
i		F $\qquad$ L	SA	LCP	BWT	$T_{\mathsf{SA}[i]}$
0	BANANA\$	\$BANANA	6	0	A	\$
1	\$BANANA	A\$BANAN	5	0	N	A\$
2	A\$BANAN	ANA\$BAN	3	1	N	ANA\$
3	NA\$BANA	ANANA\$B	1	3	B	ANANA\$
4	ANA\$BAN	BANANA\$	0	0	\$	BANANA\$
5	NANA\$BA	NA\$BANA	4	0	A	NA\$
6	ANANA\$B	NANA\$BA	2	2	A	NANA\$

Fig. 1. SA, LCP and BWT for $T = $ BANANA\$.

the BWT states that the ith occurrence of a symbol $\alpha \in \Sigma^{\$}$ in the last column L corresponds to the ith occurrence of α in the first column F.

Some other relations between the SA and the BWT are the following. It is easy to see that $L[i] = \mathsf{BWT}[i]$ and $F[i] = T[\mathsf{SA}[i]]$. Moreover, if the first symbol of $T_{\mathsf{SA}[i]}$, $T[\mathsf{SA}[i]] = \alpha$, is the kth occurrence of α in F, then j is the position of $T_{\mathsf{SA}[i]+1}$ in SA (*i.e.* j is the rank of $T_{\mathsf{SA}[i]+1}$) such that $L[j]$ corresponds to the kth occurrence of α in L.

As an example, Fig. 1 shows the circular shifts, the sorted circular shifts, the SA, the LCP, the BWT and the sorted suffixes for $T = $ BANANA\$.

The range minimum query (RMQ) with respect to the LCP is the smallest lcp value in an interval of a suffix array. We define $RMQ(i,j) = \min_{i < k \leq j}\{\mathsf{LCP}[k]\}$, for $0 \leq i < j < n$. Given a string T of length n and its LCP array it is easy to see that $lcp(T_{\mathsf{SA}[i]}, T_{\mathsf{SA}[j]}) = RMQ(i,j)$.

3   In-Place BWT

The algorithm by Crochemore *et al.* [4,5] overwrites the input string T with the BWT as it proceeds by induction on the suffix length.

Let $\mathsf{BWT}(T_s)$ be the BWT of the suffix T_s, stored in $T[s, n-1]$. The base cases are the two rightmost suffixes, for which $\mathsf{BWT}(T_{n-2}) = T_{n-2}$ and $\mathsf{BWT}(T_{n-1}) = T_{n-1}$. For the inductive step, the authors have shown that the position of \$ in $\mathsf{BWT}(T_{s+1})$ is related to the rank of T_{s+1} among the suffixes $T_{s+1}, \ldots, T_{n-1}$ (local rank), thus allowing to build $\mathsf{BWT}(T_s)$ even after $T[s+1, n-1]$ has been overwritten with $\mathsf{BWT}(T_{s+1})$. The algorithm is composed by four steps.

1. Find the position p of \$ in $T[s+1, n-1]$. Evaluating $p - s$ gives the local rank of T_{s+1} that originally was starting at position $s + 1$.
2. Find the local rank r of the suffix T_s using just symbol $c = T[s]$. To this end, sum the number of symbols in $T[s+1, n-1]$ that are strictly smaller than c with the number of occurrences of c in $T[s+1, p]$ and with s, obtaining r.
3. Store c into $T[p]$, replacing \$.
4. Shift $T[s+1, r]$ one position to the left. Write \$ in $T[r]$.

The algorithm runs in $O(n^2)$ using constant space memory. Furthermore, the algorithm is also in-place since it uses $O(1)$ additional memory and overwrites the input text with the output BWT.

4   BWT and LCP Array in Constant Space

Our algorithm computes both the BWT and the LCP array by induction on the length of the suffix. The BWT construction is the same as proposed by Crochemore *et al.* [4,5].

In a glimpse, the LCP evaluation works as follows. Suppose that $\mathsf{BWT}(T_{s+1})$ and the LCP array for the suffixes $\{T_{s+1}, \ldots, T_{n-1}\}$, denoted by $\mathsf{LCP}(T_{s+1})$, have already been built. Adding the suffix T_s to the solution requires evaluating exactly two values of *lcp*, involving the two suffixes that will be adjacent to T_s.

The first *lcp* value involves T_s and the largest suffix T_a in $\{T_{s+1}, \ldots, T_{n-1}\}$ that is smaller than T_s. Fortunately, $\mathsf{BWT}(T_{s+1})$ and $\mathsf{LCP}(T_{s+1})$ are enough to compute such value. Recall that if the first symbol of T_a is not equal to the first symbol of T_s then $lcp(T_a, T_s) = 0$. Otherwise $lcp(T_a, T_s) = lcp(T_{a+1}, T_{s+1}) + 1$ and the *RMQ* may be used, since both T_{a+1} and T_{s+1} are already in $\mathsf{BWT}(T_{s+1})$. We know that the position of T_{s+1} is p from Step 1 of the in-place BWT in Sect. 3. Then it is enough to find, in $\mathsf{BWT}(T_{s+1})$, the position of T_{a+1}, which stores the symbol corresponding to the first symbol of T_a.

The second *lcp* value involves T_s and the smallest suffix T_b in $\{T_{s+1}, \ldots, T_{n-1}\}$ that is larger than T_s. It may be computed in a similar fashion.

Basic Algorithm. Suppose that $\mathsf{BWT}(T_{s+1})$ and $\mathsf{LCP}(T_{s+1})$, have already been built, and are stored in $T[s+1, n-1]$ and $\mathsf{LCP}[s+1, n-1]$, respectively. Adding T_s, whose rank is r, to the solution requires updating $\mathsf{LCP}(T_{s+1})$: by first shifting $\mathsf{LCP}[s+1, r]$ one position to the left and then computing the new values of $\mathsf{LCP}[r]$ and $\mathsf{LCP}[r + 1]$, which refer to the two suffixes adjacent to T_s in $\mathsf{LCP}(T_s)$.

The value of $\mathsf{LCP}[r]$ is equal to the *lcp* of T_s and T_a in $\mathsf{BWT}(T_{s+1})$. The rank of T_a is r and will be $r-1$ in $\mathsf{BWT}(T_s)$ after shifting. If the first symbol of T_a is equal to $T[s]$ then $\mathsf{LCP}[r] = lcp(T_{a+1}, T_{s+1}) + 1$, otherwise $\mathsf{LCP}[r] = 0$.

We can evaluate $lcp(T_{a+1}, T_{s+1})$ by the *RMQ* function from the position of T_{a+1} to the position of T_{s+1}. We know that p is the position of T_{s+1} in $\mathsf{BWT}(T_{s+1})$. Then we have to find the position p_{a+1} of T_{a+1} in $\mathsf{BWT}(T_{s+1})$. Note that $T[p_{a+1}]$ corresponds to the first symbol of T_a. If $T[p_{a+1}] \neq T[s]$ then $lcp(T_a, T_s) = 0$, otherwise the value of $lcp(T_a, T_s)$ may be evaluated as $lcp(T_{a+1}, T_{s+1}) + 1 = RMQ(p_{a+1}, p) + 1$.

The value of $\mathsf{LCP}[r + 1]$ may be evaluated in a similar fashion. Let T_b be the suffix with rank $r+1$ in $\mathsf{BWT}(T_{s+1})$ (its rank will still be $r+1$ in $\mathsf{BWT}(T_s)$). We have to find the position p_{b+1} of T_{b+1} in $\mathsf{BWT}(T_{s+1})$ and then if $T[s] = T[p_{b+1}]$ compute $\mathsf{LCP}[r + 1] = lcp(T_s, T_b) = lcp(T_{s+1}, T_{b+1}) + 1 = RMQ(p, p_{b+1}) + 1$.

The algorithm proceeds by induction on the length of the suffix. It is easy to see that for the suffixes with length 1 and 2 the values in LCP will be always equal to 0. Let the current suffix be T_s ($0 \le s \le n - 3$). Our algorithm has new

Steps 2', 2" and 4', added just after Steps 2 and 4, respectively, of the in-place BWT algorithm as follows:

2'. Find the position p_{a+1} of the suffix T_{a+1}, such that suffix T_a has rank r in $\mathsf{BWT}(T_{s+1})$, and compute:

$$\ell_a = \begin{cases} RMQ(p_{a+1}, p) + 1 & if\, T[p_{a+1}] = T[s] \\ 0 & otherwise. \end{cases}$$

2". Find the position p_{b+1} of the suffix T_{b+1}, such that suffix T_b has rank $r+1$ in $\mathsf{BWT}(T_{s+1})$, and compute:

$$\ell_b = \begin{cases} RMQ(p, p_{b+1}) + 1 & if\, T[s] = T[p_{b+1}] \\ 0 & otherwise. \end{cases}$$

4'. Shift $\mathsf{LCP}[s+1, r]$ one position to the left, store ℓ_a in $\mathsf{LCP}[r]$ and if $r+1 < n$ then store ℓ_b in $\mathsf{LCP}[r+1]$.

Computing ℓ_a and ℓ_b. To find p_{a+1} and p_{b+1} and to compute ℓ_a and ℓ_b in Steps 2' and 2", we use the following properties.

Lemma 1. *Let T_s be the suffix to be inserted in $\mathsf{BWT}(T_{s+1})$ at position r. Let $T_a \in \{T_{s+1}, \ldots, T_{n-1}\}$ be the suffix whose rank is r in $\mathsf{BWT}(T_{s+1})$, and let p_{a+1} be the position of T_{a+1}. If $p_{a+1} \notin [s+1, p)$ then $T[p_{a+1}] \neq T[s]$.*

Proof. The local rank of T_a in $\mathsf{BWT}(T_{s+1})$ is $r - s$. We know that $T[p_{a+1}]$ corresponds to the first symbol of T_a, and it follows from LF-mapping that the local rank of $T[p_{a+1}]$ is $r - s$ in $\mathsf{BWT}(T_{s+1})$. Then $T[p_{a+1}]$ is smaller or equal to $T[s]$, since T_s has also local rank $r - s$. If $T[p_{a+1}]$ is smaller than $T[s]$, p_{a+1} must be in $[s+1, n)$. However, if $T[p_{a+1}] = T[s]$ then p_{a+1} must precede the position where $T[s]$ will be inserted, *i.e.* the position p of T_{s+1}, otherwise the local rank of T_s would be smaller than $r - s$. Then if $T[p_{a+1}] = T[s]$ it follows that $p_{a+1} \in [s+1, p)$. $\qquad\square$

We can use Lemma 1 to verify whether $T[p_{a+1}] = T[s]$ just checking if there is a symbol in $T[s+1, p-1]$ equal to $T[s]$. If no such symbol is found, $\ell_a = 0$, otherwise we have to compute $RMQ(p_{a+1}, p)$. Furthermore, if we have more then one symbol in $T[s+1, p-1]$ equal to $T[s]$, the symbol whose local rank is $r-s$ will be the last symbol found in $T[s+1, p-1]$, *i.e.* the largest symbol in $T[s+1, p-1]$ smaller than $T[s]$. Then, to find such symbol we can simply perform a backward scan in T from $p-1$ to $s+1$ until we find the first occurrence of $T[p_{a+1}] = T[s]$. One can see that we can, simultaneously, compute the minimum function for the *lcp* visited values, obtaining $RMQ(p_{a+1}, p)$ as soon as we find $T[p_{a+1}] = T[s]$.

Lemma 2. *Let T_s be the suffix to be inserted in $\mathsf{BWT}(T_{s+1})$ at position r. Let $T_b \in \{T_{s+1}, \ldots, T_{n-1}\}$ be the suffix whose rank is $r+1$ in $\mathsf{BWT}(T_{s+1})$, and let p_{b+1} be the position of T_{b+1}. If $p_{b+1} \notin (p, n-1]$ then $T[s] \neq T[p_{b+1}]$.*

The proof of Lemma 2 is similar to the proof of Lemma 1 and will be omitted. Just remember that T_b will still have rank $r+1$ in $\mathsf{BWT}(T_s)$ (after inserting T_s).

The procedure to find ℓ_b uses Lemma 2 and computes $lcp(T_{s+1}, T_{b+1})$ in a similar fashion. It scans T from $p+1$ to $n-1$ until it finds the first occurrence of $T[p_{b+1}] = T[s]$, computing the minimum function to solve the RMQ if such symbol is found.

Example. As an example, consider $T = \mathsf{BANANA\$}$ and $s = 1$. Figures 2 and 3 illustrate Steps 2' and 4', respectively. Suppose that we have computed $\mathsf{BWT}(T_2)$ and $\mathsf{LCP}(T_2)$. We then have $p = 6$ (Step 1) and the rank $r = 4$ (Step 2).

Step 2' finds the first symbol equal to $T[s]$ (A) in $T[s+1, p-1]$ at position $p_{a+1} = 5$. It represents $T_{a+1} = \mathsf{NA\$}$. In this case, the value of ℓ_a is calculated during the scan of T from $p-1 = 5$ to $s+1 = 2$, $i.e.$ $\ell_a = RMQ(p_{a+1}, p) = RMQ(5, 6) = 2$. Step 2" does not find any symbol equal to $T[s]$ (A) in $T[p+1, n-1]$. Then we know that $T[s] \neq T[p_{b+1}]$ and $\ell_b = 0$.

s	LCP	BWT	sorted suffixes
1	?	?	ANANA$
2	0	A	$
3	0	N	A$
$r \rightarrow$ 4	1	N	ANA$
$p_{a+1} \rightarrow$ 5	0	A	NA$
$p \rightarrow$ 6	2	$	NANA$

Fig. 2. After Step 2": $T = \mathsf{BANANA\$}$ and $s = 1$.

Step 3 stores $T[s]$ (A) at position $T[p]$, $p = 6$. Step 4 shifts $T[s+1, r]$ one position to the left and inserts $ at position $T[r]$, $r = 4$. The last step, 4', shifts $\mathsf{LCP}[s+1, r]$ one position to the left and sets $\mathsf{LCP}[4] = \ell_a = 3$ and $\mathsf{LCP}[4+1] = \ell_b = 0$.

s	LCP	BWT	sorted suffixes
1	0	A	$
2	0	N	A$
3	1	N	ANA$
$r \rightarrow$ 4	$\ell_a = 3$	$	ANANA$
5	$\ell_b = 0$	A	NA$
6	2	A	NANA$

Fig. 3. After Step 4': $T = \mathsf{BANANA\$}$ and $s = 1$.

In the appendix, we show a C implementation of the algorithm, that is also available at https://github.com/felipelouza/bwt-lcp-in-place.

Theoretical Costs. The worst-case time complexity is $O(n^2)$, since we only added two $O(n)$ scans over T_{s+1} (Steps 2' and 2") to compute the values of ℓ_a and ℓ_b. Step 4' shifts the LCP by the same amount that BWT is shifted. Then the time complexity remains the same as the in-place BWT algorithm. Regarding the space usage, the new algorithm needs only four additional variables to store positions p_{a+1} and p_{b+1} and the values of ℓ_a and ℓ_b, thus using constant space only.

5   Conclusion

We showed how to compute the BWT and LCP arrays using constant space. As its predecessor, our algorithm is quite simple, and although it has no practical applicability due its quadratic running time, it builds on interesting properties of the BWT and of the LCP array. As a final remark, we note that our algorithm can easily construct the suffix array using constant space, with no overhead on the running time.

Acknowledgments. FAL acknowledges the financial support of CAPES. GPT acknowledges the support of CNPq.

A   Source Code

```
1   void compute_bwt_lcp(unsigned char *T, int n, int *LCP){
2   int i, p, r=1, s, p_a1, p_b1, l_a, l_b;
3   LCP[n-1] = LCP[n-2] = 0; // base cases
4
5   for (s=n-3; s>=0; s--) {
6
7       /*steps 1 and 2*/
8       p=r+1;
9       for (i=s+1, r=0; T[i]!=END_MARKER; i++)
10          if(T[i]<=T[s]) r++;
11      for (; i<n; i++)
12          if (T[i]<T[s]) r++;
13
14      /*step 2'*/
15      p_a1=p+s-1;
16      l_a=LCP[p_a1+1];
17      while (T[p_a1]!=T[s]) // RMQ function
18          if (LCP[p_a1--]<l_a)
19              l_a=LCP[p_a1+1];
20      if (p_a1==s) l_a=0;
21      else l_a++;
22
```

```
23    /*step 2''*/
24    p_b1=p+s+1;
25    l_b=LCP[p_b1];
26    while (T[p_b1]!=T[s] && p_b1<n) // RMQ function
27       if (LCP[++p_b1]<l_b)
28          l_b=LCP[p_b1];
29    if (p_b1==n) l_b=0;
30    else l_b++;
31
32    /*steps 3 and 4*/
33    T[p+s]=T[s];
34    for (i=s; i<s+r; i++) {
35       T[i]=T[i+1];
36       LCP[i]=LCP[i+1];
37    }
38    T[s+r]=END_MARKER;
39
40    /*step 4'*/
41    LCP[s+r]=l_a;
42    if (s+r+1<n)   // If r+1 is not the last position
43       LCP[s+r+1]=l_b;
44    }
45 }
```

References

1. Adjeroh, D., Bell, T., Mukherjee, A.: The Burrows-Wheeler transform: data compression, suffix arrays, and pattern matching. Springer, Boston (2008)
2. Beller, T., Gog, S., Ohlebusch, E., Schnattinger, T.: Computing the longest common prefix array based on the Burrows-Wheeler transform. J. Discrete Algorithms **18**, 22–31 (2013)
3. Burrows, M., Wheeler, D.: A block-sorting lossless data compression algorithm. Technical report, Digital SRC Research Report (1994)
4. Crochemore, M., Grossi, R., Kärkkäinen, J., Landau, G.M.: A constant-space comparison-based algorithm for computing the Burrows–Wheeler transform. In: Fischer, J., Sanders, P. (eds.) CPM 2013. LNCS, vol. 7922, pp. 74–82. Springer, Heidelberg (2013)
5. Crochemore, M., Grossi, R., Kärkkäinen, J., Landau, G.M.: Computing the Burrows-Wheeler transform in place and in small space. J. Discrete Algorithms **32**, 44–52 (2015). stringMasters 2012 and 2013 Special Issue (Volume 2)
6. Dhaliwal, J., Puglisi, S.J., Turpin., A.: Trends in suffix sorting : a survey of low memory algorithms. In: Reynolds, M., Thomas, B.H. (eds.) Proceedings of Australasian Computer Science Conference (ACSC), CRPIT, vol. 122, pp. 91–98. Australian Computer Society, Melbourne, Australia (2012)
7. Fischer, J.: Inducing the LCP-array. In: Dehne, F., Iacono, J., Sack, J.-R. (eds.) WADS 2011. LNCS, vol. 6844, pp. 374–385. Springer, Heidelberg (2011)
8. Franceschini, G., Muthukrishnan, S.M.: In-place suffix sorting. In: Arge, L., Cachin, C., Jurdziński, T., Tarlecki, A. (eds.) ICALP 2007. LNCS, vol. 4596, pp. 533–545. Springer, Heidelberg (2007)

9. Gonnet, G.H., Baeza-yates, R., Snider, T.: New Indices for Text: PAT Trees and PAT Arrays. Prentice-Hall Inc., Upper Saddle River (1992)
10. Gusfield, D.: Algorithms on Strings, Trees, and Sequences: Computer Science and Computational Biology. Cambridge University Press, New York (1997)
11. Kärkkäinen, J., Manzini, G., Puglisi, S.J.: Permuted longest-common-prefix array. In: Kucherov, G., Ukkonen, E. (eds.) CPM 2009 Lille. LNCS, vol. 5577, pp. 181–192. Springer, Heidelberg (2009)
12. Kärkkäinen, J., Sanders, P., Burkhardt, S.: Linear work suffix array construction. ACM J. **53**(6), 918–936 (2006)
13. Kasai, T., Lee, G.H., Arimura, H., Arikawa, S., Park, K.: Linear-time longest-common-prefix computation in suffix arrays and its applications. In: Amir, A., Landau, G.M. (eds.) CPM 2001. LNCS, vol. 2089, p. 181. Springer, Heidelberg (2001)
14. Kim, D.K., Sim, J.S., Park, H., Park, K.: Constructing suffix arrays in linear time. J. Discret. Algorithms **3**(2–4), 126–142 (2005)
15. Ko, P., Aluru, S.: Space efficient linear time construction of suffix arrays. J. Discret. Algorithms **3**(2–4), 143–156 (2005)
16. Manber, U., Myers, E.W.: Suffix arrays: a new method for on-line string searches. SIAM J. Comput. **22**(5), 935–948 (1993)
17. Manzini, G.: Two space saving tricks for linear time LCP array computation. In: Hagerup, T., Katajainen, J. (eds.) SWAT 2004. LNCS, vol. 3111, pp. 372–383. Springer, Heidelberg (2004)
18. Nong, G.: Practical linear-time O(1)-workspace suffix sorting for constant alphabets. ACM Trans. Inform. Syst. **31**(3), 1–15 (2013)
19. Nong, G., Zhang, S., Chan, W.H.: Two efficient algorithms for linear time suffix array construction. IEEE Trans. Comput. **60**(10), 1471–1484 (2011)
20. Ohlebusch, E.: Bioinformatics Algorithms: Sequence Analysis, Genome Rearrangements, and Phylogenetic Reconstruction. Oldenbusch Verlag, Bremen (2013)
21. Ohlebusch, E., Gog, S., Kügel, A.: Computing matching statistics and maximal exact matches on compressed full-text indexes. In: Chavez, E., Lonardi, S. (eds.) SPIRE 2010. LNCS, vol. 6393, pp. 347–358. Springer, Heidelberg (2010)
22. Okanohara, D., Sadakane, K.: A linear-time Burrows-Wheeler transform using induced sorting. In: Karlgren, J., Tarhio, J., Hyyrö, H. (eds.) SPIRE 2009. LNCS, vol. 5721, pp. 90–101. Springer, Heidelberg (2009)
23. Puglisi, S.J., Smyth, W.F., Turpin, A.H.: A taxonomy of suffix array construction algorithms. ACM Comp. Surv. **39**(2), 1–31 (2007)

EERTREE: An Efficient Data Structure for Processing Palindromes in Strings

Mikhail Rubinchik and Arseny M. Shur$^{(\boxtimes)}$

Ural Federal University, Ekaterinburg, Russia
mikhail.rubinchik@gmail.com, arseny.shur@urfu.ru

Abstract. We propose a new linear-size data structure which provides a fast access to all palindromic substrings of a string or a set of strings. This structure inherits some ideas from the construction of both the suffix trie and suffix tree. Using this structure, we present simple and efficient solutions for a number of problems involving palindromes.

1 Introduction

Palindromes are one of the most important repetitive structures in strings. During the last decades they were actively studied in formal language theory, combinatorics on words and stringology. Recall that a palindrome is any string $S = a_1 a_2 \cdots a_n$ equal to its reversal $S^{\leftarrow} = a_n \cdots a_2 a_1$.

There are a lot of papers concerning the palindromic structure of strings. The most important problems in this direction include the search and counting of palindromes in a string and the factorization of a string into palindromes. Manacher [12] came up with a linear-time algorithm which can be used to find all maximal palindromic substrings of a string, along with its palindromic prefixes and suffixes. The problem of counting and listing distinct palindromic substrings was solved offline in [6] and online in [10]. Knuth, Morris, and Pratt [9] gave a linear-time algorithm for checking whether a string is a product of even-length palindromes. Galil and Seiferas [4] asked for such an algorithm for the k-*factorization* problem: decide whether a given string can be factored into exactly k palindromes, where k is an arbitrary constant. They presented an online algorithm for $k = 1, 2$ and an offline one for $k = 3, 4$. An online algorithm working in $O(kn)$ time for the length n string and any k was designed in [11]. Close to the k-factorization problem is the problem of finding the *palindromic length* of a string, which is the minimal k in its k-factorization. This problem was solved by Fici et al. in $O(n \log n)$ time [3]. In this paper we present a new tree-like data structure, called eertree[1], which simplifies and speeds up solutions to search, counting and factorization problems as well as to several other palindrome-related algorithmic problems. This structure can also cope with Watson–Crick palindromes [8] and other palindromes with involution and may be interesting for the RNA studies along with the affix trees [13] and affix arrays [17].

---

[1] This structure can be found, with the reference to the first author, in a few IT blogs under the name "palindromic tree". See, e.g., http://adilet.org/blog/25-09-14/.

© Springer International Publishing Switzerland 2016
Z. Lipták and W.F. Smyth (Eds.): IWOCA 2015, LNCS 9538, pp. 321–333, 2016.
DOI: 10.1007/978-3-319-29516-9_27

In Sect. 2 we first recall the problem of counting distinct palindromic sub-strings in an online fashion. This was a motive example for inventing eertree. This data structure contains the digraph of all palindromic factors of an input string S and supports the operation add(c) which appends a new symbol to the end of S. Thus, the number of nodes in the digraph equals the number of distinct palindromes inside S. Maintaining an eertree for a length n string with σ distinct symbols requires $O(n \log \sigma)$ time and $O(n)$ space (for a random string, the expected space is $O(\sqrt{n\sigma})$). After introducing the eertree we discuss some of its properties and simple applications.

In Sect. 3 we study advanced questions related to eertrees. We consider joint eertree of several strings and name a few problems solved with its use. Then we design two "smooth" variations of the algorithm which builds eertrees. These variations require at most logarithmic time for each call of add(c) and then allow one to support an eertree for a string with two operations: appending and deleting the last symbol. Using one of these variations, we design a fast backtracking algorithm enumerating all *rich* strings over a fixed alphabet up to a given length. (A string is rich if it contains the maximum possible number of distinct palindromes.) Finally, we show that eertree can be efficiently turned into a persistent data structure.

The use of eertrees for factorization problems is described in Sect. 4. Namely, new fast algorithms are given for the k-factorization of a string and for computing its palindromic length. We also conjecture that the palindromic length can be found in linear time and provide some argument supporting this conjecture.

Definitions and Notation. We study finite strings, viewing them as arrays of symbols: $w = w[1..n]$. The number of distinct symbols of the processed string is denoted by σ. We write ε for the empty string, $|w|$ for the length of w, $w[i]$ for the ith letter of w and $w[i..j]$ for $w[i]w[i+1]\ldots w[j]$, where $w[i..i-1] = \varepsilon$ for any i. A string u is a *substring* of w if $u = w[i..j]$ for some i and j. A substring $w[1..j]$ (resp., $w[i..n]$) is a *prefix* [resp. *suffix*] of w. If a substring (prefix, suffix) of w is a palindrome, it is called a *subpalindrome* (resp. *prefix-palindrome, suffix-palindrome*).Throughout the paper we do not count ε as a palindrome.

Trie is a rooted tree with some nodes marked as terminal and all edges labeled by symbols such that no node has two outgoing edges with the same label. Each trie represents a finite set of strings, which label the paths from the root to the terminal nodes.

2   Building an Eertree

Motive Problem: Distinct Subpalindromes Online. Well-known online linear-time Manacher's algorithm [12] encodes all subpalindromes of a string in a special way. Another problem is to find and count all *distinct* subpalindromes. Groult et al. [6] solved this problem offline in linear time and asked for an online solution. Such a solution in $O(n \log \sigma)$ time and $O(n)$ space was given in [10], based on Manacher's algorithm and Ukkonen's suffix tree algorithm [18]. As was also proved in [10], this time is optimal over a general ordered alphabet. In spite of a good asymptotics, this algorithm is based on two rather "heavy" data structures. In is natural to try finding a lightweight structure for solving the analyzed problem with the same asymptotics. Such a data structure is described below. Its further analysis revealed that it is useful for many algorithmic problems involving palindromes.

Eertree: Structure, Interface, Construction. The basic version of eertree supports a single operation add(c), which appends the symbol c to the processed string (from the right), updates the data structure respectively, and returns the number of new palindromes appeared in the string. According to the next lemma, add(c) returns 0 or 1.

Lemma 1 ([2]). *Let S be a string and c be a symbol. The string Sc contains at most one palindrome which is not a substring of S. This new palindrome is the longest suffix-palindrome of Sc.*

From inside, eertree is a directed graph with some extra information. Its nodes numbered with positive integers starting with 1 are in one-to-one correspondence with subpalindromes of the processed string. Below we denote a node and the corresponding palindrome by the same letter. We write eertree(S) for the state of eertree after processing the string S symbol by symbol, left to right.

Remark 1. To report the number of distinct subpalindromes of S, just return the maximum number of a node in eertree(S).

Each node v stores the length len[v] of its palindrome. For the initialization purpose, two special nodes are added: with the number 0 and length 0 for the empty string, and with the number -1 and length -1 for the "imaginary string".

The edges of the graph are defined as follows. If c is a symbol, v and cvc are two nodes, then an edge labeled by c goes from v to cvc. The edge labeled by c goes from the node 0 (resp. -1) to the node labeled by cc (resp., by c) if it exists. This explains why we need two initial nodes. The outgoing edges of a node v are stored in a dictionary which, given a symbol c, returns the edge to[v][c] labeled by it. Such a dictionary is implemented as a binary balanced search tree.

An unlabeled *suffix link* link[u] goes from u to v if v is the longest proper suffix-palindrome of u. By definition, link[c] = 0, link[0] = link[-1] = -1. The resulting graph, consisting of nodes, edges, and suffix links, is the eertree; see Fig. 1 for an example. The following lemma is straightforward.

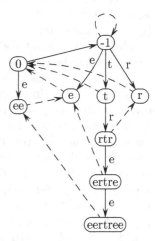

Fig. 1. Eertree of *eertree*. Edges are black, suffix links are dashed.

Lemma 2. *A node of positive length in an eertree has exactly one incoming (labeled) edge.*

Proposition 1. *The eertree of a string S of length n is of size $O(n)$.*

Proof. The eertree of S has at most $n+2$ nodes, including 0, -1 (by Lemma 1), at most n edges (Lemma 2), and at most $n+2$ suffix links (one per node).   □

Proposition 2. *For a string S of length n, eertree(S) can be built online in $O(n \log \sigma)$ time.*

Proof. We start defining eertree(ε) as the graph with two nodes (0 and -1) and two suffix links. Then we make the calls add$(S[1]), \ldots,$ add$(S[n])$ in this order. By Lemma 1 and the definition of add, after each call we know the longest suffix-palindrome maxSuf(T) of the string T processed so far. We support the following invariant: after a call to add, all edges and suffix links between the existing nodes are defined. In this case, adding a new node u one must build exactly one edge (by Lemma 2) and one suffix link: any suffix-palindrome of u is its prefix as well, and hence the destination node of the suffix link from u already exists.

Consider the situation after i calls. We have to perform the next call, say add(a), to $T = S[1..i]$. We need to find the maximum suffix-palindrome P of Ta. Clearly, $P = a$ or $P = aQa$, where Q is a suffix-palindrome of T. Thus, to determine P we should find the longest suffix-palindrome of T preceded by a. To do this, we traverse the suffix-palindromes of T in the order of decreasing length, starting with maxSuf(T) and following suffix links. For each palindrome we read its length k and compare $T[i-k]$ against a until we get an equality or arrive at the node -1. In the former case, the current palindrome is Q; we check whether it has an outgoing edge labeled by a. If yes, the edge leads to $aQa = P$, and P is not new; if no, we create the node P of length $|Q| + 2$ and the edge (Q, P). In the latter case, $P = a$; as above, we check the existence of P in the graph from the current node (which is now -1) and create the node if necessary, together with the edge $(-1, P)$ and the suffix link $(P, 0)$.

It remains to create the suffix link from P if $|P| > 1$. It leads to the second longest suffix-palindrome of Ta. This palindrome can be found similar to P: just continue traversing suffix-palindromes of T starting with the suffix link of Q.

Now estimate the time complexity. During a call to add(a), one checks the existence of the edge from Q with the label a in the dictionary, spending $O(\log \sigma)$ time. The path from the old to the new value of maxSuf requires one transition by an edge (from Q to P) and $k \geq 0$ of transitions by suffix links, and is accompanied by $k+1$ comparisons of symbols. In order to estimate k, follow the position of the first symbol of maxSuf: a transition by a suffix link moves it to the right, and a transition by an edge moves it one symbol to the left. During the whole process of construction of eertree(S), this symbol moves to the right by $\leq n$ symbols. Hence, the total number of transitions by suffix links is $\leq 2n$. The same argument works for the second longest suffix-palindrome, which was used to create suffix links. Thus, the total number of graph transitions and symbol comparisons is $O(n)$, and the time complexity is dominated by checking the existence of edges, $O(n \log \sigma)$ time in total.   □

Some Properties of Eertrees. We call a node *odd* (*even*) if its palindrome has odd (resp., even) length. A path consisting of suffix links is a *suffix path*.

Lemma 3. (1) *Nodes and edges of an eertree form two weakly connected components: the tree of odd (resp., of even) nodes rooted at −1 (resp., at 0).*
(2) *The tree of even (resp., odd) nodes is precisely the trie of right halves of even-length palindromes (resp., the trie of right halves, including the central symbol, of odd-length palindromes).*
(3) *Nodes and inverted suffix links form a tree with a loop at its root −1.*

Remark 2. Tries are convenient data structures, but a trie built from the set of all suffixes (or all factors) of a length n string is usually of size $\Omega(n^2)$. For a linear-space implementation, such a trie should be compressed into a more complicated and less handy structure: suffix tree or suffix automaton (DAWG). On the other hand, eertrees are linear-size tries and need no compression. Moreover, the size of an eertree is usually much smaller than n, because the expected number of distinct palindromes in a length n string is $O(\sqrt{n\sigma})$ [15]. This fact explains high efficiency of eertrees in solving different problems.

Remark 3. A θ-*palindrome* is a string $S = a_1 \cdots a_n$ equal to $\theta(a_n \cdots a_1)$, where θ is a symbol-to-symbol function and θ^2 is the identity (see, e.g., [8]). Clearly, an eertree containing all θ-palindromes of a string can be built in the way described in Proposition 2 (the comparisons of symbols should take θ into account).

First Applications. We demonstrate the performance of eertrees on two test problems taken from student programming contests. The first problem is Palindromic Refrain [19, Problem A], stated as follows: for a given string S find a palindrome P maximizing the value $|P| \cdot \mathrm{occ}(S, P)$, where $\mathrm{occ}(S, P)$ is the number of occurrences of P in S.

Proposition 3. Palindromic Refrain *can be solved by an eertree with the use of $O(n)$ additional time and space.*

Proof (Idea). The occurrence of P ending at position i is either the longest suffix-palindrome of $S[1..i]$ or the longest suffix-palindrome of another suffix-palindrome of $S[1..i]$. The number of occurrences of the first type can be computed during the construction of eertree(S); after that, $\mathrm{occ}(S, P)$ for all subpalindromes P can be found by traversing all nodes of eertree(S) in decreasing order (the last created node first). □

The second problem is Palindromic Pairs [20, Problem B]: for a string S, find the number of triples i, j, k such that $1 \le i \le j < k \le |S|$ and the strings $S[i..j]$, $S[j+1..k]$ are palindromes.

Proposition 4. Palindromic Pairs *can be solved by an eertree with the use of $O(n \log \sigma)$ additional time and $O(n)$ space.*

Proof (Idea). The number of triples for a fixed j is the number of suffix-palindromes of $S[1..j]$ times the number of prefix-palindromes of $S[j+1..n]$. The former value can be computed during the construction of eertree(S) and stored in the node maxSuf($S[1..j]$). The latter value is computed in a symmetric way from eertree($S^\leftarrow$) (this is why the time bound is $O(n\log\sigma)$). $\square$

3   Advanced Modifications of Eertrees

Joint Eertree for Several Strings. When a problem assumes the comparison of two or more strings, it may be useful to build a joint data structure. For example, a variety of problems can be solved by joint ("generalized") suffix trees, see [7]. The *joint eertree* of a set of strings, denoted by eertree($S_1,\ldots,S_k$), is built as follows. We build eertree(S_1) in a usual fashion; then reset the value of maxSuf to 0 and proceed with the string S_2, addressing the add calls to the currently built graph; and so on, until all strings are processed. Each created node stores a k-element boolean array flag. After each call to add, we update flag of the current maxSuf node, setting its ith bit to 1, where S_i is the string being processed. As a result, flag$[v][i]$ equals 1 if and only if v is contained in S_i. Some problems easily solved by a joint eertree are gathered below.

Problem	Solution
Find the number of subpalindromes, common to all k given strings.	Build eertree($S_1,\ldots,S_k$) and count the nodes having only 1's in the flag array
Find the longest subpalindrome contained in all k given strings.	Build eertree($S_1,\ldots,S_k$). Among the nodes having only 1's in the flag array, find the node of biggest length
For strings S and T find the number of palindromes P having more occurrences in S than in T.	Build eertree(S,T), computing occ$_S$ and occ$_T$ in its nodes (see the proof of Proposition 3). Return the number of nodes v such that occ$_S[v] >$ occ$_T[v]$
For strings S and T find the number of triples (i,j,k) such that $S[i..i+k]=T[j..j+k]$ is a palindrome.	Build eertree(S,T), computing the values occ$_S$ and occ$_T$ in its nodes. The answer is $\sum_v$ occ$_S[v]\cdot$ occ$_T[v]$

Coping with Deletions. In the proof of Proposition 2, an $O(n\log\sigma)$ algorithm for building an eertree is given. Nevertheless, in some cases one call of add requires $\Omega(n)$ time, and this kills some possible applications. For example, we may want to support an eertree for a string which can be changed in two ways: by appending a symbol on the right (add(c)) and by deleting the last symbol (pop()). Consider the following sequence of n calls:

$$\underbrace{\mathsf{add}(a),\ldots,\mathsf{add}(a)}_{n/3\ \text{times}},\underbrace{\mathsf{add}(b),\mathsf{pop}(),\mathsf{add}(b),\mathsf{pop}(),\ldots,\mathsf{add}(b),\mathsf{pop}()}_{n/3\ \text{times}}$$

Since each add(b) requires $n/3$ suffix link transitions, the algorithm from Proposition 2 will process this sequence in $\Omega(n^2)$ time independent of the implementation of the operation pop(). Below we describe two algorithms which build eertrees in a way providing an efficient solution to the problem with deletions.

Searching Suffix-Palindromes with Quick Links. Consider a pair of nodes v, link[v] in an eertree and the symbol $b = v[|v| - |\text{link}[v]|]$ preceding the suffix link[v] in v. In addition to the suffix link, we define the *quick link*: let quickLink[v] be the longest suffix-palindrome of v preceded in v by a symbol different from b.

Lemma 4. *As a node v is created,* quickLink[v] *can be computed in $O(1)$ time.*

Proof. The two longest suffix-palindromes of v are $u = \text{link}[v]$ and $u' = \text{link}[\text{link}[v]]$. Assume that v has suffixes bu and cu'. If $c \neq b$, then quickLink[v] = u' by definition. If $c = b$, then quickLink[v] = quickLink[u]. $\square$

Recall that appending a letter c to a current string S, we scan suffix-palindromes of S to find the longest suffix-palindrome Q preceded by c; then maxSuf(Sc) = cQc. (If cQc is a new palindrome, then this scan continues until link[cQc] is found.) The use of quick links reduces the number of scanned suffixes as follows. When the current palindrome is v, we check both v and link[v]. If both are not preceded by c, then all suffix-palindromes of S longer than quickLink[v] are not preceded by c either; so we skip them and check quickLink[v] next.

Example 1. Let us call add(b) to the eertree of the string $S = aabaabaaba$. The longest suffix-palindrome of S is the string $v = abaabaaba$. Since the symbols preceding v and link[v] = $abaaba$ in S are distinct from b, we jump to quickLink[v] = a, skipping the suffix-palindrome aba preceded by the same letter as link[v]. Now quickLink[v] is preceded by b, so we find maxSuf(Sb) = bab. Note that v "does not know" which symbol precedes its particular occurrence. So there is no way to avoid checking the symbol preceding link[v].

Constructing an eertree with quick links, on each step we add $O(1)$ time and space for maintaining these links and possibly reduce the number of processed suffixes. So the time and space bounds from Proposition 2 are in effect. The bound on the number of operations per step is based on the following proposition.

Proposition 5. *In* eertree(S), *any path of quick links has length $O(\log n)$.*

Corollary 1. *The algorithm constructing an eertree using quick links spends $O(\log n)$ time and $O(1)$ space for any call to* add.

Using Direct Links. Now we describe the fastest algorithm for constructing an eertree which, however, uses more than $O(1)$ space for creating a node. Still, the space requirements are quite modest, so the algorithm is highly competitive:

Proposition 6. *There is an algorithm which constructs an eertree spending $O(\log \sigma)$ time and $O(\min(\log \sigma, \log \log n))$ space for any call to* add.

Proof. For each node we create up to σ *direct links*: directLink$[v][c]$ is the longest suffix-palindrome of v preceded in v by c.

Let Q be the longest suffix-palindrome of a string S, preceded by c in S. Then either $Q = $ maxSuf(S) or $Q = $ directLink$[$maxSuf$(S)][c]$, and the longest suffix-palindrome of Q, preceded by c, is directLink$[Q][c]$. Thus, we scan suffixes in constant time, and the time per step is dominated by the $O(\log \sigma)$ search for an edge in the dictionary plus the time for creating direct links for a new node. The arrays directLink$[v]$ and directLink$[$link$[v]]$ coincide for all symbols except for the symbol c preceding link$[v]$ in v. Hence, creating a node v we first find link$[v]$, then copy directLink$[$link$[v]]$ to directLink$[v]$ and assign directLink$[v][c] = $ link$[v]$. Storing or copying direct links explicitly would cost a lot of space and time; so we do this using fully persistent balanced binary search tree (*persistent tree* for short; see [1]). The persistent tree provides full access to any of its *versions*, which are balanced binary search trees ordered by the time of their creation. An update of any version results in creating a new version, which is also fully accessible; the updated version remains unchanged. Such an update as adding a node or changing the information in a node takes $O(\log k)$ time and space, where k is the size of the updated version.

We store direct links from all nodes of the eertree in a single persistent tree, one version per node. Direct links directLink$[v][c]$ in a version v form a search tree, with c serving as the key for sorting (assuming an ordered alphabet). Creation of a version for a node v is an update of the version for link$[v]$. The size of a single search tree is at most σ by definition and is $O(\log n)$ by Proposition 5. Thus, the update time and space is $O(\min(\log \sigma, \log \log n))$, as required. $\square$

Comparing Different Implementations. The three methods of building an eertree are gathered in the following table.

Method	Time for n calls	Time for one call	Space for one node
Basic	$\Theta(n \log \sigma)$	$\Omega(\log \sigma)$ but $O(n)$	$\Theta(1)$
QuickLink	$\Theta(n \log \sigma)$	$\Omega(\log \sigma)$ but $O(\log n)$	$\Theta(1)$
DirectLink	$\Theta(n \log \sigma)$	$\Theta(\log \sigma)$	$O(\min(\log \sigma, \log \log n))$

The basic version is the simplest one and uses the smallest amount of memory. Quick and direct links work somewhat faster, but their main advantage is that any call can be reversed without much pain. Thus, one can maintain an eertree for a string with both operations add(c) and pop$()$. Indeed, let add(c) push to a stack the node containing $P = $ maxSuf(Sc) and, if P is a new palindrome, the node containing Q such that $P = cQc$. This takes $O(1)$ additional time and space. Then pop$()$ reads this information from the stack and restores the previous state of the eertree in $O(1)$ time. The table above also suggests

Question 1. Is there an online algorithm which builds an eertree spending $O(\log \sigma)$ time and $O(1)$ space for any call to add?

Enumerating Rich Strings. By Lemma 1, the number of distinct subpalindromes in a length n string is at most n. Such strings with exactly n palindromes are called *rich*. Rich strings possess a number of interesting properties; see, e.g., [2,5]. The sequence A216264 in the OEIS [16] is the growth function of the language of binary rich strings, i.e., the nth term of this sequence is the number of binary rich strings of length n. J. Shallit computed this function up to $n = 25$, thus enumerating several millions of rich strings. Using an eertree with direct links, we raised the upper bound to $n = 60$, enumerating several *trillions* of rich strings in 10 hours on an average laptop. The new numerical data shows that this sequence grows much slower than it was expected before. Our implementation is available at http://pastebin.com/4YJxVzep. Proposition 7 below serves as the theoretic basis for such a breakthrough in computation.

Proposition 7. *Suppose that R is the number of k-ary rich strings of length $\leq n$, for some fixed k and n. Then the trie built from all these strings can be traversed in time $O(R)$.*

Proof (Idea). We recursively traverse the set of rich strings depth first. In a call for string S, for each letter c we call add(c), then make a recursive call if Sc is rich (i.e., add(c) = 1), and then call pop(). Over a constant-size alphabet, add(c) can be computed with direct links in $O(1)$ time. $\square$

Persistent Eertrees. Earlier in this section we have built an eertree supporting deletions from a string. A natural generalization of this approach leads to *persistent* eertrees. Recall that a persistent data structure is a set of "usual" data structures of the same type, called *versions* and ordered by the time of their creation. A call to a persistent structure asks for the access or update of any specific version. Existing versions are neither modified nor deleted; any update creates a new (latest) version.

Consider a *tree of versions* $\mathcal{T}$ whose nodes, apart from the root, are labeled by symbols. The tree represents the set of versions of some string S: each node v represents the string read from the root to v. Recall that we denote a node of a data structure by the same letter as the string related to it. Note that some versions can be identical except for the time of their creation (i.e., for the number of a node). The problem we study is maintaining an eertree for each version of S. More precisely, the function addVersion(v, c) to be implemented adds a new child u labeled by c to the node v of $\mathcal{T}$ and computes eertree(u). The data structure which performs the calls to addVersion, supporting the eertrees for all nodes of $\mathcal{T}$, will be called a *persistent eertree*. Surprisingly enough, this complicated structure can be implemented efficiently in spite of the fact that the current string cannot be addressed directly for symbol comparisons. Due to space constraints, we give the following proposition without proof.

Proposition 8. *The persistent eertree can be implemented to perform each call to* addVersion(v, c) *in $O(\log |v|)$ time and space.*

4   Factorizations into Palindromes

The k-factorization problem can be solved online in $O(kn)$ time for any length n string. Here we are aimed at solving this problem in time independent of k. This is motivated by the fact that the expected palindromic length of a random string is $\Omega(n)$ [14], so the $O(kn)$ asymptotics can be quite bad. On the positive side, the palindromic length of a string S, which is the minimum k such that a k-factorization of S exists, can be found in $O(n \log n)$ time [3].

Palindromic Length vs. k-Factorization. Due to the following lemma, the k-factorization problem is reduced in linear time to two similar problems: factor a string into the minimum possible odd (resp. even) number of palindromes. These two problems can be solved similar to the palindromic length problem.

Lemma 5. *Given a k-factorization of a length n string S, it is possible, in $O(n)$ time, to factor S into $k+2t$ palindromes for any $t \in \mathbb{N}$ such that $k+2t \leq n$.*

Proposition 9. *Using an eertree, the palindromic length of a length n string can be found online in time $O(n \log n)$.*

Proof (Idea). For a string S we compute online the array ans such that ans$[i]$ is the palindromic length of $S[1..i]$. Any k-factorization of S can be obtained by appending a suffix-palindrome $S[j+1..n]$ of S to a $(k-1)$-factorization of $S[1..j]$. Thus,

$$\mathsf{ans}[n] = 1 + \min\{\mathsf{ans}[j] \mid S[j + 1..n] \text{ is a palindrome}\}. \tag{1}$$

A naive algorithm computes the minimum in (1) in $O(n)$ time, scanning all suffix-palindromes of S starting at maxSuf(S) and following suffix links. In our fast algorithm we split the suffix path from maxSuf(S) into $O(\log n)$ blocks ("series of palindromes") in a special way and compute the minimum inside each block in $O(1)$ time. This gives us ans$[n]$ in $O(\log n)$ time and thus proves the proposition.

The announced split is based on two additional parameters of the nodes of eertree(S): *difference* diff$[v] = $ len$[v] - $ len$[\mathsf{link}[v]]$ and *series link* seriesLink$[v]$,

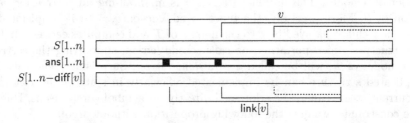

Fig. 2. Series of a palindrome v in $S[1..n]$ and of link$[v]$ in $S[1..n-\mathsf{diff}[v]]$. "Heads" of the next series are shown by dash lines. The function getMin(v) returns the minimum of the values of ans in the marked positions, plus one.

which is the longest suffix-palindrome u of v such that $\mathsf{diff}[u] \neq \mathsf{diff}[v]$. (Series links are not the same as quick links!) Both $\mathsf{diff}[v]$ and $\mathsf{seriesLink}[v]$ are computable in $O(1)$ time on the creation of the node. Series of palindromes consist of nodes with the same difference and the series link points to the "head" of the next block. It is nothing to compute for 1-element series; for longer series the situation is shown in Fig. 2: to compute the minimum for the series headed by v we compare such a minimum computed $\mathsf{diff}[v]$ steps before for the series headed by $\mathsf{link}[v]$ with one new value of ans. The proof of this property requires some combinatorics on words. The code below computes $\mathsf{ans}[n]$, using an auxiliary array $\mathsf{dp}[v]$ for storing precomputed minimums for the use in later steps.   □

```
int getMin(v)
        dp[v] = ans[n - (len[seriesLink[v]] + diff[v])]
        if (diff[v] == diff[link[v]])
            dp[v] = min(dp[v], dp[link[v]])
    return dp[v] + 1
ans[n] = /*@$\infty$@*/
for (v = maxSuf; len[v] > 0; v = seriesLink[v])
    ans[n] = min(ans[n], getMin(v))
```

Remark 4. Series links can replace quick links in the construction of eertrees. Each step requires at most $t+1$ comparisons of letters, where t is the number of series of suffix-palindromes of S (we omit the details).

Remark 5. Let t_i be the maximum length of a path of series links for the eertree of the string $S[1..i]$. Our computation of palindromic length (for an implementation, see http://ideone.com/xE2k6Y) performs, on each step, the following operations. For the eertree: at most t_i+1 symbol comparisons (Remark 4) and one $(\log \sigma)$-time access to a dictionary. For palindromic length: t_i calls to getMin, which fills one cell in dp and one cell in ans.

The algorithm by Fici et al. [3, Fig. 8] on each step fills $9t_i$ cells. So, our algorithm should work significantly faster.

Now we return to the k-factorization problem.

Proposition 10. *Using an eertree, the k-factorization problem for a length n string can be solved online in time $O(n \log n)$.*

Proof. The above algorithm for palindromic length can be easily modified to obtain both minimum odd number of palindromes and minimum even number of palindromes needed to factor a string. Instead of ans and dp, one can maintain in the same way four parameters: ans_o, ans_e, dp_o, dp_e, to take parity into account. Now ans_o (resp., ans_e) uses dp_e (resp., dp_o), while dp_o (resp., dp_e) uses ans_o (resp., ans_e). The reference to Lemma 5 finishes the proof.   □

Towards a Linear-Time Solution. Can palindromic length be found faster than in $O(n \log n)$ time? This is indeed the worst case time bound for our algorithm, since a length n string can have $\Theta(n \log n)$ series of palindromes [3]. On the other hand, the bit compression technique (the so-called *method of four Russians*) is capable of reducing the complexity of many problems by a $\log n$ factor. In particular, in [11] an $O(kn \log n)$ algorithm for k-factorization was transformed into a $O(kn)$ algorithm using bit compression. That algorithm produced a $k \times n$ bit matrix (showing whether a jth prefix of the string is i-factorable), so the speed up by grouping bits in packs of $\log n$ size was natural. In our case we work with integers, so the direct application of a bit compression is impossible. However, we have the following property.

Lemma 6. *If S is a string of palindromic length k and c is a symbol, then the palindromic length of Sc is $k-1, k,$ or $k+1$.*

By Lemma 6, if we have a $n \times n$ bit matrix M such that $M[i,j] = 1$ iff $S[1..j]$ is i-factorable, then, when filling the jth column, we have to compute just $M[k-1,j]$ and $M[k,j]$, where k is the palindromic length of $S[1..j-1]$. For each entry we should apply the OR operation to $\log n$ bit values, to the total of $2n \log n$ bit operations. If we will be able to arrange these operations in groups of size $\log n$, we will use the bit compression to obtain the palindromic length in just $O(n)$ operations. But to get an overall linear time, we need to build an eertree in $O(n)$ time as well; fortunately, this is possible.

Proposition 11. *The eertree of a length n string over the alphabet $\{1, \ldots, n\}$ can be built offline in $O(n)$ time.*

The proof is omitted due to space constraints. Finally, we formulate

Conjecture 1. Palindromic length of a string can be found in $O(n \log \sigma)$ time online and in $O(n)$ time offline using Lemma 6, eertree and bit compression.

5   Conclusion

In this paper, we proposed a new tree-like data structure, named eertree, which stores all palindromes occurring inside a given string. The eertree has linear size (even sublinear on average) and can be built online in nearly linear time. We proposed some advanced modifications of the eertree, including the joint eertree for several strings, the version supporting deletions from a string, and the persistent eertree.

Then we provided a number of applications of the eertree. The most important of them are the new online algorithms for k-factorization, palindromic length, the number of distinct palindromes, and also for computing the number of rich strings up to a given length. For further research we formulated a conjecture on the linear-time factorization into palindromes and an open problem about the optimal construction of the eertree.

Acknowledgement. The authors thank A. Kul'kov, O. Merkuriev and G. Nazarov for helpful discussions.

References

1. Driscoll, J.R., Sarnak, N., Sleator, D.D., Tarjan, R.E.: Making data structures persistent. J. Comput. Syst. Sci. **38**(1), 86–124 (1989)
2. Droubay, X., Justin, J., Pirillo, G.: Episturmian words and some constructions of de Luca and Rauzy. Theor. Comput. Sci. **255**, 539–553 (2001)
3. Fici, G., Gagie, T., Kärkkäinen, J., Kempa, D.: A subquadratic algorithm for minimum palindromic factorization. J. Discrete Algorithms **28**, 41–48 (2014)
4. Galil, Z., Seiferas, J.: A linear-time on-line recognition algorithm for "Palstar". J. ACM **25**, 102–111 (1978)
5. Glen, A., Justin, J., Widmer, S., Zamboni, L.: Palindromic richness. Eur. J. Comb. **30**(2), 510–531 (2009)
6. Groult, R., Prieur, E., Richomme, G.: Counting distinct palindromes in a word in linear time. Inf. Process. Lett. **110**, 908–912 (2010)
7. Gusfield, D.: Algorithms on Strings, Trees and Sequences. Computer Science and Computational Biology. Cambridge University Press, Cambridge (1997)
8. Kari, L., Mahalingam, K.: Watson—Crick palindromes in DNA computing. Nat. Comput. **9**, 297–316 (2010)
9. Knuth, D.E., Morris, J., Pratt, V.: Fast pattern matching in strings. SIAM J. Comput. **6**, 323–350 (1977)
10. Kosolobov, D., Rubinchik, M., Shur, A.M.: Finding distinct subpalindromes online. In: Proceedings of the Prague Stringology Conference, PSC 2013, pp. 63–69. Czech Technical University, Prague (2013)
11. Kosolobov, D., Rubinchik, M., Shur, A.M.: Pal^k is linear recognizable online. In: Italiano, G.F., Margaria-Steffen, T., Pokorný, J., Quisquater, J.-J., Wattenhofer, R. (eds.) SOFSEM 2015-Testing. LNCS, vol. 8939, pp. 289–301. Springer, Heidelberg (2015)
12. Manacher, G.: A new linear-time on-line algorithm finding the smallest initial palindrome of a string. J. ACM **22**(3), 346–351 (1975)
13. Mauri, G., Pavesi, G.: Algorithms for pattern matching and discovery in RNA secondary structure. Theor. Comput. Sci. **335**, 29–51 (2005)
14. Ravsky, O.: On the palindromic decomposition of binary words. J. Automata Lang. Comb. **8**(1), 75–83 (2003)
15. Rubinchik, M., Shur, A.M.: The number of distinct subpalindromes in random words [math.CO] (2015). arXiv:1505.08043
16. Sloane, N.J.A.: The on-line encyclopedia of integer sequences. http://oeis.org
17. Strothmann, D.: The affix array data structure and its applications to RNA secondary structure analysis. Theor. Comput. Sci. **389**, 278–294 (2007)
18. Ukkonen, E.: On-line construction of suffix trees. Algorithmica **14**(3), 249–260 (1995)
19. Problems of Asia-Pacific Informatics Olympiad 2014 (2014). http://olympiads.kz/apio2014/apio2014_problemset.pdf
20. Problems of the MIPT Fall Programming Training Camp 2014. Contest 12 (2014). https://drive.google.com/file/d/0B_DHLY8icSyNUzRwdkNFa2EtMDQ

On the Zero Forcing Number
of Bijection Graphs

Denys Shcherbak, Gerold Jäger$^{(\boxtimes)}$, and Lars-Daniel Öhman

Department of Mathematics and Mathematical Statistics,
University of Umeå, 901-87 Umeå, Sweden
{denys.shcherbak,gerold.jaeger,lars-daniel.ohman}@math.umu.se

Abstract. The zero forcing number of a graph is a graph parameter based on a color change process, which starts with a state, where all vertices are colored either black or white. In the next step a white vertex turns black, if it is the only white neighbor of some black vertex, and this step is then iterated. The zero forcing number $Z(G)$ is defined as the minimum cardinality of a set S of black vertices such that the whole vertex set turns black.

In this paper we study $Z(G)$ for the class of bijection graphs, where a bijection graph is a graph on $2n$ vertices that can be partitioned into two parts with n vertices each, joined by a perfect matching. For this class of graphs we show an upper bound for the zero forcing number and classify the graphs that attain this bound. We improve the general lower bound for the zero forcing number, which is $Z(G) \geq \delta(G)$, for certain bijection graphs and use this improved bound to find the exact value of the zero forcing number for these graphs. This extends and strengthens results of Yi (2012) about the more restricted class of so called permutation graphs.

Keywords: Zero forcing set · Zero forcing number · Bijection graph

1 Introduction

The *zero forcing number* is a well-studied graph parameter that was introduced in [2]. Let each vertex of a graph be colored either black or white. The *color change rule* says that if a vertex u is the only white neighbor of a black vertex v, then the color of u is turned to black. In this case, we say v forces u and denote it by $v \to u$. A set of black vertices S is said to be a *zero forcing set* if all vertices of the graph turn to black after finitely many applications of the color change rule. The zero forcing number of a graph G, denoted by $Z(G)$, is the minimum cardinality over all zero forcing sets S.

The zero forcing number has been used to bound the minimum rank of a graph, which is the smallest possible rank over all symmetric real matrices that are prescribed by the graph. The connection between the zero forcing number and the minimum rank problem has been considered in [3,4]. Furthermore, the

© Springer International Publishing Switzerland 2016
Z. Lipták and W.F. Smyth (Eds.): IWOCA 2015, LNCS 9538, pp. 334–345, 2016.
DOI: 10.1007/978-3-319-29516-9_28

zero forcing process and its variants have independently been studied in quantum mechanics [5,6,11] and in power networks [9,15]. For more details about computing the zero forcing set and the behavior of the zero forcing number, see [7,10,13,14]. The determination of the zero forcing number is $\mathcal{NP}$-hard [13]. Furthermore, no good bounds for the zero forcing number are known. For a given minimum and maximum degree of a graph G, $\delta(G)$ and $\Delta(G)$, respectively, it holds that $\delta(G) \leq Z(G) \leq \frac{n\Delta(G)}{\Delta(G)+1}$, see [1,4].

In this paper we introduce a new class of graphs, called *bijection graphs*, and consider the behavior of the zero forcing number on it. A bijection graph G is a graph that can be presented as the union of two subgraphs H_0, H_1, each with n vertices and joined by a *perfect matching*. Studying the zero forcing number of bijection graphs in relation to the zero forcing number of the two subgraphs is of interest as a possible reduction step in determining the zero forcing number of a general graph: If a general graph G on $2n$ vertices has the structure of a bijection graph, then the question of the zero forcing number of G can be informed by studying the properties of the two component graphs, having n vertices each. We present a general lower and upper bound for the zero forcing number of a bijection graph (Theorem 1). The class of bijection graphs includes the more restricted class of *permutation graphs* (also called *prism graphs*) defined by Harary and Chartrand, see [8][1]. The zero forcing number of permutation graphs was considered by Yi [14]. Note that since a permutation graph is a special case of a bijection graph, all results for bijection graphs hold also for permutation graphs.

Since the determination of the zero forcing number is $\mathcal{NP}$-hard, one fundamental question is how to optimize the process of finding a minimum zero forcing set. Our first result in this direction describes the structure of a zero forcing set of a bijection graph (Proposition 1). For this we use the specific structure of a bijection graph and the fact that the zero forcing number depends on the minimum degree.

Another way to simplify the task of finding a minimum zero forcing set S is to get information about the content of S. In particular we show which subsets of vertices of a bijection graph cannot be a zero forcing set (Lemma 2). As a consequence, for certain bijection graphs we improve the lower bound for $Z(G)$ (Theorems 2 and 3).

As the main result of the paper we show conditions for subgraphs H_0, $H_1 \subset G$, when the zero forcing number of a bijection graph reaches the upper bound (Theorem 4). This result is important, as in this case $Z(G)$ doesn't depend on the bijection f, and a minimum zero forcing set can be found without any computation. Applying this result to permutation graphs leads to an improvement of a result of Yi [14] (Theorem 5).

The paper is organized as follows. We give basic notation and preliminaries in Sect. 2. We show general rules and principles of the zero forcing process on bijection graphs and get a general lower and upper bound for the zero forcing

---

[1] Note that the term permutation graph is also used in a different way in literature, see [12].

number in Sect. 3. For certain types of graphs we improve the lower bound in Sect. 4 and consider graphs where the zero forcing number is equal to the upper bound n.

2   Notation and Preliminaries

Let $G = (V(G), E(G))$ be a simple and undirected graph of order $|V(G)| = n$. The *degree* $\deg_G(v)$ of a vertex $v \in V(G)$ is the number of edges incident with v in G. The *minimum degree* over all vertices of a graph G is denoted by $\delta(G)$ and the *maximum degree* by $\Delta(G)$.

Definition 1. Let H_0 and H_1 be two disjoint copies of a graph, and let $\sigma : V(H_0) \to V(H_1)$ be a permutation. A permutation graph $G_\sigma = (V, E)$ consists of the vertex set $V(G_\sigma) = V(H_0) \cup V(H_1)$ and the edge set $E(G_\sigma) = E(H_0) \cup E(H_1) \cup \{\{v, u\} \mid u = \sigma(v), v \in V(H_0), u \in V(H_1)\}$.

Definition 2. Let $H_0 = (V, X)$ and $H_1 = (U, Y)$ be graphs, where $|V| = |U|$, and let $f : V \to U$ be a bijection. A bijection graph $G = (W, Z)$ consists of the vertex set $W = V \cup U$, and the edge set $Z = X \cup Y \cup \{\{v, u\} \mid u = f(v), v \in V, u \in U\}$

We refer to the vertex subset $V(H_0)$ of a bijection graph $G = (H_0, H_1, f)$ as the H_0-side of G and denote it by $V_G(H_0)$, and the vertex subset $V(H_1)$ as the H_1-side of G and denote it by $V_G(H_1)$ (see Fig. 1). We refer to the graphs H_0 and H_1 as the components of the bijection graph G. For brevity we write H_{i+1} instead of $H_{(i+1) \bmod 2}$ for $i = 0, 1$. This is convenient when we consider a parameter of the "other side" of G. For instance, if we consider a vertex in one side (in H_0 or in H_1) and need to use the minimum degree of the graph of the other side, we may write "$v \in H_i$ has at least $\delta(H_{i+1})$ neighbors".

The *color change rule* turns the color of a white vertex v_2 to black, if it is the only white neighbor of some black vertex v_1. We denote this change by $v_1 \to v_2$. A sequence $v_1 \to v_2 \to \ldots \to v_t$, obtained through an iterative application of the color change rule is called a *forcing chain*. Notice that the set of black vertices changes after each iteration of the color change rule. We call the set of black vertices obtained after each iteration of the color change rule the

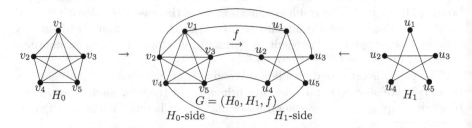

Fig. 1. Example for a bijection graph.

current set of black vertices. The *derived set* of black vertices $F(S)$ is the result of applying the color change rule starting with an initial set S, until no more changes are possible. If the derived set $F(S)$ includes all vertices of a graph G, then S is called a zero forcing set. The minimum cardinality over all zero forcing sets is called the zero forcing number of G, denoted by $Z(G)$.

It may happen that more than one vertex can force at the same time, as illustrated by Example 1. This is the reason why the color change process is not unique. However, the derived set of black vertices $F(S)$ is independent of the order of the color change process [2]. We will need the following two definitions.

Definition 3. Let G be a graph and let M be a current set of black vertices. A vertex $v \in M$ is called an *active vertex* of M if v has only one white neighbor in $G \setminus M$. The union of all active vertices of M is called *active subset* of M.

Definition 4. Let G be a graph and let M be a current set of black vertices. The forcing of all active vertices of M is called a step of the color change process, denoted by $M \to M'$, where M' is the set of vertices turned to black.

Example 1. Consider Fig. 2a, where we have a graph G with an initial set S. The vertices v_1 and v_3 are active vertices of S and can force. Since it does not matter which vertex v_1 or v_3 forces first, we assume that they force simultaneously. Thus, the first step of the color change process is $S \to S'$, where $S' = \{v_5, v_6\}$. After the first step of the color change process the current set of black vertices is $S_1 = S \cup S' = \{v_1, v_2, v_3, v_5, v_6\}$ (see Fig. 2b).

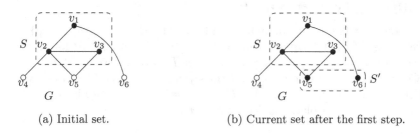

(a) Initial set.   (b) Current set after the first step.

Fig. 2. One step of a zero forcing process.

Note that if we have more than one vertex that forces the same vertex, then we also assume that all of them force this vertex. The zero forcing process can thus be presented as a nested sequence of black vertices $S = S_0 \subsetneq S_1 \subsetneq \cdots \subsetneq S_k$, where $S_{j+1} = S_j \cup S_j'$, $S_j \to S_j'$ for $j = 0, 1, \ldots, k-1$. If S is a zero forcing set, we obviously have $S_k = V(G)$.

3   The Zero Forcing Process on Bijection Graphs

Consider an initial set S of a bijection graph $G = (H_0, H_1, f)$ with an active vertex $w \in V(H_i)$, where $i \in \{0, 1\}$. Let w force w_1, i.e., $w \to w_1$. There are

338   D. Shcherbak et al.

two cases for the location of w_1. If w_1 lies at the same side, i.e., $w_1 \in V_G(H_i)$, then only one neighbor of w in $V_G(H_i)$ is white. Thus, $V_G(H_i)$ contains at least $(\deg_{H_i}(w) - 1) + 1 = \deg_{H_i}(w)$ black vertices. If w_1 lies at the other side, i.e., $w_1 \in V_G(H_{i+1})$, then all neighbors of w in $V_G(H_i)$ are black, Thus, $V_G(H_i)$ contains at least $\deg_{H_i}(w) + 1$ black vertices. In addition, if S has p black vertices at the H_i-side, any black $v \in V_G(H_i)$ cannot have more than $p-1$ black neighbors in $V_G(H_i)$, and maybe one black neighbor in $V_G(H_{i+1})$. Therefore, for an active vertex w it holds that $\deg_{H_i}(w) \leq p$. The following observation summarizes these results.

Observation 1. *Let $G = (H_0, H_1, f)$ be a bijection graph and S be an initial set. Then the following holds for $i = 0, 1$:*

(a) *If $w \in V_G(H_i)$ is an active vertex of S, then S must contain at least $\deg_{H_i}(w)$ black vertices at the H_i-side.*

(b) *If S contains p black vertices in $V_G(H_i)$, then an active vertex w fulfills $\deg_{H_i}(w) \leq p$.*

Sometimes it is important to know where the vertices whose color has currently been changed are placed. We now prove a lemma which gives information about this issue.

Lemma 1. *Let H_0 and H_1 be graphs of order n, and let $f : V(H_0) \to V(H_1)$ be a bijection. For $i = 0, 1$ the following holds: If an initial set S of the corresponding bijection graph $G = (H_0, H_1, f)$ has an active vertex in $V_G(H_i)$, then there is an initial set T with $|T| = |S|$, such that $F(S) = F(T)$ and each active vertex of T in $V_G(H_i)$ forces a vertex in $V_G(H_{i+1})$.*

Proof. Let $i = 0, 1$ be fixed, S be an initial set, $M := \{v_1, v_2, \ldots, v_p\}$ be an active subset of S, and suppose that $M \subseteq V_G(H_i)$. If each $v_j \in M$ forces $f(v_j) \in V_G(H_{i+1})$ for $j = 1, 2, \ldots, p$, then we may set $T := S$ and we are done.

Consider the case where some vertices $\{w_1, w_2, \ldots, w_t\} \subseteq M$ force vertices in $V_G(H_i)$, namely $w_j \to w_{j,1} \in V_G(H_i)$ for $j = 1, 2, \ldots, t$ (see Fig. 3a). After w_1 has forced, the set of black vertices is $S \cup \{w_{1,1}\}$.

(a) Initial set S.   (b) Initial set $T(w_1)$.

Fig. 3. Two initial sets with the same derived set.

Consider a new initial set $T(w_1) = S \cup \{w_{1,1}\} \setminus \{f(w_1)\}$ (see Fig. 3b), i.e., we change the only white neighbor of the vertex w_1 from $w_{1,1}$ to $f(w_1)$. In this case, after w_1 has forced, the set of black vertices is $T(w_1) \cup \{f(w_1)\}$ (see Fig. 3b). It is easy to see that $S \cup \{w_{1,1}\} = T(w_1) \cup \{f(w_1)\}$, and thus, $F(S) = F(T(w_1))$. Therefore, for the new initial set $T(w_1)$ we have $|S| = |T(w_1)|$ and $F(S) = F(T(w_1))$ and by construction of $T(w_1)$ we receive $w_1 \rightarrow f(w_1) \in V_G(H_{i+1})$.

We repeat this process of switching vertices for each new initial set and for all active vertices in $V_G(H_i)$. This process terminates, because the number of black vertices in $V_G(H_{i+1})$ decreases after each step. Thus, after a finite number of steps we get an initial set T with $|T| = |S|$, where each active vertex of T in $V_G(H_i)$ forces a vertex in $V_G(H_{i+1})$, and $F(T) = F(S)$. $\qquad\square$

Note that for a bijection graph $G = (H_0, H_1, f)$, $\delta(G) = \min\{\delta(H_0), \delta(H_1)\} + 1$ holds. It is easy to see that the initial set of black vertices $V(H_0)$ (or $V(H_1)$) is a zero forcing set of G. Using the well-known bound for the zero forcing number $Z(G) \geq \delta(G)$ [4], we receive the following theorem:

Theorem 1. *Let H_0 and H_1 be graphs of order n, and let $f : V(H_0) \rightarrow V(H_1)$ be a bijection. Then for the bijection graph $G = (H_0, H_1, f)$ the following holds:*

$$\min\{\delta(H_0), \ \delta(H_1)\} + 1 \ = \ \delta(G) \leq Z(G) \leq n.$$

Note that for the case $\delta(H_0) \leq \delta(H_1)$, it holds that $\delta(H_0) + 1 \leq Z(G) \leq n$. The following proposition establishes a necessary condition for a zero forcing set of a bijection graph.

Proposition 1. *Let H_0 and H_1 be graphs of order n, and let $f : V(H_0) \rightarrow V(H_1)$ be a bijection. Then for any zero forcing set S of the bijection graph $G = (H_0, H_1, f)$ at least one of the following conditions holds: $|S \cap V_G(H_0)| \geq \delta(H_0)$ or $|S \cap V_G(H_1)| \geq \delta(H_1)$.*

Proof. Let $S \subseteq G$ be a zero forcing set of G, and set $b_0 := |S \cap V_G(H_0)|$ and $b_1 := |S \cap V_G(H_1)|$. Assume that $b_0 < \delta(H_0)$ and $b_1 < \delta(H_1)$. Since $V_G(H_0)$ contains b_0 black vertices, by Observation 1(b), for an active vertex $v \in V_G(H_0)$ it holds that $\deg_{H_0}(v) \leq b_0$, which is a contradiction. Thus, there is no active vertex of S in $V_G(H_0)$. The same is true for $V_G(H_1)$. Overall, there is no active vertex in S, i.e., S cannot be a zero forcing set. Thus, $b_0 \geq \delta(H_0)$ or $b_1 \geq \delta(H_1)$ holds. $\qquad\square$

We combine Lemma 1 and Proposition 1 to describe the structure of a zero forcing set of a graph with $Z(G) = \delta(G)$.

Proposition 2. *Let H_0 and H_1 be graphs of order n, and let $f : V(H_0) \rightarrow V(H_1)$ be a bijection. If the zero forcing number of the corresponding bijection graph $G = (H_0, H_1, f)$ is $Z(G) = \delta(G)$, then there is a minimum zero forcing set which is not divided between the H_0-side and H_1-side of G.*

Proof. Let S be a zero forcing set with $|S| = \delta(G)$. If $S \subseteq V_G(H_0)$ or $S \subseteq V_G(H_1)$, there is nothing to prove. Let S be divided between the sides. For definiteness and without loss of generality we assume that $\delta(H_0) \leq \delta(H_1)$. By

Proposition 1, one of the sides of the bijection graph must contain at least $\delta(H_0) = \delta(G) - 1$ vertices of S. In this case, the other side contains one vertex of the initial set.

Obviously, for an active vertex v of S it holds that $\deg_G(v) = \delta(G)$, and all other black vertices of S are its neighbors. Since one of the black neighbors of v is at the other side, v forces a vertex at the same side (see Fig. 4). By Lemma 1, we can get a zero forcing set T with $|T| = \delta(G)$, where v forces $f(v)$. In this case all black vertices are at the same side, and thus, T is not divided. $\square$

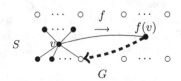

Fig. 4. Replacing the black vertex to get an initial set which is not divided.

4   The Zero Forcing Number of Bijection Graphs

Lemma 2. *Let H_0 and H_1 be graphs of order n and $\delta(H_i) = n - r_i$, $r_i \geq 1$ for $i = 0,1$. Furthermore, let $f : V(H_0) \rightarrow V(H_1)$ be a bijection, and $G = (H_0, H_1, f)$ be the corresponding bijection graph. Then for $i = 0,1$ the following holds: If $S \subsetneq V_G(H_i)$ is an initial set, it follows that:*

(a) *If $S \rightarrow S'$, then $|S'| \leq r_i - 1$.*
(b) *If $\delta(H_{i+1}) \geq r_i$, then S is not a zero forcing set.*

Proof. Let $i = 0,1$ be fixed. Note that, if an initial $S \subsetneq V_G(H_i)$ cannot force more than $r_i - 1$ vertices, then consequently, neither can any subset $K \subset S$. So, if we prove the statement for an initial set $S \subsetneq V_G(H_i)$ with $|S| = n - 1$, it is true for any initial set $K \subsetneq V_G(H_i)$. Thus, we may assume that $|S| = n - 1$.

(a) As $|S| = n - 1$, $V_G(H_i)$ contains only one white vertex v_1. Writing the degree of v_1 as $\deg_{H_i}(v_1) = n - p$, we receive $n - p = \deg_{H_i}(v_1) \geq \delta(H_i) = n - r_i$ and thus, $p \leq r_i$. Each black vertex $v \in V_G(H_i)$ which is adjacent to v_1 cannot force at this step of the forcing process, because it is also adjacent to the white vertex $f(v) \in V_G(H_{i+1})$ (see Fig. 5a).

On the other hand, exactly $p - 1$ vertices in S are not adjacent to v_1, and thus, none of them has a white neighbor in $V_G(H_i)$, and each of them has only one white neighbor in $V_G(H_{i+1})$. Thus, each of these $p - 1$ vertices forces at the first step of the forcing process (see Fig. 5b). As $p \leq r_i$, S cannot force more than $r_i - 1$ vertices. This proves (a).

(b) We continue the zero forcing process in G. After all active vertices of S have forced, i.e., $S \rightarrow S'$, the current set of black vertices is $S_1 = S \cup S'$. Note

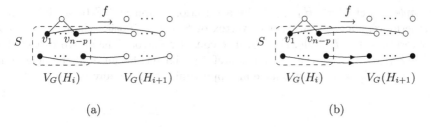

$V_G(H_i)$ $V_G(H_{i+1})$ $V_G(H_i)$ $V_G(H_{i+1})$

(a)   (b)

Fig. 5. The forcing of the initial set $S \subsetneq V_G(H_i)$.

that the remaining black vertices $v_1, \ldots, v_{n-p}$ at the H_i-side still cannot force, because they didn't receive any new black neighbor (see Fig. 5b). Therefore, for continuation of the zero forcing process, S' must contain at least one active vertex. By Observation 1(a), $V_G(H_{i+1})$ must contain at least $\delta(H_{i+1})$ black vertices. We have shown in a) that $|S'| \leq r_i - 1$. Consequently, if $\delta(H_{i+1}) \geq r_i$, then there is no active vertex in $S_1 = S \cup S'$. Thus, the forcing process stops, and S is not a zero forcing set. This proves (b). □

For the following results we need the following equivalence statement.

Remark 1. Let H_0 and H_1 be graphs of order n, and let $\delta(H_i) = n - r_i$, where $r_i \geq 1$ for $i = 0, 1$. Then $\delta(H_0) + \delta(H_1) \geq n$ is equivalent to $\delta(H_{i+1}) \geq r_i$ for $i = 0, 1$.

Proof. "⇒": Let $\delta(H_0) + \delta(H_1) \geq n$. By $\delta(H_0) = n - r_0$ we get $\delta(H_1) \geq r_0$. Analogously, by $\delta(H_1) = n - r_1$ we get $\delta(H_0) \geq r_1$.
"⇐": Let $\delta(H_0) \geq r_1$ and $\delta(H_1) \geq r_0$. By $\delta(H_1) = n - r_1$ we get $\delta(H_0) + \delta(H_1) \geq n$. □

As a consequence we get an improved lower bound on the zero forcing number for the bijection graphs from Lemma 2.

Theorem 2. *Let H_0 and H_1 be graphs of order n, which are not both complete graphs, and let $\delta(H_0) + \delta(H_1) \geq n$. Furthermore, let $f : V(H_0) \to V(H_1)$ be a bijection. Then for the corresponding bijection graph $G = (H_0, H_1, f)$ it holds that $Z(G) > \delta(G)$.*

Proof. Assume that $Z(G) = \delta(G)$, and let S be a zero forcing set with $|S| = \delta(G)$. By Proposition 2, we may assume that S is not divided between the sides, i.e., $S \subseteq V_G(H_i)$, where $i \in \{0, 1\}$. As H_0 and H_1 are not both complete graphs, it follows that $S \subsetneq V_G(H_i)$. By Lemma 2(b) and Remark 1, $S \subsetneq V_G(H_i)$ cannot be a zero forcing set. The result follows. □

Our aim is to get a better lower bound for $Z(G)$ of a bijection graph G, where the minimum degree of its components satisfies $\max\{\delta(H_0), \delta(H_1)\} \geq \frac{n}{2}$. To do this we use the technique shown in the following example.

Example 2. Let $G = (H_0, H_1, f)$ be a bijection graph and S be an initial set. Furthermore, suppose each active vertex of S at the H_0-side changes the color of a vertex at the H_1-side. For instance consider Fig. 6a. The active vertex of S in $V_G(H_0)$ is v_4. Note that S is a wasteful initial set, because all the neighbors of v_5 are black, so it doesn't have an opportunity to force any vertex in G.

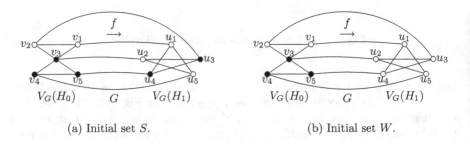

(a) Initial set S.   (b) Initial set W.

Fig. 6. Comparison of the forcing processes of two initial sets.

Consider an initial set W which we get from S by removing all black vertices at the H_1-side (see Fig. 6b). We can see that v_4 is an active vertex of W, as well as of S, because it doesn't have a black neighbor in $V_G(H_1) \cap S$, and so the coloring of its neighbors is the same for S and W. On the other hand, v_5 is an active vertex only for W and not for S. We can also observe that the number of active vertices at the H_0-side for S is less than the number of active vertices for W.

As in this simple example we get the following general conclusion.

Observation 2. *Consider a bijection graph* $G = (H_0, H_1, f)$ *with two initial sets S and W such that the following holds:*

(i) *Each active vertex of S in $V_G(H_0)$ forces a vertex in $V_G(H_1)$.*
(ii) *W is the result of removing all black vertices of S which lie in $V_G(H_1)$.*

Then all active vertices of S at the H_0-side are active vertices of W, too.

Now, we determine a lower bound of the zero forcing number $Z(G)$ of a bijection graph $G = (H_0, H_1, f)$ with $\delta(H_0) + \delta(H_1) \geq n$.

Theorem 3. *Let H_0 and H_1 be graphs of order n, and let $\delta(H_0) + \delta(H_1) \geq n$. Assume that $\delta(H_0) \leq \delta(H_1)$. Furthermore, let $f : V(H_0) \to V(H_1)$ be a bijection. Then for the corresponding bijection graph $G = (H_0, H_1, f)$ it holds that*
$$Z(G) \geq \min\{n, 2\delta(H_0) + \delta(H_1) - n + 2\}.$$

Proof. Let $\delta(H_0) = n - r_0$, $\delta(H_1) = n - r_1$, where $r_0, r_1 \geq 1$. Let S be a minimum zero forcing set of the bijection graph G. By Lemma 2(b) and Remark 1, an initial set $W \subsetneq V_G(H_i)$, where $i \in \{0, 1\}$, is not a zero forcing set. Thus, S must either fulfill $|S| = n$, or be divided between two sides. Since the case

$|S| = n$ satisfies the statement, we consider the case where S is divided and $|S| < n$.

Let $V_G(H_i)$ be a side of G with active vertices, where $i \in \{0, 1\}$. Denote by $M = \{v_1, v_2, \ldots, v_p\}$ the set of active vertices of S in $V_G(H_i)$, and set $b_0 := |S \cap V_G(H_i)|$ and $b_1 := |S \cap V_G(H_{i+1})|$. By Lemma 1, we may assume that each active vertex $v_j \in M$ forces a vertex $f(v_j) \in V_G(H_{i+1})$ for $j = 1, 2, \ldots p$ (see Fig. 7a). Consider an active vertex $v \in M$. As v forces the vertex $f(v) \in V_G(H_{i+1})$, all neighbors of v in $V_G(H_i)$ are black. Thus, $V_G(H_i)$ contains at least $\deg_{H_i}(v) + 1$ black vertices, as v and all of its neighbors must be black. In other words, $b_0 \geq \deg_{H_i}(v) + 1$, and consequently

$$b_0 \geq \delta(H_i) + 1. \tag{1}$$

By assumption,

$$Z(G) = b_0 + b_1 < n \tag{2}$$

$$\Rightarrow b_1 \overset{(1),(2)}{<} n - \delta(H_i) - 1 = n - (n - r_i) - 1 = r_i - 1 \overset{Remark\ 1}{\leq} \delta(H_{i+1}) - 1.$$

Thus, each black vertex at the H_{i+1}-side has less than or equal to $\delta(H_{i+1}) - 2$ black neighbors, i.e., at least two white neighbors. It follows that there are no active vertices in $V_G(H_{i+1})$. In other words, at the first step of the color change process only black vertices in $V_G(H_i)$ can force. By assumption, each active vertex of S in $V_G(H_i)$ changes the color of a vertex in $V_G(H_{i+1})$. Therefore, we can write the set of black vertices after the first step of the color change process as $S_1 = S \cup S'$, where $S \to S'$ and $S' \subset V_G(H_{i+1})$.

By Observation 2, each active vertex of S is an active vertex of $W = S \cap V_G(H_i)$, too. Using these facts we can write $|W'| \geq |S'|$, where $S \to S'$, $W \to W'$. By Lemma 2(a), for such a type of bijection graph it holds that $|W'| \leq r_i - 1$.

$$\Rightarrow |S'| \leq r_i - 1. \tag{3}$$

Therefore, after the first step of the zero forcing process we have b_0 black vertices at the H_i-side, and $b_1 + |S'|$ black vertices at the H_{i+1}-side (see Fig. 7b). The vertices in $V_G(H_i)$ didn't receive any new black neighbor except the active vertices of S. Thus, there is no vertex in $V_G(H_i)$ which can continue the color change process. In this case, active vertices of S_1 lie in $V_G(H_{i+1})$. Since the H_{i+1}-side has an active vertex, by Observation 1(a), the number of black vertices in $V_G(H_{i+1})$ must be at least $\delta(H_{i+1})$. It follows:

$$|b_1| + |S'| \geq \delta(H_{i+1}) \tag{4}$$

$$\Rightarrow \qquad b_1 \overset{(4)}{\geq} \delta(H_{i+1}) - |S'| \overset{(3)}{\geq} \delta(H_{i+1}) - (r_i - 1) \tag{5}$$

$$\Rightarrow b_0 + b_1 \overset{(1),(5)}{\geq} (\delta(H_i) + 1) + (\delta(H_{i+1}) - (r_i - 1)) = 2\delta(H_i) + \delta(H_{i+1}) - n + 2.$$

Using the fact that $\delta(H_1) \geq \delta(H_0)$ we receive that $|S| \geq 2\delta(H_0) + \delta(H_1) - n + 2$. Thus, the cardinality of a minimum zero forcing set S satisfies:

(a) $|S| \geq 2\delta(H_0) + \delta(H_1) - n + 2$, if it is divided between the sides,
(b) $|S| = n$, if it is equal to one of the sides.

We can rewrite this to $Z(G) \geq \min\{n, 2\delta(H_0) + \delta(H_1) - n + 2\}$. ☐

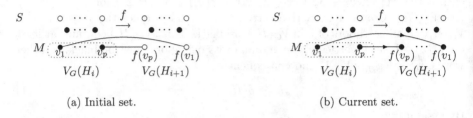

(a) Initial set.   (b) Current set.

Fig. 7. First step of the zero forcing process.

In the following we will show bounds on $Z(G)$ for some special n and $\delta(H_0)$, $\delta(H_1)$.

Theorem 4. *Let H_0 and H_1 be graphs of order n, and let $\delta(H_0) + \delta(H_1) \geq n$. Assume that $\delta(H_0) \leq \delta(H_1)$. Furthermore, let $f : V(H_0) \to V(H_1)$ be a bijection. Then for the corresponding bijection graph $G = (H_0, H_1, f)$ the following holds:*

(a) $Z(G) \geq \min\{n, \delta(H_0) + 3\}$, *if $\delta(H_0) + \delta(H_1) > n$,*
(b) $Z(G) = n$, *if $\delta(H_0) \geq \frac{2n-2}{3}$.*

Proof. (a) Let $\delta(H_0) + \delta(H_1) > n$.

$$\Rightarrow 2\delta(H_0) + \delta(H_1) - n + 2 = \delta(H_0) + \delta(H_0) + \delta(H_1) - n + 2 \geq \delta(H_0) + 3.$$

Thus, (a) follows from Theorem 3.
(b) Let $\delta(H_0) \geq \frac{2n-2}{3}$. This is equivalent to $3\delta(H_0) + 2 \geq 2n$. It follows that

$$2\delta(H_0) + \delta(H_1) - n + 2 \geq (3\delta(H_0) + 2) - n \geq 2n - n = n.$$

Thus, (b) follows from Theorems 1 and 3. ☐

Remark 2. From the assumption $\delta(H_0) + \delta(H_1) \geq n$, it follows that

$$\max\{\delta(H_0), \delta(H_1)\} \geq \frac{n}{2}.$$

Yi proved that for a permutation graph G_σ with n vertices and $\delta(G) = n - 2$ it holds that $Z(G_\sigma) = n$ [14]. Theorem 4(b) improves on this theorem. We just need to use the fact that for a permutation graph G_σ the minimum degree of its components is the same and receive the following statement.

Theorem 5. *Let H_0 and H_1 be two disjoint copies of a graph of order n, and let $\delta(H_0) = \delta(H_1) \geq \frac{2n-2}{3}$. Furthermore, let $\sigma : V(H_0) \to V(H_1)$ be a permutation. Then for the corresponding permutation graph G_σ it holds that $Z(G_\sigma) = n$.*

References

1. Amos, D., Caro, Y., Davila, R., Pepper, R.: Upper bounds on the k-forcing number of a graph. Discrete Appl. Math. **181**, 1–10 (2015)
2. Barioli, F., et al.: (AIM minimum rank special graphs work group): zero forcing sets and the minimum rank of graphs. Linear Algebra Appl. **428**(7), 1628–1648 (2008)
3. Barioli, F., et al.: Zero forcing parameters and minimum rank problems. Linear Algebra Appl. **433**(2), 401–411 (2010)
4. Berman, A., et al.: An upper bound for the minimum rank of a graph. Linear Algebra Appl. **429**(7), 1629–1638 (2008)
5. Burgarth, D., Giovannetti, V.: Full control by locally induced relaxation. Phys. Rev. Lett. **99**, 100–501 (2007)
6. Burgarth, D., Maruyama, K.: Indirect hamiltonian identification through a small gateway. New J. Phys. **11**, 103019 (2009)
7. Chilakamarri, K.B., Dean, N., Kang, C.X., Yi, E.: Iteration index of a zero forcing set in a graph. Bull. Inst. Combin. Appl. **64**, 57–72 (2012)
8. Harary, F., Chartrand, C.: Planar permutation graphs. Annales de l'I.H.P. (B) Probabilités et Statistiques **3**(4), 433–438 (1967)
9. Haynes, T.W., Hedetniemi, S.M., Hedetniemi, S.T., Henning, M.A.: Domination in graphs applied to electric power networks. SIAM J. Discrete Math. **15**(4), 519–529 (2002)
10. Row, D.D.: A technique for computing the zero forcing number of a graph with a cut-vertex. Linear Algebra Appl. **436**(12), 4423–4432 (2012)
11. Severini, S.: Nondiscriminatory propagation on trees. J. Phys. A: Math. Theor. **41**, 482001 (2008)
12. Skiena, S.: Implementing Discrete Mathematics: Combinatorics and Graph Theory with Mathematica. Addison-Wesley, Reading, MA (1990)
13. Trefois, M., Delvenne, J.-C.: Zero forcing sets, constrained matchings and strong structural controllability. Linear Algebra Appl **484**, 199–218 (2015). arXiv:1405.6222v1
14. Yi, E.: On zero forcing number of permutation graphs. In: Lin, G. (ed.) COCOA 2012. LNCS, vol. 7402, pp. 61–72. Springer, Heidelberg (2012)
15. Zhao, M., Kang, L., Chang, G.J.: Power domination in graphs. Discrete Math. **306**(15), 1812–1816 (2006)

The k-Leaf Spanning Tree Problem Admits a Klam Value of 39

Meirav Zehavi[1($\boxtimes$)]

Department of Computer Science, Technion IIT, 32000 Haifa, Israel
meizeh@cs.technion.ac.il

Abstract. The *klam value* of an algorithm that runs in time $O^*(f(k))$ is the maximal value k such that $f(k) < 10^{20}$. Given a graph G and a parameter k, the k-LEAF SPANNING TREE (k-LST) problem asks if G contains a spanning tree with at least k leaves. This problem has been extensively studied over the past three decades. In 2000, Fellows *et al.* [FSTTCS'00] asked whether it admits a klam value of 50. A steady progress towards an affirmative answer continued until 5 years ago, when an algorithm of klam value 37 was discovered. Our contribution is twofold. First, we present an $O^*(3.188^k)$-time parameterized algorithm for k-LST, which shows that the problem admits a klam value of 39. Second, we rely on an application of the bounded search trees technique where the correctness of rules crucially depends on the *history* of previously applied rules in a non-standard manner, encapsulated in a "dependency claim". Similar claims may be used to capture the essence of other complex algorithms in a compact, useful manner.

1 Introduction

Given an undirected graph $G = (V, E)$ and a parameter k, the k-LEAF SPANNING TREE (k-LST) problem asks if G contains a spanning tree with at least k leaves. Due to its general nature, k-LST has applications in a variety of areas, including, for example, the design of ad-hoc sensor networks (see [3,19,21]) and computational biology (see, e.g., [18]). Furthermore, k-LST is tightly linked to the classic k-CONNECTED DOMINATED SET (k-CDS) problem, which asks if a given graph $G = (V, E)$ contains a connected dominating set of size at most k.[1] On the one hand, given a spanning tree T with t leaves, the set of internal vertices of T forms a connected dominating set of size $|V| - t$. On the other hand, given a connected dominating set S of size t, one can construct a spanning tree T with at least $|V| - t$ leaves (attach the vertices in $V \backslash S$ as leaves to a tree spanning the subgraph of G induced by S).

Even in restricted settings, it has long been established that k-LST is NP-hard (see, e.g., [9]). Thus, over the past three decades, k-LST has been extensively studied in the fields of Parameterized Complexity, Exact Exponential-Time Computation and Approximation. We focus on *fixed-parameter tractable*

---

[1] A connected dominating set is a subset $U \subseteq V$ such that the subgraph of G induced by U is connected and every vertex in $V \backslash U$ is a neighbor of a vertex in U.

© Springer International Publishing Switzerland 2016
Z. Lipták and W.F. Smyth (Eds.): IWOCA 2015, LNCS 9538, pp. 346–357, 2016.
DOI: 10.1007/978-3-319-29516-9_29

Table 1. Known FPT algorithms for k-LST.

Reference	First published	Klam value	Running time
Fellows et al. [14]	1988 [14]	0	FPT
Bodlaender et al. [4]	1989 [4]	1	$O^*((17k^4)!)$
Fellows et al. [11]	1995 [10]	5	$O^*((2k)^{4k})$
Fellows et al. [15]	2000 [15]	17	$O^*(14.23^k)$
Bonsma et al. [5]	2003 [5]	20	$O^*(9.49^k)$
Estivill-Castro et al. [13]	2005 [13]	22	$O^*(8.12^k)$
Bonsma et al. [6]	2008 [6]	24	$O^*(6.75^k)$
Kneis et al. [17]	2008 [16]	33	$O^*(4^k)$
Daligault et al. [8]	2008 [7]	35	$O^*(3.72^k)$
Binkele-Raible et al. [2]	2010 [20]	37	$O^*(3.46^k)$
This paper	2015	39	$O^*(3.19^k)$

(FPT) algorithms, which run in time $O^*(f(k))$ for some function f, where O^* hides factors polynomial in the input size.

Table 1 presents a summary of known FPT algorithms for k-LST. The *klam value* of an algorithm that runs in time $O^*(f(k))$ is the maximal value k such that $f(k) < 10^{20}$. In 2000, Fellows et al. [15] asked whether k-LST admits a klam value of 50. A steady progress towards an affirmative answer continued until 2010, when an algorithm of klam value 37 was discovered [20].

We develop a deterministic polynomial-space FPT algorithm for k-LST that runs in time $O(3.188^k + \text{poly}(|V|))$. Thus, our contribution it twofold: we show that k-LST admits a klam value of 39, and we explicitly define the notion of a "dependency claim" (see below). Our result, like previous algorithms for k-LST, is based on the bounded search trees technique (see Sect. 2): Essentially, when applying a branching rule, we determine the "role" of a vertex in G—i.e., we decide whether it should be, in the constructed tree, a leaf or an internal vertex (which, in turn, may determine roles of other vertices). Also, along with the constructed tree (to be completed to a spanning tree), we maintain a list of "floating leaves"—vertices in G that are not yet attached to the constructed tree, but whose roles as leaves has been already determined.

The difference between our result and previous algorithms for k-LST lies in the encapsulation of dependencies among nodes (of a search tree) in a "dependency claim". In our application of the bounded search trees technique, nodes depend on the *history* of nodes that precede them in a non-standard manner— the correctness of many of our reduction and branching rules crucially relies on formerly executed branching rules, particularly on the fact that certain branches considered by them could not lead to the construction of a solution. More precisely, for certain vertices whose roles are to be determined, our decision will rely on the fact that there is no solution in which their parents are leaves.[2] Indeed, the crux of our algorithm is the explicit definition and demonstration of the usefulness of a "dependency claim", which describes that dependency of the nodes (of a search tree) on the nodes that precede them. We consider this claim as the foundation on which we build the rules of our algorithm. More precisely, our rules are carefully designed to preserve the correctness of the dependency claim along the search tree, while, most importantly, allowing us to exploit its implications. Generally, the use of similar dependency claims may capture the dependencies implicitly exploited by other complex algorithms, which are based on the bounded search trees technique, in a compact, useful manner. This can lead to the development of faster algorithms that might not be more compact, but whose essence and "bigger picture", presented by the claims, will be clearer.

Organization. First, Sect. 2 gives necessary information on bounded search trees, along with standard definitions and notation. Then, Sect. 3 presents our algorithm, including our measure and the dependency claim. We only present a brief overview of the entire set of rules of the algorithm, after which we discuss in detail two rules which capture its spirit. The complete set of rules can be found at http://www.arxiv.org/abs/1502.07725. Finally, Sect. 4 concludes the paper.

2   Preliminaries

Bounded Search Trees. Bounded search trees is a fundamental technique in the design of recursive FPT algorithms (see [12]). To apply it, one defines a list of rules of the form Rule X. [condition] action, where X is the number of the rule in the list. At each recursive call (i.e., a node in the search tree), the algorithm performs the action of the first rule whose condition is satisfied. If by performing an action, the algorithm recursively calls itself at least twice, the rule is a *branching rule*, and otherwise it is a *reduction rule*. We only consider polynomial-time actions that increase neither the parameter nor the size of the instance.

The running time of the algorithm can be bounded as follows (see, e.g., [1]). Suppose that the algorithm executes a branching rule where it recursively calls itself ℓ times, such that in the ith call, the current value of the parameter decreases by b_i. Then, $(b_1, b_2, \ldots, b_\ell)$ is called the *branching vector* of this rule. We say that α is the *root* of $(b_1, b_2, \ldots, b_\ell)$ if it is the (unique) positive real root of $x^{b^*} = x^{b^* - b_1} + x^{b^* - b_2} + \ldots + x^{b^* - b_\ell}$, where $b^* = \max\{b_1, b_2, \ldots, b_\ell\}$. If $b > 0$ is the initial value of the parameter, and the algorithm (a) returns a result when (or before) the parameter is negative, and (b) only executes branching rules whose roots are bounded by a constant c, then its running time is bounded by $O^*(c^b)$.

Standard Definitions and Notation. Given a graph $G = (V, E)$ and a vertex $v \in V$, let N(v) denote the set of neighbors of v (in G, which will be clear from

---

[2] Problematic vertices in whose examination we cannot rely on such a fact will be handled by a "marking" approach—we will be able to consider our treatment of them as better than it is, since we previously considered the treatment of the vertices that marked them as worse than it is.

context). Given subsets $S, U \subseteq V$, let $\text{Paths}(S, v, U)$ denote the set of paths that start from a vertex in S and end at the vertex v, whose *internal* vertices belong to U (only). Given a rooted tree $T = (V_T, E_T)$, let $\text{Lea}(T)$, $\text{Int}(T)$ and $\text{Chi}_i(T)$ denote the leaf-set, the set of internal vertices and the set of vertices with exactly i children in T, respectively. Clearly, $\text{Lea}(T) = \text{Chi}_0(T)$. Given a vertex $v \in V_T$, let $\text{par}(v)$ and $\text{Sib}(v)$ denote the parent and set of siblings of v (in T, which will be clear from context), respectively.

3 The Algorithm

Our algorithm, Alg, is based on the bounded search trees technique, in which we integrate the ideas mentioned in the introduction.

Intermediate Instances. Each call to Alg is associated with an instance $(G = (V, E), T = (V_T, E_T), L, M, F, k)$. Since G and k are always the graph and parameter given as part of the (original) input, we simplify the notation to (T, L, M, F). This corresponds to:

- A rooted subtree T of G.
- L ("fixed leaves") and M ("marked leaves") are disjoint subsets of $\text{Lea}(T)$.
- F ("floating leaves") is a subset of $(\text{Lea}(T) \backslash L) \cup (V \backslash V_T)$.

Informally, T is a tree that we try to extend to a solution (that is, a spanning tree of G with at least k leaves) by attaching new vertices to its leaves; L contains leaves in T that should be leaves in the solution (thus, we cannot attach new vertices to them); M contains leaves in T that other vertices have "marked", thus when their roles are decided, the measure (defined below) is decreased by a value large enough for our purpose; F contains vertices in G that are outside L, but whose roles as leaves have been already determined. We note that once a floating leaf becomes a leaf in T (which possibly belongs to M), the algorithm applies a reduction rule (Rule 4) that inserts it into L and removes it from F and M (if necessary). For the sake of clarity, denote $N = \text{Lea}(T) \backslash (L \cup M)$. That is, N is the set of leaves in T whose roles as leaves (in the solution) has not yet been determined, and which are not marked. Roughly speaking, the separation between M and N will reflect the different means using which we decide whether their vertices should be internal vertices or fixed leaves: most of the vertices in M will be handled using exhaustive rules whose analysis will rely on the fact that $|M|$ appears in the measure, while vertices in N will be handled using rules that are not necessarily exhaustive (to ensure that they result in good branching vectors), and whose correctness will rely on the dependency claim (defined below). An example of an instance (T, L, M, F) is illustrated in Fig. 1.

Goal. Our goal is to accept the (intermediate) instance *iff* G contains a spanning tree $S = (V_S, E_S)$ with at least k leaves that *complies with* $(T, L \cup F)$—i.e., (1) T is a subtree of S, (2) the vertices in $L \cup F$ are leaves in S, and (3) the neighbor set of each internal vertex in T is the same as its neighbor set in S. An example of such a spanning tree S is illustrated in Fig. 1.

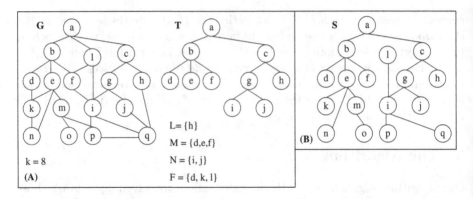

Fig. 1. (A) An instance associated with a node of the search tree; (B) A spanning tree with 10 leaves that complies with $(T, \{d, h, k, l\})$.

By calling Alg with $(T = (\{r\}, \emptyset), \emptyset, \{r\}, \emptyset)$ for all $r \in V$, and accepting *iff* at least one of the calls accepts, we clearly solve k-LST in time that is bounded by O^* of the running time of Alg.

Measure. To ensure that the running time of Alg is bounded by $O^*(3.188^k)$, we propose the following measure:

Measure: $2k + \dfrac{1}{4}|M| - (|L| + |F| + \sum_{i \geq 2}(i - 1)|\,\mathrm{Chi}_i(T)|)$.

Clearly, the measure is initially $2k + \frac{1}{4}$. Moreover, as we will prove in this paper, this measure was carefully chosen to ensure that Alg can return a correct decision when the measure drops to (or below) 0, and that the roots of the branching vectors associated with the branching rules are bounded by $3.188^{0.5}$.

We note that at this point, where we have not yet presented our rules, it is already easy to see that the measure makes sense in the following manner: (1) Marking a vertex (i.e., inserting it to M) increases the measure, so when the vertex is "handled" and thus removed from M, its treatment is considered to be better than it actually is, and (2) Determining the role of a vertex as a leaf (i.e., inserting it to $L \cup F$) or an internal vertex with at least two children (i.e., inserting it to $\mathrm{Chi}_i(T)$ for some $i \geq 2$) decreases the measure by a significant value (at least 1). When determining the role of a vertex as an internal vertex with one child, we can avoid decreasing the measure, since this decision will be made either in a reduction rule or in a branching rule where the role of another vertex, which decreases the measure, is determined.

The Dependency Claim. To ensure the correctness of our rules, we will need to preserve the correctness of the *dependency claim* (defined below), which describes the dependency of a node in the search tree on the nodes preceding it. This claim supplies information that is relevant only to vertices in N, and allows us to handle them as efficiently as we handle marked vertices. More precisely, in some branching rules that determine the roles of vertices in N, the dependency claim will justify why certain branches are either omitted or present but also determine the roles of other vertices. Formally, for each $v \in N$, the dependency claim supplies the following information:

1. $|\operatorname{Sib}(v)| \leq 1$.
2. Let (T', L', M', F') be the instance associated with the (unique) ancestor node (in the search tree) in which $p \triangleq \operatorname{par}(v)$ was inserted to T as an internal vertex. Then,
   (a) There is no solution that complies with $(\widetilde{T}, L' \cup F' \cup \{p\})$, where $\widetilde{T}$ is the tree T' from which we remove the descendants of p.[3]
   (b) If there is $s \in \operatorname{Sib}(v)$, then
      i. $s \notin M \cup (\bigcup_{i \geq 2} \operatorname{Chi}_i(T))$.
      ii. $\operatorname{Paths}(\operatorname{Lea}(\widetilde{T}) \backslash (L' \cup F' \cup \{p\}), s, V \backslash (V_{\widetilde{T}} \cup L' \cup F')) \neq \emptyset$.

Roughly speaking, for each vertex $v \in N$, the dependency claim states that we *must* have determined (at an ancestor node) that the parent p of v is an internal vertex since otherwise there is no solution (**item 2(a)**). Moreover, it states that if v has a sibling s, then v does not have another sibling (**item 1**), the sibling s is neither marked nor has more than one child in T (**item 2(b)i**), and p was not the only vertex from which we could have reached s when we determined the role of p (**item 2(b)ii**).

We note that at this point, where we have not yet presented our rules, it is already possible to see that the dependency claim is potentially useful when handling vertices in N. Indeed, suppose that we can use **items 1, 2(b)i** and **2(b)ii** to show in some "problematic situations", where the role of at least one vertex in N is determined, that the existence of a solution S where the roles of certain vertices x, y and z are a, b and c implies that there also exists a solution S' which contradicts **item 2(a)**. Then, we can avoid (in advance) setting the roles of x, y and z as a, b and c. Having less options to consider implies that Alg might examine a smaller search tree, and thus it would be faster.

Observe that initially (i.e., in an instance of the form $(T = (\{r\}, \emptyset), \emptyset, \{r\}, \emptyset)$), the dependency claim is correct since the root is inserted to M.

Result. We will show how to devise a set of 39 rules that preserve the correctness of the dependency claim, solve the problem in polynomial time when the measure drops to (or below) 0, and such that the roots of the branching vectors associated with the branching rules are bounded by $3.188^{0.5}$. We thus obtain that Alg runs in time $O^*(3.188^k)$ and uses polynomial-space. Estivill-Castro *et al.* [13] proved that in polynomial time, given an instance $(G = (V, E), k)$ of k-LST, one can compute another instance $(G' = (V', E'), k)$ of k-LST such that $|V'| = 3.75k$, and (G, k) is a yes-instance *iff* (G', k) is a yes-instance. Therefore, by first running the kernelization algorithm in [13], and then calling Alg on the instance it computes, we obtain the following result.

---

[3] For example, if $T = (\{r, p, v, u, a, b, c\}, \{(r, p), (p, v), (p, u), (r, a), (a, b), (b, c)\})$, and exactly before (after) p was inserted to T as an internal vertex, the tree was $T_1 = (\{r, p, a\}, \{(r, p), (r, a)\})$ $(T_2 = (\{r, p, v, u, a, b\}, \{(r, p), (p, v), (p, u), (r, a), (a, b)\}))$, then $T' = T_2$ and $\widetilde{T} = (\{r, p, a, b\}, \{(r, p), (r, a), (a, b)\})$.

Theorem 1. *The k-LST problem is solvable in deterministic time* $O(3.188^k + \text{poly}(|V|))$, *using polynomial-space.*

3.1   A Brief Overview of the Rules

Alg starts by examining three (reduction) rules that identify cases where the instance can be solved in polynomial time. First, Rule 1 rejects the instance when there is a vertex outside the constructed tree T that cannot be attached to T via a path that starts at a vertex in $M \cup N$. Then, Rule 2 shows that when the measure drops to (or below) 0, we have that $k \leq \max\{|\text{Lea}(T)|, |L \cup F|\}$, in which case the instance is necessarily a yes-instance. Now, Alg considers the case where all the vertices in G are already contained in T—then it concludes (since Rule 2 was not applied) that the instance should be rejected.

Next, Alg examines six (reduction) rules that identify cases where the instance, although not necessarily solvable in polynomial time, is still simple in the sense that we can currently decrease its measure or add a vertex to T without branching. First, Rule 4 turns a floating leaf that is a leaf in T into a fixed leaf. Rule 5 turns a vertex outside T that does not have any neighbor outside T into a floating leaf. Then, Rules 6 and 7 handle *certain* situations where there are two vertices such that the neighbor set outside T of one of them is a subset of the neighbor set of the other. In these situations, Alg turns one of the two vertices into a floating leaf. Rule 8 handles the case where there is a vertex $v \in M \cup N$ and a vertex u outside T such that v is the only vertex in $M \cup N$ from which we can reach u (without using vertices whose roles have been already determined). In this case, Alg determines that v is an internal vertex (while maintaining the correctness of the dependency claim). Rule 9 shows that at this point, if there is a vertex $v \in M \cup N$ that has only one neighbor, u, outside the tree, and u also has only one neighbor outside the tree, then v can be turned into a fixed leaf.

Rule 10 considers the case where there is a vertex $v \in M \cup N$ that has exactly two neighbors outside T, and these neighbors belong to F. It examines two branches to determine the role of v (i.e., to determine whether v should be an internal vertex or a fixed leaf), while relying on the dependency claim to show that if v is an internal vertex, the neighbors of its neighbors in F should be floating leaves.

Now, Rules 11–18 determine the roles of all of the remaining vertices in M. First, Rule 11 handles the case where a vertex $v \in M$ has at least three neighbors outside T; then, in the branch where it decides that v is an internal vertex, it inserts its children to M. Rule 12 handles the case where v has two neighbors outside T; then, in the branch where it decides that v is an internal vertex, it shows that its children can be inserted to N without contradicting the dependency claim.[4] Afterwards, Rule 13 handles the case where v has only one neighbor, u, outside T, and u has at least three neighbors outside T. This rule is

---

[4] In this context, recall that we need to avoid marking vertices when it is possible, since each marked vertex increases the measure.

similar to Rule 11, where upon deciding that v is internal vertex, u should also be an internal vertex. Rules 14–18 handle the remaining situations, where v has only one neighbor, u, outside T, and u has two neighbors outside T. Here we do not simply perform one rule whose action is similar to the one of Rule 12, since to preserve the correctness of the dependency claim, more delicate arguments are required.

Rules 19–23 handle the situations where there is a vertex $v \in N$ that does not have a sibling in N. In the branches of these rules where Alg decides that v is a fixed leaf, it relies on the dependency claim to show that it can insert the neighbors outside T of v into F. Rule 19 (Rule 20) examines the case where v has at least three (exactly two) neighbors outside T. Only when v has exactly two neighbors outside T, Alg inserts them to N rather than M in the branch where it decides that v is an internal vertex. Then, Rule 21 handles the case where v has only one neighbor, u, outside T, and u has at least three neighbors outside T, while Rules 22 and 23 handle the case where u has two neighbors outside T. More precisely, Rule 22 assumes that there is no vertex outside T that can be reached (from a vertex in N) only via paths that traverse u, while Rule 23 assumes that such a vertex exists. This separation allows Alg to perform different actions in Rules 22 and 23 in order to maintain the dependency claim.

Then, Rules 24–28 determine the roles of all the vertices in N that have a sibling in N, and this sibling has only one neighbor outside T. These rules are similar to rules 19–23, except that now, in order to preserve the correctness of the dependency claim, we also need to handle the sibling, which is simply done by inserting it to M. Observe that after Rule 28, we are necessarily handling a situation where there is a vertex $v \in N$ with a sibling $s \in N$, and both v and s have at least two neighbors outside T. In most of the following rules, the roles of both v and s are determined together.

Rules 29 and 30 handle the case where there is a vertex u outside T that can be reached only from v and s. More precisely, Rule 29 (Rule 30) considers the case where v has at least three (exactly two) neighbors outside T. Roughly speaking, these situations require special attention, since upon deciding the roles of v and s, it might be problematic to insert their children to N rather than M (for some intuition why this action causes a problem, we refer the reader to item 2(b)ii of the dependency claim). However, rather than considering the situation in these rules as a disadvantage, Alg actually exploits it; indeed, for intuition why this is possible, observe that if v is a (fixed) leaf, s must be an internal vertex, since otherwise it is not possible to connect the vertex u to T.

Next, Rule 31 handles the case where v and s have a common neighbor outside T, and this neighbor is not a floating leaf. The efficiency of this rule relies on an observation (whose correctness is based on the dependency claim) that upon deciding that v is a leaf, Alg can also safely decide that s is an internal vertex. After this rule, Alg examines two rules, Rules 32 and 33, that handle the (remaining) cases where v has at least three neighbors outside T. The separation between Rules 32 and 33 is done according to the number of neighbors in F that v has outside T; the case where there are at least two such neighbors is simple, while the case where there is at most one such neighbor requires (in Rule 33) to rely on the dependency claim.

Finally, in Rules 34–39, Alg handles the remaining cases, where both v and s have exactly two neighbors outside T. More precisely, Rules 34–36 handle such cases where v and s have a common neighbor outside T, and Rules 37–39 handle such cases where the sets of neighbors of v and s outside T are disjoint. The inner distribution of situations corresponding to these cases between Rules 34–36 and between Rules 37–39 is quite delicate, and all of these rules perform actions whose correctness crucially relies on the dependency claim. The following subsection discusses one of these rules in detail.

3.2   Examples of Central Rules

In this subsection, we present a reduction rule, as well as a branching rule, that capture the spirit of our algorithm. In particular, Rule 37 demonstrates the power of the dependency claim. We note that each rule is followed by an illustration.

Reduce 8. [There are $v \in M \cup N$ and $u \in V\backslash V_T$ such that Paths($(M \cup N)\backslash\{v\}, u, V\backslash(V_T \cup F)) = \emptyset$] Let $X = \mathrm{N}(v)\backslash V_T$.

1. If $|X| = 1$: Return Alg($T' = (V_T \cup X, E_T \cup \{(v,w) : w \in X\}), L, M\backslash\{v\}, F$).
2. Return Alg($T' = (V_T \cup X, E_T \cup \{(v,w) : w \in X\}), L, (M\backslash\{v\}) \cup (\mathrm{Sib}(v) \cap N) \cup X, F$).

In this rule, there is a vertex $v \in M \cup N$ and a vertex u outside the constructed tree T such that v is the only vertex in $M \cup N$ from which we can reach u (via a path whose internal vertices are neither floating leaves nor belong to T). Therefore, if there is a solution S (which, in particular, means that S is a spanning tree that complies with $(T, L \cup F)$), it contains v as an internal vertex. Moreover, the vertices in X are not ancestors of v (since $X \cap V_T = \emptyset$ and $v \in M \cup N$); thus, we can disconnect each of them from its parent in S and attach it to v as a child, obtaining a solution with at least as many leaves as S. This implies that we can safely turn v into an internal vertex such that the vertices in X are its children.

In the first case, the measure clearly does not increase. In the second case, the measure both decreases by at least $(|X|-1)$ (since v is inserted to $\mathrm{Chi}_{|X|}(T)$) and increases by at most $\frac{1}{4}(|X|+1)$ (since $X \cup (\mathrm{Sib}(v) \cap N)$ is inserted to M, where by the dependency claim, $|\mathrm{Sib}(v) \cap N| \leq 1$); thus, the measure decreases by at least $\frac{3}{4}|X| - \frac{5}{4} \geq \frac{1}{4}$. Observe that in the second branch, we need to insert $\mathrm{Sib}(v) \cap N$ to M, since otherwise we might have a vertex in N whose sibling, v, has at least two children in T, which contradicts item 2(b)i of the dependency claim.

Branch 37. [There are $v, s \in N$ such that $s \in \mathrm{Sib}(v) \cap N$, $|X| = |Y| = 2$, $X \cap Y = \emptyset$ and $|(X \cup Y) \cap F| \leq 1$, where $X = \mathrm{N}(v)\backslash V_T$ and $Y = \mathrm{N}(s)\backslash V_T$. Moreover, there is no $u \in V\backslash V_T$ such that Paths($N\backslash\{v,s\}, u, V\backslash(V_T \cup F)) = \emptyset$]

1 If Alg($T, L \cup \{v,s\}, M, F \cup X \cup Y$) accepts: Accept.
2 Else if Alg($T' = (V_T \cup X, E_T \cup \{(v,u) : u \in X\}), L \cup \{s\}, M, F$) accepts: Accept.

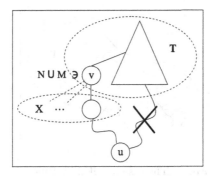

Fig. 2. The case handled by Rule 8.

3 Else if $\mathsf{Alg}(T' = (V_T \cup Y, E_T \cup \{(s, u) \colon u \in Y\}), L \cup \{v\}, M, F)$ accepts: Accept.

4 Return $\mathsf{Alg}(T' = (V_T \cup X \cup Y, E_T \cup \{(v, u) \colon u \in X\} \cup \{(s, u) \colon u \in Y\}), L, M, F)$.

The rule is exhaustive in the sense that we try all four options to determine the roles of v and s. Also, recall that once a vertex is determined to be an internal vertex, we can attach each of its neighbors outside T as a child (as explained in Rule 8). Thus, to prove the correctness of the rule, it suffices to show that in the first branch, inserting the vertices in $X \cup Y$ to F is safe (i.e., if $(T, L \cup \{v, s\}, M, F)$ is a yes-instance, then $(T, L \cup \{v, s\}, M, F \cup X \cup Y)$ is also a yes-instance). Let S be a solution to $(T, L \cup \{v, s\}, M, F)$. Suppose that there is a vertex $u \in X \cap \mathrm{Int}(S)$. Then, we can disconnect (in S) the leaf v from its parent and reattach it to u, obtaining a spanning tree S' with the same number of leaves as S (since $u \in \mathrm{Int}(S)$). Next, we disconnect the leaf s and reattach it (in S') to another neighbor $w \in V \backslash (\mathrm{Int}(\widetilde{T}) \cup L' \cup F')$, where $\widetilde{T}$, L' and F' are defined as in the dependency claim (the existence of w is guaranteed by the dependency claim). We thus obtain a solution S'' with at least as many leaves as S', in which the parent of v and s in T is a leaf. By our construction, S'' complies with $(\widetilde{T}, L' \cup F')$ (since as we progress in a certain branch, we only extend the sets $\mathrm{Int}(T)$ and $L \cup F$). This contradicts the dependency claim. Thus, there is no vertex $u \in X \cap \mathrm{Int}(S)$. Symmetrically, there is no vertex $u \in Y \cap \mathrm{Int}(S)$. Thus, S is also a solution to $(T, L \cup \{v, s\}, M, F \cup X \cup Y)$.

Next, we argue that the dependency claim holds in all branches. In the first branch, this is clearly correct. Denote $X = \{x_1, x_2\}$ and $Y = \{y_1, y_2\}$. Now, consider the second branch. There is no solution for $(T, L \cup \{v, s\}, M, F)$, since otherwise Alg would have accepted in the first branch. Moreover, $x_1, x_2 \in \mathrm{Lea}(T') \backslash M$ (where T' is defined in the second branch), and $\mathrm{Paths}(N \backslash \{v, s\}, x_1, V \backslash (V_T \cup F))$, $\mathrm{Paths}(N \backslash \{v, s\}, x_2, V \backslash (V_T \cup F)) \neq \emptyset$ (this follows from the condition of the rule). Therefore, the claim holds in the second branch. Symmetrically, the claim holds in the third branch. Similarly, noting that Alg did not accept in the second *and* third branches, the claim holds in the fourth branch.

Finally, the branching vector is at least as good as $(5, 2, 2, 2)$ since in the first branch, at least five vertices are inserted to $L \cup F$, in each of the second and

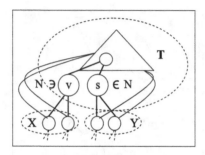

Fig. 3. The case handled by Rule 37.

third branches, one vertex in inserted to $\text{Chi}_2(T)$ and one vertex is inserted to L, and in the fourth branch, two vertices are inserted to $\text{Chi}_2(T)$. The root of this branching vector is at most $3.188^{0.5}$.

4   Conclusion

In this paper, we developed an $O^*(3.188^k)$-time parameterized algorithm for the k-LST problem, which implies that it admits a klam value of 39. To this end, we employed the classic bounded search trees technique, in whose application we integrated an interesting claim that captures dependencies between nodes in a compact and useful manner. It is natural to ask whether our approach can be further improved to obtain a faster algorithm for k-LST, which, in turn, might prove that k-LST admits a klam value better than 39. Although a more refined set of rules might result in a better running time, in order to make significant progress while using our approach, we suggest to devise a more powerful dependency claim. The most problematic cases in our analysis occur when we handle vertices in N. The worse root that we obtain is associated with Rule 30; however, optimizing this rule will not be very useful, since our algorithm contains other rules (which handle vertices in N) whose roots are quite close to it. However, a dependency claim that is easier to *both* exploit and maintain might lead to the desired improvement. The new dependency claim should capture more delicate relationships between decisions made along the branches of a search tree, which might both speed-up and simplify our algorithm.

References

1. Binkele-Raible, D.: Amortized analysis of exponential time and parameterized algorithms: Measure & conquer and reference search trees. Ph.D. Thesis, Universität Trier (2010)
2. Binkele-Raible, D., Fernau, H.: A parameterized measure-and-conquer analysis for finding a k-leaf spanning tree in an undirected graph. Discrete Math. Theor. Comput. Sci. **16**(1), 179–200 (2014)

3. Blum, J., Ding, M., Thaeler, A., Cheng, X.: Connected dominating set in sensor networks and MANETs. In: Du, D.-Z., Pardalos, P.M. (eds.) Handbook of Combinatorial Optimization B, pp. 329–369. Springer, New York (2005)
4. Bodlaender, H.L.: On linear time minor tests and depth-first search. In: Dehne, F., Sack, J.-R., Santoro, N. (eds.) Algorithms and Data Structures. LNCS, vol. 382, pp. 577–590. Springer, Berlin (1989)
5. Bonsma, P.S., Brueggemann, T., Woeginger, G.J.: A faster FPT algorithm for finding spanning trees with many leaves. In: Rovan, B., Vojtáš, P. (eds.) MFCS 2003. LNCS, vol. 2747, pp. 259–268. Springer, Heidelberg (2003)
6. Bonsma, P.S., Zickfeld, F.: Spanning trees with many leaves in graphs without diamonds and blossoms. In: Laber, E.S., Bornstein, C., Nogueira, L.T., Faria, L. (eds.) LATIN 2008. LNCS, vol. 4957, pp. 531–543. Springer, Heidelberg (2008)
7. Daligault, J., Gutin, G., Kim, E.J., Yeo, A.: FPT algorithms and kernels for the directed k-leaf problem. In: CoRR abs/0810.4946 (2008)
8. Daligault, J., Gutin, G., Kim, E.J., Yeo, A.: FPT algorithms and kernels for the directed k-leaf problem. J. Comput. Syst. Sci. **76**(2), 144–152 (2010)
9. Douglas, R.J.: NP-completeness and degree restricted spanning trees. Discrete Math. **105**(1–3), 41–47 (1992)
10. Downey, R.G., Fellows, M.R.: Parameterized computational feasibility. In: Feasible Mathematics, vol. II, pp. 219–244 (1995)
11. Downey, R.G., Fellows, M.R.: Parameterized Complexity. Springer, New York (1999)
12. Downey, R.G., Fellows, M.R.: Fundamentals of Parameterized Complexity. Springer, Heidelberg (2013)
13. Estivill-Castro, V., Fellows, M.R., Langston, M., Rosamond, F.A.: FPT is P-time extremal structure I. In: Proceedings of ACiD, pp. 1–41 (2005)
14. Fellows, M.R., Langston, M.: On well-partial-order theory and its applications to combinatorial problems of VLSI design. Technical report, CS-88-188, Department of Computer Science, Washington State University (1988)
15. Fellows, M.R., McCartin, C., Rosamond, F.A., Stege, U.: Coordinatized kernels and catalytic reductions: an improved FPT algorithm for max leaf spanning tree and other problems. In: Kapoor, S., Prasad, S. (eds.) FST TCS 2000. LNCS, vol. 1974, p. 240. Springer, Heidelberg (2000)
16. Kneis, J., Langer, A., Rossmanith, P.: A new algorithm for finding trees with many leaves. In: Hong, S.-H., Nagamochi, H., Fukunaga, T. (eds.) ISAAC 2008. LNCS, vol. 5369, pp. 270–281. Springer, Heidelberg (2008)
17. Kneis, J., Langer, A., Rossmanith, P.: A new algorithm for finding trees with many leaves. Algorithmica **61**, 882–897 (2011)
18. Milenković, T., Memišević, V., Bonato, A., Pržulj, N.: Dominating biological networks. PLOSone **6**(8), e23016 (2011)
19. Rai, M., Verma, S., Tapaswi, S.: A power aware minimum connected dominating set for wireless sensor networks. J. Network. **4**(6), 511–519 (2009)
20. Raible, D., Fernau, H.: An amortized search tree analysis for k-leaf spanning tree. In: van Leeuwen, J., Muscholl, A., Peleg, D., Pokorný, J., Rumpe, B. (eds.) SOFSEM 2010. LNCS, vol. 5901, pp. 672–684. Springer, Heidelberg (2010)
21. Thai, M.T., Wang, F., Liu, D., Zhu, S., Du, D.Z.: Connected dominating sets in wireless networks with different transmission ranges. IEEE Trans. Mob. Comput. **6**, 721–730 (2007)

Author Index

Printed in the United States
By Bookmasters